terra australis 23

Terra Australis reports the results of archaeological and related research within the south and east of Asia, though mainly Australia, New Guinea and island Melanesia — lands that remained *terra australis incognita* to generations of prehistorians. Its subject is the settlement of the diverse environments in this isolated quarter of the globe by peoples who have maintained their discrete and traditional ways of life into the recent recorded or remembered past and at times into the observable present.

Since the beginning of the series, the basic colour on the spine and cover has distinguished the regional distribution of topics as follows: ochre for Australia, green for New Guinea, red for South-East Asia and blue for the Pacific Islands. From 2001, issues with a gold spine will include conference proceedings, edited papers and monographs which in topic or desired format do not fit easily within the original arrangements. All volumes are numbered within the same series.

List of volumes in *Terra Australis*

Volume 1: Burrill Lake and Currarong: coastal sites in southern New South Wales. R.J. Lampert (1971)

Volume 2: Ol Tumbuna: archaeological excavations in the eastern central Highlands, Papua New Guinea. J.P. White (1972)

Volume 3: New Guinea Stone Age Trade: the geography and ecology of traffic in the interior. I. Hughes (1977)

Volume 4: Recent Prehistory in Southeast Papua. B. Egloff (1979)

Volume 5: The Great Kartan Mystery. R. Lampert (1981)

Volume 6: Early Man in North Queensland: art and archaeology in the Laura area. A. Rosenfeld, D. Horton and J. Winter (1981)

Volume 7: The Alligator Rivers: prehistory and ecology in western Arnhem Land. C. Schrire (1982)

Volume 8: Hunter Hill, Hunter Island: archaeological investigations of a prehistoric Tasmanian site. S. Bowdler (1984)

Volume 9: Coastal South-west Tasmania: the prehistory of Louisa Bay and Maatsuyker Island. R. Vanderwal and D. Horton (1984)

Volume 10: The Emergence of Mailu. G. Irwin (1985)

Volume 11: Archaeology in Eastern Timor, 1966–67. I. Glover (1986)

Volume 12: Early Tongan Prehistory: the Lapita period on Tongatapu and its relationships. J. Poulsen (1987)

Volume 13: Coobool Creek. P. Brown (1989)

Volume 14: 30,000 Years of Aboriginal Occupation: Kimberley, North-west Australia. S. O'Connor (1999)

Volume 15: Lapita Interaction. G. Summerhayes (2000)

Volume 16: The Prehistory of Buka: a stepping stone island in the northern Solomons. S. Wickler (2001)

Volume 17: The Archaeology of Lapita Dispersal in Oceania. G.R. Clark, A.J. Anderson and T. Vunidilo (2001)

Volume 18: An Archaeology of West Polynesian prehistory. A. Smith (2002)

Volume 19: Phytolith and starch research in the Australian-Pacific-Asian regions: the state of the art. D. Hart and L. Wallis (2003)

Volume 20: The Sea People: late-Holocene maritime specialisation in the Whitsunday Islands, central Queensland. B. Barker (2004)

Volume 21: What's changing: population size or land-use patterns? The archaeology of Upper Mangrove Creek, Sydney Basin. V. Attenbrow (2004)

Volume 22: The Archaeology of the Aru Islands, Eastern Indonesia. S. O'Connor, M. Spriggs and P. Veth (2005)

Volume 23: Pieces of the Vanuatu Puzzle: Archaeology of the North, South and Centre. S. Bedford (2006)

terra australis 23

Pieces of the Vanuatu Puzzle:
Archaeology of the North, South and Centre

Stuart Bedford

ANU
THE AUSTRALIAN NATIONAL UNIVERSITY

E PRESS

ANU
E PRESS

This edition © 2006 ANU E Press

The Australian National University
Canberra ACT 0200 Australia
Email: anuepress@anu.edu.au
Web: http://epress.anu.edu.au

National Library of Australia Cataloguing-in-Publication entry

Bedford, Stuart.

Pieces of the Vanuatu puzzle: archaeology of the north, south and centre.

Bibliography.

ISBN 1 74076 093 X (pbk).
ISBN 978 1 921313 03 5

1. Lapita culture — Vanuatu. 2. Archaeology — Vanuatu. 3. Vanuatu. I. Title. (Series : Terra Australis ; 23).

995.95

Cover: Area A excavation, Ponamla, Erromango (photo: S. Bedford).
Back cover map: *Hollandia Nova.* Thevenot 1663 by courtesy of the National Library of Australia.
Reprinted with permission of the National Library of Australia.

Series Editor: Sue O'Connor

Typesetting and design: Emily Brissenden

Foreword

STUART BEDFORD was born in New Zealand in 1960. He received his BA and MA degrees from the University of Auckland in 1982 and 1994 respectively. In between he gained valuable archaeological field experience as Assistant Archaeologist with the Clutha Valley Archaeology Project, Cromwell, New Zealand and then as Archaeologist and finally Supervisory Archaeologist with the Museum of London in the UK. I first met Stuart in 1994 at the notorious World Archaeological Congress in New Delhi. He had already applied for a PhD scholarship at ANU and got the thumbs up when he returned from his Indian travels. He arrived in the then-Division of Archaeology and Natural History in the Research School of Pacific and Asian Studies in 1995.

He arrived at ANU at just the right moment. The 1985 to 1994 research ban in Vanuatu had just been lifted under the enlightened policies of new Vanuatu National Museum Director, Ralph Regenvanu. There were many archaeological questions to answer there concerning whether Vanuatu had been occupied in the Pleistocene period, on the stratigraphy and pottery sequence at the Mangaasi type-site on Efate, the cultural sequence of the ethnographically-rich (but little known archaeologically) island of Malakula in the north, the biological anthropology of the Vanuatu population, and the stylistic analysis and external connections of Vanuatu rock art, among a range of other topics. Clearly, with all these projects and potential projects there were numerous opportunities for PhD and other thesis projects and I set about recruiting suitable candidates. Stuart was among the first crop, which also included Nicola van Dijk and Meredith Wilson.

Stuart's family had earlier had business contacts in Vanuatu and so he had a ready-made social circuit of family friends in Port Vila to help him settle in to the rigours of fieldwork. Very soon after arriving he met his future wife, Caroline Brunet, daughter of a long-established French family in Vanuatu. Suitably distracted he began fieldwork on Erromango in mid-1995, investigating the site of Ponamla, an immediately post-Lapita pottery site found by one of the National Museum *filwokas*, Sempit Naritantop. During this trip he quickly picked up the lingua franca of Vanuatu, *Bislama*, and clearly displayed exactly the right attitudes and manner to fit him for independent fieldwork in local conditions.

Museum Director Ralph Regenvanu and I then accompanied Stuart to Northwest Malakula to introduce him to a new fieldwork area. After an initial misunderstanding by our party which left both Ralph and I embarrassed and necessitated a custom 'fine' and elaborate kava ceremony in recompense, we left Stuart to fend for himself. This he did admirably in what — as we had already discovered to our cost — was a very politically and culturally-sensitive environment. His collaboration with local *filwoka* Jimmyson Sanhanbath was a major key to his success in this regard, as I know he would be the first to admit.

In 1996 Stuart carried out field work on Erromango and Malakula, and that year we had also begun the re-excavation of the Mangaasi-Arapus site on Efate Island. This project lasted, with only one year without fieldwork, from 1996 to 2003. Stuart also undertook the analysis of the pottery and other artefact sequences from that site. Only some of those analyses appeared in the PhD and in this monograph resulting from it, as the PhD was awarded in 2000 when excavations were still underway.

In 2000 Stuart took up the job of Regional Archaeologist for Auckland and Northland at the New Zealand Historic Places Trust. He was able to persuade the Trust to allow him a period of leave-without-pay each year to pursue his Vanuatu researches. Apart from jointly-directing the continuing Mangaasi-Arapus excavations, he also developed and gained funding from the Sasakawa Pacific Island Nations Fund and other sources for a project on the Small Islands off NE Malakula which has produced spectacular results in terms of finding Lapita settlements — of which he had found only a hint in his PhD research in NW Malakula — on several of the islands investigated, and in the recovery of painted Lapita vessels. He has subsequently extended his Malakula research to the Maskelyne Islands to the SE of Malakula.

Early in 2004 Stuart was involved in the location of a major Lapita site at Teouma on the island of Efate, just outside the capital Port Vila by staff of the Vanuatu National Museum. The Museum asked Stuart and I to secure funding for a salvage excavation of the site, which had been impacted by quarrying activities. We were able to secure funding from the Pacific Biological Foundation and again Stuart's legendary 'finder's luck' has come into play with the excavation of what has turned out to be the oldest cemetery site yet found in the Pacific.

Also in 2004 Stuart and I put together an ARC application for a comprehensive program of research in Northern Vanuatu to investigate inter-island and inter-archipelago relations during prehistory. This was successful and has allowed Stuart to return to the ANU early in 2005 as an ARC Postdoctoral Fellow and to complete turning his PhD into a *Terra Australis* Series monograph.

Given current time and financial restrictions on PhD students a project of the scope of Stuart's PhD would no longer be contemplated today. But in the aftermath of the lifting of the research ban in Vanuatu there just seemed so much to learn about the archipelago's prehistory and a feeling of the need to 'catch up' with the kind of research coverage and results starting to be produced from other parts of Island Melanesia and New Guinea. A comparison of cultural sequences from North, South and Central Vanuatu was a tall order, given that before Stuart began the only reasonably detailed sequence we had was from the Centre through José Garanger's work, and basically there were no cultural sequences from the North or South. Stuart, however, seemed completely unfazed by all this, and as you can read in this *Terra Australis* he came through with the goods famously.

Much, of course, remains to be done, particularly with the poorly-understood middle period of Malakula's history, but a firm foundation has been laid in this monograph for all future work on the archaeological investigation of Vanuatu's cultural history. I have greatly enjoyed collaborating with Stuart in our various joint Vanuatu projects and in hearing about his own projects. These projects have always been greatly enhanced by our long-standing collaboration with the staff and *filwokas* of the Vanuatu National Museum, prominent among whom at various times should be mentioned the Director Ralph Regenvanu, Jacob Kapere, Marthe Yamsiu, Willie Damelip, Richard Shing, Jennifer Toa, Marcelin Abong, Sero Kauatonga, Jimmy Kauatonga, Douglas Meto, Jimmy Sanhanbath, Numa Fred, Silas Alben, Philip Tepahae, and 'the pugilist' Jerry Taki.

Matthew Spriggs
May 2005

Acknowledgements

THIS MONOGRAPH is largely based on research carried out as PhD student at The Australian National University (Bedford 2000a). Acknowledgements firstly go to the three archaeologists who pioneered archaeological research in Vanuatu namely José Garanger, Mary Elizabeth Shutler and Richard Shutler Jr. Without their trailblazing efforts, research in Vanuatu would have posed a much more daunting challenge. Professor Matthew Spriggs was the principal supervisor of my PhD dissertation and was inspirational throughout with his seemingly boundless energy and enthusiasm for the archaeology of the Pacific. At times we arrived at diametrically opposed interpretations of the archaeological data but rather than attempting to stifle dissent he simply emphasised that whatever my view was it had to be robustly argued.

Archaeologists at the Australian National University who freely shared their knowledge of the Pacific, often over lunch at the Fellows Gardens, included Jim Allen, Atholl Anderson, Wallace Ambrose, Geoff Clark, Jack Golson, Jean Kennedy and Glenn Summerhayes. Glenn Summerhayes reviewed the thesis for publication and provided valuable constructive comment. Thesis examiners Roger Green and Dave Burley also provided constructive comment.

Others who have provided comment and support over the years have included Harry Allen, Judith Cameron, William Dickinson, Barry Fankhauser, Matthew Felgate, Jean-Christophe Galipaud, Joe Gyngell, Roger Green, Geoff Hope, Christophe Sand, Jim Specht and Dave Steadman, Darrell Tryon, Lynley Wallis and Meredith Wilson.

Much of the funding for fieldwork and other support was provided by the Department (former Division) of Archaeology and Natural History, Research School of Pacific and Asian Studies, Australian National University, Canberra. The excavations at the Mangaasi site on Efate were also funded by grants to the Vanuatu National Museum, co-ordinated by Professor Yosi Sinoto of the Bishop Museum in Honolulu as part of the pilot archaeological training program of the Sasakawa Pacific Island Nations Fund. In 1996 the South Pacific Cultures Fund of the Australian Government also made a contribution. In 1999 the excavation at the Arapus site on Efate was funded by the Japanese Government via a grant to the Pacific Islands development Program at the East-West Center in Honolulu administered by Professor Yosi Sinoto. The vast majority of the radiocarbon dates were provided by the Quaternary Dating Research Centre of the Australian National University, while the AMS dates were processed at the ANTARES AMS Facility, Australian Nuclear Science and Technology Organisation (ANSTO). AMS funding was provided by the Australian Institute of Nuclear Science and Engineering (AINSE) through a grant to Matthew Spriggs.

Accumulated time in the field, carrying out excavations and surveys in Vanuatu was made possible through the assistance of literally hundreds of individuals. Support from Ralph Regenvanu, Director of the Vanuatu Cultural Centre, and staff of the Vanuatu Historical and

Cultural Sites Survey and the Vanuatu Cultural Centre (Marcelin Abong, Willy Damelip, Peter Kolmas, Richard Shing, Martha Yamsiu and the late Jean-Paul Batick) was crucial. Research permission was granted by the Vanuatu Cultural Council. A special thanks goes to all the landowners and chiefs who supported the work and to Vanuatu National Museum fieldworkers Jerry Taki (Erromango), Douglas Meto (Efate) and Jimmy Sanhanbath (Malekula) and Kalisak Masing and Numa Fred, former and current directors of the Malekula Cultural Centre.

On a more personal note I acknowledge a great debt to my parents Hugo and Helen, to whom this publication is dedicated, who have always been supportive, especially during what seemed like, to both them and me, an interminable period as a student. Les familles Brunet, Colardeau, De Gaillande, Dufus, Russet and Traverso have all been very supportive during any time spent in Port Vila. Finally, but by far from last, my love and thanks to my children, Ysée Hélène Marie-Thérèse, Henri Victor and Louis Frederick and my wife, Caroline Brunet, who have all variously contributed to either prolonging the completion of my thesis or this publication but have also helped me firmly place it in a much wider perspective.

Contents

List of Figures

List of Tables

Tables in Appendices

1

Introduction and Research Design

'What is needed now is the same attention to the form and decoration of the other early pottery of the Southwest Pacific, and to the archaeology of its field occurrences, as has been devoted over the last 20 or so years, with such rewarding results, to the Lapita phenomenon…' (Golson 1992:165).

Research background

It is now more than fifty years since Father Patrick O'Reilly and Jacques Avias (Avias 1950:131) simultaneously and independently recognised that a distinctively decorated ceramic, later to become known as Lapita, could be shown to have direct parallels over vast areas of the Pacific (Watom and New Caledonia). Since that time a full 'cultural complex' has been identified, the geographical spread of which has now been extended from Aitape on the north coast of New Guinea to Samoa and Tonga in the east along with many of the islands in between (Anderson *et al.* 2001; Kirch 1997:55).

The 1950s and 1960s was a golden age for scientific archaeology in the Pacific (Fig. 1.1) with much of the attention being focused on Lapita sites in Remote Oceania (Kirch 1988a, 1997). Nascent theories as to what Lapita represented began to develop and these were inevitably tied up with the search for explanations which might account for the biological, linguistic and cultural diversity that is found across the contemporary Pacific and more specifically the Melanesian-Polynesian divide. This remains a central issue in Pacific archaeology today and one which has intrigued and perplexed observers and researchers since the first European explorers appeared on the horizon over 400 years ago and is still very far from being entirely understood.

Lapita clearly spanned the Melanesian-Polynesian divide and was seen at least beyond the end of the Solomons chain (later to be known as Remote Oceania [Green 1991a]) to represent the founding population which either pre-dated the arrival of 'Melanesian' cultures as in the case of New Caledonia or was seen as ancestral to the Polynesian populations east of Fiji (Golson 1961; Green 1963). But it was soon recognised, as the pace of excavations increased, that the picture might not be so simple, particularly as other ceramic traditions began to be identified which were thought

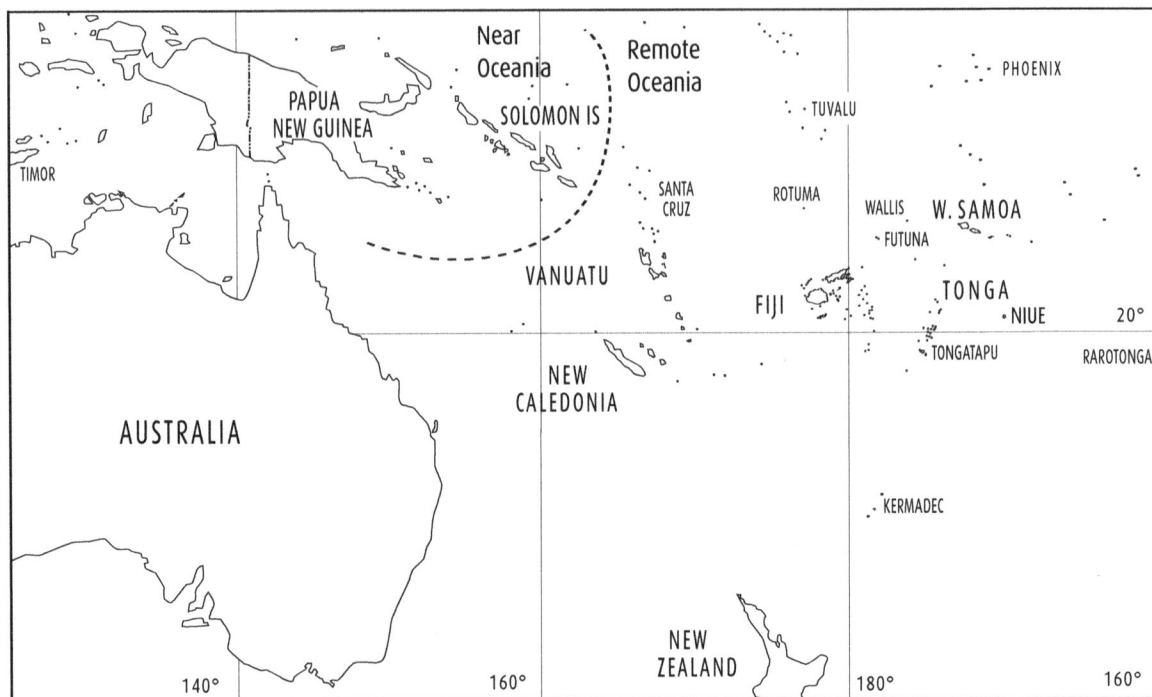

Figure 1.1 South West Pacific

to be either contemporary with or even pre-dated Lapita, i.e. Paddle Impressed and Incised and Applied Relief traditions (Garanger 1971, 1972; Golson 1968). With the added ingredient of increasing evidence for Pleistocene settlement in mainland New Guinea (White 1971) and later in the Bismarck Archipelago and the Solomons (Gosden *et al.* 1989) a whole raft of other issues were raised and variables added to the increasingly complex picture that was beginning to emerge.

The origins of the Lapita Cultural Complex with its associated trappings (pottery, horticulture, pigs, dogs and chickens and an array of shell ornaments and distinctive adzes) were widely debated, and two opposing camps began to form in the 1970s. One (the Melanesian indigenists) argued that these developments could have largely occurred within the Bismarck Archipelago with little requirement for any migration or other significant input from Southeast Asia (White, Allen and Specht 1988). The other viewpoint argued for wholesale migration into the Bismarcks area from Southeast Asia (the intrusive model) (Bellwood 1978; Shutler and Marck 1975). The origin question was tackled more seriously with the intensive fieldwork program during the mid to late 1980s, entitled The Lapita Homeland Project, which was centred on the Bismarck Archipelago (Allen and Gosden 1991). Although far from resolving the question of origins the project did encourage some modification of the opposing views. Lapita origins remains a continuing dispute but the more recent fine tuning of the chronology in the Bismarcks (Specht and Gosden 1997) tends to confirm it was a rather sudden and significant event which spread rapidly east and not one that can easily accommodate indigenist arguments of a process that occurred over a long period (Allen and Gosden 1996:193). The 'hot source of debate' (Bellwood and Koon 1989) has little chance of being further resolved in the near future without additional research focusing both in Island Southeast Asia on sites dating to a similar period and on immediate pre-Lapita sites in the Bismarcks both of which have proved so far to be somewhat elusive. The most parsimonious explanation, or at least an outline of potential processes, remains Green's (1991b:298; 1994:35) model of intrusion, interaction and innovation or Triple I model, although the proposed chronology (long duration) must now be substantially shortened which might arguably weaken somewhat the integration component (Spriggs 1997:91).

Over 50 years of scientific archaeological research, dating from Gifford's pioneering work in Fiji in 1947 (Gifford 1951), is now coming to fruition and the Lapita Cultural Complex and what its represents is being more fully comprehended, the detailed elucidation of which can be found in numerous recent publications (Best 2002; Burley 1998; Clark *et al.* 2001; Green 2003; Green and Kirch 1997; Kirch 1997, 2000; Sand 1995, 2003; Spriggs 1997; Summerhayes 2000). The continued research focus on Lapita has radically altered many earlier perceptions and theories and, as with much of the rest of Pacific archaeology, has reached a pace where theories may be wholly or partially obsolete soon after being published (Spriggs 1997:13).

If Lapita research has made such great strides over the last 50 years, what of the other ceramic traditions first identified in the 1960s (paddle impressed and incised and applied relief ware) that had initially challenged theories which argued for Lapita being the primary ceramic tradition in the region? Most researchers nowadays, whether they favour the largely indigenous, the 'Southeast Asian fast train' or the 'compromise model' scenario (i.e. Triple I) for the origins of the Lapita Cultural Complex, tend to agree that it can be accepted as an entity (biological and cultural) associated with the initial colonisation and settlement of Remote Oceania, across to at least Tonga and Samoa. There is still however a lingering reluctance amongst some researchers in accepting this scenario. This reluctance is notably associated with those who either lean towards a greater indigenist input for the Lapita Cultural Complex and believe there is still a possibility that pre-Lapita evidence might be found in Remote Oceania (Allen and Gosden 1996; Gosden 1991) and others who argue more positively that there is indeed definitive evidence of pre-Lapita settlement or at least contemporary cultural groups settling Remote Oceania, and more specifically Vanuatu (Galipaud 1996a, 1996b; Gorecki 1992, 1996).

It must be said that these claims have neither been developed in a void nor were they particularly outlandish but can be partly traced to the initial research in both New Caledonia and Vanuatu which laid the foundations of doubt regarding the primacy of Lapita settlement in these archipelagos. This is coupled with the fact that there has been little subsequent research until recently in Vanuatu which has challenged earlier assertions. The anomalies initially identified in New Caledonia have been the specific focus of detailed research which has provided some clarification. The mysterious tumuli have now been assigned to the birds (Green 1988) and the paddle-impressed or Podtanean pottery, once seen as representing a separate cultural group (Green and Mitchell 1983), is now interpreted as the domestic component of the Lapita assemblages (Galipaud 1990; Sand 1995; Spriggs 1997:145).

Another complication which added further to an increasingly crowded group of ceramic entities was the initial identification, associated with a number of ceramic assemblages that were almost exclusively plainwares, of other separate but contemporary cultural groups to Lapita (Green 1985; Kirch and Rosendahl 1973; McCoy and Cleghorn 1988). However, with the further refinement of radiocarbon dates and many more excavations this scenario has been generally discounted and the plainwares, that are found throughout the Pacific from the Bismarcks across to Tonga, have been more convincingly ascribed to the Lapita Cultural Complex (Kirch 1997:146–150; Spriggs 1997:108–150). They are identified as the domestic component of Lapita assemblages which continued in use after dentate stamping had dropped out, an argument put forward long ago by Golson (1971).

What then of the discordant evidence in Vanuatu? This related principally to the pioneering research carried out in the 1960s by both Garanger (1972) and the Shutlers (M.E. and R. Shutler 1965, 1968) which had focused on the establishment of cultural sequences for the central and southern islands of Vanuatu (see Chapter 2 for a detailed discussion). From their work in the south the Shutlers proposed a tentative culture history that argued that horticulturists, accompanied by the pig, dog and chicken, had arrived in the islands some 3000 years ago (Shutler 1969:137). But most significantly there did not appear to be any evidence of ceramics in the archaeological record.

Garanger, on the other hand, recovered quite different evidence in central Vanuatu particularly in terms of ceramic remains. His excavations revealed the Mangaasi ceramic tradition, present on Efate and the Shepherd Islands, which he argued was in existence for 2000 years, was associated with a cultural group unrelated to Lapita and possibly pre-dated Lapita, at least in Vanuatu. Garanger noted that the Mangaasi tradition seemed likely to be related to a number of other incised and applied relief traditions that were found throughout the Southwest Pacific (Golson 1968; Specht 1969) and pointed to New Guinea as a possible source for its origin (Garanger 1972:124–125).

The apparently aceramic southern region of Vanuatu once again became the focus of research when Spriggs undertook fieldwork on Erromango in 1983 as part of the Tafea Culture History Project (Spriggs and Wickler 1989). Hypothesising that ceramics could be found on the uplifted eastern coast of the island Spriggs targeted a number of reef passages and fresh water sources. Ceramics were indeed identified at a number of sites and they were described as comprising principally a regional variant of Mangaasi with a minor Lapita component (Spriggs and Wickler 1989:82). The results from Erromango, along with a summary of artefact forms and a consideration of what he termed 'transitional' sites, inspired Spriggs to argue that rather than representing two separate pottery traditions and cultural groups there was a developmental sequence from Lapita to Mangaasi. In this he was following a hypothesis first canvassed as a Melanesia-wide phenomenon by Specht (1969) and Kennedy (1982). Spriggs (1984:217) also argued that there was evidence of widespread similarities in the form and decoration of the post-Lapita ceramic sequences, indicative of broadly synchronous change across the Southwest Pacific, which would have required a 'continuing communication network'. This too was was an extension of earlier theories that had argued for tentative connections across the Southwest Pacific that could be identified through the incised and applied relief tradition (Garanger 1972; Golson 1968; Specht 1969).

This revamped version of earlier syntheses initially had little effect on the status of the Mangaasi tradition in Vanuatu. Despite earlier challenges by Ward (1979) over its proposed chronology and now its relationship to Lapita (Spriggs 1984), the Mangaasi tradition continued to remain generally accepted as the unchanged entity first outlined by Garanger (1971, 1972).

Green (1985:222) questioned Spriggs' assertions of cultural continuity in the ceramic records and argued that it had yet to be demonstrated. This was still a valid criticism ten years later (Green 1997:8), certainly at least in the case of Vanuatu. But with further archaeological research in the region the claim that the Mangaasi ceramic tradition was unrelated to Lapita and represented a separate cultural group contemporary with Lapita, began to be seen as increasingly problematic (Spriggs 1996a). Resolution of this issue, however, was put on hold in Vanuatu when archaeological and other social science research was banned by the government from 1984 until 1994. During the same period of the ban in Vanuatu the Lapita Homeland Project was undertaken in the Bismarck Archipelago (Allen and Gosden 1991). The spectacular results indicated that human occupation extended well back into the Pleistocene across the Bismarck Archipelago. Those results, combined with Wickler's later demonstration of a 29,000 year prehistory for the Solomon Islands (Wickler 1995; Wickler and Spriggs 1988), again raised the question of whether Vanuatu too might have been settled pre-Lapita (Gosden *et al.* 1989).

During the period of the ban there was also renewed interest in Garanger's original assertions of a possible New Guinean connection with Mangaasi (Garanger 1972:124), when ceramics were recovered from several sites in New Guinea claimed to be associated with dates of c. 5000 BP (Gorecki *et al.* 1991; Swadling *et al.* 1989). The reliability of the assertions of ceramics dating to such an early period in New Guinea has been challenged at some length (Spriggs 1996a, 1996b) and certainly in the case of the Ramu-Sepik sites (Swadling *et al.* 1989) the original dates associated with ceramics have been contradicted by further determinations (Swadling *et al.* 1991). Putative Mangaasi connections with these very early New Guinea ceramics have only been seriously argued by Gorecki (1992, 1996) and Galipaud (1996a, 1996b). While these claims might seem a little far-fetched, Gorecki quite rightly pointed out that 'answers to these questions can only be provided by archaeological investigation in Vanuatu' (Gorecki 1996:64).

This background sets the research context and outlines a number of salient research issues that relate to the puzzle that comprises the linguistic, biological and cultural milieu that is contemporary Vanuatu. If further understanding of the underlying processes which have influenced the development of this diversity, found both in Vanuatu and across the Pacific, is to be realistically achieved through archaeology then the establishment and detailed comparison of regional sequences is essential (Clark 1999:252; Kirch and Hunt 1988; Hunt 1987:330; Weisler 1997:7). However in Vanuatu, right up to the mid-1990s fundamental questions relating to the initial colonisation and settlement of the archipelago and the succeeding changes which took place were still largely unanswered. Much of the country remained an archaeological *terra incognita*, the sum of the chronological information aptly described by Kirch and Hunt (1988:28) as 'dreadful'. When the research ban was lifted in 1994, Matthew Spriggs received research permit number one and the Australian National University-Vanuatu National Museum Archaeological Project began (Bedford 2000b; Bedford *et al.* 1998, 1999; Wilson 2002). While this monograph focuses primarily on research carried out during the period of the above project (1995–1999), aspects of more recent significant research (2001–2003), carried out on a number of islands, particularly in northern Malekula (Bedford 2003) but also Efate (Bedford *et al.* 2004) will also be briefly mentioned.

Research objectives

This research focuses on the archaeology of Vanuatu (Fig. 1.2) and more specifically the timing and nature of initial colonisation and settlement and the later cultural transformations which ensued. Once some semblance of a regional sequence can be established for Vanuatu wider issues pertaining to the region in general can then be more closely scrutinised. One of the more pertinent issues and one which Vanuatu has played a central role, vis à vis the Mangaasi ceramic tradition, is the nature of the cultural change which occurred during the post-Lapita period which ultimately led to the conspicuous diversity that is found in the region. A widely accepted explanation for these changes in the Southwest Pacific, has been that it was related to a secondary wave or at least continued contact with Non-Austronesian populations further west, which contributed to the 'Melanesianisation' of the region as far east as Fiji (Bellwood 1979; Golson 1961; Green 1963; Spriggs 1984). It has been argued that its most visible manifestation archaeologically is to be found in the ceramic record with some authors claiming that there is evidence of a Melanesia-wide Incised and Applied Relief tradition which demonstrated synchronous change from the post-Lapita period (Spriggs 1984, 1997; 2000) possibly lasting right up to 800 BP (Wahome 1997, 1999). Once detailed ceramic sequences from Vanuatu are established a re-appraisal of the ceramic assemblages which have been used to support the above claims can be carried out in order to further assess the validity of such a scenario.

Dating from initial arrival in Vanuatu, human populations have experienced a metamorphosis over the last 3000 years from an arguably single broad ancestral Austronesian language and cultural complex to 110 distinct languages (Tryon 1996) and a profusion of cultural forms. The archaeological record in Vanuatu has in the past contributed both confusion and clarification to our understanding of that archipelago's history and the wider region. A set of inter-related research objectives and strategies were established prior to the commencement of fieldwork and continued to develop throughout its progress. They were as follows:

1. Testing for evidence of pre-Lapita settlement. This followed on from the work of Spriggs on Erromango in 1994 where a series of caves in areas affected by tectonic uplift had been

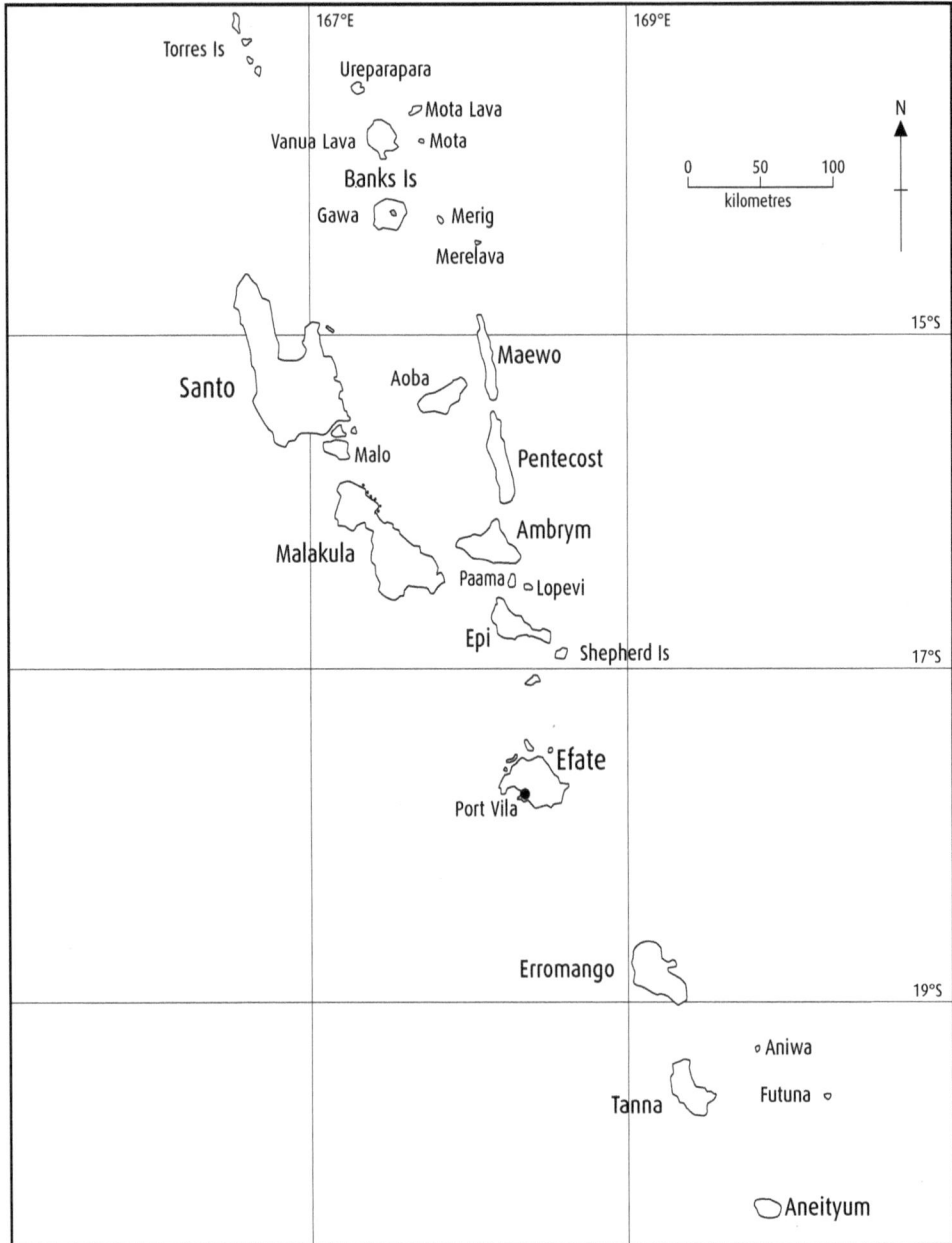

Figure 1.2 Vanuatu

targeted for excavation. The northern island of Malekula (the second largest in Vanuatu at some 2024 km^2) and more specifically the Northwest area was chosen as an ideal region to test for evidence of pre-Lapita settlement. It is an island that experiences periodic tectonic uplift with some of the highest rates in the whole of Vanuatu, particularly in the Northwest where uplift is estimated to be some 3m per 1000 years (Taylor *et al*. 1980:5369). The coastal landscape comprises a series of uplifted coral terraces riddled with caves and shelters. At the completion of an initial survey some fifty caves/shelters or overhangs were recorded at varying altitudes with fifteen being targeted for excavation.

2. Clarification of the pioneering work of the 1960s and 1970s (Garanger 1972; Hedrick nd; Ward 1979). This objective was tied up largely with establishing the timing and nature of initial settlement and succeeding transformations. The relationship of the Lapita and Mangaasi ceramic traditions was one of the key issues. This required the identification and

excavation of a number of sites which possessed lengthy cultural sequences dating from initial settlement. The two sites of Ponamla and Ifo on Erromango provided such data. Initially a similar research strategy was employed on Malekula during the 1995-1999 research period but no sites that had lengthy continuous cultural sequences were located. Excavations instead revealed two ends of the cultural sequence, one associated with first arrival on the northwest coast and the other with the last 500 years leading up to European contact in the eighteenth century. During subsequent research (2001–2003) on the small islands of northeast Malekula, sites with lengthy cultural sequences were identified and they have greatly increased our knowledge particularly of Lapita settlement in that area (Bedford 2003). A return to the eponymous Mangaasi site on Efate was also undertaken in light of both Ward's initial (1979) questioning of the chronology of the site and its ceramics and new interpretations inspired by the ceramics originally recovered from Erromango in 1983 and in much greater quantity in the excavations of 1995 and 1996 (Bedford 1999; Spriggs and Wickler 1989). Seven seasons of excavation at areas adjacent, immediately inland of and along the coast from Garanger's original site, carried out between 1996 and 1999 and again from 2001 to 2003, have provided a detailed picture of settlement pattern and subsistence and a lengthy well dated cultural sequence.

3. Establish the basic outlines of prehistoric subsistence patterns in Vanuatu. Information regarding prehistoric subsistence activities in Vanuatu has been largely restricted to research in the Banks Islands by Ward (1979) and on Aneityum by Spriggs (1981) who focused exclusively on the evidence for horticultural intensification. The prehistoric faunal record for Vanuatu was very poorly known. An attempt to close this large gap in the record and highlight any changing trends was a priority of this research.

4. The clarification of the history of settlement pattern in Vanuatu. By combining the results of earlier research with those from a new range of site types and varied geographical locations on different islands it was expected that at least provisional conclusions could be drawn regarding the nature of the archipelago's settlement history.

5. An assessment of the evidence for a Melanesia-wide incised and applied relief tradition. Early theories of the existence of such a tradition (Golson 1968; Specht 1969) still retain considerable influence today (Spriggs 1997, 2001; Wahome 1999 [see Clark 2003 for an historical review]). These have often utilised Garanger's Mangaasi tradition as a point of reference. Once ceramic sequences had been established for a number of different islands in Vanuatu, evidence for inter-archipelago and/or inter-regional interaction as evidenced by homologous ceramic traits could be more accurately assessed.

In theoretical terms much of the archaeological research that has been carried out in the Pacific to date can be broadly positioned within the culture-historical paradigm. It is an approach that in the past has been criticised for producing 'just so stories' (Clark and Terrell 1978) and one that lacks any detailed consideration or explanation of the processes that lead to cultural or social change. But as eloquently stated by Sharp, 'criticising a program of constructing accurate cultural sequences as 'mere' culture history is something of a luxury that can be indulged in only after culture history has been done, and done well' (Sharp 1991:326). Moreover the culture history methodology that has been criticised relates more to the paradigm as practiced in the first half of the twentieth century (Lyman *et al.* 1997) but which has since that time seen significant transformation. No longer do culture histories comprise simple artefact inventories. They are now more often utilised in a complementary or supportive fashion in conjunction with newer approaches in archaeology (Green 1982:17; Lyman *et al.* 1997:231) that incorporate a wide range of variables which can include environmental and socio-economic factors and evolutionary theory.

Theoretical models constructed without detailed examination of empirical evidence or in the absence of such evidence can be shown to be fundamentally flawed (Green 1982). In fact either without the other can lead to less than secure conclusions. Vanuatu is a classic case where early theories based on pioneering fieldwork remained unchallenged because little further archaeological research focusing on cultural sequences was carried out. As outlined above, Vanuatu's archaeology has remained stuck in the pioneering phase for many years, lacking basic empirical data to facilitate further progress. This fundamental gap in the archaeological record of Vanuatu has necessitated that this research is heavily empirically orientated, in an attempt to redress the deficiency of fundamental data which ultimately provides the background to a more informed engagement of the theoretical issues.

Much of the archaeological research that has been carried out in the Pacific to date has been data-driven, largely because so little is actually known of the region. With the extraordinary volumes of new information that have been generated over the last 10-15 years, it has often been simply a matter of attempting to make some sense of the information as we endeavour to further understand the deep human history of the region. Theories are often outmoded soon after publication, or cannot be tested due to a total lack of pertinent basic data (Spriggs 1997:13). Simply collecting new data however is not in itself justification for research. Researchers need to be continually reassessing previously collected data in the light of new information and it must be continually integrated into new and changing interpretative frameworks (Denham and Ballard 2003:132; Felgate 2003:21; Gosden 1991:260).

Outline

This monograph comprises another ten chapters with related figures and the appendices. Chapter 2 outlines the history of archaeological research in Vanuatu and how it has contributed to the understanding of the archipelago's *longue durée* (Braudel 1980). A number of conclusions which were gleaned from the pioneering research are detailed along with later reassessments that have been partly inspired by subsequent work. All the radiocarbon dates presented through the text include the uncalibrated date followed by the laboratory number and then the calibrated date at two standard deviations using the Calib program REV 4.1.2 of Stuiver *et al.*, 1998 with delta R as 0 for marine samples (see Appendix 1 for a detailed list of radiocarbon and AMS dates associated with this research).

Chapter 3 outlines the excavation strategy and details the individual site stratigraphies and chronologies for the excavations at Ponamla and Ifo on Erromango, the series of cave sites and open areas on Malekula, and Mangaasi and Arapus on Efate. At a number of the sites both areal and test pit excavations were undertaken. Data from the individual test pits from the sites of Ponamla, Ifo, Mangaasi and Arapus are presented in detail in Appendix 2. Some assessment of regional settlement patterns is also presented at the conclusion of this chapter.

Chapter 4 deals with the methodology utilised for the analysis of the recovered ceramics which primarily focused on a combination of vessel form and decoration and to a lesser degree on fabric analysis through petrography. Various petrographic reports completed by Professor William Dickinson relating to this research are to be found in Appendix 3. Chapter 5 discusses the recovered ceramics from the sites of Ponamla and Ifo on Erromango. Chapter 6 outlines in detail the recovered ceramics from the recent excavations (1996-1999) at the Mangaasi site on Efate, while Chapter 7 focuses on the excavated and surface collected ceramics from Malekula. A full inventory of all the distinctive decorative motifs identified from the recovered ceramics that were assigned a numeric is given in Appendix 4.

Chapter 8 presents a synthesis of the proposed ceramic sequences from the various islands and makes further inter-island Vanuatu comparisons and inter-regional/archipelago comparisons. Sites and ceramic assemblages outside Vanuatu that were specifically targeted were those that have been argued previously as having parallels to the Vanuatu assemblages in terms of homologous ceramic traits.

Chapter 9 deals with the non-ceramic items of material culture recovered from the excavations. Attention is focused both on a detailed discussion of the composition of the recovered non-ceramic items and an attempt to determine if there were any changing temporal trends through time. Chapter 10 presents the faunal remains which, as noted, have to date been poorly documented. Emphasis is again focused on presenting a basic outline of the recovered faunal remains which is then used to establish patterns of subsistence. Any temporal change or regional variation is highlighted. Chapter 11 presents the conclusions of the study and discusses its implications and points to potentially productive avenues for future research.

2

Archaeology of Vanuatu

'...another incident in a long series of incidents, a chaotic chain of events in a chaotic chain of islands, coralline or volcanic restless' (Harrisson 1937:110).

Introduction

Although the above epigraph refers to the long series of outside contacts that have occurred throughout Vanuatu's history it also has some relevance when reviewing the history of archaeological research in the archipelago. This chapter concentrates on a chronological review of that research and in doing so attempts to place it and the current research in historical and archaeological context. The review avoids delving in the minutiae but rather highlights general research themes and results. More comprehensive discussion is included for archaeological research which has not been fully published. On the islands where this current research has been carried out, namely Erromango, Efate and Malekula, more detail will be included in relevant chapters. In an attempt to retain contextual consistency the former colonial name for Vanuatu, the New Hebrides/Nouvelles Hébrides will be used in discussion of earlier research which took place prior to 1980. In addition the terms 'Incised and Applied Relief' (Golson 1968:12) and/or 'Mangaasi' (Garanger 1971) will also be used throughout the review when referring to ceramic traditions as, although they have recently been shown to be somewhat imprecise and all encompassing (Bedford et al. 1998; Bedford 2000a, 2000b), it was within these frameworks that much of the earlier research and discussion was conducted. Earlier researchers have also somewhat freely used and often intermixed terms such as style, ware, phase and tradition when referring to ceramic assemblages and this will again be reflected throughout this review[1]. All radiocarbon dates are presented in the following format; date, sample number and calibrated age at two standard deviations using the Calib. program REV 4.1.2 of Stuiver et al., 1998 with delta R as 0 for marine samples.

---

1. The more recently excavated Vanuatu ceramic assemblages will be referred to in the following terms. The term 'style' is used in a generic sense when discussing assemblages that lack detailed well-defined chronologies. Once a proposed ceramic chronology is outlined in detail (Chapter 8) the term 'ware' is used to refer to a distinctive class of pottery which shares a number of well defined attributes (Rice 1987:484). The term 'phase' is used to define homogeneous and distinct sections of a dated ceramic sequence (Rice 1987:484).

It is more than forty years since archaeological research began in Vanuatu and it must be seen in its historical context. Forty years ago and for much of the succeeding period the geology of many of the islands was virtually unknown, and archaeologically they were a complete blank as were many other areas of the Pacific. The idea that dentate stamped ceramics (later universally known as Lapita) were somehow related across different island groups and evidence of some 'community of culture' first became accepted only in the 1960s (Golson 1961). The full implications continue to be vigorously debated today. Other ceramic traditions and theories relating to them have been established and dismantled during the same period. Radiocarbon dating with its associated complexities and potential pitfalls, of which we have only recently become more fully aware and begun to examine in detail in the Pacific (Kirch and Hunt 1988; Spriggs 1990a, 1996a; Spriggs and Anderson 1993), was for much of the history of archaeological research in Vanuatu regarded as the definitive scientific tool which could single-handedly resolve the chronology of sites. Progress in the understanding of the Pacific *longue durée* will only continue through the juxtaposition of results from previous research with those from new ventures. Such is the case with this research endeavour.

The dominant theoretical framework that has characterised archaeological research in the Pacific, briefly discussed in the introductory chapter, has been that of the culture historical approach. In those areas of the Pacific where ceramics occur, they have been the major component in establishing cultural sequences. Vanuatu is no exception and consequently this review will reflect that. But although ceramics have contributed greatly to the elucidation of the archipelago's archaeology, there are large bodies of research that have dealt with aceramic periods or ceramics were not recovered, as was the case with Garanger's excavations of the burials on Retoka and Tongoa and the Shutler's work in Southern Vanuatu (1968). In the case of Spriggs' (1981) study of the intensification of agriculture on Aneityum there was not a single sherd in sight.

The first arrival of Europeans in Vanuatu and their increasing involvement and influence on the archipelago over time have been documented in a number of both general (Corris 1973; Howe 1984; Scarr 1990; Spriggs 1997) and more specific studies (Adams 1984; Guiart 1983; McClancy 1981; Shineburg 1967). The early ethnographic and historic records that survive come from a typically eclectic group ranging from the earliest explorers and observers, to ethnographers, whalers, missionaries, traders, travelers and scholars. They can provide valuable baselines from which archaeologists can either work back or forward but they become increasingly detached when dealing with deeper time. Brief mention only will be made here of a number of key figures whose passing observations have aided or influenced archaeological research strategies.

The first group of Europeans to visit Vanuatu were the crews of the Spanish ships *San Pedro y Pablo, San Pedro* and *Los Tres Reyes* (Kelly 1966:26) under the leadership of Don Pedro Fernandez de Quiros. Sailing from the northeast, they first sighted Mere Lava (27th April) in the Banks Islands before landing on Gaua on April the 28th 1606. Clearly Gaua was not their idea of a 'great southern continent' and they sailed further south reaching Santo and entering Big Bay the next day. The crews of the boats were to spend almost two months in Big Bay expending huge energy in first establishing and then suddenly abandoning the 'capital' of Terra Austrialia del Espiritu Santo. The Spanish accounts written at the time which were later published in various forms (Kelly 1966; Markham 1904) are a valuable ethnographic record for the Big Bay area in particular both for the fact of being a record of first contacts and that they took place some 400 years ago. It is from these accounts that the first mention of pottery was made, 'The natives make from the black clay some very well-worked pots, large and small, as well as pans and poringers in the shape of small boats. It was supposed that they made some beverage, because in the pots were found certain sour fruits' (Kelly 1966:215). Had Quiros and company ventured further south to Efate and beyond they may well have commented on the changing cultural landscape including the absence of pottery on these other islands.

It was not until some 162 years later that other European visitors began to appear. It was firstly the French in 1768 under Bougainville who visited northern Vanuatu, and shortly later in

1774 Cook arrived visiting the islands in the north, centre and south. Cook named the island group, the New Hebrides (Beaglehole 1969), a name that remained in place until independence in 1980. Again valuable ethnographic information for the period can be gleaned from the early records but the contacts were marked by their brevity and were often less than amicable. Greatly increased European contact dates from the early nineteenth century, largely through the activities of whalers and traders (particularly sandalwood traders, see Shineberg 1967) and later on the 'labour recruitment' or 'blackbirding' voyages (Corris 1973). Again many of these contacts were often peripheral and rarely recorded, particularly those of the earlier nineteenth century. It was not until the arrival of the missionaries from the mid-nineteenth century that more detailed long-term ethnographic observations were recorded along with occasional mention of archaeological stratigraphy and the possibility that people had occupied the islands for some period of time.

Reverend Oscar Michelsen, the first missionary on the island of Tongoa, arrived in 1879. He was the earliest writer to mention the Kuwae eruption, recording oral traditions relating to the event (Michelsen 1893). Kuwae was said to be a large island which was blown apart during a violent volcanic eruption, leaving the present Shepherd Islands as testimony to its existence. Michelsen calculated that the volcanic eruption had occurred around A.D. 1540 (Michelsen 1893:13). In relation to the phenomenon he noted the presence of pottery sherds and bone in layers some depth below volcanic ash. The same site was later to be visited by Hébert (1965:91) and excavated by Garanger (1972). The volcanic event which has since been more precisely dated to AD 1452 (Eissen *et al.* 1994) was one of the worlds ten largest eruptions of the last 10,000 years (Robin *et al.* 1994).

In a similar stratigraphic situation the French traveller and sometime archaeologist Glaumont (1899) was to note pottery under volcanic ash on Ambae. The sherds were described as coarse without design or engraving and ascribed to a people living at an undetermined but clearly earlier period unrelated to the present inhabitants. Glaumont further speculated that the earlier 'primitive' inhabitants were probably absorbed by the current population on Ambae who were principally made up of Polynesians (Glaumont 1899:66). There are numerous other references throughout the early literature referring to presence of pottery on the ground surface, particularly in the north and centre of the country, along with the fact that it was no longer manufactured (Douceré 1922; Etheridge 1917; Lawrie 1892; Schurig 1930).

The Swiss ethnographer Speiser provides us with the first, published at least, accounts of excavations and the associated results from Vanuatu (Speiser 1996[1923]:83-6). He carried out a series of excavations on different islands in a search for evidence of an earlier 'Palaeolithic culture'. These investigations included a series of 'trenches' on the islands of Santo (Port Olry), Aore, Ambrym, Malekula (Vao) and Lelepa. Very little was noted in any of the excavations and Speiser was to conclude that evidence of an earlier Palaeolithic population did not seem to exist. On several occasions he recorded the recovery of sherds from the test trenches, namely on Vao, Ambrym and Lelepa. Speiser described these sherds along with a number presented to him by locals on the islands of Iririki (Efate) and Vao (Malekula), commenting on their appearance and whether or not they appeared similar to the contemporary Santo material. In a more detailed discussion (along with photographs) of the contemporary ceramic traditions then extant on the west coast of Santo, Speiser was reluctant to accept that the numerous sherds found throughout the northern and central islands were made on those islands. Rather he ascribed the remains as ultimately having originated, through a series of exchange networks, from Santo (Speiser 1996:230).

Examples of the west coast Santo traditions were further reviewed by MacLachlan (1939) in his article on the *Native Pottery of the New Hebrides*. The studied materials, held by New Zealand museums, were illustrated and subjected to a series of seven measurements. Although understandably dominated by the recent whole pots from Santo his study also included an analysis of sherds collected by the missionary Milne from the islands of Emae, Nguna and Emai. A total of 21 sherds are illustrated and their colour and decoration are described in detail. This article was to remain the primary source of information regarding ceramic remains from Vanuatu until further work was carried out in the 1960s. Some of the same sherds were to be used by Golson (1968, 1972) in his definition of the 'Incised and Applied Relief' tradition.

Nascent planning

When Golson published a wide ranging and detailed review of the archaeology of the South Pacific in 1959 he noted that although the practice of archaeology had arrived late in the Pacific, great progress had occurred in the previous ten years, dating from Gifford's pioneering systematic archaeological work in Fiji in 1947 (Golson 1959). The New Hebrides was referred to in two sentences. One mentioned the similarities of incised ceramic material from New Caledonia to that of the New Hebrides, as noted by Gifford and Shutler (1956:93) who in turn had been forced to rely on the only source available, MacLachlan's article of 1939. The second, appearing under 'Perspectives', noted that similar pottery traditions (soon after identified as evidence for a 'community of culture' [Golson 1961:176]) had been identified in New Caledonia, Fiji and Tonga and as far away as New Britain, but that there was a complete lack of archaeological knowledge for areas between these two geographic regions. Golson stressed the need for archaeological work to be carried out in the Solomons and the New Hebrides (Golson 1959:49). Clearly then, the archaeology of Vanuatu, up to the start of the 1960s, was a veritable *terra incognita*, a status (along with the term) further emphasised by two of the researchers who contributed to it being put on the archaeological map in the 1960s and 1970s (Garanger 1966:59; Ward 1979:1–11). It was a status that some islands continued to be noted for even thirty years after research began in Vanuatu (Allen and Gosden 1996; Gorecki 1992, 1996; Kirch and Hunt 1988; Spriggs 1990a, 1993, 1997).

At the 10th Pacific Science Congress in Hawaii in 1961 the need for coordinated Pacific archaeological research was further recognised. Pacific archaeologists attending the Congress presented a petition to that effect which was adopted by the Congress and led to the formulation of the Pacific Area Archaeological Programme (PAAP). Initially a number of aims were presented in the form of a resolution followed by several meetings which decided on a number of research areas which were top priority in Melanesia, Polynesia and Micronesia (Green 1961:478). The New Hebrides was seen as one of the five crucial areas of research in Melanesia, along with Fiji, New Caledonia, New Guinea and the New Guinea North Coast and off-shore islands (Solheim 1961:72). The PAAP prospectus for the New Hebrides, which qualified the islands as a crucial area of research, again emphasised the fact that no previous archaeological work had been carried out there, their favourable geographical position relative to Melanesia and Polynesia and that a wealth of ethnographic information was available (Solheim 1961:72). The push for archaeological research in the islands galvanised Richard Shutler and Jean Guiart who proposed a joint Franco-American expedition at the Congress (Garanger 1972:9). The archipelago was divided in two, the Americans were to carry out research in the south and the French the north, with overlapping interests on Efate (Shutler 1998 pers. comm.). Shutler who had intimate knowledge of the archaeology of New Caledonia saw the New Hebrides as a key region for the elucidation of the history of colonisation and settlement of the region and the ever present puzzle of the relationship of Melanesians and Polynesians (M.E. and R. Shutler 1968:157), a theme also emphasised by Garanger (1966:60). The French ethnologist Guiart had carried out many years of fieldwork throughout Vanuatu from 1948 onwards gaining intimate knowledge of the many and varied contemporary customs and social systems and collecting oral traditions, including those related to the arrival and influence of Polynesians in the central islands (Guiart 1956, 1963, 1973). Guiart was acknowledged by the Shutlers as inspiring their work in Southern Vanuatu (M. E. and R. Shutler 1968:158) and was regularly mentioned by Garanger as also having been instrumental in guiding his research. The oral traditions collected by Guiart had great influence over Garanger's chosen areas of research and his interpretation of the archaeology (Garanger 1972).

Before embarking on overviews of the research of the major players of the 1960s, mention must be made of Bernard Hébert, the French Overseas Territories Administrator, who also had some influence on the work of the Shutlers and Garanger. Hébert spent six years in Vanuatu during which time he carried out a number of surveys visiting archaeological sites and surface collecting ceramics and other artifactual material on the main island of Efate and some of its smaller offshore islands and on the Shepherd Islands. He also visited a number of cave sites recording art, made note of various stone features, and identified archaeological deposits on Tongoa beneath the ash of the Kuwae volcanic eruption (Hébert 1965). The ceramics illustrated largely comprise of 'incised and applied relief' ware although at the site of Erueti he recovered four distinct and quite different sherds. By gleaning information from publications of New Caledonian work (Avias 1950; Lenormond 1948; Gifford and Shutler 1956) he was able to identify the sherds as Lapita, the first for Vanuatu, assigning them the pre-Lapita label of Vao-Lapita-Vuatom, 'je me permets de désigner par le groupe *Vao-Lapita-Vuatom*' (Hébert 1965:78). Garanger acknowledged the value of the Hébert's article (Garanger 1972:16,18) which guided his initial archaeological program on Efate and the Shepherd Islands.

The Shutlers in the South, Centre and North

The New Hebrides portion of PAAP began in 1963 with Richard and Mary Shutler arriving on Aneityum in November. The Shutlers were to carry out a series of extensive surveys along with some excavation, principally in cave sites, on the islands in the south. The fieldwork which lasted almost a year has only ever been briefly outlined in a number of articles (M. E and R. Shutler 1965, 1968; R. and M.E. Shutler 1968; Shutler *et al.* 2002) and further used in general syntheses attempting to establish cultural chronologies for Vanuatu (M. E. Shutler and R. Shutler 1967; Shutler R. 1969; R. and M. E. Shutler 1975). It was soon recognised that pottery did not appear to be present (although see below) in the southern islands and the strategy of the research then concentrated on identifying other artefacts which could be used as chronological markers. The publications associated with this initial fieldwork generally comprise a listing of artefact types and associated radiocarbon dates (see Hoffman 2003 and Shutler *et al.* 2002 for a detailed summary).

The results of the Shutlers' fieldwork in the southern islands are briefly outlined below. Working on Aneityum between November 1963 and January 1964, the Shutlers reported 20 sites, namely 11 caves/rockshelters, seven old village sites and two petroglyhs. The extensive remains of abandoned agricultural terraces were also noted. Five rockshelters were partially excavated from which a series of relatively late dates were recovered. From the lowest cultural layer in two rockshelters, namely AtRS1 and AtRS3, dates of 470±80 BP (UCLA-693) 646-315 BP and 850±120 BP (WSU-140) 971-560 BP respectively were recovered.

The Shutler's work on Tanna from February to April of 1964 was restricted to the south of the island. They noted numerous small middens, a large village site and five rockshelters, two of which were excavated. The rockshelter TaRS1, measuring some 24 by 54 feet (7.3 by 16.4m) was completely excavated and returned a number of dates. One date of 2370±90 BP (UCLA-734) 2737–2154 BP came from the lowest cultural level of the site while all other dates, generally from the upper layers of the site were late 645±80 BP (UCLA-1295A) 725–517 BP, 1095±80 BP (UCLA-1295B) 1225–796 BP, 192±45 BP (P-1188) 308–0 BP, 767±47 BP (P-1190) 760–652 BP, 518±46 BP (P-1189) 629–503 BP and 567±46 BP (P-1191) 651–513 BP. A number of bone, stone and shell artefacts were recovered but the disturbed nature of the site with European material being found up to mid-way down through the stratigraphy must cast some doubt on the reliability of the conclusions including the claimed date (1095 BP or earlier) for the earliest appearance of pig (Shutler 1969:136).

Short visits were made to both Aniwa, a Polynesian 'outlier', and Erromango where a number of archaeological sites were recorded, but in both cases bad weather prevented a return to excavate. On Aniwa 15 village sites and one cave were recorded while on Erromango six caves with cultural remains were noted. A piece of shell money (*navela*) was also collected from Erromango (Shutler *et al.* 2002: Plate 7g).

During April and May 1964 the Shutlers concentrated on Futuna, another Polynesian 'outlier', recording 19 rockshelters and 19 open midden sites. Excavations were undertaken on one open site and seven rockshelters which were largely unproductive. One site that did prove worthwhile was the rockshelter FuRS12, measuring some 45 feet long by 12 feet wide (13.7 by 3.6m). The entire site was excavated to bedrock down to a maximum depth of 48 inches (88.8cm). A large earth oven dated to 905±190 BP (WSU-184) 1258–536 BP was recorded along with fifteen burials and associated grave goods which were dated (seven dates on human bone) to between 1650±100 BP (GaK-757) 1818–1314 BP and 510±90 BP (GaK-763) 658–323 BP. The time span in the dates for the burials apparently indicated that the site had been used for burial over a very long period (R. and M.E. Shutler 1968:16). However with recent questions surrounding the reliability of both the dating of human bone and dates in general from the Gakushuin Laboratory (Kirch 1975, 1984; Spriggs 1990a) these conclusions must now be regarded as somewhat unreliable.

Although it was initially stated that pottery was not found in the southern islands (R. and M.E. Shutler 1968:17), an accepted 'fact' repeated by a number of authors when discussing settlement by aceramic groups (Bellwood 1978; Green 1979), the statement is somewhat modified later when Richard Shutler mentions 'the *almost* total lack of pottery in the southern New Hebrides' (Shutler 1969). In the earliest reports (Shutler and Shutler 1965; M.E. and R. Shutler 1968[1966]) a few sherds are mentioned as being recovered from a midden site on Aneityum (At1) and a sherd from Tanna is also noted as being included in a collection analysed by Dickinson and Shutler (1979). In their 1975 publication the Shutlers mention that 'very little pottery' had been recovered from the south (R. and M. E. Shutler 1975:69). It should be noted that the French geologist Aubert de la Rüe (1945:174) mentioned finding pottery on a number of islands including Aneityum.

The apparent partitioning of the group between the French and Americans seemed a somewhat relaxed arrangement as demonstrated by the Shutlers final fieldwork of 1964. In June of 1964 they worked in the Vila harbour area recording four village sites and eight rockshelters. Excavations were carried out on three of the open area sites and two rockshelters. The two midden areas (Ef1 and Ef2) on the Efate mainland opposite Fila island proved to be relatively unproductive despite the large areas excavated (12 six by three foot squares (1.8 by .9m) and 17 six by three foot squares respectively) as were the two rockshelters (EfRS6 and 7) on the Efate mainland. On Fila island itself (Ef3) a grand total of 63, 6 by 3 foot (1.8 by .9m) test pits were dug at three different locations recovering a wide range of artefacts, including a 'considerable collection' of pottery decorated with relief, appliqué and incision (Shutler and Shutler 1965; M. E. and R. Shutler 1968). Three dates from different test pits and different levels were obtained, 815±180 BP (WSU-200) 1064–510 BP, 1020±130 BP (WSU-199) 1258–673 BP and 1090±140 BP (WSU-198) 1291–709 BP. It is however, difficult to determine the significance of these dates in relation to the excavated pottery without further details. The Shutlers also managed to survey east of Vila along the coast as far as Forari (R. Shutler 1998 pers. comm.). The Shutlers visited the site of Erueti with Garanger, the location of which had been outlined by Hébert, but were not initially successful in their search for Lapita sherds. On a second visit the Shutlers recovered 'five small Lapita sherds' (Garanger 1972:27).

In 1966 and 1967 the Shutlers embarked on a reconnaissance of the northern islands of Vanuatu (Shutler 1970). The fieldwork was largely concentrated on Santo (and adjacent small islands) with surveys of areas of the east coast noting a number of open sites with pottery. An excavation was carried out in a rockshelter near Hog Harbour recovering pottery from throughout the 2 foot (44cm) deep stratigraphy. The nearby island of Aore was also visited and a number of mound features containing pottery were excavated. Araki, another small island off the south coast of Santo, was also visited and a number of sites with abundant pottery were recorded. Encouraged by Bill Camden, the missionary on the island of Tangoa, excavations were carried out there on two former village sites. Abundant pottery and shell implements along with some stone and bone artefacts were recovered from the 3 foot (90cm) deep stratigraphy. It was during this period of fieldwork that Mary Shutler made her initial visits to the west coast of Santo recording ethnographic data on the last remaining potters of Vanuatu, at the villages of Wusi and Olpoe (Shutler M.E. 1968, 1973).

All visits to other northern islands were very brief. On Malekula numerous sites with pottery were recorded on the north east coast and offshore islands. Surface collections of pottery were made but no excavations were carried out. Further brief visits were made to Ambae where pottery was noted, as it was on Pentecost. On Ambrym and Paama it was recognised that the regular volcanic eruptions and associated ash falls explained a lack of archaeological evidence. A number of former village sites were recorded on the Banks Islands but no pottery was in evidence.

In summing up the results of the various expeditions Richard Shutler noted that, 'The nature of most of the New Hebridean archaeological sites i.e., shallow and often mixed with European material, makes it extremely difficult to attempt even a relative chronology, so that the application of the C-14 method of dating to the New Hebrides is of the greatest importance' (Shutler 1969:135). Ironically, however, no account seems to have been taken made of the possibility that the samples submitted for radiocarbon dating might also come from mixed contexts. In the end they were left with an eclectic group of artefacts and dates which led the Shutlers to conclude that 'no single artefact or artefact complex appears to be chronologically significant' (Shutler 1969:136). This it seems, along with a lack of deeply stratified undisturbed sites, essentially inhibited the establishment of a cultural sequence for the islands, but the Shutlers were confident enough to suggest that 'a tentative culture history of the New Hebrides indicated that people had been living in Vanuatu for 3000–4000 years with the pig, dog and chicken and practiced shifting cultivation supplemented with shellfish collection' (Shutler 1969:137). Although it was felt that establishing a cultural sequence had greater prospects in the central and northern islands where pottery had been recorded it remained an unrealised goal. The intensive radiocarbon dating strategy that had been carried out in the southern islands was not repeated in the north, possibly due to the above notion regarding ceramics.

Despite the enormous amount of archaeological fieldwork carried out by the Shutlers, the contemporary lack of geomorphological and archaeological knowledge and other associated difficulties of primary research, frustrated the return of conclusive results. The heavy reliance on (the then recently developed) radiocarbon dating as being able to almost single-handedly establish cultural sequences also proved to be somewhat over emphasised. Although at least 56 dates, the largest number for any area of the western Pacific at the time (Shutler 1969, 1973; Shutler *et al.* 2002), were collected between the Shutlers and Garanger, their interpretation remained problematic. In a recent review of this earlier work (Shutler *et al.* 2002) recognition of the disturbed nature of these sites and the unreliability of a whole number of radiocarbon dates has been highlighted. What has emerged from this reassessment, decades after the initial fieldwork and conclusions is that the research and collections from southern Vanuatu in fact represent the most comprehensive data set on the last 500–1000 years of prehistory for that region (Shutler *et al.* 2002). Other profitable research was the ethnographic study carried out by M. E. Shutler on the potters of the west coast of Santo. In addition, the Shutlers also established type collections of Vanuatu pottery which were sent to a number of Museums in the Pacific. Other researchers to benefit from the Shutlers encouragement and supervision were Caroline Leaney who carried out investigations on Malekula and John Hedrick on Malo, the results of which are further discussed below.

Garanger on Efate and the Shepherds

The French component of the PAAP in the New Hebrides began in 1964 with the arrival of José Garanger. Garanger, at the time employed by the Centre National de la Recherche Scientifique (CNRS), carried out a series of extensive surveys and excavations on Efate and the Shepherd Islands from April to October 1964 (Garanger 1966). A further twelve months fieldwork carried out between October 1966 and October 1967 was again concentrated on the Shepherds and Efate and also included the offshore island of Retoka. The large areas that were excavated, the spectacular finds and their detailed publication have resulted in Garanger's seminal work remaining, even after some thirty years, synonymous with the 'Archéologie des Nouvelles Hebrides' and more specifically with both the Mangaasi ceramic tradition and the communal burials on the small island of Retoka (Garanger 1972, 1982). The influence and contribution of his research remain central to any discussion of archaeology in Vanuatu. Only a very general summary is outlined here and will be supplemented further in Chapters 3 and 6 where the more recent research at the Mangaasi site is outlined.

The initial research on Efate and the Shepherds in 1964 proved to be successful both in terms of the archaeology and in the attempts to relate the archaeology to oral traditions (Garanger 1966). On Efate, surveys around the island and on the smaller offshore islands of Nguna, Mau and Moso, returned large collections of 'incised and applied relief' ceramics (later named Mangaasi) along with shell and stone artefacts. Along with the identification of this distinctive widespread 'incised and applied relief ware' was the apparent recovery of a limited number of distinctive sherds from the surface surveys of the Mele plain. These cord-marked sherds were recognised as a separate ceramic tradition (Garanger 1966). They have since been identified as Japanese Jomon pottery dating to between 5500–3500 BP, the presence of which still remains somewhat of a mystery (Dickinson *et al.* 1999). A number of test pits were also excavated at different locations on Efate but generally proved to be of little interest (Garanger 1972). More detailed excavation and survey was carried out on the small offshore islands of Mele and Lelepa. The excavations on the tiny Mele Island uncovered a total of 18 burials that were dated to the seventeenth century. Also noted was a series of occupation floors constructed with coral gravel, along with the fact that no ceramics were recovered.

On Lelepa Garanger excavated a total of 28 test pits both in caves and open sites, the results of which proved to be somewhat disappointing (Garanger 1972:45). Again a number of burials were recorded along with the extensive painted and engraved art within the caves. A total of almost 10,000 sherds were recovered during surface collections, all of which belonged to the 'incised and applied relief tradition'. None of the excavated test pits however revealed undisturbed stratigraphic layering and therefore proved unhelpful in the establishment of a ceramic chronology (Garanger 1972:45). Later research in 1966/67 on the west coast of Efate directly opposite Lelepa was to prove much more productive in that respect.

Shifting to the Shepherd Islands, Garanger was to have much greater success with deeply layered stratigraphy, and he was also able to confirm that the 'incised and applied relief tradition' found on Efate appeared there also. Armed with the earlier observations of both Michelsen and Hébert regarding archaeological deposits buried beneath volcanic ash, Garanger's aim was to research further the pre- and post-volcanic remains (ceramic and aceramic) and attempt to verify the local traditions regarding the cataclysmic Kuwae eruption and the resettlement of the area by Ti Tongoa Liseiriki (Garanger 1972:84).

Nine excavations were undertaken on Tongoa, largely located on the east coast of the island. An extraordinary 360m^2 in total was excavated over two seasons, but many of the excavations (particularly those associated with pre-volcanic deposits), although providing detailed geomorphological information, proved to be largely sterile. Several however did prove to be worthwhile, namely Aknau (TO-22), Euta (TO-11) and Lamalake (TO-23). All the sites contained varying numbers of 'incised and applied relief' sherds along with another quite distinct style. The recovered ceramics from TO-22 provided the largest sample of this distinctive tradition, later named Aknau ware which was characterised by internal incision on pots that had plain exteriors (Garanger 1972:87). It appeared to post-date the 'incised and applied relief ware', being dated on Tongoa and Makura to around 1000–900 BP. The earliest date for the 'incised and applied relief ware' recovered from Tongoa was 2460±80 BP (B 740) 2758–2347 BP at TO-23. A series of excavations were also carried out on a number of ceremonial structures and burial sites associated with detailed oral traditions relating to the re-settlement of Tongoa after the Kuwae eruption. Various oral traditions were able to be confirmed including that associated with the Kuwae eruption. A re-interpretation of these archaeological remains and the excavated burials on Retoka Island and how they relate to oral traditions can be found in Spriggs (1997:207–218).

The results from excavations further south on the island of Makura tended to corroborate those from Tongoa. A total of nine test pits were excavated, three of the more productive all located near the sea, were extended into trenches. One of the trenches reached a depth of 2m and the basal cultural level was dated to 2540±110 BP (GX-0223) 2852–2345 BP. The pottery from the lowest levels of the test pits was again characterised by 'incised and applied relief ware' (Garanger 1972:81). Aknau ware was recovered from the upper levels and appeared to have disappeared by 1000 BP.

Returning to Vanuatu in 1966 Garanger was again on the trail of oral traditions (now regarded as somewhat controversial), this time those associated with the 'foreign chief' Chief Roy Mata who was said to have transformed the political structure of Efate introducing a stabilising peace ceremony at a time of major conflict. When he died, he and a number of his supporters were buried on the island of Retoka. The excavations and the oral traditions appeared to coincide to some extent and were published in detail (Garanger 1972), although as mentioned above have been recently subjected to some re-assessment (Spriggs 1997). Initially the burials were rather uncertainly dated to the 12th century AD but more recent research indicates that the burials took place sometime during the 17th century (Bedford *et al.* 1998:187). Although no ceramics were associated with the burials, 'incised and applied relief ware' was recovered on the surface of the island and from a deep test pit adjacent to the burial area.

After being told that the location of Roy Mata's village lay on the mainland of Efate Garanger shifted his research to the site of Mangaasi. Garanger excavated 118m² at Mangaasi where some 17000 sherds were recovered, approximately 3500 of which were diagnostic. They were of the same form and decoration as those that had been surface-collected and excavated on Efate and the Shepherds. Garanger then decided the term Mangaasi was a more appropriate and accurate term, than 'incised and applied relief', to describe the tradition in Vanuatu (Garanger 1971:54). The ceramics were characterised by globular incurving pots decorated with incised and or applied relief. Garanger argued that the ceramics represented a tradition that appeared around 2600 BP ([2445±80 BP (GX-0963) 2749–2334 BP], [2595±95 BP (GX-0964) 2915–2358 BP]) had survived for up to 2000 years, and although exhibiting aspects of conservatism in form and decoration could be divided into 'Early' and 'Late' Phases (Garanger 1971:54). Garanger noted some difficulty in interpreting the stratigraphy of the site and this is reflected in both his summing up of the ceramic chronology and the definitions of his Early and Late Phases.

Mangaasi was thought to represent a separate and distinct cultural tradition that was contemporary with or possibly pre-dated Lapita in Vanuatu (Bellwood 1979; Garanger 1972; Green 1979; M.E and R. Shutler 1967). Garanger (1972:124) speculated that its origins lay in mainland Papua New Guinea an argument taken up more recently by other authors (Galipaud 1996a, 1996b; Gorecki 1992, 1996).

Many of Garanger's assertions regarding the Mangaasi tradition have been questioned, notably the validity of the Early to Late sequence and the proposed termination date for pottery production and use. Graeme Ward (1979, 1989) highlighted these problems after his own excavations in the Banks Islands to the north. Ward argued that the stratigraphy at Mangaasi appeared somewhat disturbed, many of the sherds did not seem to be in primary deposition and the termination date for ceramic production and use was more likely to be around 2000 BP. In a recent article Garanger (1996b:70) stated that he now accepts Ward's criticisms. The more recent excavations at the Mangaasi site from 1996 to 1999 (Bedford *et al.* 1998; Bedford 2000b; Bedford and Spriggs 2000), outlined in more detail below, were explicitly designed to clarify the stratigraphy of the site and the chronology of the ceramic material in the light of the criticisms by Ward and others.

The other important site excavated by Garanger on Efate was that of Erueti. The site is located on the south coast of Efate some 400m from the current beach at the rear of a bay fronted by a coral reef and lagoon. Fresh water springs are found at the rear of the site. As noted earlier the site had initially been recorded by Hébert, and later the Shutlers had collected further Lapita dentate stamped sherds from the surface of the site. As it was at the time the only place in Vanuatu where Lapita sherds had been recovered, Garanger needed little further encouragement to excavate. He excavated a single area measuring 35m² down to a depth of 80cms. Garanger noted that the stratigraphy seemed to be disturbed and the excavation was carried out in arbitrary 20cm spits (Garanger 1972:27). The recovered pottery from the site was dominated by plainware sherds. Some 5000 sherds were recovered from Erueti of which 96% were undecorated, a radical contrast to other sites where Mangaasi style decoration was in a ratio, decorated to plain of 50:50 (Garanger 1971:61). The vessels from Erueti were globular with outcurving rims. The distinctive lips were wide and flat and often notched on the outer edge, along with occasional decoration on the flat lip itself, including incision, punctation and in one case dentate stamping. Several carinated sherds were also recovered as were six Lapita dentate-stamped sherds. A smaller component of the collection displayed linear incision and only six sherds of Mangaasi-like appliqué were noted (Garanger 1972:27). The classification system applied to the rim and lip forms from the site was used, with some modification, for the entire ceramic collections recovered by Garanger from Vanuatu (1972:fig. 22).

The pottery of Erueti was initially identified by Golson (1971), and accepted by Garanger (1972:29), as being part of the Lapitoid tradition. A single date of 2300±95 BP (GX-1145) 2710–2069 BP was obtained from the 60cm below surface level at the site. As mentioned a number of Mangaasi-style sherds were found throughout the excavation, including some in the lowest 20cm spit. Garanger argued (1971:61), based on a comparison of radiocarbon dates from Mangaasi and Erueti, that the Lapitoid ware appeared some time later than the Mangaasi material in Central Vanuatu, that it was unrelated to it, and had a restricted influence in Vanuatu. With the benefit of recent archaeological research both in Vanuatu and the wider Pacific, along with improved knowledge of site formation processes, a re-assessment of the sites excavated by Garanger and the related conclusions can be made and are discussed in detail in Chapter 6.

Leaney on Malekula

Having briefly broken away from the chronological review of research carried out in northern Vanuatu we return to the work of Caroline Leaney. Leaney had a BA in archaeology from Cambridge and was initially based on Tanna with her husband who was the British District Agent for the Southern New Hebrides. She worked with the Shutlers during their investigations on the island in 1964. Later that same year the Leaneys shifted to Malekula taking 'charge' of the Northern New Hebrides. It was during this time that it was suggested by Richard Shutler that Leaney carry out a series of surveys on the island. The research objectives were of a general reconnaissance nature and the work was essentially consisted of a series of surveys recording present and prehistoric village sites, caves and rockshelters and rock art sites. Collections of pottery and artefacts were also carried out along with preliminary test pitting. The work carried out between November 1964 and April 1965 was sponsored by the Bishop Museum and supervised from a distance by Richard Shutler. The only record of the work is an unpublished typescript (Leaney 1965).

Leaney's research was concentrated in a number of discrete coastal areas namely around Lakatoro, along the coast and the small offshore islands in the north east, north and north west coast and in the south and south west including, Port Sandwich/Lamap, the Maskelyne Islands and round to Toman Island. Many of these areas were visited only briefly, due to time constraints and inaccessibility. The test pit excavations carried out by Leaney are reported with the barest of details (Leaney 1965) and largely proved to be unproductive. They were all single test pits carried out within the vicinity of Lakatoro and included a rockshelter north of Lakatoro, another rockshelter north of Litzlitz (near Lakatoro) and a ceremonial structure on Uripiv Island.

When discussing the archaeological possibilities of the island the first point that Leaney made was 'one of the most striking things about Malekula is the enormous quantity of potsherds to be found on the surface' (Leaney 1965:7), a point regularly highlighted throughout the descriptions of the village and associated ceremonial sites. Leaney was the first archaeologist to record the large caves, Yalo A and B on the northwest coast, and their impressive collections of rock art which have recently been the focus of more detailed study (Bedford *et al.* 1998; Wilson 2002). The report is useful for its descriptions of population density and location in the 1960s and often emphasises the evidence for dramatic depopulation which in some areas of Malekula has only really been turned around in the 1990s.

Hedrick on Malo

Slightly further north and commencing in 1968 under the guidance of Mary Elizabeth Shutler, John Hedrick, a masters student at San Diego State College, carried out survey and excavation work on the island of Malo, just south of Santo (Hedrick and Shutler 1969:262-265) where he located evidence for a 'Lapita-style' site (Fig. 2.1). The site (NHMa-7) was situated near the village of Avunatari on the northwest coast some 10 metres above sea level. Hedrick initially commented that the site was stratified and undisturbed (Hedrick and Shutler 1969:262) and having received one radiocarbon date of 2020±60 BP (UCLA-1412) 2146–1826 BP from the lowest level of the excavation suggested Lapita settlement at the site dated from this period. A coral slab platform was also suggested to be related to the Lapita occupation.

In a later publication however Hedrick discussed the stratigraphy in more detail noting several inconsistencies largely indicated by two more radiocarbon dates (940±80 BP [LJ-1907] 1046–686 BP and 1200±200 BP [LJ-1906] 1520–694

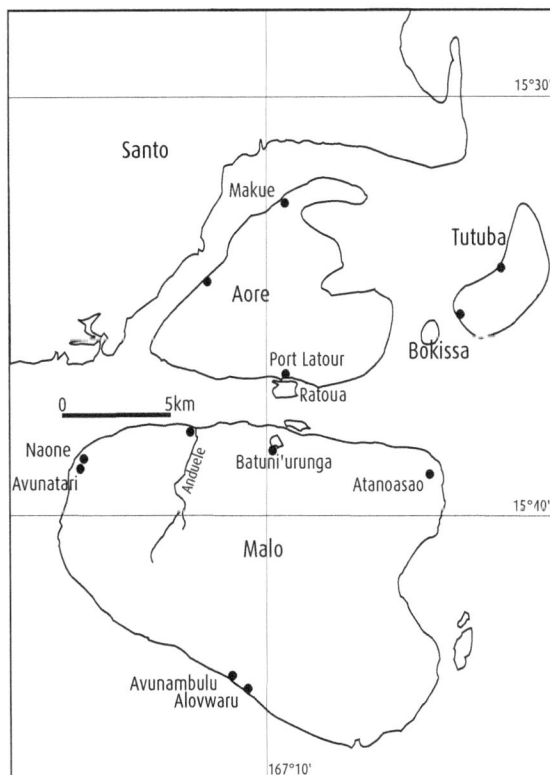

Figure 2.1 Malo and surrounding islands

BP) he had received (Hedrick 1971:13). In his re-analysis of the site Hedrick stated that the 1000 year occupation of the site, as suggested by the dates, seemed very unlikely particularly as the style of the pottery remained consistent throughout the layers. The late date for the arrival of Lapita at Malo was also difficult to explain (Hedrick 1971:18). Hedrick emphasised the preliminary nature of the results and stressed the need for more work which was in fact carried out by him between July 1972 and July 1973. The results of the work from this period have only ever been published in a popular article where the archaeology is barely noted (Hedrick and Hedrick 1975). Much more detail can be found in a partial PhD draft produced in the early 1980s (Hedrick nd). The PhD was never completed.

During the year of fieldwork Hedrick recorded 19 Lapita sites, 3 of which he excavated. Some of the sites were quite extensive and all were located in the north and east of the island up to several hundred metres inland on a former shoreline some 10-12 metres above the present sea level, which has prograded since initial settlement due to uplift. He first revisited Avunatari (NHMa-7) excavating a 3 by 3 metre area adjacent to his earlier test pit. The later excavation clarified the stratigraphy at the site and Hedrick recognised that the *in situ* Lapita material was restricted to the lowest levels of the site, below the coral slab platform and the Lapita recovered from the mound was not associated with the very late dates. This possibility had been earlier suggested to him by a number of people including Palmer, Green, Golson and Specht (Hedrick nd) and was hinted at by Ward (1979:1–17).

Some 200 metres north of NHMa-7 Hedrick excavated another Lapita site again located on the remains of a former beach ridge. At Naone (NHMa-8) he excavated a 5 by 5 metre area identifying four layers. Again the upper levels of the sites were disturbed. From the lowest levels he obtained radiocarbon dates on marine shell of 2980±70 BP (ANU-1134) 2880–2656 BP and 3150±70 BP (ANU-1135) 3138–2759 BP. Accepting these dates and the 2020±60 BP from NHMa-7 Hedrick concluded that 'classic' Lapita settlement in the area appeared at around 3000 BP and lasted to around 2000 BP.

The other Lapita site excavated by Hedrick was located on the north coast of Malo at Batuni'urunga. The site actually stretched some 3 kilometres east to west but Hedrick concentrated in the general area of a fresh-water spring, NHMa-101, excavating at three discrete locations. The sites were all somewhat disturbed and the two dates from the site of 620±85 BP (Gak-4567) 691–509 BP and 999±49 BP (P-2087) 1046–790 BP were, as Hedrick noted, not very enlightening. Hedrick also made extensive 'highly selective' surface collections in the area (including recovering the large pieces of a single pot from Paoancarai Lagoon) which he described as being very representative of the 'entire (Lapita) pottery corpus' found on Malo (Hedrick nd).

In the search for an undisturbed well stratified site which would contribute to the establishment of a cultural chronology Hedrick moved to a cave site, Avunamatala NHMa-9, located inland of the Lapita sites NHMa-7 and 8. A 1 by 3.6m trench was excavated in the cave and two levels were identified. The remains indicated sporadic short-term use of the cave and periods of abandonment. Pottery recovered from the cave included a variety of 'incised and applied relief' material. No dentate stamped Lapita was recovered from the cave. One date of 195±35 BP (P-2089) 304-0 BP from the upper level of the cave certainly indicated it had been used up to very recent times. Although this review has largely concentrated on a discussion of the Lapita ceramics on Malo, as did the research of Hedrick, 'incised and applied relief' ceramics were also recovered from both the excavations and surface collections, particularly at NHMa-9 and NHMa-101. Hedrick noted similarities between some of the material with that recovered by Garanger on Efate and the Shepherds, while other material seemed more similar to material from Malekula (Hedrick nd). However, being unable to locate sites that were less disturbed he was unable to establish the relationship of the different ceramic styles. Hedrick returned briefly to Malo in 1983 but as yet no report on his survey work of that year has appeared.

Despite the disturbed nature of the sites investigated and the lack of published data the research carried out by Hedrick on Malo has contributed significantly to the understanding of Lapita settlement pattern in general and more specifically in Vanuatu. It established that initial settlement on Malo was concentrated around lagoonal complexes and a preference for smaller islands was also suggested. On the evidence to date Malo might appear to be somewhat of an enigma in Vanuatu, a sort of 'Lapita metropolis' or an initial central place from where other islands may have been visited or colonised (Bedford *et al.* 1999:21). Recent research, however, in Malekula (Bedford 2003) and Aore (Galipaud pers. comm. 2002) where a series of Lapita sites have been identified, suggest that factors of site visibility are playing a significant role in the currently identified Lapita settlement pattern.

The work of Hedrick remained the largest record by far of Lapita ceramics for Vanuatu for another thirty years. He was able to define a large number of different Lapita vessel forms along with a whole suite of designs. He incorporated this information into the Donovan (1973) classification system and this enabled later researchers to include Malo within both studies of Lapita design and decoration (Anson 1983; Mead *et al.* 1975; Spriggs 1990b) and more general syntheses of the Lapita cultural complex (Green 1979; Kirch 1997; Kirch and Hunt 1988). Evidence for Malo initially being part of an exchange network or at least in some form of interaction with other Lapita settlements further afield was found in the recovered exotic materials. These included obsidian that came from the Banks Islands, Talasea and Lou (Ambrose 1976) and at least one of the Lapita sherds which was sourced to New Caledonia (Dickinson and Shutler 1979:1696). Preliminary results from more recent work on Malo and Aore, carried out by Jean-Christophe Galipaud, are discussed below.

Groube on Aneityum and the Banks

Some ten years after the Shutler's expedition, the southern islands of Vanuatu once again became a focus for investigation. In 1972 Les Groube from the Australian National University carried out a short survey on Erromango and spent a much longer period on Aneityum. There he surveyed various areas of the island concentrating on the extensive abandoned agricultural terracing systems (Groube 1972, 1975). Groube chose the Imkalau valley as a focus of investigations where the deep flood plains had been downcut by the Imkalau river exposing a four metre deep section with apparent evidence of agricultural activities throughout the stratigraphy. At the base of the deposits was a thick black soil rich in charcoal which returned a date of around 2000 BP (Groube 1975). Mapping of the surface features was completed with the help of Norma McArthur who further extended the research to other agricultural and historic sites (McArthur 1967, 1974). Groube also excavated a habitation site nearby named the 'White Walls' which was dated to within the last 200 years (Groube 1975:29). In a popular article of 1975 Groube summed up by describing the terraced agricultural remains on Aneityum as 'the Easter Island of Melanesian agriculture' (Groube 1975:30), an evocative phrase which was to inspire later researchers.

In October and November of 1972 Groube also spent six weeks in the Banks Islands visiting three islands in the northern part of the group, recording a variety of surface features and making surface collections. The recovered pottery was identified as relating to the 'incised and applied relief tradition' seen further south. On the small islet of Pakea off Vanua Lava, Groube test excavated a low mound feature to a depth of 1.5 metres defining four layers (Ward 1979: Appendix IV-2, 4–4). Abundant faunal material was recovered along with a limited number of sherds. Two dates from charcoal samples collected between 650 and 1250 mm below the surface returned ages of 1230±70 BP (ANU-1150) 1291-970 BP and 1380±140 BP (ANU-1151) 1544–973 BP.

Ward in the Banks

It was to the Banks Islands that Graeme Ward, who was carrying out fieldwork for a PhD, headed in 1973. Encouraged both by Groube's results and the recently published research on other islands of Vanuatu, namely Efate and the Shepherds (Garanger 1972), and the Southeast Solomon Islands (Green 1973) (a project in which Ward participated) which pointed to considerable change in the prehistoric record in the region, Ward targeted the Banks Islands citing their strategic location in Island Melanesia, as potentially providing some solutions to developing archaeological questions (Ward 1979:2). The main focus of Ward's research was the study of the development of subsistence activities, particularly marine-related strategies on small islands in the Pacific. The establishment of a chronological sequence and how that related to comparable sites in Vanuatu and nearby archipelagos was a further aim of the research. The initial fieldwork of 1973 (May to August) concentrated on environmental and archaeological surveys. Ward managed to carry out surveys on most of the islands/islets that make up the Banks Group including Mota Lava, Rah, Mota, Ureparapara, Rowa, Gaua, Vanua Lava, Pakea and Nawila. On all the islands except Ureparapara pottery was recorded. The sherds were generally small, somewhat worn, and when decorated exhibited characteristics of the incised and applied relief tradition (Ward 1979 Appendix III-1 3-12). A second field season from June 1974 to April 1975 focused on archaeological excavation and the collection of ethnographic data.

Although Ward excavated a number of sites on different islands in the Banks Group, the principal excavation and the only one described in any detail in his PhD was that of the open site on Pakea Islet earlier tested by Groube. The Pakea site comprised an area of low mounds covering some 160,000 square metres, about a tenth of the total area of the islet. The mounds seemed to have been formed by midden dumping over the previous 1500 years. Ward identified three major cultural layers at the site of which only the lowest (Layer 3) was seen as an *in situ* deposit while the upper layers of the excavation appeared to be somewhat mixed (Ward 1979:5–19). The lowest layer revealed evidence of short term occupation, possibly seasonal visits. Although pottery was found throughout the stratigraphy Ward believed it was intrusive in the upper layers. The earliest date from the site was 2600±130 BP (ANU-1711) 2955–2348 BP but the concentrated cultural material was associated with dates at the site of around 2000 BP (2300±80 BP [ANU-1874] 2706–2122 BP, 2000±120 BP [ANU-1822] 2310–1632 BP, 1890±70 BP [ANU-1813] 1991–1629 BP and 2240±70 BP [ANU-1710] 1994–1685 BP). The upper layers dating between 1300 and 650 BP were principally made up of shell midden and according to Ward could be seen as evidence of more permanent settlement. They also contained abundant evidence for the manufacture of shell tools and ornaments. Evidence for subsistence activities suggested that initially there was a predominance of fish over shellfish and that pig was also an important component in the diet. A change over time in shellfish collection strategies was also noted, from initially utilising a wide range of species to a more restricted range, possibly due to targeting and/or depletion.

The pottery excavated by Ward was highly fragmented and worn and his proposed sequence has been disputed (Kirch and Yen 1982:204, 206). Based on the relatively small sample of 168 diagnostic sherds including 4 with evidence of appliqué, 28 with incision and 31 with punctation, Ward slotted the material into the 'incised and applied Mangaasi tradition', although he noted some difficulty assigning it to Early or Late Mangaasi (Ward 1979:7–22). He believed pottery use and manufacture ceased some time after 2000 BP (Ward 1979:7–43). Ward dismissed the idea that the Pakea ceramics had any connection with Lapita or the Lapitoid plainware found at Erueti. As mentioned earlier Ward also suggested that the termination date for the Mangaasi ceramics of Central Vanuatu seemed more likely to correspond to a similar 2000 BP date.

Kirch and Yen (1982: 204, 206) were later to question Wards termination date for ceramic use in the Banks and his rejection of a connection to Lapitoid plainware. Ward's termination date of c.

2000 BP clearly did not fit with Kirch and Yen's proposed importation of Mangaasi-like pottery (Sinapupu ware) from Vanuatu into Tikopia from 2000 BP until 750 BP. Kirch and Yen (1982) argued that the Pakea site was mixed and therefore the dates could not be relied upon. With the benefit of hindsight and recent research results from other sites in Vanuatu a reassessment of the ceramics recovered by Ward can be made. They are characterised by plainware globular vessels along with occasional carinated vessels with a high percentage of outcurving rims and notching on the lip. A smaller component of incised and punctate material and very occasional appliqué was also present. The material would appear to show greater affinity with the ceramics from the Erueti site than those of Garanger's Mangaasi tradition. The plainware pots with outcurving notched rims would fit with Ward's date of c. 2600 BP and are likely to have been succeeded by the incised material also present and reminiscent of the early decorated material from Erueti. The applied relief and punctate material appears at the end of the sequence around 2000 BP or later. There is consistency here at least with Wards chronological interpretation although no doubt further excavations to refine the end of the sequence are required.

Spriggs on Aneityum and Erromango

After Ward had completed his fieldwork in Vanuatu, Matthew Spriggs, another student from the Australian National University, arrived in 1978 to begin fieldwork for his PhD. As a student at Cambridge he had been inspired by some of Les Groube's lectures on the 'Easter Island of Melanesian agriculture' terracing remains that covered much of the small island of Aneityum. Spriggs' fieldwork, carried out between 1978 and 1980, was concentrated on Aneityum, the southernmost inhabited island of the archipelago, investigating agricultural intensification and human impact on the environment. Spriggs' fieldwork also included an ethnoarchaeological study of irrigation systems on the northern island of Maewo (Spriggs 1981). Spriggs took the view that in certain respects prehistoric human-accelerated erosion was beneficial, creating the large coastal plains on which much of the population of Pacific Islands live (see Spriggs 1984, 1986). As part of that project, the first pollen analysis for Vanuatu was carried out by Geoff Hope. The results revealed vegetation clearance on Aneityum on a massive scale at about 3000 BP (Hope and Spriggs 1982). Hope and Spriggs argued that these changes were human induced and the results certainly suggest that people first arrived in the area around the time Lapita colonists might be expected to have appeared.

Rapid large scale erosion over an extended period led to extensive valley in-filling and coastal progradation creating whole new areas of coastal plain for settlement and agriculture. The newly created landscapes were occupied by 950 BP with people initially practicing dry land agriculture. By 400 BP there is evidence for more intensive agricultural activity with irrigated agriculture becoming increasingly prevalent. By AD 1830 these agricultural systems were at their peak, the valley floors were covered with irrigated gardens as were many hillsides with extensive irrigated agricultural terracing. Spriggs argued that the human-accelerated erosion which created the rich alluvial coastal plains provided the potential for greatly expanded agricultural intensification and associated social stratification (Spriggs 1986).

Despite and because of the fact that evidence for large-scale human induced erosion and subsequent valley infilling dating from c. 3000 BP was able to be identified, artefactual material that could have contributed to the establishment of cultural sequences was not recovered. There were no archaeological remains or land surfaces in the alluvial sections older than 2000 years, meaning that at least 1000 years of the island's history was missing. Despite the fact that valuable information had been gleaned regarding the history of landscape change and sociopolitical machinations dating from likely first human arrival of around 3000 BP, information on the cultural sequences for at least the first 1000 years of Southern Vanuatu were still non-existent and at best very sparse for the remaining 2000 years.

In 1983 Spriggs returned to Aneityum for further archaeological and palaeoenvironmental research. He then shifted his research focus further north when he commenced a project on Erromango, assisted by Vanuatu National Museum fieldworker Jerry Taki who had earlier worked with Les Groube. The explicit aim was to search for history of Southern Vanuatu that was earlier than 2000 years old. The uplifted coral reef terraces on the east coast of the island were not covered by alluvial deposition and it was hypothesised that Lapita and other early pottery sites would be found there. Locations near reef passages and freshwater sources at rivermouths were targeted, following the model of Lapita site location developed by Frimigacci (1980) for New Caledonia. The series of surveys and several excavations located the first *in situ* pottery recovered from southern Vanuatu. Pottery was recovered from a total of seven sites, two of which were excavated, namely Ifo and Naen. The recovered pottery was described as largely comprising of a regional variant of the Mangaasi tradition with a smaller Lapita component, dating to around 2300 BP (Spriggs and Wickler 1989). The excavated ceramics and other materials from the excavations on Erromango were analysed in detail by Wickler (1985). Further discussion of the 1983 fieldwork and ceramics is included in Chapters 3 and 5. The site of Ifo, which was described as being 'transitional' between Lapita and Mangaasi, provided Spriggs with further data to challenge the idea that Lapita and Mangaasi were two separate pottery traditions (Spriggs 1984). Further survey was carried out by Spriggs on Erromango in 1988 in preparation for the establishment of the Erromango Kauri Reserve (Spriggs 1988). Later, this work and all the earlier surveys and excavations which had been carried out on Erromango were synthesised in a cultural resources study of the island (Spriggs and Roe 1989).

The moratorium which was placed on archaeological research from 1984 clearly restricted research potential but with the establishment in 1990 of the Vanuatu Cultural and Historic Sites Survey (VCHSS), headed by David Roe and Jean-Christophe Galipaud, extensive survey and limited research re-commenced. The work of the VCHSS involved extensive surveys on many of the islands of Vanuatu along with the training of local personnel. To date over 40 development impact reports have been produced. These reports are designed to record archaeological, cultural and historic sites which may be threatened by development projects. However until 1994, when the research ban was lifted, no significant excavation was part of the VCHSS program.

Galipaud on Malo, Torres and Santo

In 1995 Galipaud was freed from his duties with the VCHSS and he became the resident ORSTOM archaeologist in Vanuatu. The ban on archaeological research had been lifted and over a four year period (Galipaud left Vanuatu in 1998) he carried out a series of surveys and excavations, principally on the islands of Santo, Malo, and the Torres.

Earlier surveys of Malo by Galipaud were followed up in July and August of 1997 with limited excavations at two sites on the island (see Fig. 2.1). The earlier surveys on the island, showed that Lapita deposits could now be identified as being located along much of the east and north coasts. Two sites were targeted for excavation in 1997, Avunatari and Atanoasao. Two sites at Avunatari (located in the same area where Hedrick had excavated earlier) proved to be disappointing as the deposits were very disturbed. Atanoasao was somewhat different, the Lapita deposit there being sealed by up to a metre of sediment. Several test pits uncovered a cooking and dumping area where the remains of ovens and oven rakeout were recorded along with an array of dentate stamped ceramics along with frequent shellfish and turtle bone. Several shell arm ring fragments were also recovered. The site appeared to represent the remains of a short term colonising occupation with dates of 2830±100 BP (Beta-110143) 3244–2752 BP, 2900±50 BP (Beta-110144) 3209–2872 BP and

2830±60 BP (Beta-110146) 3158–2781 BP, all being recovered from charcoal samples from the lowest level of the site associated with Lapita dentate stamped ceramics. Further details of the excavations are outlined in Galipaud (1998a), Noury (1998) and Pineda and Galipaud (1998).

Galipaud's work in the Torres Group was concentrated on the islands of Tegua and Toga (Galipaud 1998b) where in some areas pottery was described as being relatively frequent on the surface. Excavations were carried out at an open site and a cave site (Woga) on Tegua and an open site at the village of Kurvot on Toga where mound features were noted. At the open site of Litetona three test pits were excavated to a depth of over one metre but returned few artefacts. Numerous surface collected materials were however recovered in the vicinity of the test pits, including shell adzes, arm rings, volcanic glass and worn pottery. Galipaud (1998b:163) indicated that the pottery was Mangaasi-style from the form and colour, although only one recovered sherd was decorated. At the cave site of Woga a test pit was dug to a depth of 2.5m. Artefactual material was very sparse with shellfish and several pieces of worked stone making up the total. No pottery was recovered. A date from charcoal collected from the very base of the test pit returned a date of 2100±60 BP (Beta-100138) 2305–1902 BP, perhaps indicating that pottery by that date was no longer a component of the material culture on the island.

The excavations on the island of Toga were concentrated at the village of Kurvot. Seven test pits were excavated along an east-west axis beginning behind the village and continuing at intervals of 30m. Artifactual material was again relatively sparse. Two phases of occupation were noted. *In situ* pottery was only found in the earlier levels and appeared to be largely a plainware which Galipaud compared to Kiki ware from Tikopia and early material from Pakea (Galipaud 1998b:166). Two dates from charcoal were recovered from the lower levels of TP 3, one AMS date of 2470±40 BP (Beta-118605) 2736-2354 BP and one standard C14 date of 2420±70 BP (Beta-118606) 2739–2333 BP, indicated the earliest occupation of the area.

During his tenure with the VCHSS Galipaud had also carried out a series of surveys and limited excavation on the island of Santo (Galipaud 1996c:27). Further survey and excavation was completed in 1996 from Wusi on the south west coast of Santo right up to the northern point of the island. A total of 66 sites were recorded and a series of test pits were excavated in both open sites and cave sites. The objectives of the research were to establish basic chronologies related to the human settlement of Santo and compare the results with those already known from the Banks and Malo. The test pit results proved to be rather disappointing with little artifactual material being recovered from what were largely relatively recent deposits. Collections of surface pottery were made and a basic typology of decorative styles was illustrated along with three distinct traditions being identified.

Two cave sites excavated at the northern tip of Santo, Malsosoba 1 and 2, are reported in some detail, but again were not particularly rich in artifactual material. Two radiocarbon dates were reported from charcoal recovered from the lowest levels of the caves, 1150±80 BP (Beta-98570) 1264-927 BP (Malsosoba 1) and 350±60 BP (Beta-97558) 498-310 BP (Malsosoba 2) (Galipaud 1996c).

As noted by Galipaud (1996c:27) the west coast has proved to be a difficult area to find archaeological sites with any depth of stratigraphy and consequently the results are somewhat tentative. The complex and varied ceramic remains relating to human colonisation and settlement, which seem likely to have a 3000 year history on Santo, as yet require a great deal of further research.

Commencement of the ANU-VNM Archaeological Project

Also following the lifting of the ban on research in 1994 The Australian National University-Vanuatu National Museum archaeological project was commenced. Spriggs again targeted Erromango, this time testing for evidence of pre-Lapita settlement. A series of rockshelters/caves were selected for excavation along the west coast at a range of altitudes from 5 to 125 metres where

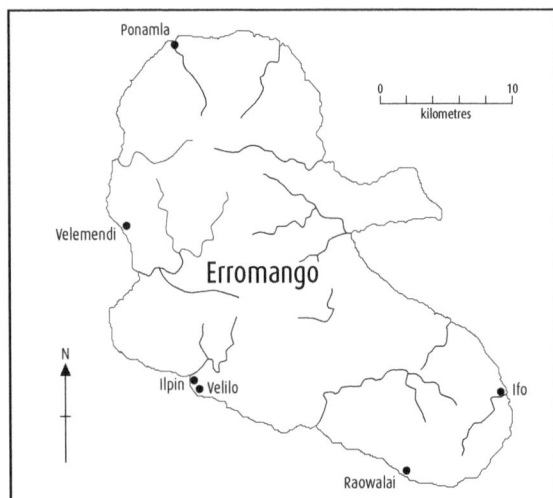

Figure 2.2 Erromango

early Holocene and Pleistocene shorelines were preserved. Four sites were excavated in 1994, namely Velemendi, Velilo, Raowalai and Ilpin (Fig. 2.2). The earliest cultural deposit from any of the cave sites occurred at 5-10 metres above sea level at Velilo shelter. It consisted of a shallow trench and several postholes, associated with a charcoal date of 1220±130 BP (ANU-9709) 1350–916 BP. A date on marine shell taken from the former foreshore below the cultural layer returned a date of 4290±50 BP (ANU-9710) 4521–4257 BP. The only other excavation which provided dates earlier than the last few hundred years was at Raowalai Cave where charcoal and shell associated with burials was dated to 810±80 BP (ANU-9703) 923–571 BP and 910±70 BP (ANU-9705) 631–427 BP respectively. The next use of the site was domestic in the form of a large stone oven which gave a carbon 14 determination of 200±60 BP (ANU-9702) 425–0 BP. The other cave sites returned radiocarbon determinations no earlier than the last few hundred years (Bedford *et al.* 1998). The lack of evidence for early use of these sites would be surprising if there was widespread occupation of the island before the Lapita expansion. The only pottery found on Erromango, with the exception of a single sherd in secondary, surface deposition in a cave, comes from village sites situated at prime settlement locations at river mouths and associated reef passages. Two of these, namely Ponamla and Ifo which were excavated in 1995 and 1996 respectively, are discussed in detail in Chapters 3 and 5. Spriggs also returned to Aneityum in 1996 with Geoff Hope and Brad Pillans to undertake further palaeoenvironmental research.

Summary

This review clearly highlights the somewhat piecemeal nature of archaeological research in Vanuatu to 1995 and the pressing need for further work. Up to 1994 Vanuatu could still be described as largely an archaeological *terra incognita* or certainly a somewhat confused archaeological terrain. The large scale pioneering projects of the 1960s and 1970s which produced spectacular results provided a basic platform from which to proceed but for a number of reasons further progress was unrealised.

In many respects the archaeology of Vanuatu had been stuck in a pioneering phase. Many of the results from the earlier research remained unpublished or unconfirmed and had become increasingly enigmatic over time when compared to results from more recent archaeological programs throughout the Pacific. The Australian National University-Vanuatu National Museum Archaeological Project was commenced in 1994 to move specifically beyond the lingering pioneering phase.

Fieldwork post 2000

Although not outlined in any great detail here it is necessary to briefly mention a number of other, largely as yet, unpublished research projects which will be referred to in the text that have been undertaken since 2000. A number of them have particular relevance to subsequent discussions particularly in relation to Lapita settlement across the archipelago.

Small Islands of Malekula

In 2001 a three year archaeological research and training program was commenced on the small offshore islands of northeast Malekula (Bedford 2003). The program was specifically designed to rectify the dearth of information on Lapita sites in Vanuatu. The focus of the project was a search for Lapita on smaller islands and more specifically those on small islands not too distant from Lapita-rich Malo. Those islands chosen for study were a group located on the northeast coast of Malekula, namely Vao, Atchin, Wala and Uripiv. Lapita sites were found on the uplifted former beach terraces, on the sheltered western side of all these islands. The sites are generally very well sealed and preserved through a combination of tephra deposits, cyclonic sand deposits and accumulated debris from later habitation in the same area.

Aore and Tutuba

Since 2000 Jean-Christophe Galipaud has been carrying out survey and limited excavation on Aore Island ($60km^2$). He reports the presence of at least two Lapita sites, Makue in the north and Port Latour in the south (Fig. 2.1), with some ephemeral evidence in the west. He also suspects that it is highly likely that Lapita will be found on the small offshore islet of Ratoua in the south (Bedford 2003; Galipaud pers comm.). Only limited information regarding the sites is available to date but one of the significant features of the northern site of Makue is the relatively large number of obsidian flakes sourced to the west (Galipaud pers. comm.). Galipaud has also more recently confirmed the presence of two Lapita sites on the smaller adjacent island of Tutuba ($14km^2$). The Lapita sites on Malo and Aore which have obsidian from the west are likely to provide some of the earliest dates for Lapita arrival in Vanuatu. This more recent research has also demonstrated the widespread nature of Lapita settlement across Vanuatu.

Arapus and Teouma, Efate

In 1999 the Arapus site, located directly southwest of Mangaasi was identified and some of the results from the research of that year are included in this publication. Further research at the site was carried out over separate field seasons in 2001-2003. This subsequent research has defined in detail the pattern of settlement at the site and extent of the archaeological deposits. Further fine dating of the site and detailed analysis and publication is forthcoming but the accumulated evidence now indicates that the Arapus site is associated with first settlement on the west coast of Efate but that it post-dates initial settlement of the island by at least some decades if not longer. The recent discovery of the Teouma site on the south coast of Efate further strengthens this argument (Bedford *et al.* 2004). Teouma is a Lapita site that was very recently uncovered during construction work for a prawn farm. The dentate-stamped motifs and vessel forms identified amongst the scattered ceramics indicate that the site dates to the earliest part of Vanuatu's settlement with a number of sherds showing clear affiliation to sherds from Mussau and Reef Santa Cruz Lapita sites. Excavations at the site were undertaken in mid-2004 and 2005.

Rock-art

Another area of research that has seen major advances over the last five years is that of rock-art in Vanuatu which is now in many respects is one of the better known collections of the Pacific. In 2002 Meredith Wilson completed her doctoral research on the rock art of Vanuatu (Wilson 2002) and then subsequently in collaboration with Bruno David carried out one of the most intensive rock art dating programs in the world concentrating on the rock-art of northwest Malekula (Wilson and David in prep.).

Over the last ten years archaeological research in Vanuatu has shifted significantly from the pioneering phase in which it had for so long remained. An outline and some detail of the various advances in research and knowledge will be demonstrated in the following pages.

3

Excavated sites: plans, stratigraphy and dating

'The focus should be on large islands within regions where rockshelter sites have never been seriously investigated: the Solomons, Vanuatu and New Caledonia. Such an approach would certainly put to the test most of the ideas presented in this essay. Whatever the outcome of such a campaign, it has to provide a more complete and more realistic picture of Melanesian prehistory' (Gorecki 1992:44).

This chapter outlines in detail the various sites where excavations were carried out and includes descriptions and illustrations of stratigraphy along with site plans. Descriptions of individual layers include colour (Munsell Soil Color Charts 1975), consistency and form along with general summaries of any recovered artifacts and other midden debris. More detailed information, summary tables and discussion regarding midden remains are presented in relevant chapters and appendices. A full list of all radiocarbon dates associated with this research is presented in Appendix 1. Dates that have been interpreted as being anomalous are identified with an asterisk*.

All of the excavations were carried out using trowels and hand shovels. Apart from some areas of the Ponamla and Arapus sites (see below) all excavated material was dry sieved then wet sieved through 2mm screens and sorted into separate categories on site. Datums were generally established adjacent to the excavated areas and depths were consistently measured from those points. The relative position of the datum points to the surrounding topography was calculated in relation to their height above sea (asl) which in all cases except for Mangaasi was an approximation of the mean high tide level. In the case of Mangaasi the datums were related to their height above a live coral datum (mean sea level) (see Appendix 2). Sites were initially excavated in spits of 10 or 20 cm until distinctive stratigraphic layers could be identified. Thicker stratigraphic layers were excavated in spits facilitating the identification of finer temporal change within those layers. The layer depths for the various test pits and excavated areas are given as centimetres below datum (cm bd). The chapter content is divided by island i.e. Erromango, Efate and Malekula.

Erromango

Two sites were excavated on the island of Erromango, namely Ponamla in the north and Ifo in the southeast (see Fig. 2.2). As noted in Chapter 2, both sites had been previously investigated, although somewhat briefly in the case of Ponamla, by Matthew Spriggs and Jerry Taki.

Ponamla 1994–5

This site was initially identified in the last few days of Spriggs' 1994 field season. The potential of the site had been recognised and brought to Spriggs' attention by former Vanuatu Cultural Centre fieldworker Sempit Naritantop, who identified pottery brought to the surface during posthole digging for a fence around the hamlet of Ponamla. The one by one metre test pit excavated in 1994 confirmed the rich nature of the site's deposits. In 1995 Matthew Spriggs and the author returned to the site for over five weeks of excavation. The work was again carried out with the assistance of Jerry Taki, the incumbent Vanuatu Cultural Centre fieldworker for Erromango, and a crew of local landowners.

Ponamla is a bay at the northern end of Erromango facing the island of Efate. It is a prime location for settlement with its sheltered bay facilitating canoe access and reliable freshwater supplied by the Ponamla River. The main site area appears to be a remnant Pleistocene alluvial terrace, subject to talus slope encroachment from the limestone hillslope at its eastern edge. Ponamla is a relatively undisturbed settlement site with cultural deposits dating from c. 2800 to 2500 BP. An areal excavation, Area A (Fig. 3.1), revealed what appeared to be a former cooking area, possibly also associated with pottery production. The presence of ash, charcoal and cooking stones, mixed with shellfish and faunal material indicated that the remains from hearths and ovens were being deposited in this area. Thousands of sherds, along with an assortment of other artefacts such as *Tridacna* adzes and arm rings, *Conus* shell rings, shell beads, a drilled shark tooth, bone needles, scoria abraders and stone flakes were recovered. Structural features were also recorded at the site. These were stone terraces which appear to have been constructed to form flat areas for the construction of houses and/or activity areas (Fig. 3.2). At least three levels of structural features were identified within the almost two metre deep cultural deposit. These features and the associated stratigraphy have been discussed in more detail by Spriggs (1999). The consistent and tight range of radiocarbon dates and the evidence for stylistic change in pottery within the deposit confirmed the stratigraphic integrity of the site.

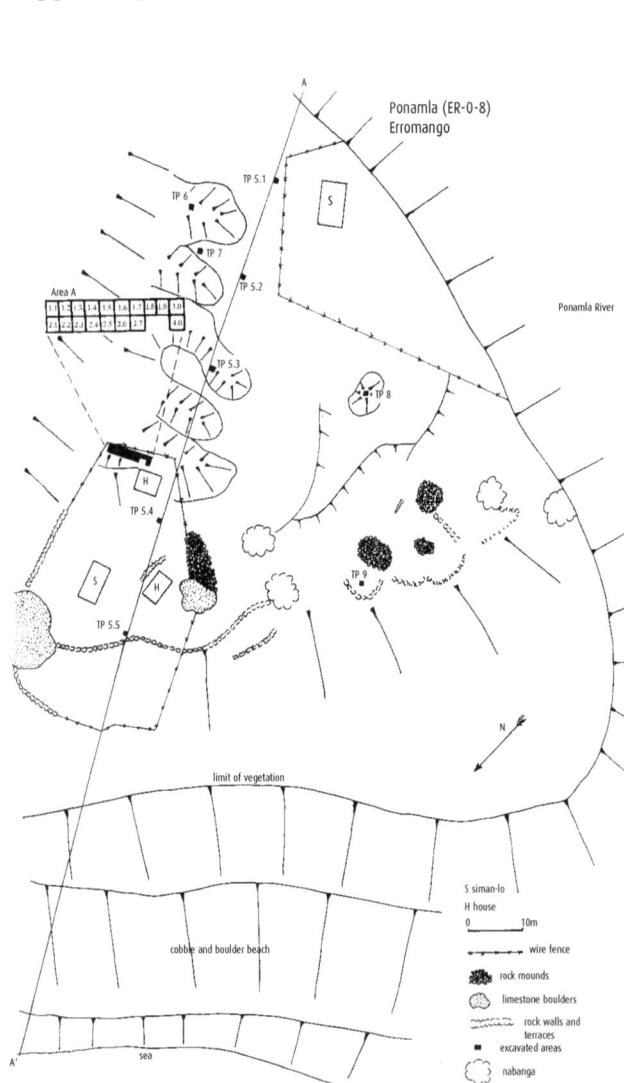

Figure 3.1 Ponamla, Erromango

Ponamla ER-0-8   Area A South Section

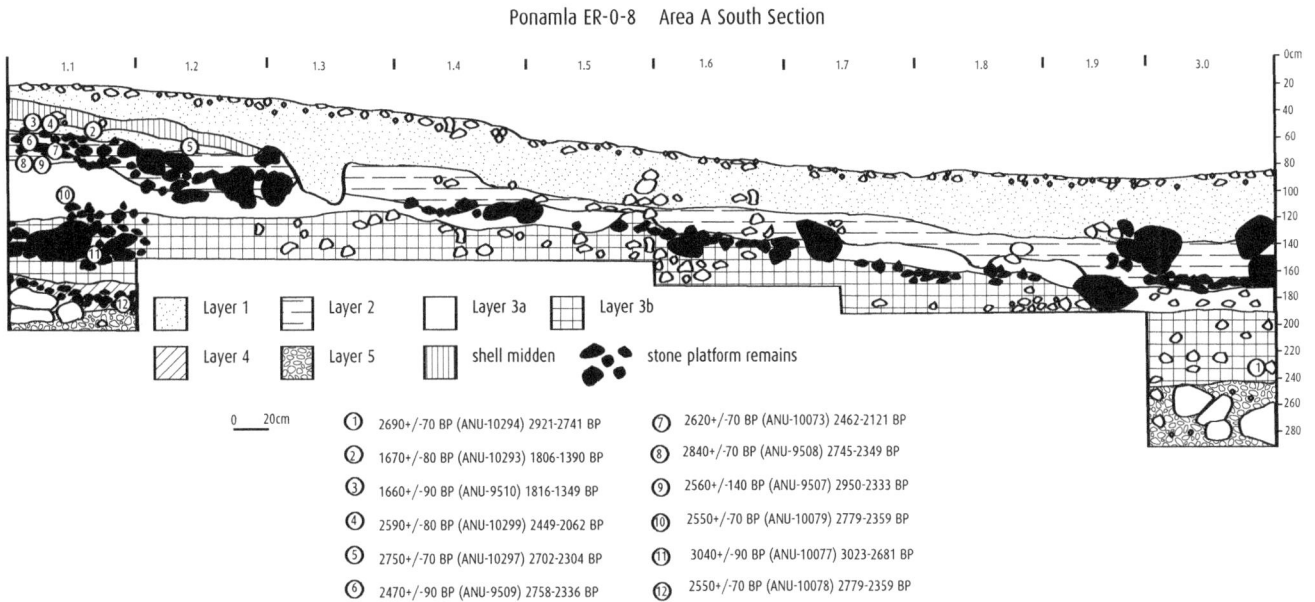

Figure 3.2 Ponamla Area A south section

A series of test pits (TP) were also excavated across the site (see Fig. 3.1). These were designed to determine the extent of the site, any spatial or temporal variation and to identify any distinct activity areas. The stratigraphy of the test pits (Fig. 3.3) is outlined in detail in Appendix 2. They were spread across the entire site and therefore often demonstrated quite different stratigraphies. Although several corresponding layers can be identified between the various test pits, the stratigraphic layers are labeled separately and discussed as per each test pit. Initially five test pits (TPs 5.1–5.5) were excavated across a north-south transect A–A', followed by another four test pits in various locations (TPs 6–9). Only the stratigraphy of Area A will be discussed in detail here.

Area A (12.64m asl, datum at south eastern corner). The areal excavation which covered an area of some 20m^2, ran along the central ridge of the most northerly mound feature (see Fig. 3.1). Five distinct layers were identified, although in only three square metres was the stratigraphy excavated to the basal sterile, namely the eastern and western ends of the excavation area (see Fig. 3.2). The excavated material from much of the central section of Area A (TP1.3/2.3–1.9) was only partly sieved. Concentrated *in situ* midden along with the associated stone platforms were contained within Layers 2, 3 and 4. Full details of the recovered midden materials are presented in other relevant chapters.

Layer 1 consisted primarily of a black (10YR 2/1) humic soil with frequent basalt cobbles. It has been formed after the intensive early occupation of the site through the agencies of slopewash and humic accumulation. The cobbles seem likely to have been associated with clearance activities, perhaps for gardening. The recovered artifactual material from this layer, nearly all associated with the initial settlement of the area, appeared largely to be in secondary deposition. Shellfish were sparse as were charcoal and ash deposits. Much later, more ephemeral use of the site was indicated at the western end of the excavation by a lens of shell midden (see Fig. 3.2) which would appear to correspond to a period dating to around 1600 BP (1660±90 BP [ANU-9510] 1816–1349 BP and 1670±80 BP [ANU-10293] 1806–1390 BP. An early date of 2590±80 (ANU-10299) 2449–2062 BP from a marine shell recovered from Layer 1 suggests it was in secondary deposition and that the date is more likely to relate to Layer 2. Layer 1 covered all of Area A, varying in thickness from 20–45cms. The very different matrices of Layer 1 and 2 provided a sharp delineation between the two layers.

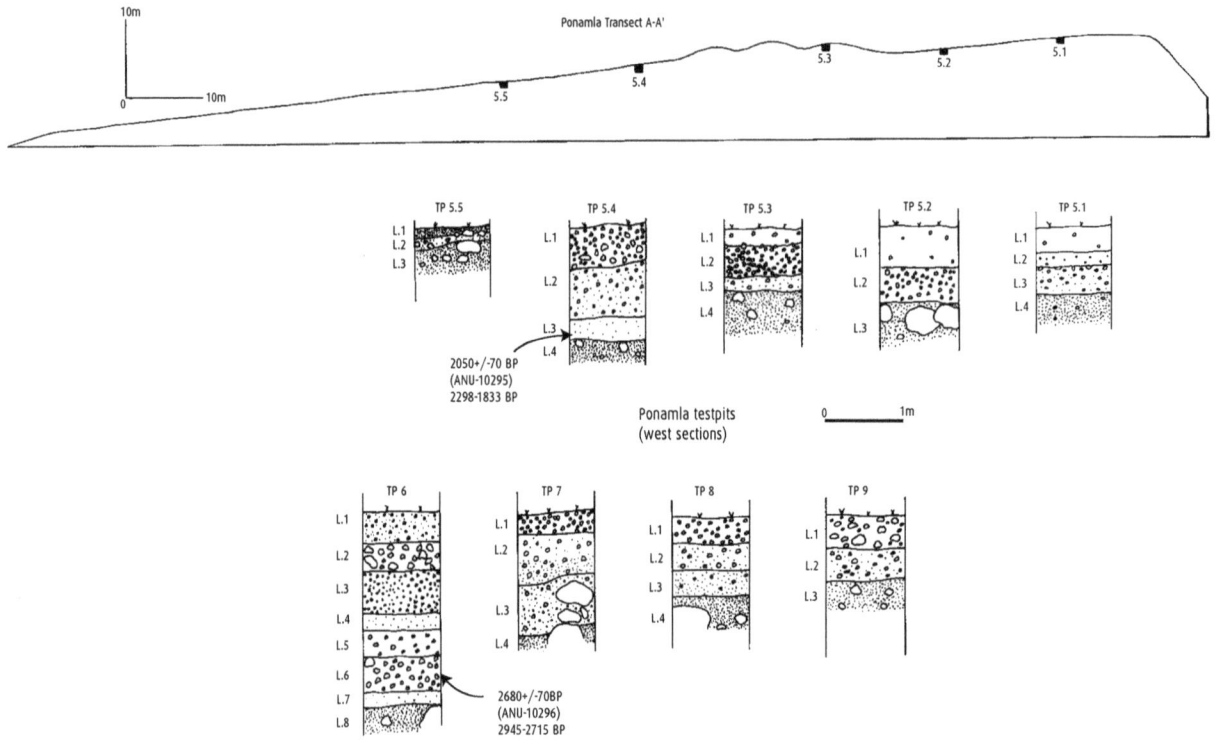

Figure 3.3 Ponamla Transect A-A' and testpit sections

At the interface of Layers 1 and 2, two marine shell samples [2620±70 BP (ANU-10073) 2462–2121 BP; 2750±70 BP (ANU-10297) 2702–2304 BP] provided a *terminus ad quem* date for the early occupation site.

Layer 2 marked the appearance of the *in situ* cultural remains. Large coral boulders which delineated the structural remains (platforms) were recorded in association with concentrated basalt and coral cobbles which provided an infill. Grayish brown (10YR 5/2) sandy silt and concentrated midden were also present. The pattern of boulders and cobbles was patchy across much of Area A at this level reflecting the presence or absence of stone platforms. Three radiocarbon determinations were returned from this layer, two from charcoal samples, namely 2560±140 BP (ANU-9507) 2950–2333 BP and 2470±90 BP (ANU-9509) 2758–2336 BP, and one from a marine shell 2840±70 BP (ANU-9508) 2745–2349 BP.

Layer 3a lay directly beneath the upper level of stone platforms and consisted of a friable very dark grayish brown (10YR 3/2) sediment, with large quantities of oven rake-out and associated midden debris up to 20cm in thickness. Ash lenses and charcoal concentrations were recorded throughout the layer. The build-up of the layer can be largely attributed to midden dumping. A single radiocarbon determination on a charcoal sample returned a date of 2550±70 BP (ANU-10079) 2779–2359 BP.

Layer 3b was very similar to above but the sediment was a darker brown (very dark brown 10YR 2/2). These layers could have been combined but the colour variation and the appearance of another platform at the eastern end of the excavation justified their separation. The delineation also provided a greater control on any temporal change through the somewhat thicker layer. The two dates gleaned from marine shell and charcoal samples respectively at either ends of the excavation area [3040±90 BP [ANU-10077] 3023–2681 BP; 2690±60 BP [ANU-10294] 2921–2741 BP] provided further indication of the rapid nature of the stratigraphic accumulation.

Layer 4 again comprised a very dark brown (10YR 2/2) sediment along with concentrated basalt cobbles. It was only recorded in the western end of the excavation and appears to be associated with an earlier stone terrace feature. This layer sat directly on top of sterile alluvium which provided a distinct boundary. A charcoal sample from the lowest level of this layer returned a radiocarbon date of 2550±70 BP (ANU-10078) 2779–2359 BP.

Layer 5 comprised of loosely compacted sterile very pale brown (10YR 8/3) river sand and silt with occasional limestone boulders and basalt cobbles. No cultural material was recovered from this layer. It represents a probable Pleistocene-age river terrace (Spriggs 1999) which at the time of human arrival provided a suitable flood-free environment for settlement. A 50cm deep sondage was excavated into this basal layer, confirming its sterile nature.

Discussion

The test pitting program enabled the extent of early settlement to be pinpointed. *In situ* midden and associated structural features were largely restricted to the mound features located on the eastern area of the river terrace. Stone terraces and platforms were constructed above the flood zone of the river and sea. These features became the focus for habitation and cooking activities and areas where refuse accumulated. After the abandonment of the site these features also became foci for the buildup of slopewash and humic accumulation. In areas away from the mound features the midden appeared to be generally in secondary deposition, the pottery was often worn in appearance through exposure and had been shifted around over long periods of time through various post-depositional processes. The recovered pottery also indicated that there does not appear to be any great temporal variation across the site. No doubt there were distinct activity areas within the site but the limited nature of the test pitting failed to reveal any evidence of this. Those test pits closer to the sea or river displayed limited stratigraphic depth before the sterile former beach or river terrace were reached. These areas appear to have only been utilised in the relatively recent past, either because of a drop in relative sea level or because tidal or river action has removed early deposits.

The post-occupational buildup was most dramatically demonstrated in TP 6 where over 160cm of overburden lies on top of the *in situ* deposit. It was not possible to determine the eastern extent of the site due to the increasing depth of overburden nearer the hillside.

Initial human occupation appears to have been relatively short-term and intensive with the site being abandoned after a few hundred years. The area appears to be a secondary colonising settlement on Erromango, perhaps 2–300 years after it was first settled by Lapita colonists possessing the full suite of dentate-stamped ceramics (see Ifo below). People arrived to colonise the area c. 2800 BP at a time when dentate-stamped Lapita was generally no longer being produced. Initially plainware dominated but over time fingernail and incised ceramics appeared. The ceramics are culturally transitional between Lapitoid plainware and a fingernail and incised tradition. People moved into a pristine environment and commenced an intensive exploitation of the local fauna and marine resources, highlighted by the rapid stratigraphic accumulation and overlap at two standard deviations of most of the radiocarbon dates from Layers 2 and 4. There is some indication of abandonment after 2500 BP, probably due to dual factors of resource depletion and the attraction of other readily available pristine environments. The first indication of a return to the area are dates of around 1600 BP associated with a shell midden suggesting ephemeral use of the area. People left with the ceramic tradition intact and returned without.

Ifo 1983 and 1996

This site is located on the southeast coast of Erromango (see Fig. 2.2), a coast that comprises an extensive area of recently raised coral reef known as the Imponkor Limestone (Colley and Ash

1971:48–49). Although tectonic uplift has increased the chances for the preservation of the earliest settlement sites, the raised coral reef along much of the south eastern coast presents today a very hostile environment in terms of canoe access. It may well be that this was not the case in the initial settlement period prior to this uplift.

Ifo is located near a reef passage and river outlet, a few hundred metres from the shore, on the north bank of the Ifo river which provides canoe access. The site is concentrated on a series of linear mound formations (Fig. 3.4). A number of these ridges run parallel to the river and appear to be former beach ridges while others run at right angles and are primarily made up of cultural material. Scattered cultural debris was noted over an area covering approximately 60 by 80m. The site is some 7–8m above sea level (mean high tide).

As noted earlier Spriggs and Taki recorded and tested the site in 1983 (Spriggs and Wickler 1989). The site was re-visited by the author in 1996 for six weeks in June and July during which time, again with the assistance of Jerry Taki along with a team of local landowners, a series of test pits and larger areal excavations were completed (Figs 3.4 and 3.5). The results from this period of excavation are outlined below. After a test pitting program (13, 1 by 1m test pits) to determine the area of the site and any temporal/spatial variance (see Appendix 2 for test pit details), it was found that the most productive and undisturbed area at Ifo was the ridge that Spriggs and Taki had tested in 1983. Upon completion of the test-pitting program, a larger area on this same east-west aligned ridge was excavated. Two parallel trenches some five metres apart and on either side of the ridge were excavated from the edge of the ridge into the centre and connected by a trench along the spine of the ridge (Trenches B, C and D), a total area of sixteen square metres (Fig. 3.4). This strategy was employed to gain information on the structure of the ridge and how it had been formed, along with the added goal of establishing the cultural sequence that was to be found within it.

Trenches B, C and D

These trenches accounted for the bulk of the excavated area at Ifo (16m^2) along with the greatest quantity of recovered midden remains. This excavated ridge consisted of a central core of flattish

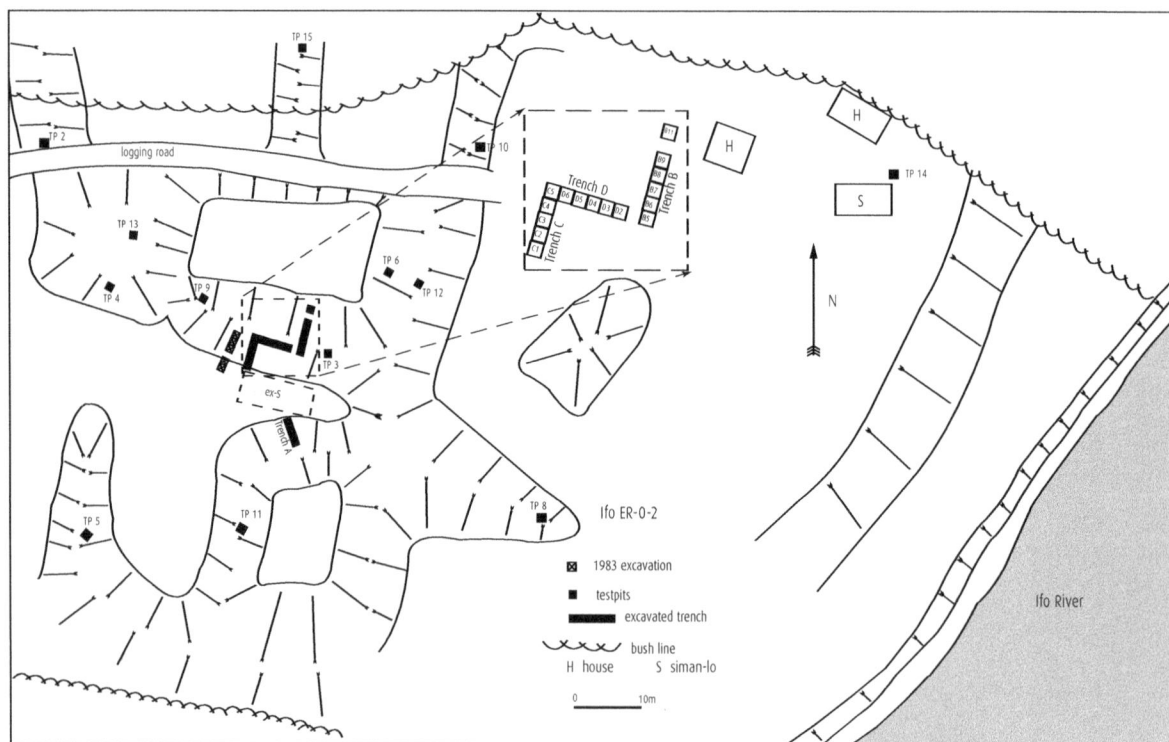

Figure 3.4 Ifo, Erromango

coral blocks on top of a relatively level degraded coral terrace (Fig. 3.6). This accumulation of coral blocks appears to have been the direct result of people clearing the area for settlement on first arrival at the site. Similar scattered coral blocks in uncleared areas can be seen further towards the coast. Once these linear piles of coral had been formed they appear to have served as a focus for the dumping of cooking debris and refuse, a practice that is still seen across Vanuatu and many parts of the Pacific today. Living areas are cleaned daily and over time mound features consisting of dumped material are built up. There is no permanent village in the vicinity of the site today but as recently as 1983 there were several houses and a s*iman-lo* (communal cook house) located on the flat areas adjacent to the linear mounds (Spriggs and Wickler 1989). No excavation was carried out on the flat areas of the site due to the very shallow stratigraphy noted by Spriggs and Wickler (1989).

The excavated ridge has thus been formed from a series of relatively undisturbed *in situ* dumping layers with a maximum accumulation of cultural material of up to 1.5m. The nature of the depositional processes involved at the site make it inevitable that some mixing of deposits would have occurred. Evidence of this is highlighted by the inversion of a number of the radiocarbon dates from the upper levels of the site. Despite this, the recovered midden remains and the clearly delineated layering of the mound features tend to indicate overall stratigraphic integrity. The datum for these trenches was located adjacent to B5 and was 9.07m above the high tide mark.

The most detailed stratigraphic record was found only towards the centre of the mound features and became less complete, particularly evidence of the earliest layers, towards the periphery (Fig. 3.7). Concentrated midden remains also tended to be located in the central spine of the mounds. Five distinct layers were identified:

Layer 1 consisted of concentrated water worn coral cobbles and firecracked basalt cobbles within a black (10YR 2/1) humic sediment. Recovered materials included frequent pottery, shellfish, bone and occasional artefacts. At the lowest level of this layer, two radiocarbon determinations, on charcoal and marine shell respectively, returned dates of 2690±70 BP (ANU-

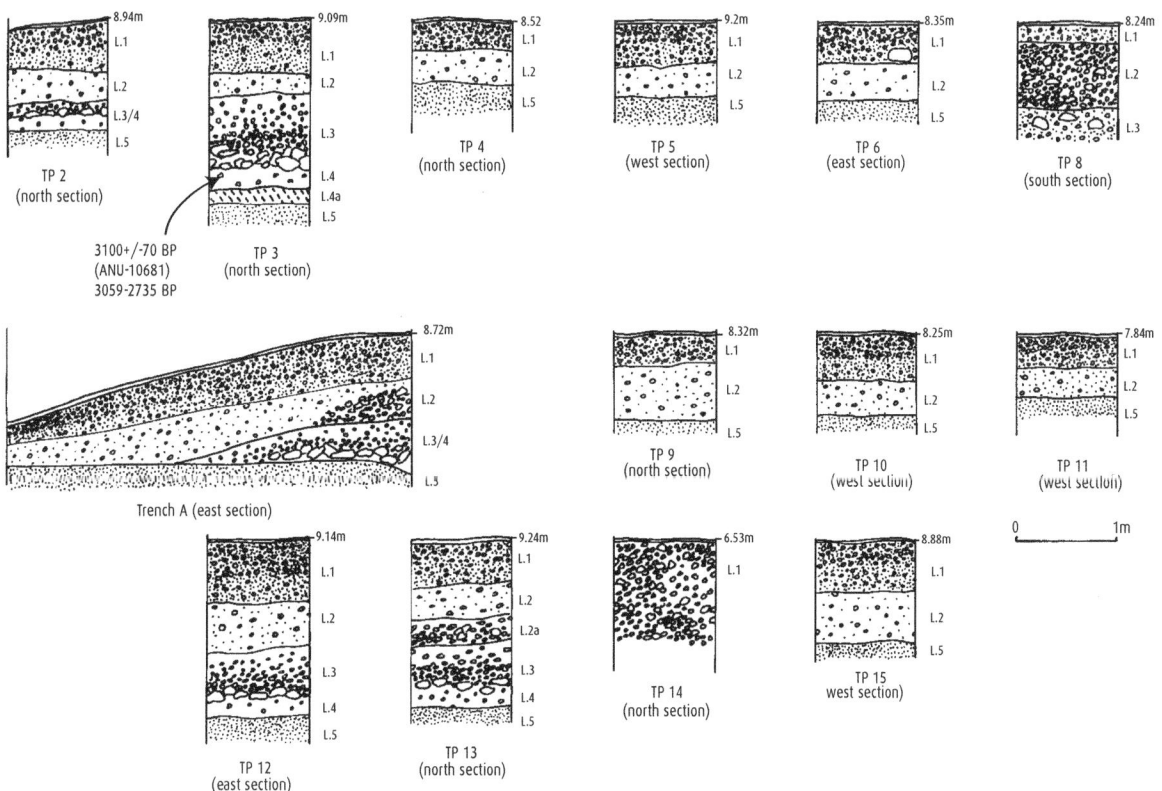

Figure 3.5 Ifo Trench A and testpit sections

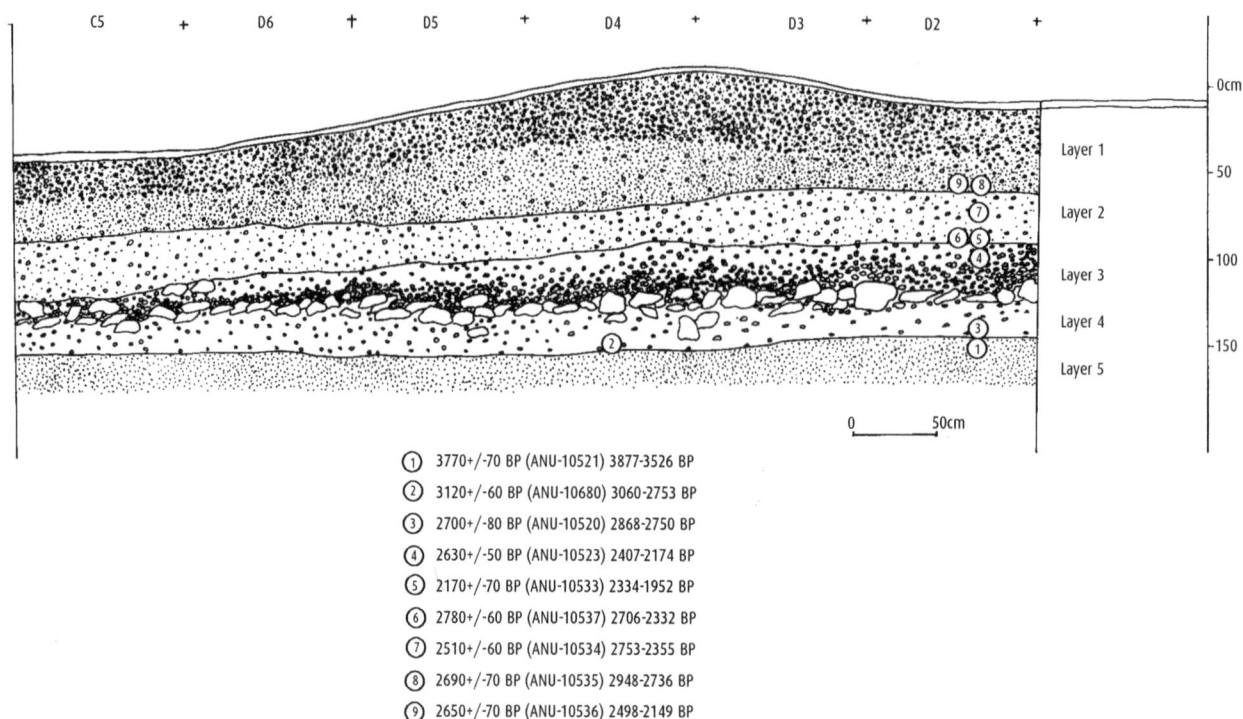

① 3770+/-70 BP (ANU-10521) 3877-3526 BP
② 3120+/-60 BP (ANU-10680) 3060-2753 BP
③ 2700+/-80 BP (ANU-10520) 2868-2750 BP
④ 2630+/-50 BP (ANU-10523) 2407-2174 BP
⑤ 2170+/-70 BP (ANU-10533) 2334-1952 BP
⑥ 2780+/-60 BP (ANU-10537) 2706-2332 BP
⑦ 2510+/-60 BP (ANU-10534) 2753-2355 BP
⑧ 2690+/-70 BP (ANU-10535) 2948-2736 BP
⑨ 2650+/-70 BP (ANU-10536) 2498-2149 BP

Figure 3.6 Ifo, Trench D north section

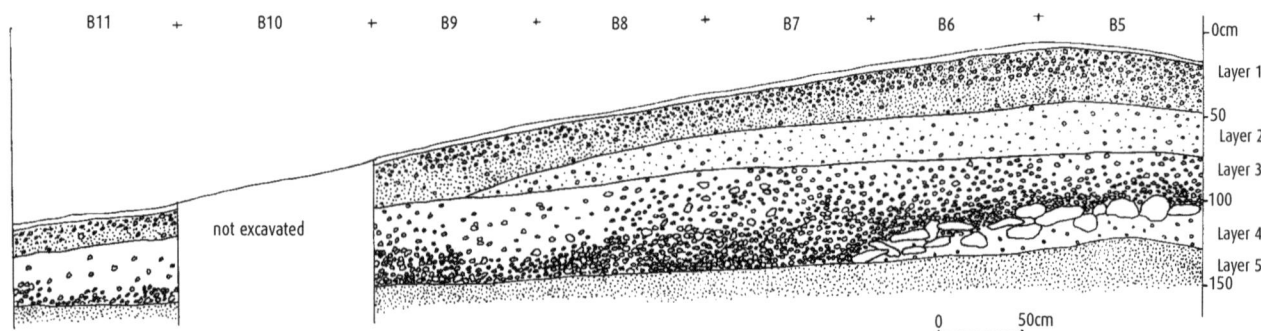

Figure 3.7 Ifo, Trench B east section

10535) 2948–2736 BP* and 2650±70 BP (ANU-10536) 2498–2149 BP. The date on the charcoal (ANU-10535) at least would seem, on the grounds of stratigraphic position and artefact association, to be somewhat anomalous. The recovered ceramics from this layer which are dominated by fingernail decoration associated with a single vessel form tend to support overall stratigraphic integrity. This layer graded into the layer below as the cobbles became less frequent with depth.

Layer 2 was a black (10YR 2/1) silty sediment with only occasional coral and basalt cobbles. Midden material was again frequent with patches of ash and charcoal also being recorded throughout the layer. Fingernail decorated ceramics continued to dominate. From the upper level of the layer a charcoal sample returned a date of 2510±60 BP (ANU-10534) 2753–2355 BP while at the lower level charcoal and marine shell samples respectively returned dates of 2170±70 BP (ANU-10533) 2334–1952 BP and 2780±60 BP (ANU-10537) 2706–2332 BP.

A distinctive change in the composition of the matrix signalled the presence of Layer 3. It consisted of concentrated basalt cobbles and coral boulders and flattish blocks amongst a very dark gray (10YR 3/1) sandy sediment. This layer was associated with early human activity in the area where level surfaces were being cleared of miscellaneous coral debris to facilitate settlement. Linear alignments of coral blocks and boulders then became foci for the regular dumping of midden. Frequent midden remains were encountered and it was from this layer that dentate-stamped and linear incised, calcareous-tempered sherds began to appear. Layer 3 was clearly defined only in the central core of the mound features. A marine shell from the upper part of this layer returned a radiocarbon determination of 2630±50 BP (ANU-10523) 2407–2174 BP.

Layer 4 lay beneath the concentrated coral debris and marked the interface of this earliest cultural layer and the sterile weathered coral terrace. It consisted of a dark yellowish brown (10YR 4/4) sandy sediment. Occasional coral cobbles and blocks were recorded. Midden materials were sparse but included a number of dentate-stamped Lapita sherds which were generally calcareously-tempered. Layer 4 was only recorded in the central core of the mound features. Marine shell and charcoal samples from the lowest level of Layer 4, returned radiocarbon determinations of 3120±60 BP (ANU-10680) 3060–2753 BP and 2700±80 BP (ANU-10520) 2868–2750 BP respectively.

Layer 5 was the cemented yellowish brown (10YR 5/4) gritty coral sand of the weathered coral terrace. A marine shell from this layer returned a radiocarbon determination of 3770±70 BP (ANU-10521) 3877–3526 BP, giving some indication of the time when the terrace was uplifted above sea level.

The above layers, where in evidence, could be directly related across the whole site (see Fig. 3.5 and Appendix 2). Water-borne pumice pebbles associated with regional volcanic activity were recorded throughout much of the excavated stratigraphy, particularly in the lower levels. In TP 3 a band of concentrated pumice pebbles (Layer 4a) lay directly on top of the coral terrace and would appear to relate to regional volcanic activity prior to human arrival at the site.

Trench A

This 4 by 1m trench was located south and opposite the areal excavation (Trench B–D) on the north facing slope of a linear mound feature. The trench ran from the centre of the mound to its edge. The datum, located at the southern end of the trench, was 8.72m above sea level (mean high tide). Five distinct layers were identified in the 130cm deep stratigraphy (Fig. 3.5). Midden remains, although sparse, were recovered from throughout the excavation.

Layer 1 consisted of a black (10YR 2/1) humic soil with concentrated water worn coral and basalt pebbles. Cobble content decreased with depth. This layer graded into the layer below. *In situ* midden remains included pottery, shellfish and bone. The recovered sherds were predominantly decorated with fingernail impression.

Layer 2 was characterised by an increasing sand and gravel content amongst the black (10YR 2/1) silty sediment with less frequent cobbles. However, toward the centre of the mound a concentration of cobbles was recorded which may represent an intensive clearance phase of use. Pottery (fingernail impressed), bone and shellfish were again recovered.

Layers 3 and 4 were recorded only in the central core of the mound and consisted of a concentration of flat coral blocks and boulders within a dark brown (10YR 3/3) sandy sediment. Several calcareously-tempered sherds were recovered along with sparse shellfish and bone. These layers, as previously noted above, are associated with the initial human settlement in the area, although evidence was somewhat ephemeral in this part of the site. Layer 5, the basal uplifted coral terrace, was encountered across the excavation at 130cm bd, at which point the excavation terminated.

Discussion

The excavation in 1996 was able to define comprehensively the nature, extent and cultural chronology of the Ifo site. Two predominant factors influenced the results of the excavation. Firstly, it was only excavation in the centre of the mound features that produced both more prolific midden remains and a complete stratigraphic history of the site. On the periphery of the mounds midden was more dispersed and the earliest cultural layers of the site were generally not present. Secondly, a number of the linear mound features were primarily made up of natural deposits and lay outside the area of settlement associated with ceramic use. These are taken to be predominantly naturally accumulated beach ridges. This was the case with TPs 8 and 14.

The remains of the initial colonising Lapita settlement, dating to c. 3000 BP, were concentrated south of the recent logging road as far as Trench A. Its most easterly extent is defined by the north-south aligned linear mound into which TP 12 was excavated (Fig. 3.4). Later occupation associated with distinctive fingernail and incised ceramics was concentrated in this area as well but was also thinly dispersed across a wider area of the site. Ceramic production may have continued up to around 2200–2100 BP (the most recent date recovered from Ifo was 2170 BP) but then disappeared from the cultural repertoire.

Efate

Mangaasi 1996–1999

The eponymous site of Mangaasi, located on the northwest coast of Efate, (Fig. 3.8) once again became the focus of archaeological investigation in August 1996 when Matthew Spriggs supervised four weeks of excavation at the site. This was the first year of the Australian National University-Vanuatu National Museum archaeological research and training program (Bedford *et al.* 1998, 1999) which continued at the site and nearby environs through to 1999 (Bedford and Spriggs 2000) and then again at the Arapus site from 2001 to 2003. The research at the site was designed to develop further the pioneering work of Garanger on the ceramic remains from the site, partly in light of the questions concerning the central Vanuatu pottery chronology raised by Ward in his 1979 thesis (see also Ward 1989). Garanger noted that he had some difficulty in interpreting the stratigraphy of the site and this is reflected in his summing up of the ceramic chronology. Researchers have questioned the validity of Garanger's Early to Late Mangaasi ceramic sequence and his proposed termination date for pottery production and use (Ward 1979, 1989; Spriggs 1997:179–81).

In 1996 four test pits (TPs 1–4) were excavated at Mangaasi both near to the earlier excavations carried out by Garanger and further inland (Fig. 3.9). This area of Efate experiences regular tectonic uplift estimated to have been some four to five metres in the last 3000 years. This was confirmed with the excavation of test pits, adjacent to the area excavated by Garanger, which reached the former reef at 3.6m below the ground surface (Fig 3.10, TP 3). This was up to some two metres below the basal levels of Garanger's earlier excavations in the same area and consisted of a series of former beach deposits. Water-worn pottery was found throughout the stratigraphy and although it was not possible to assign it to any particular style due to its condition, temper analysis carried out by Dickinson (see Appendix 3: WRD-138) on a number of waterworn sherds from the lowest levels of the test pits confirmed them as being of Efate origin. The presence of pottery in these levels suggested that people had been dumping refuse onto the beach below the high tide mark from a settlement further inland than the locations excavated in 1996. The recovered pottery from TPs 1 to 3 was broadly similar to material labelled Mangaasi by Garanger. However TP 4, some distance inland, revealed in its lowest layers (Fig. 3.11) ceramics similar to those excavated by Garanger at Erueti on the south coast of Efate, and called by him Erueti Ware.

Excavations in 1997 directed by Spriggs and Bedford extended the test pit transects further inland and parallel to the present shoreline. Relatively undisturbed, deeply-stratified (up to 2m), concentrated cultural deposits relating to an earlier occupation at the site were located and tested in 1997 through the excavation of seven 1 by 1 metre test pits (TPs 5–11) (Figs 3.10, 3.11, 3.12 and 3.13). Two of these were later expanded, namely TP 9 (2 by 2m) and TP 7 (3 by 1.5m). Pottery recovered from the more inland areas closest to the creek (primarily TP 9) consisted totally of Erueti-style ceramics. Mangaasi-style ceramics were generally recovered from the uppermost layers of the site or in test pits that were closer to the sea and thus clearly date to a later phase of occupation at the site. In a number of the test pits excavated in 1997 tephra layers were recorded. These are tephras that were initially thought (Bedford 2000b; Bedford and Spriggs 2000) to be associated with very large volcanic eruptions on Ambrym around 2000 years ago (Robin *et al.* 1993) and later of Kuwae in the Shepherds some 500 years ago (Robin *et al.* 1994). Subsequent analyses of the tephas have shown that the Ambrym tephra is in fact from nearby Nguna and the dates also suggest it was event that occurred several hundred years earlier than Ambrym at around 2200 BP. The later tephra may also be from Nguna but this has yet to be clarified. It does date to a similar period to that of the Kuwae eruption and until confirmed otherwise will continue to be referred to as such. However

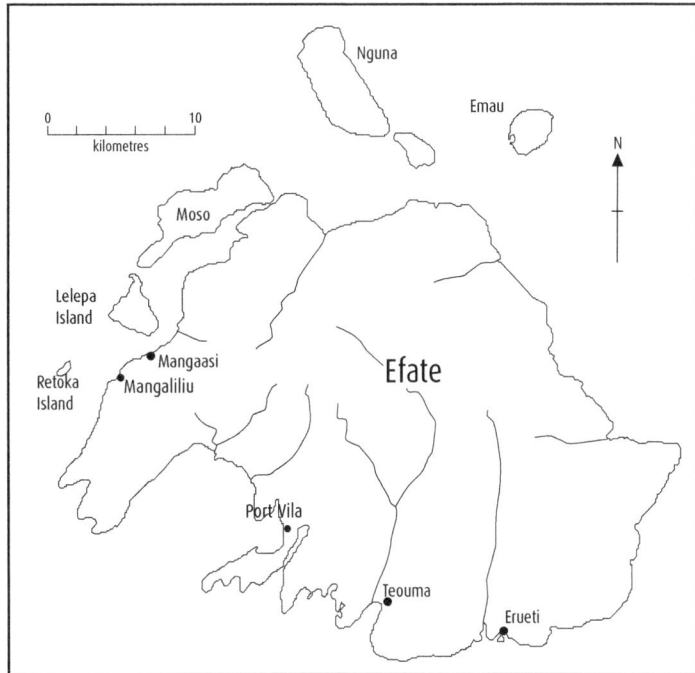

Figure 3.8 Efate

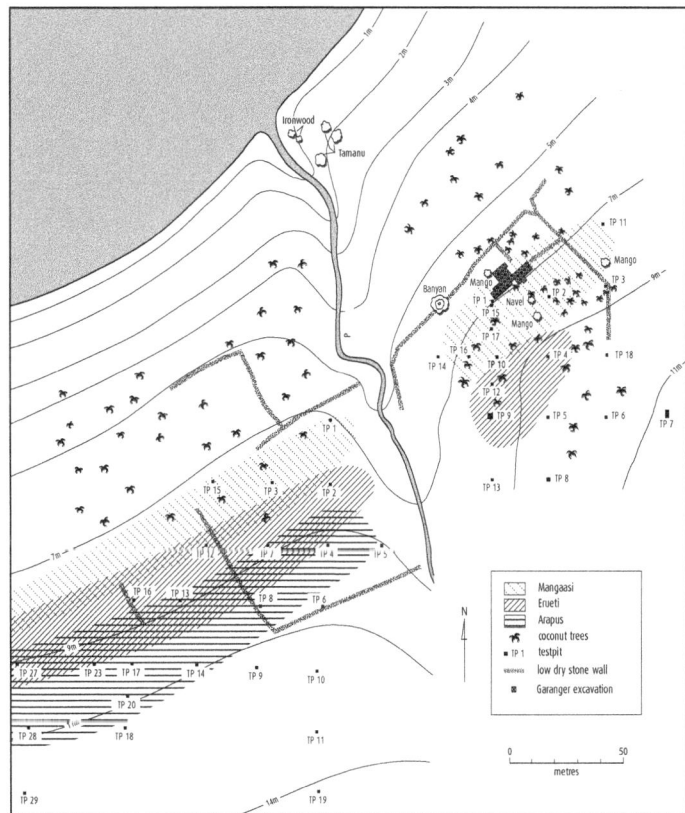

Figure 3.9 Mangaasi (right side of creek) and Arapus (left side), West Efate

whatever their origins, in test pits where these distinctive tephras were detected, they provided crisp chronological definition. When recovered from *in situ* deposits, Erueti-style pottery was found beneath the earlier Nguna tephra while the Mangaasi-style pottery was sandwiched between this tephra and that related to the Kuwae eruption (Fig. 3.12).

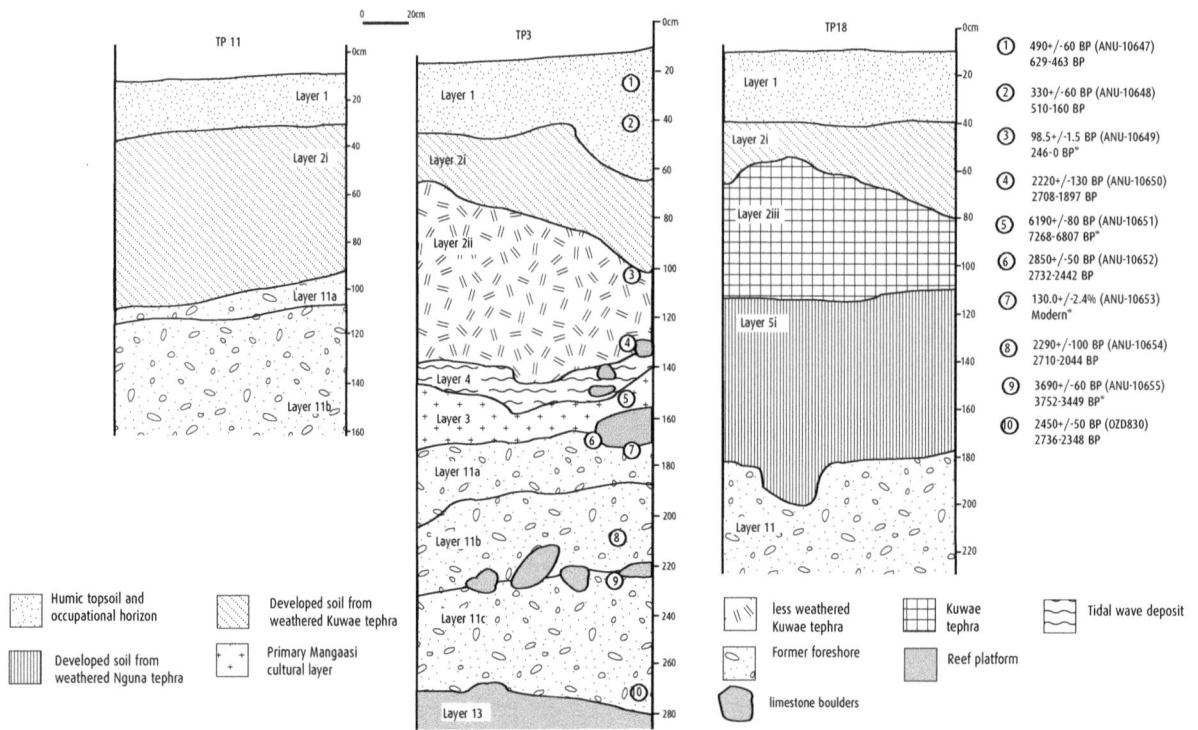

Figure 3.10 Mangaasi testpits 11, 3 and 18 east sections

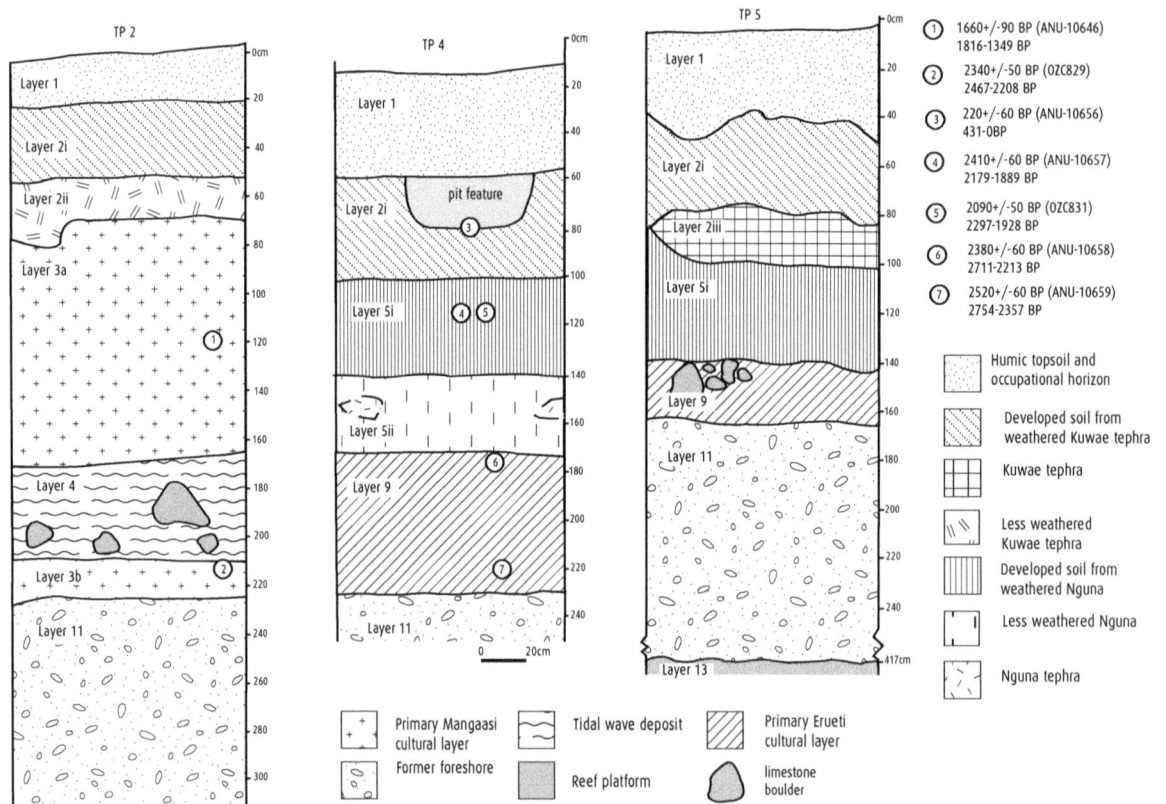

Figure 3.11 Mangaasi testpits 2, 4 and 5 east sections

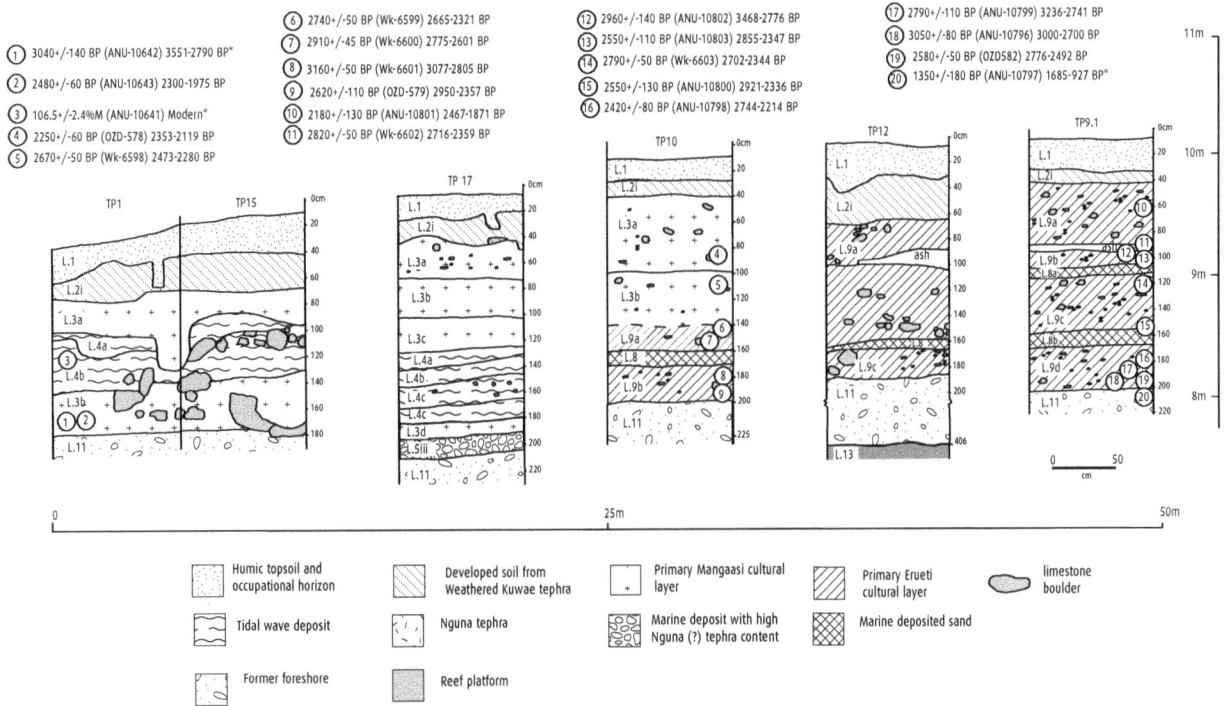

Figure 3.12 Mangaasi profile TP 1- TP 9.1 east sections

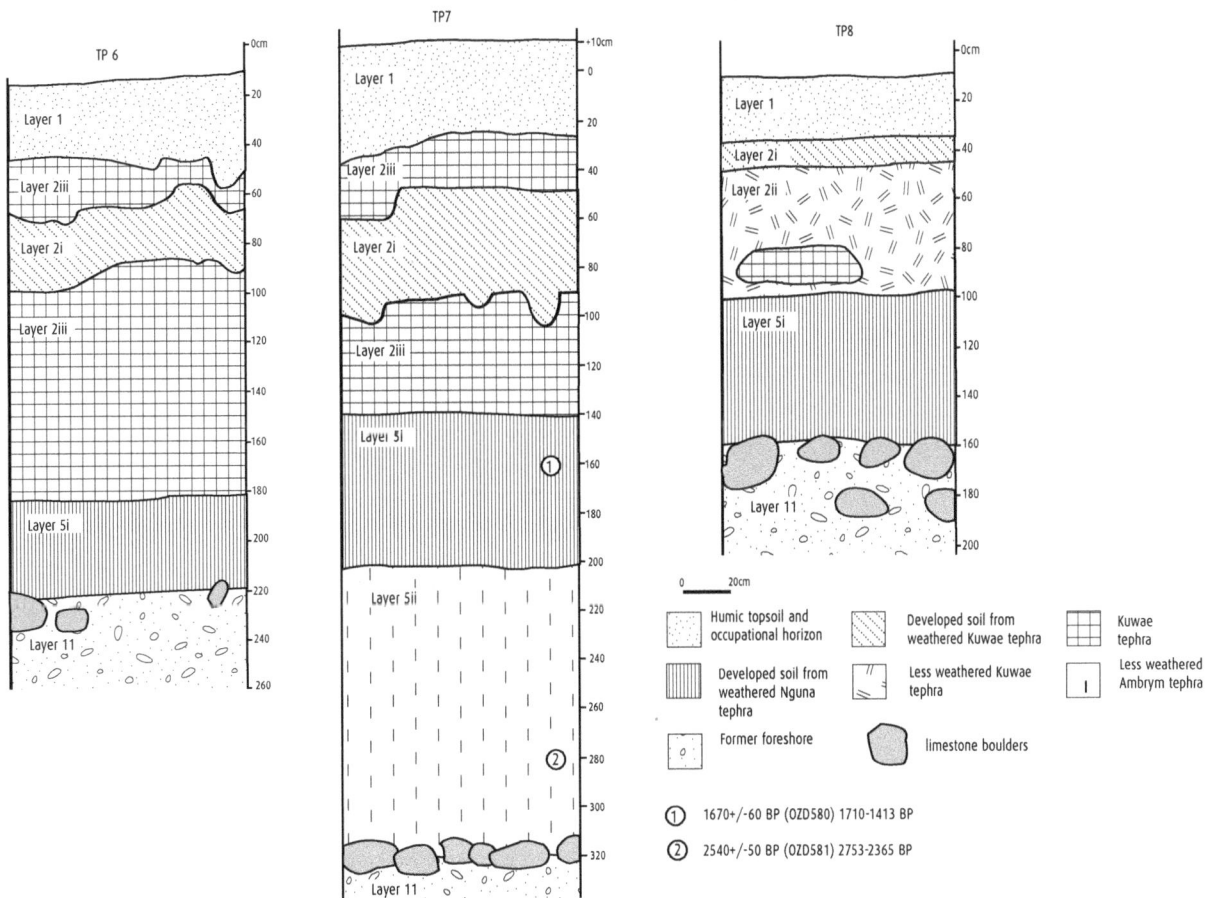

Figure 3.13 Mangaasi testpits 6, 7 and 8 east sections

In 1998 a further six 1 by 1m test pits (TPs 12–17) were excavated by Spriggs and Summerhayes primarily to define further the limits of the site and investigate the complex stratigraphy which included the evidence of tidal waves and other flood events along with the series of tephras (Fig. 3.14). These test pits confirmed the results of the previous years' work and enabled tighter delineation of the various phases of settlement. In 1999 a single 1 by 1 m test pit (TP 18) was excavated at Mangaasi to complete the grid pattern of investigation (Fig. 3.10).

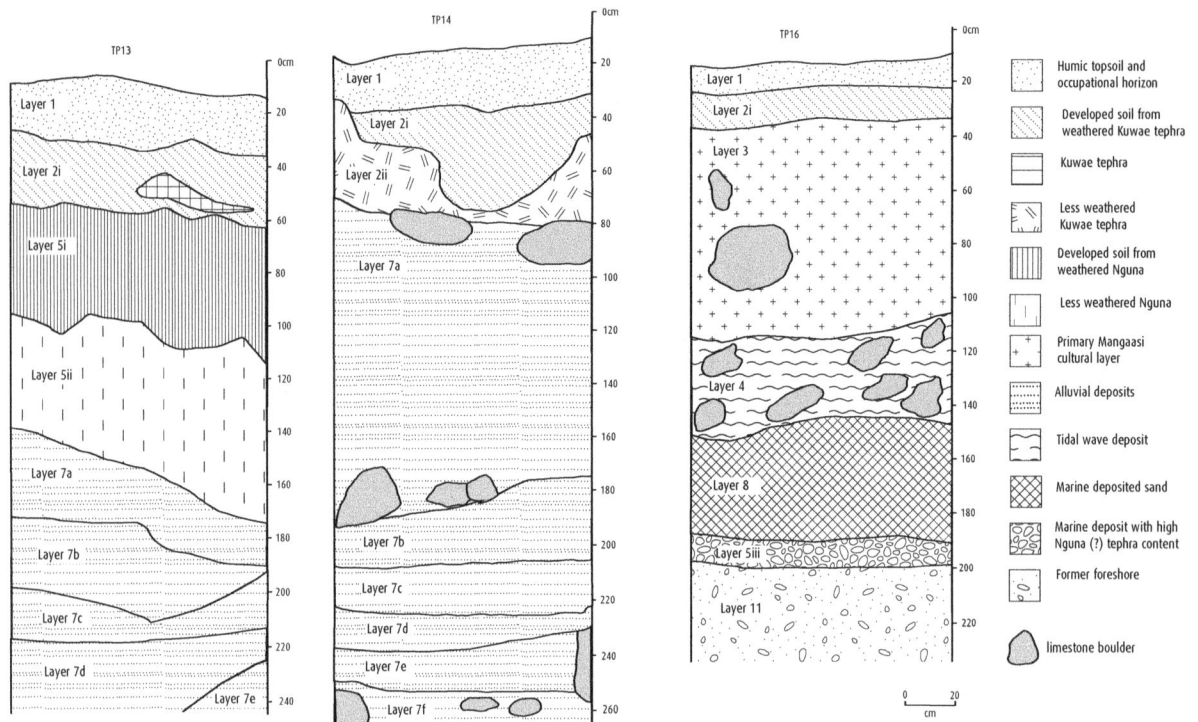

Figure 3.14 Mangaasi testpits 13, 14 and 16 east sections

Arapus 1999

The main thrust of excavations in 1999 directed by Spriggs and Bedford was the investigation of the stratigraphy southwest of the Pwanmwou Creek, on the opposite bank to the Mangaasi site. A grid of test pits orientated to the cardinal points and spaced at 25m intervals was laid across the site. This was an extension of the same grid system utilised at the Mangaasi site. As was also the case at Mangaasi, this southwest side of the river showed signs of habitation from the more recent past in the form of a number of low coral boundary walls across the site demarcating former households and gardens. A total of 24 one by one metre test pits were excavated during the six week field season (Fig. 3.9). These test pits were generally trowelled only and not sieved as the major emphasis in 1999 was to delineate the areal extent of the site.

The archaeological remains uncovered during the 1999 season proved to be quite spectacular and have greatly extended our knowledge of the settlement history of the area. Excavations on the southwest side of the Pwanmwou Creek have revealed much more extensive remains than those located at Mangaasi (Bedford and Spriggs 2000; Spriggs and Bedford 2001). These include an earlier phase of settlement than that identified at the latter site. This earlier phase of settlement appears to be associated with the first human arrival in the area and with a distinctive type of pottery cooking vessel i.e. plain globular pots with outcurving rims which are almost always notched on the lip. The pottery from this early phase has been named Arapus after one of the ancestral house areas where it was initially identified.

The test pitting program has confirmed what had earlier been indicated at Mangaasi, that the archaeological stratigraphy of the area is both horizontal and vertical. Settlements were located on or near beach ridges on the coast. However, intermittent tectonic activity during human habitation of the island had encouraged continual realignment of settlement which was required to compensate for the continually prograding shoreline. With the now well-dated and defined pottery sequences, namely Arapus, Erueti and Mangaasi, it is possible to chart both the settlement pattern in the area and the associated cultural transformations which occurred over time (Fig. 3.9). The Mangaasi site itself can now be seen as an outlier of the much larger site complex on the southwest side of the Pwanmwou creek.

At the Arapus site these various phases of settlement associated with the former beach ridges continue parallel to the present beach in a south westerly direction for at least 125 metres (established to 1999). The south western boundary of the settlement was not reached in 1999 but further research carried out in 2001–2003 has managed to establish that the site extends as least as far as the next creek to the southwest, a further 400 metres. Tephra layers related to the eruptions of Nguna and Kuwae could be identified in many of the test pits at Arapus, as was the case in the 1997 and 1998 excavations at Mangaasi (Figs 3.10–3.14). The excavations at Arapus also revealed the presence of a third, as yet unidentified tephra, directly underlying the initial human occupation of the area.

Despite the spectacular nature of the results from the excavations at Arapus, the limitations of thesis research and publication deadlines require some restriction and definition of research boundaries. Although only the results from the excavations at Mangaasi are presented here in any detail, it has been necessary to include key aspects of the stratigraphy recorded from the excavations at Arapus. Three Arapus test pits (Fig. 3.15) are outlined in some detail, namely TPs 4,

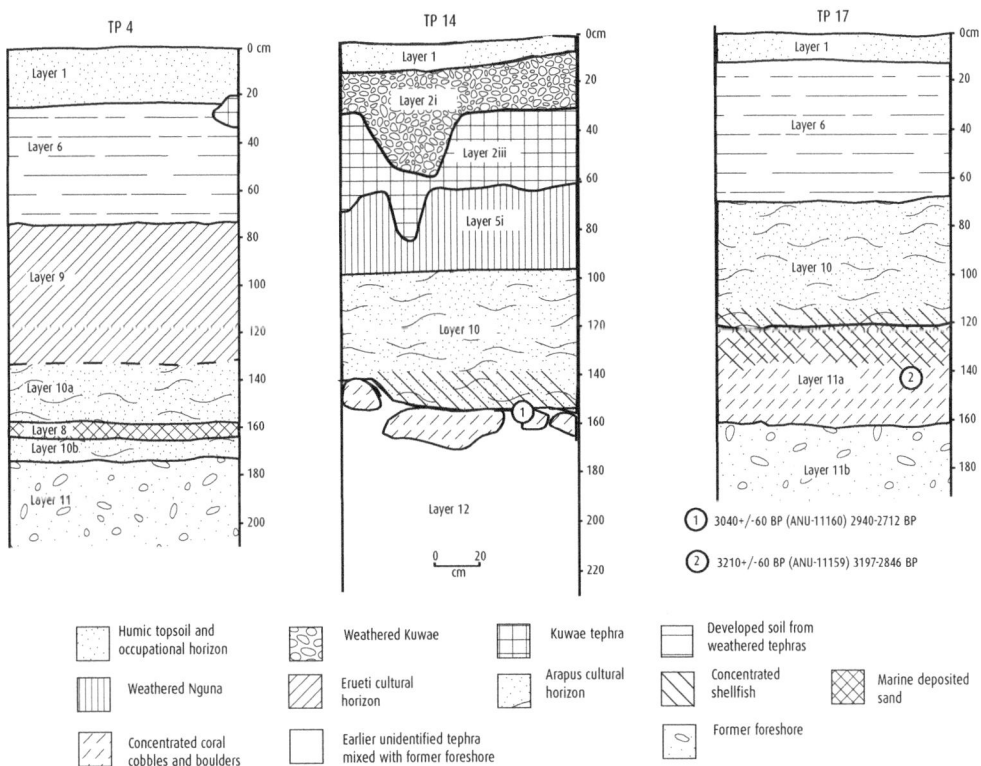

Figure 3.15 Arapus TP 4, 14, and 17 south sections

14 and 17 (see also Appendix 2). They provide evidence relating to the initial settlement of the area including deposits of the distinctive Arapus-style pottery and demonstrate the transition from Arapus to Erueti ceramic styles. Key elements of the recovered ceramics from Arapus, including an expanded repertoire of motifs and vessel forms along with chronological information have also been incorporated into relevant chapters, as have aspects of the faunal and non-ceramic artefact remains.

Mangaasi stratigraphy

Due to the complex nature of the stratigraphy and the chosen excavation strategy it was necessary to assign all layers that were found across the site a numeric label. In the case of the tephra layers (2 and 5) Roman numerals have been added to further distinguish variation. These layers only are able to be correlated across the whole site. This is in contrast to other layers where differentiation within a layer is shown by the addition of a letter to a layer's numeric designation. These letter designations are not correlated across the site but are test pit specific. The layers are not labelled serially and in some cases numeric inversion was inevitable. All 18 test pits are described in detail (Appendix 2) and are accompanied with stratigraphic illustrations (Figs 3.10–3.14). The descriptions of the test pit stratigraphy have been established through a combination of sedimentally and culturally defined layers. A total of 18 distinctive layers and/or primary cultural horizons were identified across the site (see below).

Four broad primary cultural layers were defined at the site. The most recent post-dates the eruption of Kuwae in 500 BP. Remains were sparse (ceramics were not a component of the cultural repertoire) and consisted principally of concentrated coral pebbles from former house floors and occasional burials and other features. Many of these features may relate to the settlement associated with the legendary Chief Roy Mata (Garanger 1972). The three other cultural layers, namely Mangaasi, Erueti and Arapus, were defined principally from ceramic remains. Primary deposits of these three cultural layers displayed some consistency, comprising very dark gray to black (10YR 3/1–2/1) charcoal-rich silty sediments with concentrated coral and basalt cobbles and midden remains.

Figure 3.9 shows both the layout of the test pits and the respective areas where Erueti and Mangaasi cultural horizons were concentrated. A north to south aligned transect (TP sections 1/15, 17, 10, 12 and 9) demonstrates the horizontal and vertical nature of the stratigraphy (Fig. 3.12) and the relationship between the Erueti and Mangaasi cultural horizons.

The *in situ* Erueti cultural horizon was identified in five test pits (TPs 9, 12, 10, 4 and 5) located at the rear of the site and close to the Pwanmwou Creek while the *in situ* Mangaasi cultural horizon was largely restricted to test pits closer to the sea namely 1, 2, 15, 17, 16 and 10. Only in TP 10 was the transition from the Erueti to the Mangaasi cultural horizons clearly demonstrated stratigraphically (Fig. 3.12). Seven of the excavated test pits (TPs 6, 7, 8, 11, 13, 14 and 18) lay largely outside the area of concentrated settlement. The stratigraphy of TPs 13 and 14 comprised principally of alluvial deposits (Fig. 3.14) while the stratigraphy of TPs 6, 7, 8 and 18, located on or near the base of the slope at the rear of the site, was dominated by weathered or more pure tephra deposits transported by slopewash from the hillside.

The 18 layers identified at the Mangaasi can be summarised as follows:

1) A black to dark gray (10YR 2/1 to 10YR 4/1) humic topsoil which was found across the whole site. In most cases patches of concentrated coral pebbles were found in the upper levels of this layer which are associated with former house floors from the latest phase of occupation of the site (post 500 BP).

2i) A very dark gray (10YR 3/1) developed soil derived principally from weathered Kuwae tephra. Coral pebbles were absent and midden remains were sparse and if recovered generally in secondary deposition (post 500 BP).

2ii) A dark gray (10YR 4/1) less weathered Kuwae tephra differentiated by its high tephra and low soil content ratio (post 500 BP).

2iii) Light brownish gray (10YR 6/2) sterile Kuwae tephra associated with the eruption of Kuwae in 498 BP (Robin *et al.* 1994). This layer was easily identified at test pits near or on the slope at the rear of the site where the layers were up to one metre in thickness. In a number of test pits, only patches of pure Kuwae tephra were recorded. These appeared to be the remnants of layers that had been mixed through subsequent gardening activities. An extra layer of this tephra was noted in TPs 6 and 7 which appeared to be derived from slumping due to slope failure.

3) The primary Mangaasi occupational layer dating from c. 2200–1200 BP was located in test pits further towards the sea but also concentrated near the Pwanmwou Creek. The overall matrix of this layer comprised a very dark gray to black (10YR 3/1–2/1) charcoal-rich silty sediment with concentrated coral and basalt cobbles and midden remains. Ash and charcoal lenses were also often recorded.

4) A tidal wave deposit was identified across TPs 1, 15, 2, 3, and 16. It consisted of coral debris such as sand, cobbles and coral blocks. The stratigraphic integrity of the up to 40cm thick layer indicates it was a sudden event (well before 600 BP and post 2200 BP, probably c. 1400 BP).

5i) A dark gray (10YR 4/1) developed soil derived largely from weathered Nguna (post 2200 BP) tephra.

5ii) A less weathered grayish brown (10YR 5/2) Nguna tephra, differentiated by its high tephra and low soil content ratio (post 2200 BP).

5iii) A light gray (10YR 7/1) marine deposited sand with a high (Nguna) tephra content was recorded in TPs 16 and 17. These layers lay directly on top of the foreshore deposits.

5iv) Gray (10YR 5/1) Nguna tephra (c. 2200 BP). This layer was clearly identified in test pits close to the slope at the rear of the site. As noted this tephra was previously mis-identified as deriving from Ambrym. Our dating of the site has enabled us to tentatively date this Nguna tephra-fall to around 2200 BP or slightly earlier.

6) A very dark gray (10YR 3/1) developed soil from the two mixed, weathered tephras. This tephra-rich layer occurred in areas of the site where there was no evidence of settlement between the two eruptions, nor remnants of more pure tephra, to provide a clear delineation between the Kuwae and Nguna tephra content.

7) Grouped together here simply as alluvial deposits, these layers were restricted to test pits near the Pwanmwou Creek (TPs 13, 14). The deposits demonstrated clear stratification and are described in detail in Appendix 2.

8) White to light gray (10YR 8/1–7/1) marine deposited sands derived from cyclonic events were recorded in TPs 10, 12, 9 and 4 (Figs 3.11, 3.12). These tended to be only 5–10cm thick and were restricted to the earlier period of the settlement. The marine sands confirmed the close proximity of these locations to the former foreshore.

9) The primary Erueti occupational layer dated from c. 2800 to 2200 BP and at Mangaasi was concentrated in the inland area of the site close to the Pwanmwou Creek. As noted above the matrix of the layer consisted largely of a very dark gray to black (10YR 3/1–2/1) charcoal-rich silty sediment with concentrated coral and basalt cobbles and midden remains. Ash and charcoal lenses were also often recorded.

10) The primary Arapus occupational layer dating from c. 2900 BP to 2800 BP was found only on the southwest side of Pwanmwou Creek at the Arapus site (Figs 3.9, 3.15). See above (Layer 9) for matrix description.

11) Former foreshore deposits comprised of water worn coral debris including branches, cobbles and pebbles and gravel. A number of these foreshore layers were excavated to the former reef platform. Water worn pottery was found in all test pits closer to the sea indicating that

earlier settlements were dumping midden into the inter-tidal zone. This layer was exposed in all test pits.

12) A very pale brown to white (10YR 7/3–8/2) tephra mixed with the former foreshore deposit. This layer was found in only one test pit at the Arapus site (TP 14, Fig. 3.15) and is an earlier unidentified tephra which pre-dates human arrival in the area (pre 2900 BP).

13) The former reef platform was exposed only in TPs 3, 5 and 12. It is some 7m above the live coral datum (mean sea level).

Discussion

Through the intensive test pitting program comprising a total 25.5m^2 at Mangaasi and 24m^2 at Arapus it has been established that the complex stratigraphy in the area is both horizontal and vertical and has been affected by cataclysmic events such as volcanic eruptions, tidal waves and/or cyclones along with more low key but regular activities such as gardening. The initial settlement pattern in the area was focused on beach ridges above and parallel to the foreshore which was continually prograding over time. The regular tectonic activity during human habitation of the island led to continual realignment of settlement to compensate for the shifting shoreline. The results of the excavations have further confirmed the related nature of the ceramics on Efate.

Malekula

Fieldwork on Malekula began in 1995 and continued in 1996 with the assistance of Jimmy Sanambath, the Vanuatu Cultural Centre fieldworker for northwest Malekula. The work consisted of an intensive survey along a ten kilometre stretch of the Northwest coast from Tenmiel to Tenmaru (Fig. 3.16), along with a series of transects into the interior. A number of perennial rivers drain from the interior and are often associated at their mouths with sheltered bays which represent prime areas for settlement. Malo Island with its numerous Lapita sites, can be clearly seen from the Northwest. Some 50 caves and/or rockshelters were recorded at varying altitudes. In an initial attempt to define the cultural sequence of the area some fifteen of the caves and four open sites (Malua Bay, Nuas, Fiowl and Chachara) were excavated (Fig. 3.16). The details of a number of the cave sites are presented first (in sequentially recorded order) followed by the open sites. A number of the excavated cave sites (Waprap Mk-3-30, Oochmenoch MK-3-38, Peckhara Mk-3-45, Ndavru Mk-3-46) returned what could be generously described as meagre cultural remains and they have been excluded from this publication. However a full report of all excavated cave sites can be found in Bedford 2000b. Recovered materials are presented in detail in relevant chapters.

Malekula cave sites

Woplamplam (Mk-3-26)

This cave site is located several hundred metres northeast of the present Tenmiel village some 60m above sea level. The cave is used for shelter during hurricanes. Numerous firecracked stones and fireplaces were noted on the ground surface. A 2 by 1m test pit, some 4m inside the cave entrance, was excavated to a depth of 1.05m (Fig. 3.17). The stratigraphy indicated the cave had been occupied intermittently with a series of fireplaces and cultural horizons being recorded throughout. Accumulation of sediment at the site could also be attributed to intermittent slopewash from the cliff above the entrance of the cave.

Layer 1a (10–18cm bd) consisted of a powdery ash and grayish brown (10YR 5/2) sediment. Frequent firecracked cooking stones were recorded. It was very similar to Layer 1b below but

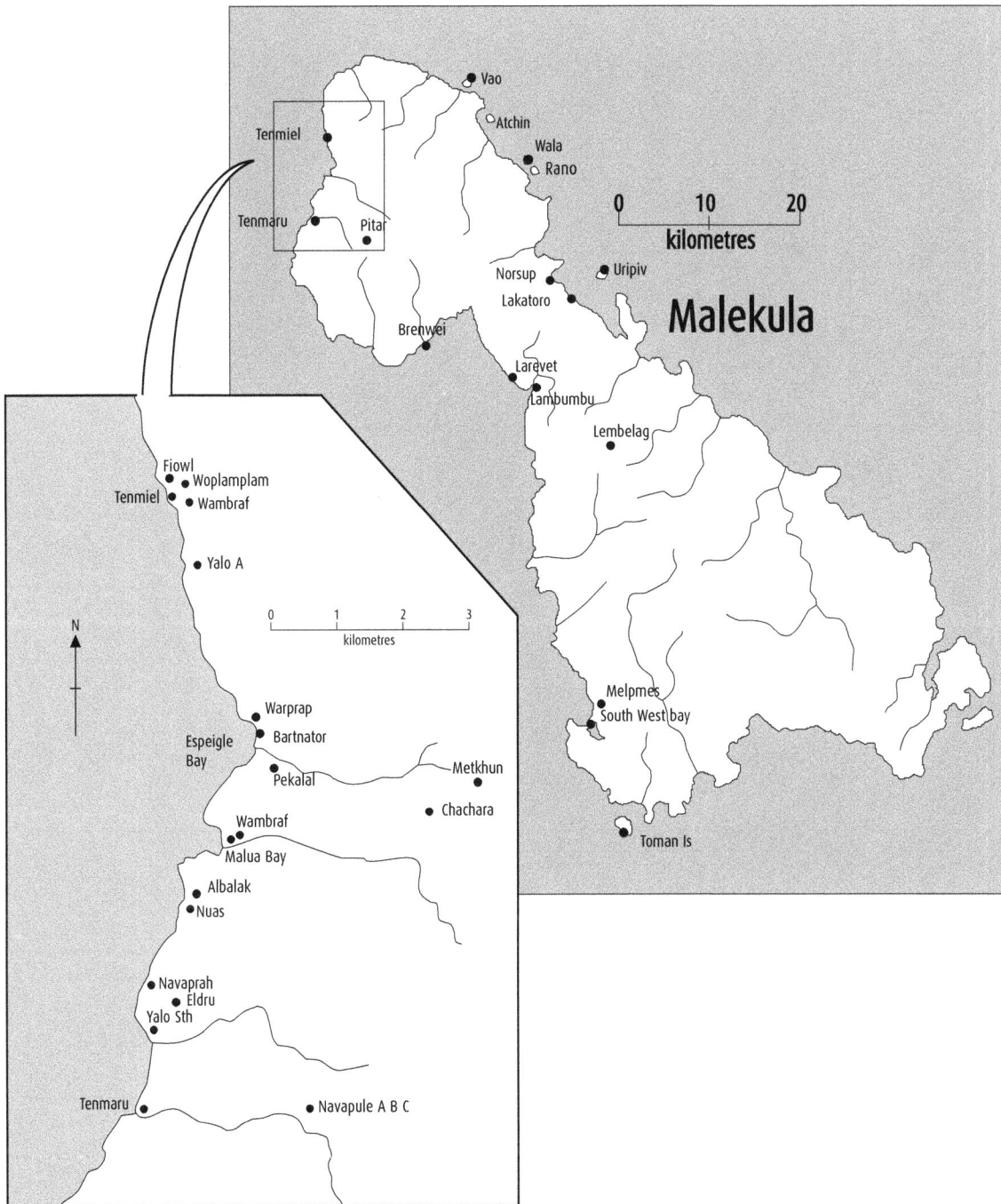

Figure 3.16 Malakula and Northwest region

separated by a thin ash lenses at 18–22cm bd which covered the whole square. Bone, pottery and shellfish were recovered.

Layer 1b (22–46cm bd) was essentially the same as that above and graded into the layer below. Frequent fire-blackened coral cobbles and occasional basalt cobbles were noted. Three firescoop hearth features were recorded within the layer. Concentrations of small intact bones suggested that much of the material was related to the cave being used as an owl roost. Fishbone was also present along with sparse shellfish. No pottery was present within this layer.

Layer 1c (46–54cm bd) was similar to above but more hard-packed and darker (dark grayish brown 10YR 4/2) due to more sediment content from slopewash. Frequent coral cobbles were noted along with continued evidence that suggested (concentrated bone) the cave had been used

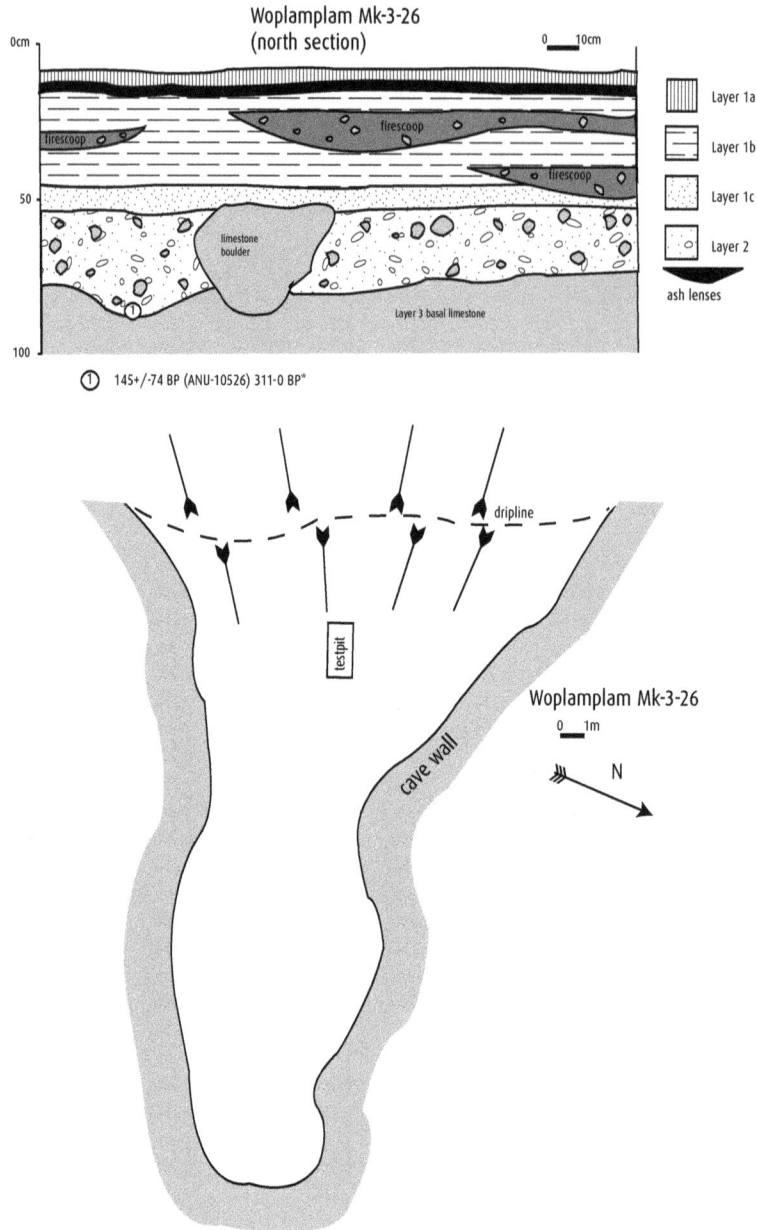

Figure 3.17 Woplamplam testpit section and plan

as an owl roost. Sparse shellfish and plain pottery (6 sherds) were also recovered. Much of Layer 1 can be attributed to accumulated debris from cooking and temporary occupation with an added component of slopewash.

Layer 2 (54-80/105cm bd) marked the basal cultural layer of the cave and comprised of a gray (10YR 6/1) powdery sediment with frequent angular coral cobbles and boulders, possibly roof-fall caused by tectonic activity. A distinct boundary between this layer and the lower layer could be identified. From the bottom level of this layer (80–105cm bd) a charcoal sample was dated (145±74 BP [ANU-10526] 311–0 BP*). The date seemed inconsistent both with the depth of stratigraphy and the ceramic remains. Fishbone was prevalent and in association with sparse shellfish remains and two calcareously-tempered sherds.

Layer 3 was the basal limestone of the cave floor. A thin largely sterile dark yellowish brown (10YR 4/6) silt lay directly on top of the basal limestone. It would appear to represent material accumulated through the natural weathering of the cave bedrock prior to human arrival. Two calcareous-tempered sherds recovered from this layer clearly derived from the layer above.

Intermittent occupation was seen throughout the stratigraphy at Woplamplam in the form of firescoops and cooking debris. Sediment accumulation can also be attributed to intermittent slopewash. The dispersed charcoal sample from the lowest cultural level of the cave seemed to be inconsistent with both the one metre of stratigraphic accumulation and the recovered midden remains. Bone, shell and pottery were recovered from throughout the three distinct layers (Fig. 3.17). Many of the bone remains which were often in nested deposits appear to be the result of owl roosting in the cave (Steadman pers. comm.). The recovered pottery comprised only (13) plain sherds, thicker in the upper layers and thinner and calcareously-tempered in the bottom layer.

Bartnator (MK-3-32)

This wave-cut coastal rockshelter is located north of Espeigle Bay some 100m above the coastal road and 30m above sea level. A freshwater creek is located nearby. A 1 by 1m test pit was excavated to a depth of 1.35m (Fig. 3.18). A total of six layers were identified.

Layer 1 (4–12cm bd) was a light gray (10YR 7/1) ashy sediment with concentrated fire-cracked stone representing more recent cooking activities. A single thick sherd of pottery which had been thoroughly burnt was recovered from amongst the stone debris and seemed likely to have been incorporated with and re-used as an oven stone. No other midden materials were recovered.

Layer 2 (12–45cm bd) consisted of a powdery dark gray (10YR 4/1) silt with frequent small-medium fire-cracked rocks. The lack of ash clearly separated this layer from that above and the firescoop below. Slopewash appeared to have contributed at least partly to the build-up of this layer. Sparse shellfish remains only were recovered.

Layer 3 (45–65cm bd) essentially comprised a substantial firescoop which covered the entire test pit. Although technically a feature, rather than a layer it was designated as a separate layer. It consisted of a light gray (5YR 7/1) ashy sediment with concentrated

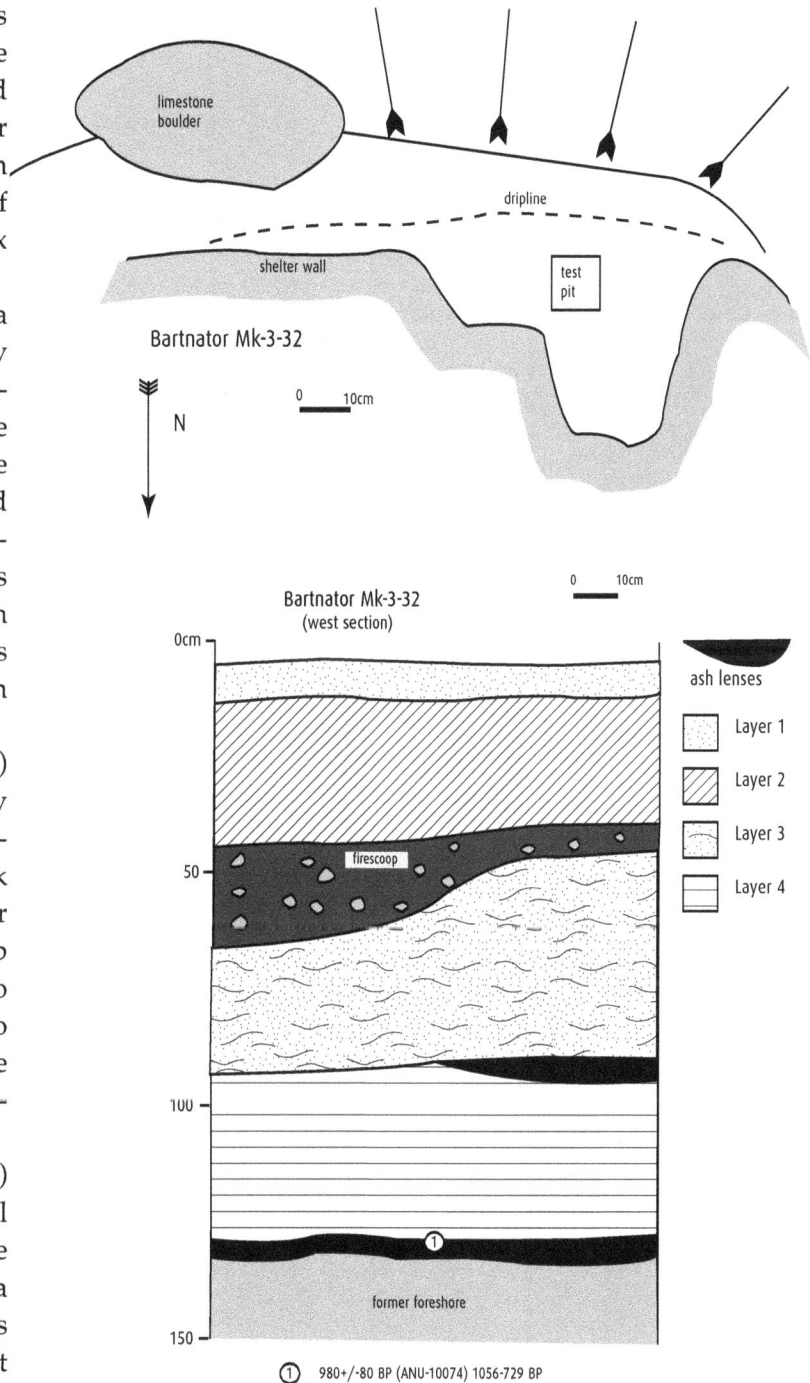

Figure 3.18 Bartnator testpit section and plan

burnt coral pebbles and cobbles, further evidence of cooking activities. Sparse shellfish remains were all that were recovered.

Layer 4 (65–85cm bd) was clearly delineated from the layer above by a change in colour and the disappearance of any ash content. It comprised a dark reddish gray (5YR 4/2) silty sediment with frequent coral cobbles and pebbles. Accumulated slopewash appeared to account for much of this layer. Again shellfish only were recovered.

Layer 5 (85–130cm bd) marked the first use of the shelter and comprised a hard-packed dark reddish brown (5YR 3/3) silty sediment with sparse coral pebbles or cobbles. A clear break could be identified between it and Layer 4. An ash lens was recorded over half the test pit at the top of this layer and another at the bottom of the layer. This lower ash lens marked both the interface of the former foreshore (Layer 6) and Layer 5 and the first use of the shelter. Shellfish were the only midden material recovered. It was from this lowest level (110–135cm) of the shelter that a dispersed charcoal sample was recovered which returned a date of 980±80 BP (ANU-10074) 1056–729 BP. Both cooking debris in the form of the ash lenses and slopewash have contributed to the stratigraphic buildup of Layer 5.

Figure 3.19 Waal testpit section and plan

Layer 6 the former foreshore consisting of sterile compacted pinkish gray (5YR 6/2) coral sand and water rolled pebbles was reached at a depth of 135cm bd.

Bartnator returned very limited amounts of midden remains, comprising principally shellfish and much lesser quantities of pottery (two sherds). Intermittent use of the shelter was identified through the presence of ash lenses and debris from cooking fires. The radiocarbon determination 1056–729 BP from the lowest cultural level of the shelter is a relatively late date for initial human use and suggests that the shelter may not have been suitable for use, due to its proximity to the sea, until the last 1000 years.

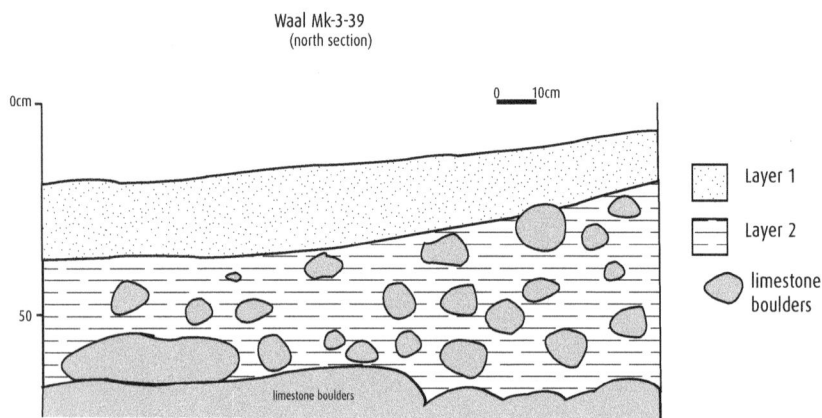

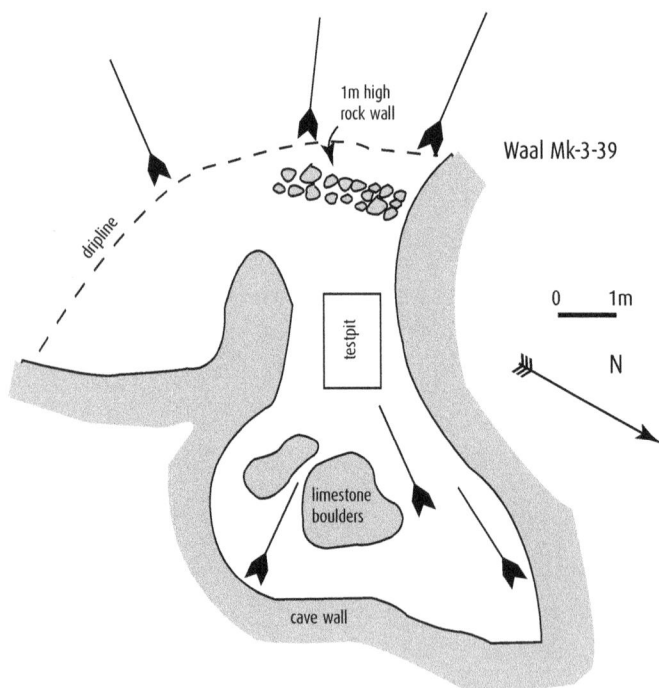

Waal (Mk-3-39)

This small cave is located some 30m above sea level behind the present Malua Bay village. The rear of the cave is almost completely infilled with slopewash. A 1m high dry stone wall marks the entrance (Fig. 3.19). On initial inspection the cave appeared to have limited prospects but the discovery of a notched applied relief sherd, similar in style to what was recognised at the time as 'late Mangaasi' pottery, on the floor surface encouraged further investigation. A test pit measuring 1 by 1.5m was excavated some 3m inside the entrance to a depth of 65cm bd. Recovered materials included shellfish, bone, pottery and a shell pendant.

Layer 1 (0–35cm bd) was characterised by a dark gray (10YR 6/1) powdery sediment with frequent small coral pebbles and cobbles. It appeared to be largely formed from slopewash which had entered the rear of the cave. Evidence of intermittent cooking activity in the form of patches of ash and charcoal was also recorded within this layer. Two sherds, sparse shellfish and bone remains were recovered.

Layer 2 (35–65cm bd) was made up of concentrated small to medium sized pebbles and cobbles amongst a dark gray (10YR 4/1) silty sediment. The radically increased cobble content provided a clear boundary between the layers. Calcareous tempered sherds were recovered from this layer along with shellfish and a shell ornament. This layer signals the first human use of the cave which appears to be characterised by sporadic short-term visits and occupation. It rests on top of a sterile layer of tightly packed small to large coral limestone cobbles and boulders.

Malua One (Mk-3-40)

This cave is located some 40m above sea level behind Malua Bay village at the northern end of Malua Bay (Fig. 3.20). The limestone cave has been subjected to substantial infill and now is rarely used due to the increasingly low roof. The entrance of the cave is some 2m high, reducing to 1.2m in the centre and only 70cm at the rear. Faint charcoal drawings were recorded at the rear of the cave at ground level. A 2 by 1m test pit was excavated 2m inside the entrance of the cave to a depth of 2.3m. Nine distinct layers were identified (Fig. 3.20).

Layer 1 (5–30cm bd) consisted of a gray (5YR 5/1) powdery sediment with frequent coral gravel inclusions. A fire-scoop

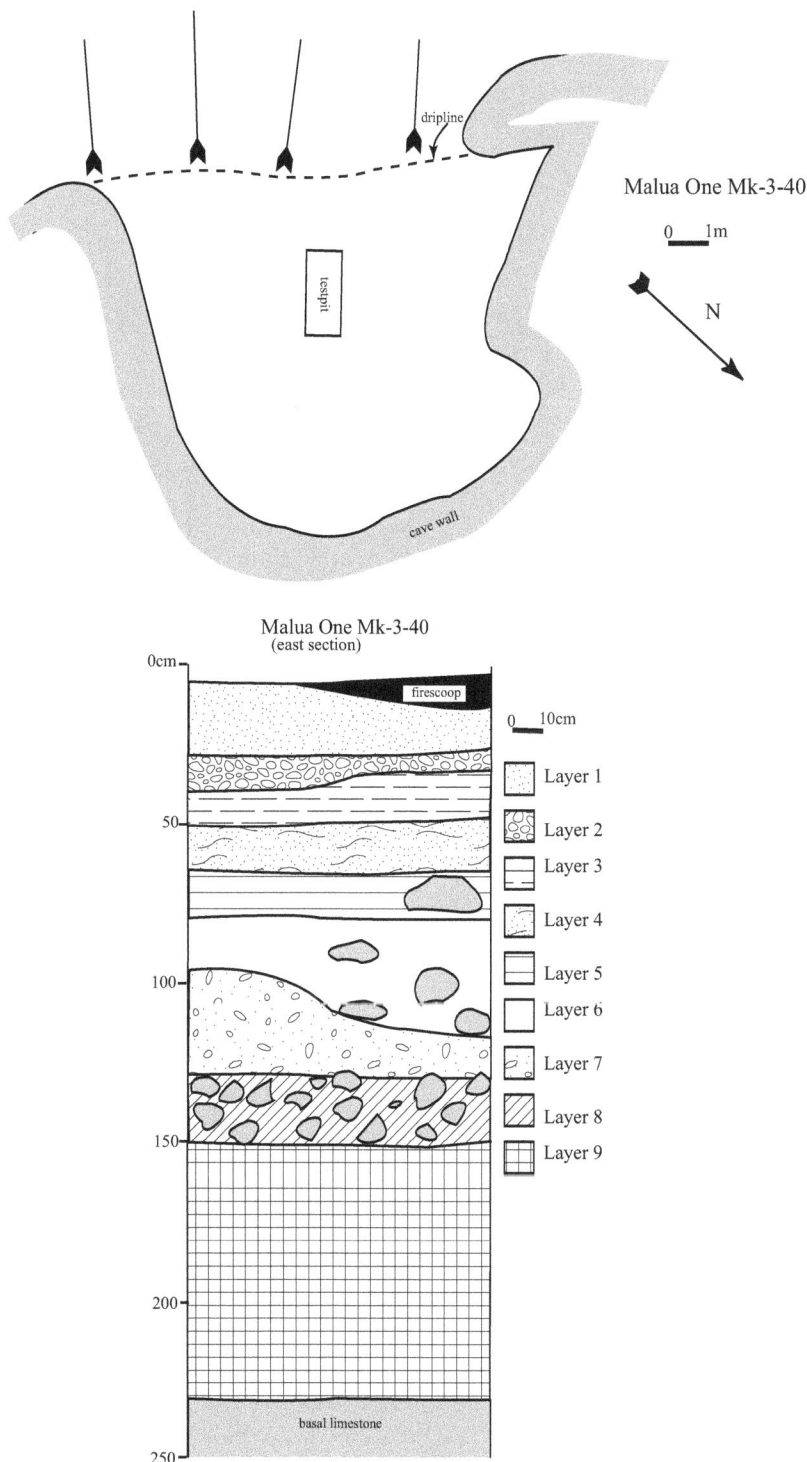

Malua One Mk-3-40

0 ___ 1m

N

Malua One Mk-3-40
(east section)

firescoop

basal limestone

Layer 1
Layer 2
Layer 3
Layer 4
Layer 5
Layer 6
Layer 7
Layer 8
Layer 9

0 ___ 10cm

Figure 3.20 Malua One testpit section and plan

filled with white ash and fire-cracked rock was exposed from the surface and was cut into the layer. Several shells and sparse bone were recovered.

Layer 2 (30–40cm bd) was easily distinguished by the frequent fire-cracked rock set amongst a light gray (10YR 7/1) ashy sediment. The layer appeared largely to have derived from rakeout from a nearby fireplace. Several shells were recovered.

Layer 3 (40–50cm bd) comprised a dark reddish brown (5YR 3/3) powdery silt with frequent gravely inclusions derived primarily from slopewash. It graded into Layer 4. Shellfish only were recovered.

Layer 4 (50–65cm bd) was a dark gray (5YR 3/1) silty sediment with frequent charcoal and gravely inclusions along with white ash patches indicating brief use of the cave. Recovered cultural materials consisted of occasional shellfish. A clearly defined boundary could be identified between this layer and the layer below.

Layer 5 (65–80cm bd) consisted of a black (5YR 2.5/1) silty sediment with frequent gritty inclusions derived from the weathering of the cave walls and roof. The occasional medium to large limestone boulder added further to the suggestion of surface weathering. The silt component of the layer appeared to be made up largely of slopewash. The layer was relatively sterile (occasional shellfish) and graded into the layer below.

Layer 6 (80–95/115cm bd) was similar to the layer above but was distinguished by more frequent cobbles and boulders and a dark reddish brown (5YR 3/3) sediment with frequent gravel inclusions. Four plain sherds and infrequent shellfish remains were recovered.

Layer 7 (95–120cm bd) was identified from a distinct change in the matrix. It consisted of a dark brown (7.5YR 3/2) silty sediment with few coral cobbles or boulders. Sparse plain sherds (3) and shellfish were again recovered.

Layer 8 (120–150cm bd) comprised a black (5YR 2.5/1) sediment with concentrated limestone cobbles and boulders. These remains would seem to be a further indication of the erosional impact caused by human use of the cave and surrounding environment. Sparse cultural materials included calcareous-tempered plain sherds (3) and shellfish.

Layer 9 (150–230cm bd) consisted of a very hard-packed black (5YR 2.5/1) sediment with frequent limestone gravel inclusions rather than cobbles. It was some 80cm thick, again indicating radically altered environmental regimes brought about by human arrival. Sparse shellfish only was recovered. Excavation terminated at a depth of 230cm bd when the basal limestone was encountered.

The stratigraphy of Malua One was characterised by a series of largely sterile layers formed through slopewash. Ephemeral use of the cave was indicated by firescoops and ash lenses and associated cultural remains. Recovered materials were sparse and included bone, shellfish and pottery (10 sherds). Initial human use of the cave is associated with a number of thin calcareous-tempered sherds collected from the bottom of the test pit. Thick non-calcareous tempered sherds were found in the upper layers of the test pit. No sherds were recovered from the first 1m of stratigraphy and as pottery was used up until the last few hundred years on Malekula this lack of remains might be a further indication of the rapid accumulation of sediment in the cave.

Woapraf (Mk-3-41)

This site comprises a 30m long 2–3m wide shelf at the base of a limestone cliff located at the back of Malua Bay some 30m above sea level. The cliff rises to a height of 75m and marks the seaward edge of a large uplifted terrace which is used extensively for gardening. The shelf is said to have been a stopping off point for those visiting the coast from the interior. At the western end of the shelf a 1 by 1 metre test pit was excavated to a depth of 2.3m (Fig. 3.21). A total of seven layers were identified.

The surface (0–12cm bd) of the test pit consisted of recent humic accumulation devoid of any midden materials.

Layer 1 (12–40cm bd) comprised a series of firescoops and associated rakeout. The matrix was characterised by a light brownish gray (10YR 6/2) ash with frequent charcoal and fire-cracked rock. The layer was trowelled without sieving and no midden materials were recorded.

Layer 2 (40–55cm bd) consisted of a concentrated gravel (weathered cliff face) and very dark gray (10YR 3/1) sediment (slopewash) which provided a distinct boundary between Layers 1 and 3. No midden materials were recovered.

Layer 3 (55–80cm bd) was characterised by concentrated firecracked rock, light brownish gray (10YR 6/2) ash and frequent charcoal. It represented a firescoop and associated rakeout which cut into Layer 4 below with the contrasting matrices providing a clear delineation. Occasional shellfish and bone were recovered.

Layer 4 (80–93cm bd) consisted of a dark

Figure 3.21 Woapraf testpit section and plan

gray (5YR 4/1) powdery sediment, with concentrated gravel. This layer returned little in terms of cultural remains and appeared to be largely made up of slopewash debris. An ashy lens, recorded at a depth of 93–110cm bd was an indication of short-term use. The first pottery (1 plain sherd) was recovered from this layer along with occasional shellfish and bone.

Layer 5 (110–115cm bd) was characterised by a dark grayish brown (10YR 4/2) silty sediment. An ash lens was recorded at a depth of 115cm bd. The layer was cut by a firescoop located on the western side of the test pit. Cultural remains were sparse with only a single plain sherd being recovered along with scattered shell. A further ash lens was recorded at a depth of between 148–150cm bd. A distinctive stratigraphic break was then identified.

Layer 6 (150–190cm bd) was a hard packed black (10YR 2/1) silty sediment, with frequent coral pebbles/gravel (weathered cliff face). Nine sherds were recovered and all were calcareously tempered. Occasional shellfish were also recovered. It was from this layer (175–200cm bd) that a

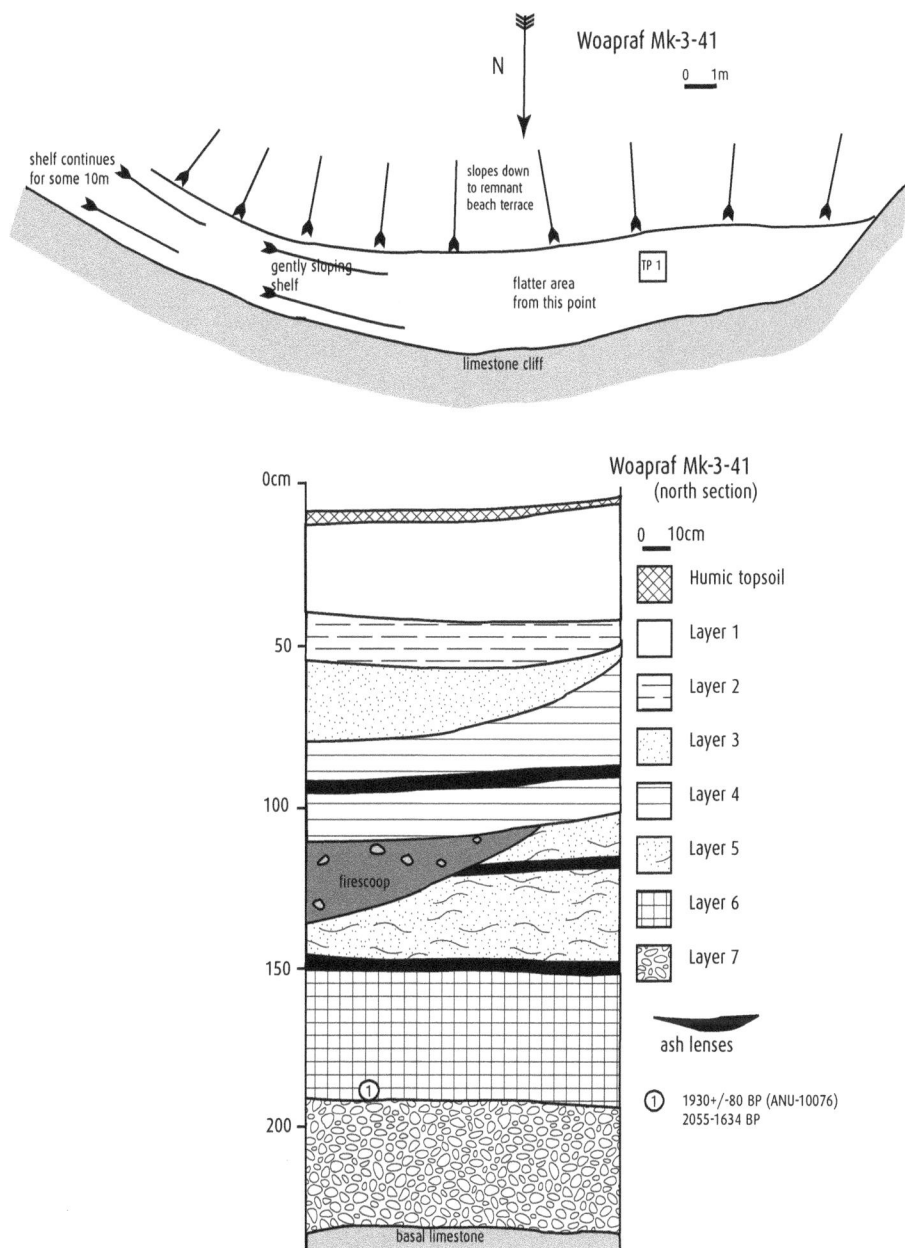

dispersed charcoal sample was dated (1930±80 BP [ANU-10076] 2055–1634 BP). The date does not seem to be consistent with the ceramic remains which at other sites have been more securely dated to c. 2700–2500 BP. This is not totally unexpected at a site of this nature where some mixing of the cultural and slopewash deposits can be expected. This layer graded into Layer 7 which became increasingly sterile with less silt content.

Layer 7 (190–230cm bd) comprised a dark brown (10YR 3/3) clay and concentrated gravel (cliff fall) with little soil content. One plain sherd along with scattered shellfish, bone and charcoal were recorded from the upper part of layer. At a depth of 2.30m the sterile basal limestone was encountered.

The stratigraphy at Woapraf revealed rapid accumulation of sediment, much of which appeared to be slopewash, but included evidence, in the form of fireplaces and rakeout, of intermittent use. Little artifactual material was recovered. The sparse collection of pottery demonstrated a consistent pattern with that of other excavated cave sites on the Northwest coast of Malekula. Thin calcareous-tempered wares were recovered from the base of the cultural stratigraphy. These are superseded by slightly thicker calcareous wares and much later by thick non-calcareous globular pots. Limited quantities of shellfish and faunal material were recovered from the test pit.

Navaprah (MK-3-47)

This cave is located near Lekhan village some 20m above sea level and adjacent to the coastal road. There is a level area located at the entrance of the cave but after several metres the floor slopes downward towards the rear. Large limestone boulders (roof-fall) dominate the rear of the cave (Fig. 3.22). In 1995 a 1 by 1.5m test pit (TP 1) was excavated some 2m in from the entrance of the cave. Further squares, namely TP 2 (3 by 1m) and TP 3 (1 by 1.5m) were excavated in 1996 (Fig. 3.22). These excavations both provided a larger ceramic sample and further confirmation of the stratigraphy and cultural sequence hinted at in the 1995 excavation. The stratigraphic sequence was made up of five distinct layers (Fig. 3.22), although their presence and appearance varied across the cave. As indicated by the recovered ceramics and associated dates there is also some internal division within layers. The four radiocarbon dates recovered from the cave site were all from charcoal samples collected from TP 1.

Layer 1 (10–70cm bd) largely comprised a series of cooking firescoops and associated rakeout which had rapidly accumulated. Several older residents of the nearby village recalled either having camped or cooked in the cave. The matrix consisted of concentrated fire-cracked stone mixed with white (10YR 8/2) through to gray (10YR 5/1) coloured ash and frequent charcoal. Shellfish and bone although sparse dominated the recovered materials. Seven plain sherds were found only in the very bottom of the layer. The layer thickness varied across the three test pits but was up to 50cm in some areas of all three test pits. The initial 30–40cm bd of TPs 2 and 3 were trowelled but not sieved. The identification of a new layer was marked by a distinctive change in the matrix.

Layer 2 (60–90cm bd) was characterised by a dark gray brown (10YR 4/2) powdery sediment with frequent charcoal and occasional rock. It would appear to indicate a less intensive period of use of the cave and some contribution to the infill is attributable to slopewash. Layer 2 was up to 25cm thick in TP 1 but petered out towards the rear of the cave, virtually disappearing in TP 2. In TP 3 it was 10–15cm thick. Shellfish and bone were recovered along with five plain sherds. A date of 510±50 BP (ANU-10540) 629–497 BP was returned from charcoal from the 50–75cm level of Layer 2 of TP1 (Fig. 3.22).

Layer 3 (60–140cm bd) consisted completely of a large oven feature filled with white (10YR 8/1) ash. The remains were clearly identified across TP 1 and 2 and were up to 80cm thick. In TP 3 however, some distance from the main oven feature the layer comprised only a thin compacted

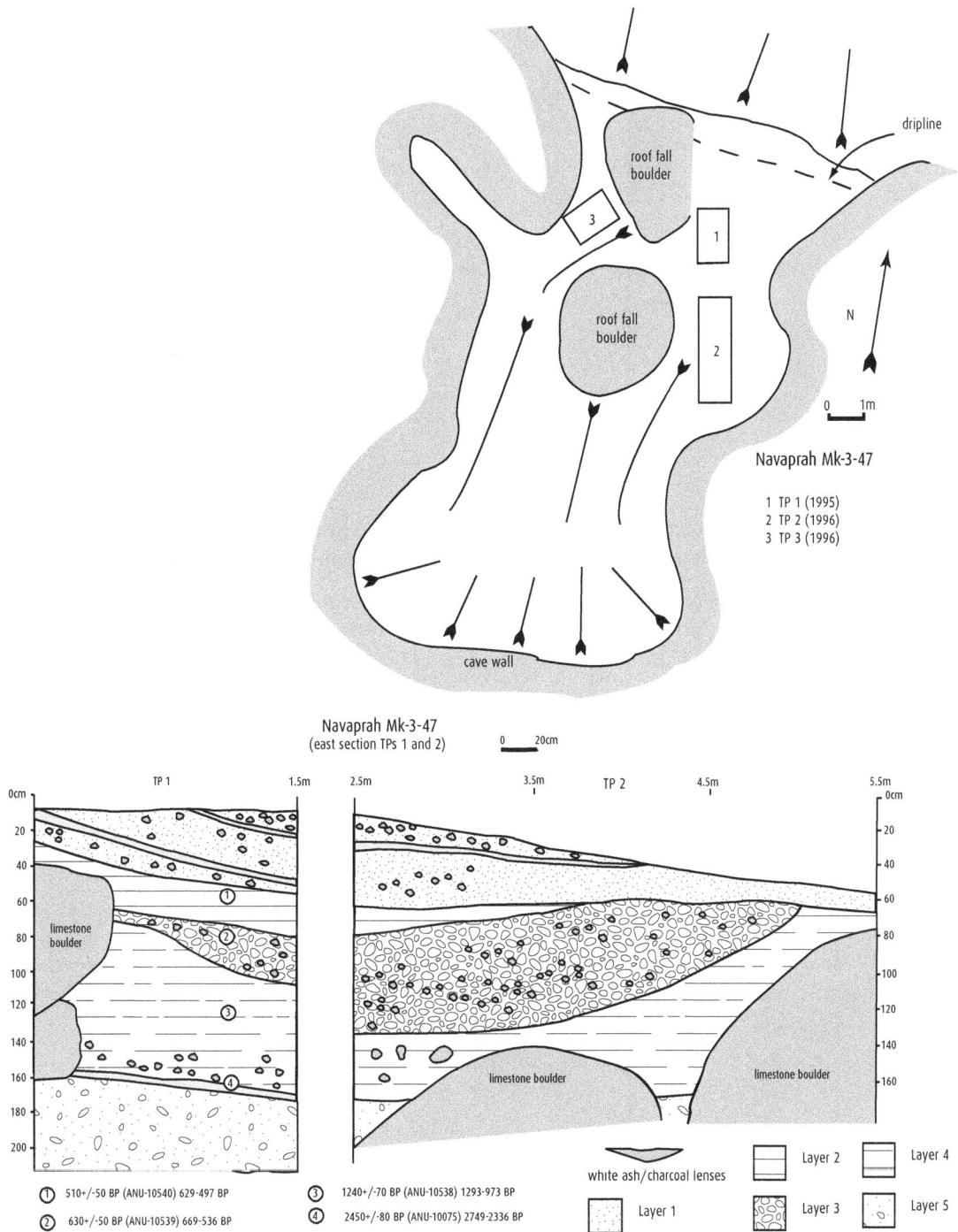

Figure 3.22 Navaprah testpit section (1 and 2) and plan

layer of white ash. Shellfish and bone were recovered from the white (10YR 8/2) ash matrix but only from TP 1. Six plain sherds were recovered from TP 1. Burnt coral and basalt cobbles were frequently present along with concentrations of charcoal. A radiocarbon date of 630±50 BP (ANU-10539) 669–536 BP was recovered from this feature in TP 1. The oven cut deeply into the lower layer and much of the earlier archaeology in TP 2 had been removed due to this event. An abrupt change in layer composition could be seen between this layer and the layer below.

Layer 4 (80–160cm bd) was the lowest cultural layer, made up primarily of a dark gray (10YR 4/1) powdery sediment with fire-cracked rock in increasing quantity towards the base. As noted above, identifying internal divisions within this layer was difficult and it appeared to be relatively homogeneous across the cave. Analysis of the ceramics and radiocarbon dates clearly demonstrates a long period of accumulation. From the upper level of Layer 4 (115–130cm) a date of 1240±70 BP (ANU-10538) 1293–973 BP was recovered. The ceramics associated with this date were large thick-walled globular pots with everted notched rims. The same material was also recovered from the lowest levels of TP 2. The earliest radiocarbon date was retrieved from a charcoal sample from the bottom 10cms of Layer 4, and dated to 2450±80 BP (ANU-10075) 2749–2336 BP. This was associated with a thin-walled plainware with calcareous temper. It was found only near the entrance of the cave in TPs 1 and 3 in the earliest cultural layers. The earliest evidence for use of the cave appears to be concentrated nearer the entrance and it is only later that larger areas of the cave were utilised. Increasing quantities and concentrations of pottery (52 sherds) and bone were noted towards the bottom of this layer. Much of the recovered bone material from the cave again appeared to be the result of owl roosting. In TP 3 concentrations of small bone were noted in all spits of Layer 4 lying against a large limestone boulder. There was an abrupt and distinctive layer change between this layer and the one below.

Layer 5 (from 160cm bd) represented the former foreshore. It consisted of a light yellowish brown (10YR 6/4) sand with concentrated water-rolled coral pebbles/branches and shellfish. A sondage was excavated in TP 1 to a depth of 230cm, confirming that the layer was culturally sterile.

Navaprah appears to have first utilised some 2700–2500 years ago. The earliest cultural layer contained relatively abundant quantities of calcareous-tempered pottery. Its restriction to the lowest level of this layer and the homogeneous nature of the ceramics suggests that it was a short-term occupation. A period of less intensive use or possible abandonment of the area is indicated by the decrease in midden content and a date in the upper level of the lowest layer of 1200 BP. Much of the upper layers of the cave were disturbed by a large oven feature but sparse pottery, bone and shellfish were recovered throughout. Pottery disappears from the record in the uppermost layer of the cave, dated to some time after 500 BP. The abundant bone recovered from the excavations can be partly attributed to owl roosting in the cave.

Yalo South (Mk-3-48)

This cave is located adjacent to the entrance to Yalo B, a large cathedral-like limestone cave with a rich collection of rock art. It is 35m above sea level and some 75m from the current shoreline. Two metres inside the entrance of this much smaller cave a 1 by 1.5m test pit was excavated to a depth of 2m (Fig. 3.23). Seven distinct layers were identified.

Layer 1 (6–12/18cm bd) was a dark grayish brown (10YR 4/2) ashy sediment with frequent fire-cracked rock. It appeared to consist of very recent (slopewash) accumulation resting on top of two fireplaces which were recorded some 10–15cm below the surface. These features had been cut into Layer 2 and comprised concentrated ash and charcoal. Shellfish and bone were recovered from Layer 1 along with a single sherd from one of the fireplaces.

Layer 2 (12–74cm bd) consisted of a dark grayish brown (10YR 4/2) silty sediment, largely sterile slopewash with occasional pebbles and cobbles of weathered limestone roof-fall. Recovered materials included pottery (at 50cm bd) bone and shellfish. The matrix of the layer had little apparent layering apart from an ash lens covering part of the surface of the square at some 50cm below the datum, indicating temporary use of the cave. A further ash lens with charcoal covering the whole square was identified at 75cm below datum.

Layer 3 (76–100cm bd) was again characterised by a dark grayish brown (10YR 4/2) silty sediment with occasional roof-fall limestone. The layer was partly divided by an ash and charcoal lens at 100cm bd and terminated at 100–115cm bd with the presence of a much thicker ash and charcoal lens covering the entire square which provided a clear division with the lower layers. Layers 2 and 3 appeared to indicate rapid accumulation in the cave although cave use appeared to be very intermittent, suggesting slope wash from human-induced landscape change was more of a

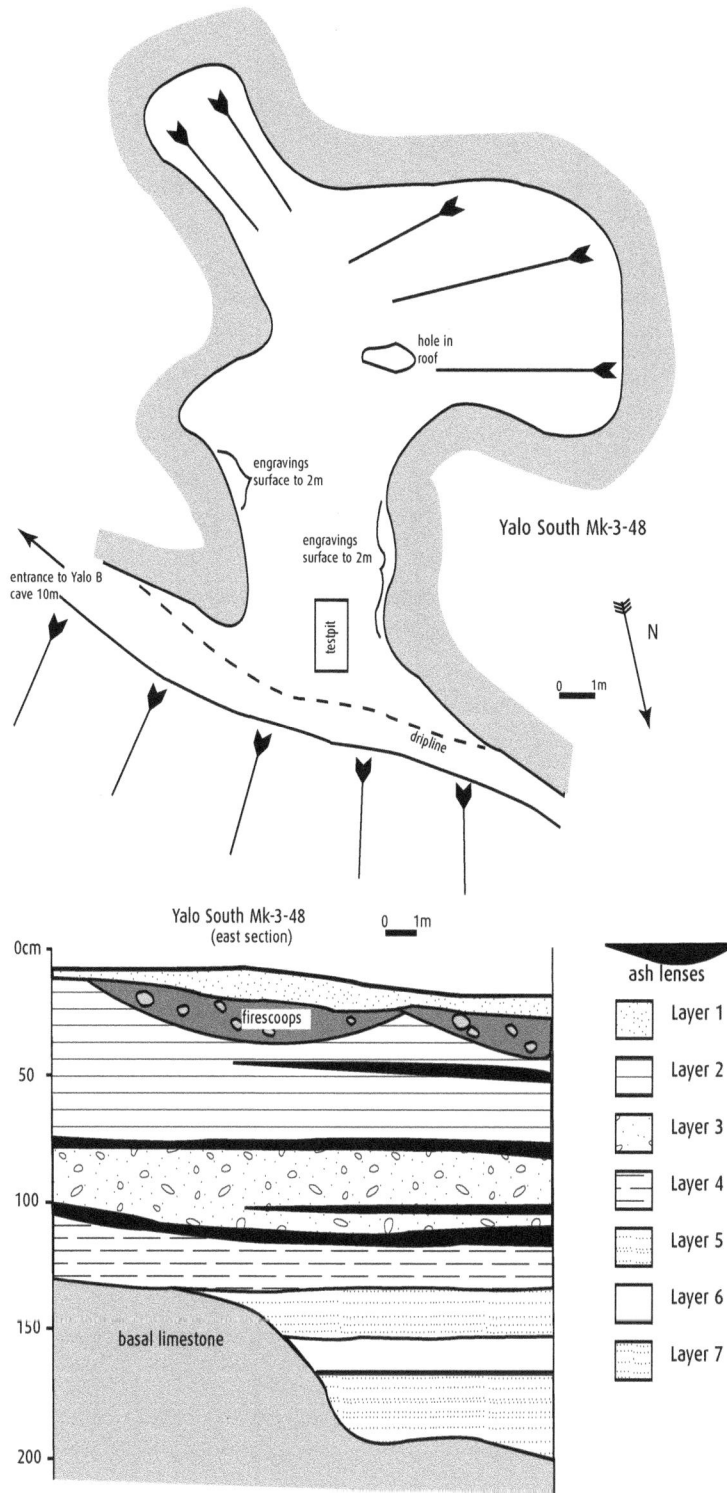

Figure 3.23 Yalo South testpit section and plan

contributing factor to the buildup. Shellfish and bone were recovered from this layer but no pottery.

Layer 4 (110–130/135cm bd) consisted of a distinctive hard-packed dark yellowish brown (10YR 4/6) sediment with some weathered limestone from the cave walls. Pottery, shell, and bone were recovered from within this layer. It appeared to mark the initial human use of the cave. The pottery was typical of the thin calcareous-tempered sherds found at the basal levels of other caves which dated to c. 2700–2500 BP.

Layer 5 (130–150cm bd) was characterised by a series of thin interspersed lightly compacted brown (10YR 5/4) sediment and white ashy lenses. It appeared to represent a tephra from volcanic activity in the region that had been washed into the cave over a short period of time. Limited worn shell and bone only were recovered which seemed likely to have derived from the layer above.

Layer 6 (150–165cm bd) consisted of a dark brown (10YR 4/4) silty sediment representing slopewash which was devoid of midden material.

Layer 7 (165–195cm bd) was the sterile basal layer which consisted of a reddish brown (5YR 4/4) sediment similar to Layer 5 which incorporated a series of thin lenses of white ash and charcoal. These had been washed into the cave and could have been related to volcanic activity on the nearby island of Ambrym. At the bottom of this layer the weathered limestone bed-rock of the cave was exposed across the square.

The excavation of Yalo South revealed a series of occupation layers punctuated by periods of non-use. Much of the accumulated stratigraphy can be attributed to slopewash. Near the base of the cultural stratigraphy thin lenses of tephra, which appear to have been washed into the cave, were recorded. Recovered artifactual material was somewhat meagre with ceramics (17 sherds) dominating. The recurring pattern of thin calcareous-tempered sherds at the base being later replaced by thicker non-calcareous tempered globular pots was again noted. The presence of the calcareous-tempered sherds in the lower layers of the site suggest a date of c. 2700–2500 BP for the initial human use of the cave. Shellfish were recovered but not in any concentrated midden deposits. The bone remains were relatively abundant. The high proportion of bone, particularly rat, in nest-like deposits, suggests much of the material is debris associated with an owl roost. The cave also has a number of rock engravings and charcoal drawings, many of which are near ground level, indicating substantial accumulation within the cave.

Wambraf (Mk-3-56)

This rockshelter is located some 40m above sea level behind the present Tenmiel village. A 2 by 1m test pit located in the middle of the shelf created underneath the boulder overhang was excavated to a depth of 1.40m at which point the former foreshore was reached (Fig. 3.24). The stratigraphy could be clearly separated into four layers.

Layer 1 (0–70cm bd) consisted of an unconsolidated light brownish gray (10YR 6/2) silty sediment with frequent angular limestone pebbles and cobbles. The top 5cm was sealed by more recent humic buildup. A firescoop filled with white ash, coral cobbles and charcoal was recorded in the southern half of the square near the surface. The edge of a further firescoop was recorded at a depth of 40–50cm below the surface and located within Layer 1. Quantities of bone, frequent shellfish and pottery (15 sherds) were recovered from the layer.

Layer 2 (70–90/110cm bd) was characterised by very concentrated coral cobbles, many of which were burnt or fire-cracked, set within a dark gray (7.5YR 4/0) sediment. This distinctive matrix provided a clear boundary between this layer and the layers above and below. A firescoop was recorded at the top of Layer 2. A large limestone boulder marked the bottom of Layer 2 in the western half of the square. Recovered materials included bone, shell and pottery (2 sherds).

Layer 3 (90–140cm bd) was made up of an unstratified gray (10YR 5/1) powdery soil with occasional coral cobbles, partly resting on top of the basal limestone and Layer 4 the sterile, former foreshore deposit. A distinct boundary between Layers 2 and 3 could be seen with the dramatic change in coral cobble content. Layer 3 signalled the initial human use of the cave. It was from the lowest part of this layer (100–120cms bd) that a charcoal sample for dating was collected (1030±70 BP (ANU-10529) 1063–788 BP). Bone, frequent shellfish and pottery (3 sherds) constituted the total recovered remains

Layer 4 was made up of the sterile dark yellowish brown (10YR 4/6) former coral foreshore which included coral sand/gravel, water worn coral pebbles, cobbles and branches.

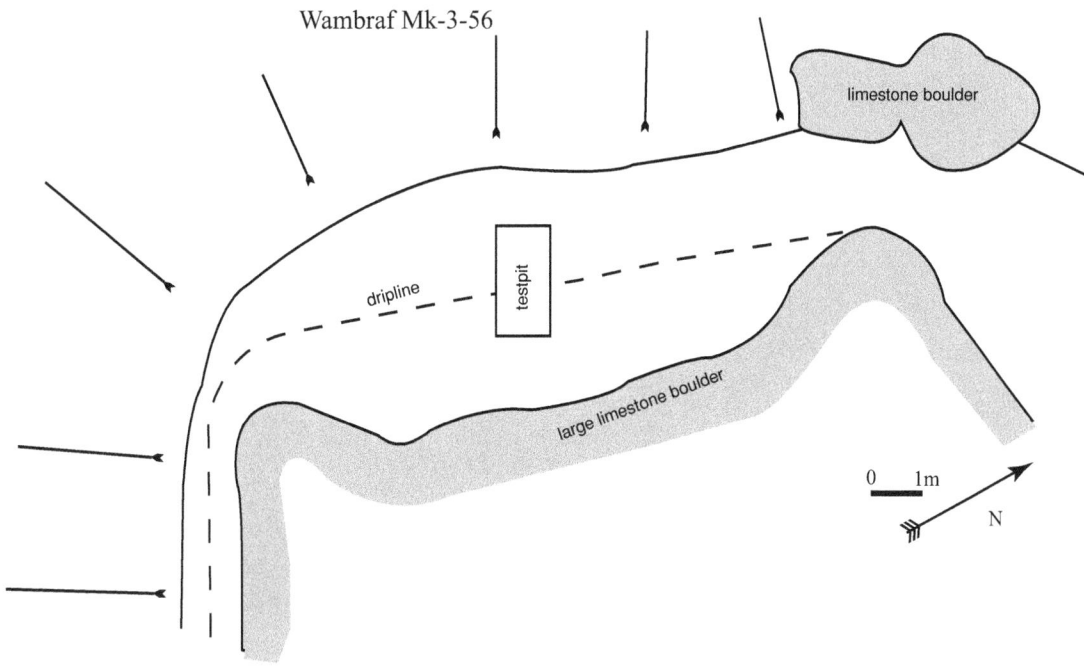

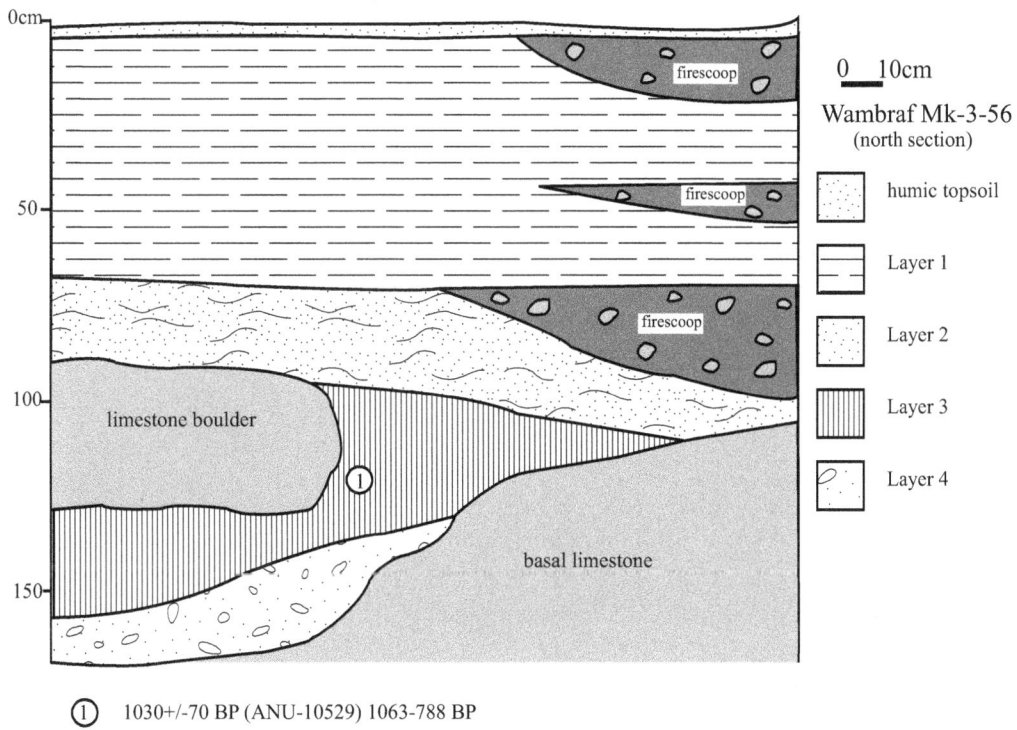

Figure 3.24 Wambraf testpit section and plan

The test pit excavation at Wambraf revealed a series of layers showing intermittent use of the shelter over the last 1000 years. Much of the accumulated stratigraphy can be attributed to slopewash but several firescoops and concentrations of fire-cracked or burnt cobbles are associated with short-term occupation and use of the site. Bone and shellfish remains were recovered along with a limited number of sherds (20) which showed little change in form or fabric from the top to the bottom of the stratigraphy.

Navepule Caves

Three caves were recorded in the area known as Navepule. The caves are located amongst the 'second step' of raised coral limestone terraces, some 1.5km inland from the coastal village of Mbenenavet at an altitude of around 200m asl. Rock art was recorded in all three caves and clear evidence of more recent use in the form of fireplaces and firecracked rock was noted on the surface. A series of 2 by 1m test pits were excavated several metres inside the cave entrances.

Navepule A (Mk-3-38)

Navepule A is a large high roofed cave measuring some 15 by 17m in area. Large limestone roof-fall boulders dominate the centre of the cave. Soil from the surrounding slope is washed into the cave in the northern part of the entrance and along the dripline. A 2 by 1m test pit was excavated to a depth of 70cm at which point the test pit was reduced to a 1 by 1m and excavated to the basal limestone at 105cm bd (Fig. 3.25). Three distinct layers were identified.

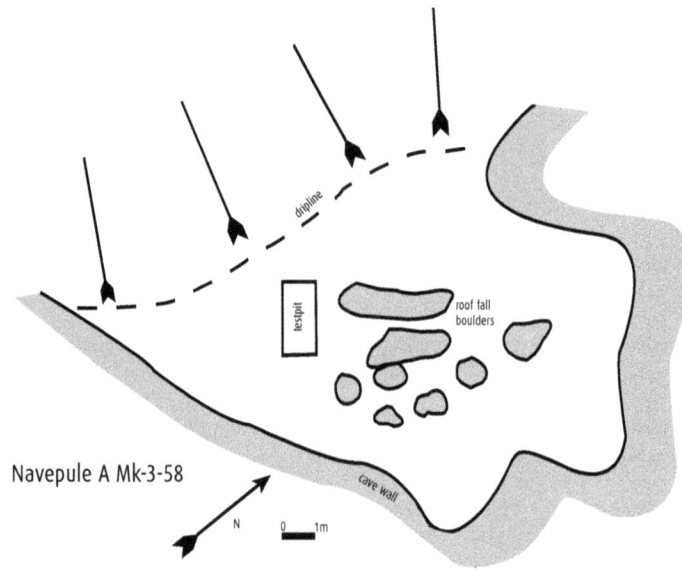

Layer 1 (5–50cm bd) was a hardpacked matrix of black (10YR 2/1) silty sediment with occasional basalt cobbles and more frequent limestone pebbles and cobbles, much of which appeared to be roof-fall. A firescoop was noted on the surface of the layer in the southern end of the test pit. A further firescoop along with an associated ashy lens was noted at 30cm bd. An abrupt change was identified between Layers 1 and 2. Recovered materials from Layer 1 included eight plain pottery sherds, sparse bone and fossilised shellfish derived from the cave walls and bed-rock.

Layer 2 (50–70cm bd) consisted of black (10YR 2/1) silty sediment that was increasingly compacted with depth and contained concen-trated limestone pebbles and boulders. These appeared not to be associated with cooking activities (i.e.; not fire black-ened, little charcoal or ash) and are more likely to be derived from roof-fall encouraged by frequent human use of the cave. A firescoop was noted in the northern part of the test

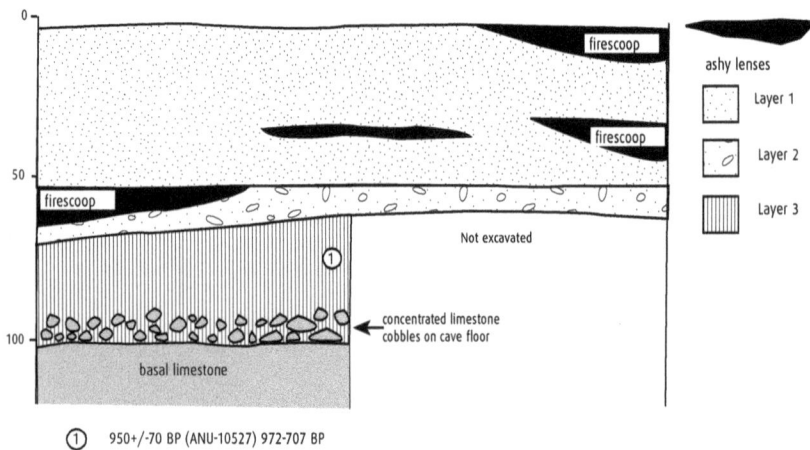

Figure 3.25 Navepule A testpit section and plan

pit. A dispersed charcoal sample from this layer (55–65cm bd) returned a date of 950±70 BP (ANU-10527) 972–707 BP. Sparse worn shellfish and pottery (3 sherds) only were recovered from this layer.

Layer 3 (70–105cm bd) was again primarily made up of a black (10YR 2/1) sterile silty sediment with frequent limestone gravel and pebbles. With increasing depth the layer graded into a lighter coloured (very pale brown 10 YR 7/3) more clay-like matrix. At the bottom of the layer concentrated limestone cobbles were noted resting on top of the basal limestone. This layer would appear to represent a largely natural, steady buildup on top of the cave floor through the agencies of slopewash and weathering of the basal limestone. Sparse charcoal throughout the layer suggests some human influence although at a much less intensive level than that which occurred later. There were no other midden materials recovered from this layer.

The excavation of the test pit at Navepule A provided evidence for the intermittent use of the cave. Firescoops and concentrations of firecracked rock were interspersed within more sterile deposits derived from slopewash. Midden remains were sparse and consisted largely of shellfish and pottery (11 plain sherds), none of which were calcareously tempered. The radiocarbon date from the lowest cultural layer suggests that increased human use of the cave and surrounding environs dates from around 1000 BP.

Navepule B (Mk-3-59)

A test pit measuring 2 by 1m was excavated some 5m inside the dripline of this cave. Recent use of the cave was indicated by a number of fireplaces on the surface. Surface indications also suggested that much of front entrance was regularly scoured out during heavy rains. The rear of the cave appeared to be subjected to regular slopewash. The somewhat less than suitable occupational environment of the cave was quickly confirmed through excavation. Only the top 20cm returned any cultural material with the remaining stratigraphy, to the basal limestone at 130cm bd, proving to be largely sterile. Twenty one sherds of pottery were scattered across the surface of the cave. Much of the material appeared to have been washed into the cave.

Layer 1 (5–20cm bd) consisted of a very dark gray (10YR 3/1) clay with frequent limestone gravel and charcoal flecks. Two firescoops were noted on the surface of the layer. This layer graded somewhat into the layer below. At the bottom of Layer 1 the test pit was reduced to a 1 by 1m square. Very occasional shellfish were recovered but all appeared to be fossilised and eroded from the walls of the cave. A single decorated sherd was also recovered.

Layer 2 (20–130cm bd) comprised an increasing clay-like matrix with depth and graded from a very dark grayish brown (10YR 3/2) to a dark yellow brown (10YR 4/6). It was virtually sterile except for the very occasional fossilised shell. This layer would appear to represent largely natural infill of the cave which has been hastened by human activity in the area. There were no recovered midden materials.

Navepule C (Mk-3-60)

The third cave of this cluster is Navepule C. It is again a relatively large cave measuring 10 by 11m in area with a high roof. It appears also to be more stable than the other two with very little evidence of roof-fall on the surface of the cave. The remains of recent cooking fires were recorded on the floor surface of the cave. A 2 by 1m test pit was excavated some 2m inside the dripline to a depth of 110cm bd (Fig. 3.26). Four distinct layers were identified.

Layer 1 (0–15cm bd) comprised a dark gray (10YR 4/1) silty sediment with occasional coral and basalt cobbles. Remnants of cooking fires were further confirmed with patches of ash and charcoal being recorded. Sparse shellfish and bone and two plain pottery sherds constituted the total recovered materials.

Layer 2 (10–50cm bd) consisted primarily of the remains of a large earth oven feature and provided a clear delineation from the layers above and below. The matrix comprised a white

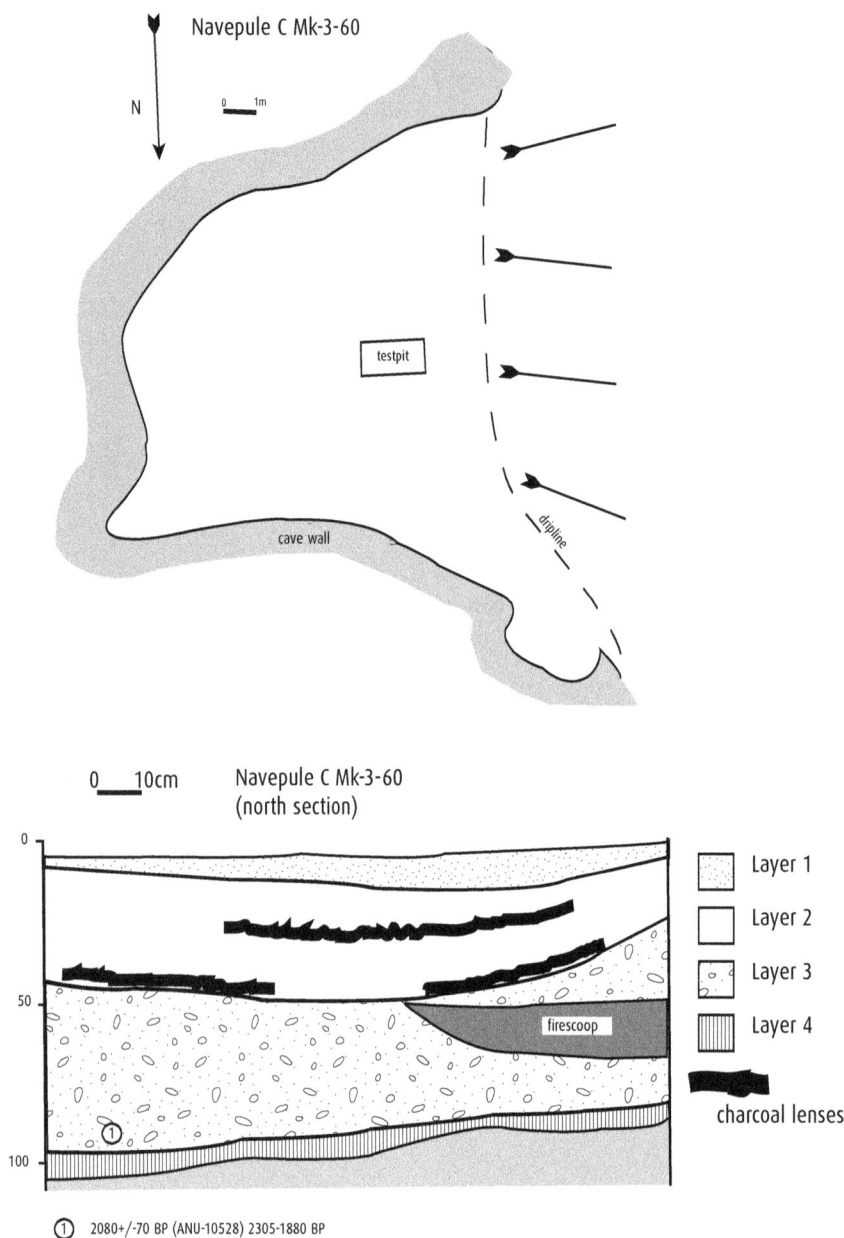

Figure 3.26 Navepule C testpit section and plan

(2.5YR N8/0–2.5 YR 8/2) ash, charcoal and concentrated burnt limestone cobbles. Recovered materials were sparse and included small amounts of shellfish, bone and pottery (3 plain sherds) remains.

Layer 3 (45–100cm bd) was made up of an unstratified pale brown (10YR 6/3) compacted sediment with frequent burnt limestone cobbles. Much of this material would appear to be slopewash accumulating at the dripline mixed with debris from intermittent occupation activities. A dispersed charcoal sample from the 80–100cm level returned a date of 2080±70 BP (ANU-10528) 2305–1880 BP which signals the initial human use of the cave.

Layer 4 (100–110cm bd) was a sterile brownish yellow silt (10YR 6/6) which appears to have largely derived from the weathering of the cave floor prior to human influence. It covered the basal limestone floor of the cave.

Excavation of the test pit at Navepule C revealed stratigraphy which indicated irregular human use of the cave dating from 2305–1880 BP. However much of the accumulation in the cave can be attributed to slopewash, as was the case in the other two Navepule caves. The recovered materials were sparse comprising largely shellfish and occasional pottery and bone. None of the recovered sherds were calcareously-tempered. Initial human use of this area appears to be very ephemeral and date to a much later period than initial coastal settlement. More frequent use of the caves and increased localised landscape change dates from around 1000 BP.

Malekula Open sites

Chachara (MK-3-43)

This is an inland site 250m above sea level, some 300m west of the modern village of Metkhun which is approximately 2km inland from the coast (Fig. 3.16). The site consists of two distinct areas (designated south and north), centred around low stone platforms some 75m apart (Figs 3.27 and 3.28). These

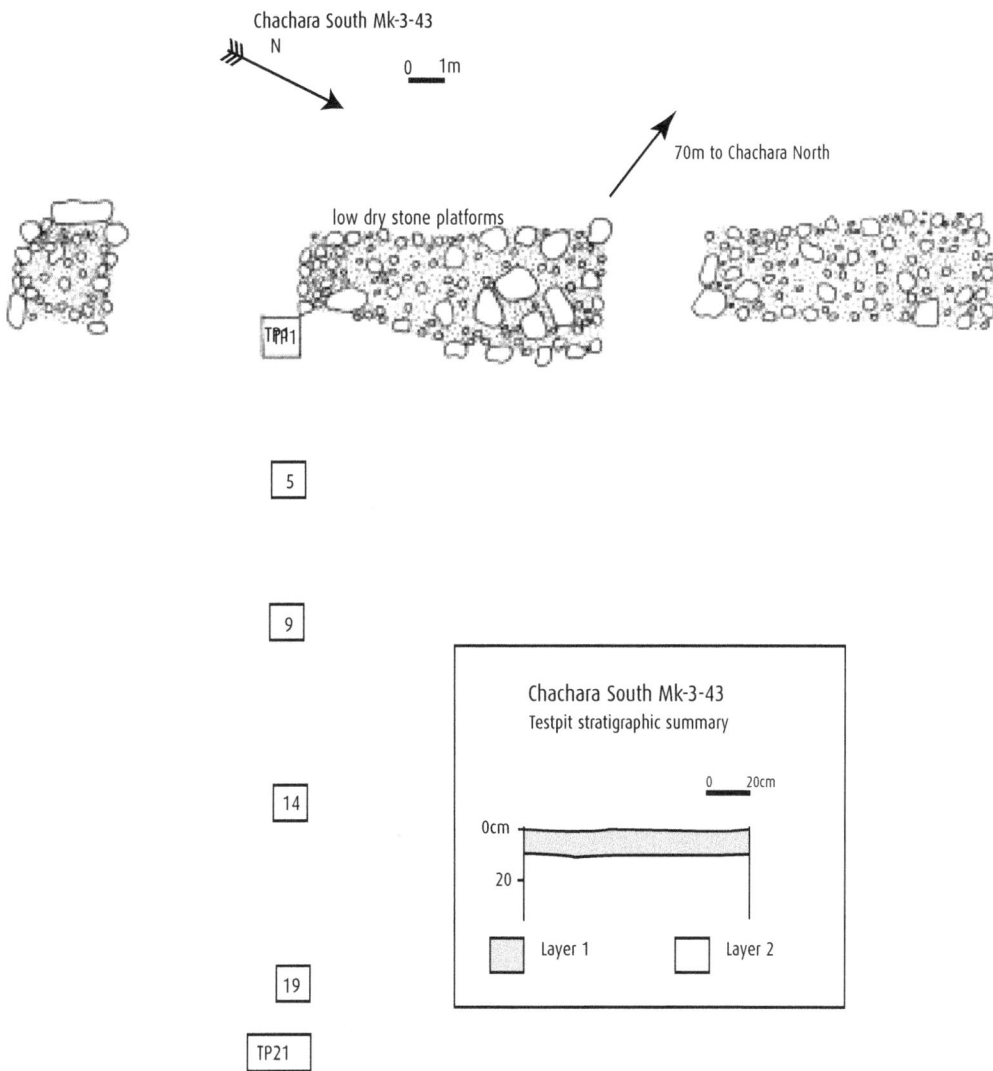

Figure 3.27 Chachara South testpit sections and plan

stone platforms, which vary greatly in size and height above ground level are a common site type in Malekula and much of Northern Vanuatu. They can also be associated with a series of upright stones set within the platform or in a linear alignment nearby. The stone platforms, known as *nasara*, are associated with ceremonial activities. Ethnographic accounts (Deacon 1934; Layard 1942; Speiser 1996[1923]) indicate that they were generally located within or adjacent to an associated village. The surface remains at the above site appeared to further confirm this. Discrete concentrations of pottery and firecracked stone, separate from the platforms were visible on the surface of the site.

The stone platforms were cleared and mapped. At Chachara South a series of test pits were excavated to clarify the stratigraphy of the site and collect a representative sample of the pottery. It very quickly became clear that the cultural material was largely restricted to the ground surface or no more than 20cm below it. Extensive surface collections of pottery were thus undertaken at Charchara North and South. At Chachara North a mound feature some 5m southwest of the platform provided more *in situ* and variable midden debris (Fig. 3.28). This feature appears to have been initially formed from the clearance of the area for settlement and later became a focus for midden dumping. The central core of the mound was made up of an assortment of limestone

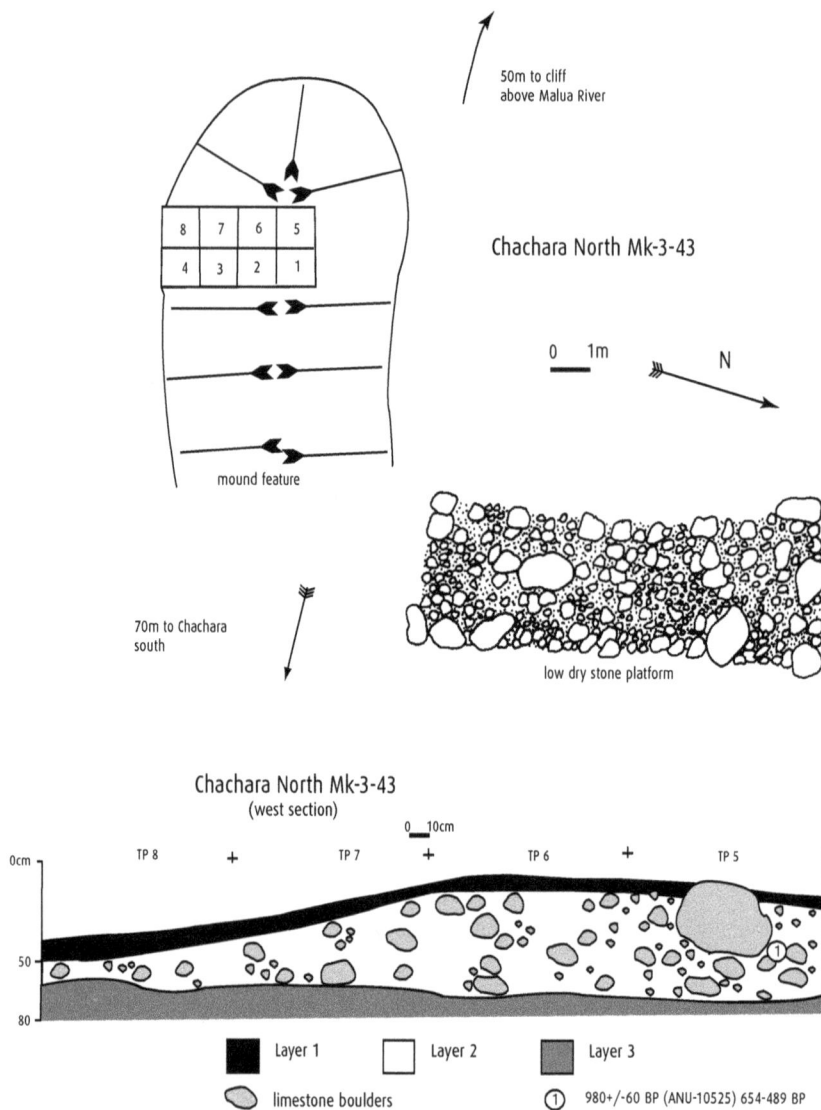

Figure 3.28 Chachara North testpit sections and plan

boulders and a black humic soil. Throughout the mound, amongst the rocks and sediment, midden remains were recovered which included shellfish, bone (largely pig), a shell adze and numerous sherds. Again, stratigraphically the mound feature appeared to represent a short term, single phase occupation with essentially only one cultural layer being identified.

All test pits excavated at Chachara South ($7m^2$) comprised two layers only (Fig. 3.27). Layer 1 was a black (10YR 2/1) humic topsoil with occasional limestone pebbles and cobbles ranging in depth from 10–15cm, which contained limited quantities of cultural material, almost exclusively pottery. This layer graded into a lighter coloured clay beneath. All test pits were 1 by 1m in area except TP 21 which was 2 by 1m.

Layer 2 was a very dark brown (10YR 2/2) clay which with depth graded in colour to a dark yellowish brown (10YR 4/6). This basal layer was completely sterile and dates to a period prior to human settlement of the area.

Chachara North is located some seventy metres to the north west of Chachara South. A 4 by 2m area was excavated across the mound feature situated adjacent to the *nasara* (Fig. 3.28). Three stratigraphic layers were identified and as noted above, concentrated midden was recovered from this feature. *In situ* cultural material was restricted to Layer 2. A radiocarbon determination from a single shell recovered from TP 1 (Layer 2, 40cm bd) returned a date of 980±60 BP (ANU-10525) 654–489 BP.

Layer 1 (0–10cm bd) consisted largely of the more recent buildup of humus and black (10YR 2/1) topsoil. Occasional sherds were recovered and frequent limestone pebbles were noted.

Layer 2 (30–70cm bd) comprised a black (10YR 2/1) silty sediment with concentrated limestone pebbles and cobbles and towards the central core of the mound some much larger limestone boulders. Occasional burnt basalt cobbles were also recorded. Concentrations of sherds were recovered from throughout this layer along with sparse shellfish remains and more frequent bone. The pottery was all of similar vessel form and fabric. A single *Tridacna* shell adze was also recovered.

Layer 3 (below 70cm bd) provided a clear delineation from Layer 2. It consisted of a sterile dark yellowish brown (10YR 4/6) clay bereft of any cultural material with an accompanying dramatic decrease in the limestone pebble and cobble content. It represents the pre-settlement subsoil of the area.

The excavations at Chachara have provided a large sample of the late period ceramics found across much of Malekula often in association with *nasara*. The homogeneous nature of the ceramics at the site suggest a relatively short-term period of occupation dating to sometime around 654–489 BP. Other midden remains included a single shell adze, bone (dominated by pig) and sparse shellfish. The surface remains indicated that there were discrete activity areas at the site.

This research, focusing on the archaeology of last 1000 years, is very preliminary in nature. The initial appearance and transformation of these ceremonial structures remains largely unknown and as yet there are no comparative excavations that have been carried out in the rest of Vanuatu. Much more detailed research into this period is required to understand this influential phase of social and cultural transformation which defined the ethnographic period and comprises the cultural heritage of much of the present populations of Northern Vanuatu.

Malua Bay School Site (Mk-3-55)

This site is located behind Malua Bay, some 12–14m above sea level on a remnant beach terrace (Fig. 3.29). The area was identified as an ideal area for a colonising settlement with its accessible all year round water source, large areas for gardening behind the beach, a sheltered bay and sandy beach which facilitates canoe access. The perennial river at the southern end of the beach has downcut through the uplifted coral terraces in a restricted area of the bay and left the site intact. This was not the case in other areas of likely early settlement such as Espeigle Bay and Tenmaru where the rivers appear to have meandered back and forth over large areas of the coastal fringe.

A total of 14m² was excavated to determine the extent and stratigraphy of the colonising site and collect a sample of the cultural material remains. The remains of the early site covered an area of some 50 by 30m (Fig. 3.29). Much later settlement appeared to cover a larger area although recovered cultural material was very sparse. Five layers were identified at the site, the depth and presence of which varied between test pits (Fig. 3.30). All test pits measured 1 by 1m except for TP 9 which was 2 by 2m. The extent of the early settlement site could be seen in the stratigraphy of TPs 3, 4, 7, 8, 9 and 10. Recovered materials from these test pits, which are presented in detail in other chapters, included concentrated sherds of calcareous-tempered plainware, shellfish, bone and occasional artefacts. One dentate-stamped Lapita sherd was also recovered from amongst this material. The stratigraphy of the excavated test pits was as follows:

Layer 1 (surface to 30cm bd) was a mixed black (10YR 2/1) topsoil representing more recent accumulation at the site. The area is now heavily gardened. Occasional basalt river cobbles, shellfish and more frequent, generally thick-walled ceramics were recovered from all but one of the excavated test pits. These remains would appear to represent a dispersed settlement, the remains of which have been greatly disturbed.

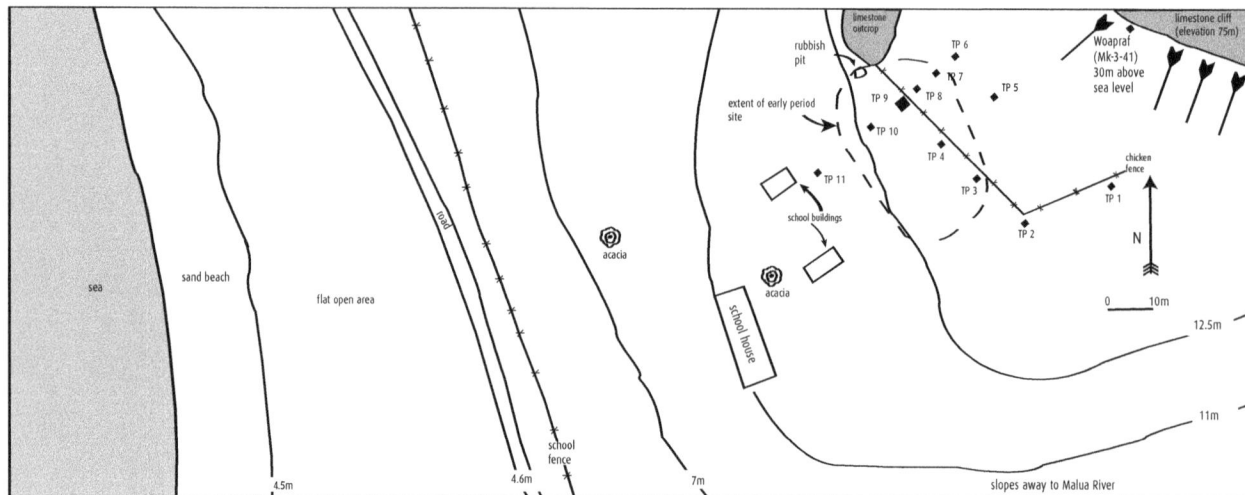

Figure 3.29 Malua Bay School site

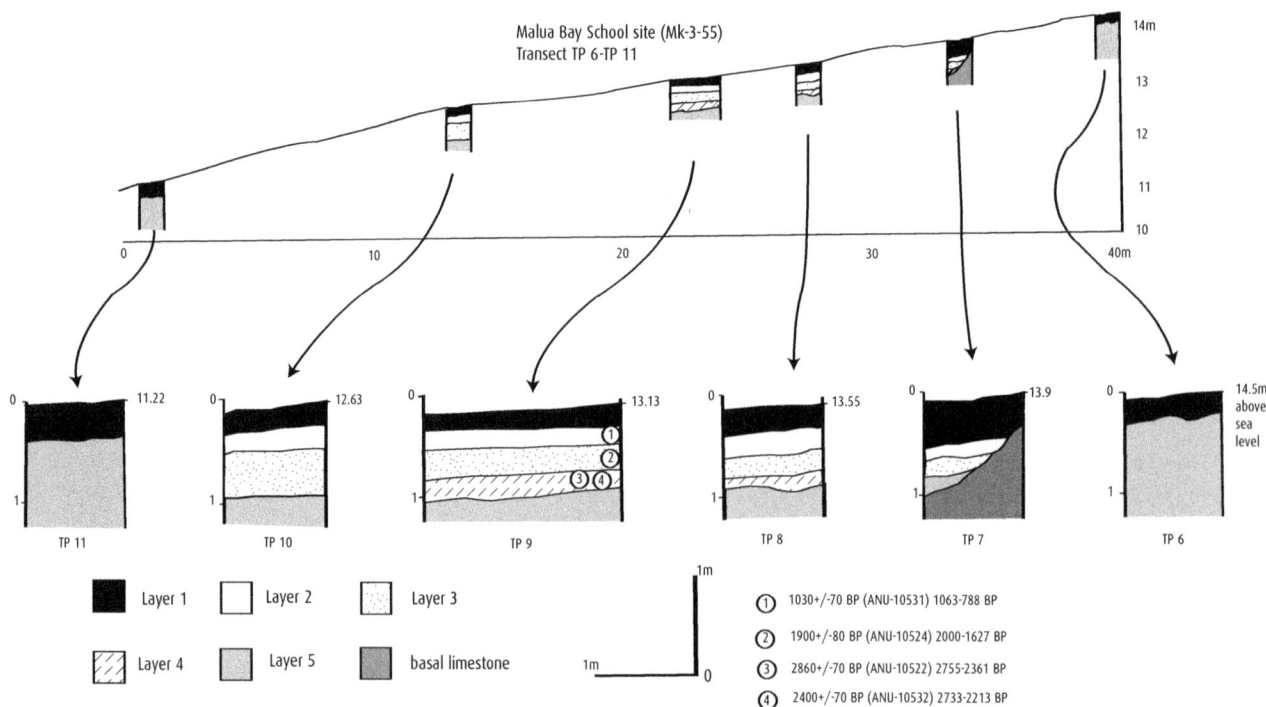

Figure 3.30 Malua Bay School site testpit sections

Layer 2 (from 30–50cm bd) consisted of a mixture of black (10YR 2/1) sandy sediment and a pale brown (10YR 6/3) tephra. It contained significantly less frequent cultural material. The tephra may be related to a particularly violent volcanic event emanating from the nearby island of Ambrym. A radiocarbon date of 1030±70 BP (ANU-10531) 1063–788 BP was recovered from the interface of this layer and the cultural layer below. The sparse cultural material recovered from Layer 2 seems more likely to relate to disturbances of either Layer 1 or 3.

Layer 3, where in evidence, was up to 50cm below the surface and comprised of a black (10YR 2/1) sandy sediment with concentrated coral gravel and frequent basalt cobbles plus cultural material which appeared to be largely cooking and frequent midden remains. The layer was very hard-packed and on average was some 25cm thick. The concentrated coral gravels look very much as if they have been transported into the area to build up a clean dry surface on top of the beach sand, a practice that is still seen at Malua Bay today. A charcoal sample was assayed from this layer and returned a date of 1900±80 BP (ANU-10524) 2000–1627 BP. This date, which suggests up to 700 years of initial occupation at the site appears somewhat inconsistent with the archaeological remains (see discussion below) and seems more likely to relate to later activity at the site.

Layer 4 represents the interface of the cultural and sterile and where present was located some 70cm below the surface. It consisted of a dark brown (10YR 3/3), to very pale brown (10YR 8/3) compacted beach sand with a decreasing sediment content along with frequent pottery and other midden remains. Shell and charcoal samples from TP 9 returned consistent dates of 2860±70 BP (ANU–10522) 2755–2361 BP and 2400±70 BP (ANU–10532) 2733–2213 BP respectively.

The recovered materials from Layers 3 and 4 included large quantities of thin plainware sherds with calcareous temper. A single dentate stamped sherd was also recovered from TP 9, Layer 3. The pots appear to be globular with outcurving rims, and some sherds have fine notching on the lip. The plainware is the same as that recovered from the basal layers of Navaprah and other coastal cave sites. Other artifactual material included part of a *Conus* shell ring, an ear pendant and a small argillite adze. Pig, sea turtle, bird, fruit bat, rat and fishbone were also recovered, along with quantities of shellfish (some of which were said to be no longer available in the area).

Layer 5 was a distinctive uncompacted very pale brown (10YR 8/3) sterile beach sand which could be identified across the whole site.

Excavations at Malua Bay revealed the remains of an initial colonising settlement site some 12–14m above the present sea level. The recovered remains included a range of both introduced and indigenous faunas, along with concentrated shellfish and pottery. The plainware pottery was calcareously-tempered with only several vessel forms being represented. A single dentate-stamped Lapita sherd was also recovered. The homogeneous nature of the ceramics from these lowest cultural layers suggests the settlement was short-term, perhaps a maximum of several hundred years or even less at c. 2700–2500 BP. It appears to post-date a period when dentate stamping was more frequent. The stratigraphy of the site also suggests that the area may have been abandoned or less intensively utilised for a period after its initial settlement. Much of Layer 2 was sterile and appeared to consist of slopewash and/or tephra.

Fiowl (Mk-3-61)

This is an open area located some 100m north of Tenmiel village and some 150m from the beach, where frequent pottery sherds were noted on the ground surface. Also in the area were a series of mounds/ridges that had been formed by clearance of the area for gardening. The area is approximately 12m above sea level. A 1 by 1m test pit was excavated on top of one of the ridge features to a depth of 80cm (Fig. 3.31). The recovered pottery was somewhat weathered due to post-depositional processes but the general pattern found throughout the excavations on Malekula, that of thicker sherds in the upper layers and thin calcareous-tempered sherds in the lowest layers, was confirmed at Fiowl. Four distinct layers were identified.

Layer 1 (8–25cm bd) comprised a very dark gray (10YR 3/1) topsoil with concentrated angular coral pebbles which graded into the layer below. Occasional very degraded shellfish were noted. Pottery (32 sherds) only was collected.

Layer 2 (25–45cm bd) was characterised by a hard-packed black (10YR 2/1) silty sediment with less frequent worn coral pebbles. Midden material became increasingly sparse with depth. At the base of this layer a distinctive lens of coral pebbles and sand was noted which may be the result of cyclone activity. Recovered materials included pottery (11 sherds) and sparse worn shellfish remains.

Layer 3 (50–80cm bd) was the earliest cultural deposit, comprising a dark gray (10YR 4/1) gravelly silt with increasing coral pebble content with depth. Again the only recovered materials

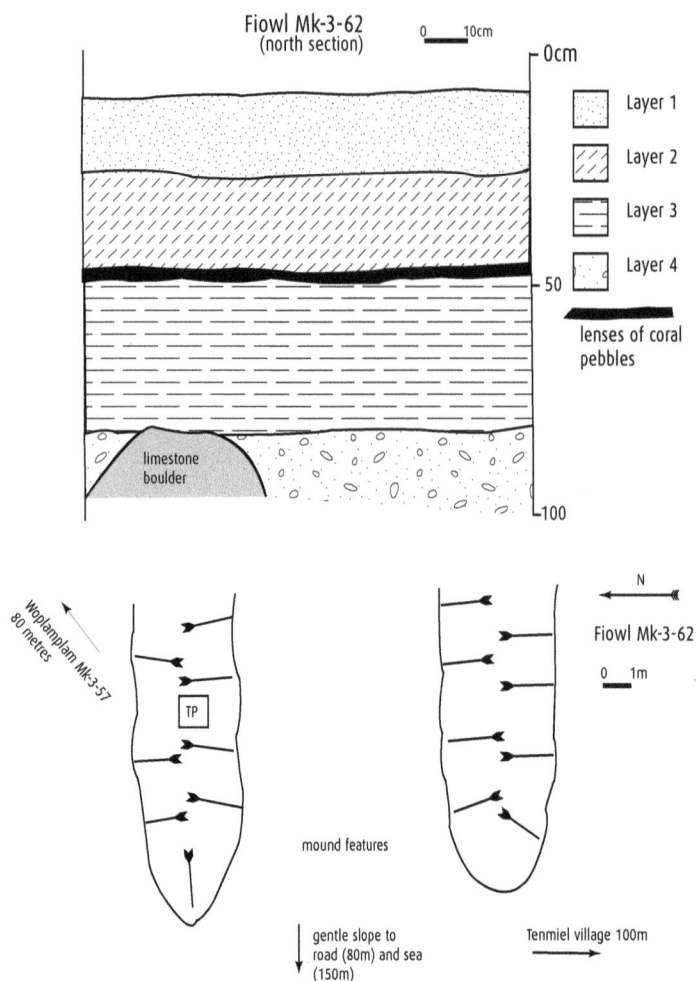

Figure 3.31 Fiowl testpit section and plan

were somewhat worn pottery (17 sherds) and sparse worn shellfish. A distinctive break was noted at 80 cms bd where Layer 4, the former foreshore was encountered.

Layer 4 (below 80cm bd) was the former foreshore consisting of a yellowish brown (10YR 5/4) matrix of gravel, coral pebbles and branch coral. It was from this lowest level that two calcareous-tempered sherds were recovered.

The testing of one of the linear mound features at Fiowl confirmed that they appeared to have been formed largely from the clearance of the area for gardening and other activities. Midden remains were found throughout the stratigraphy of the test pit but were somewhat weathered indicating they had been subjected to various post-depositional processes. The recovered ceramic remains were largely plain but a general trend of increasing thickness over time could be observed. Thin calcareous-tempered sherds were recovered from the lowest layer of the site.

Nuas (Mk-3-62)

Nuas is the name given to an open site located directly inland from the present Nuas village on the coast. The site is some 100m above sea level and 75m from the coastal cliff which backs the present village situated some 50m from the sea. The site is on a flat area at the top of a slight slope. Pottery and shellfish concentrations were noted at the site and a low platform of coral cobbles indicated the remains of a former *nasara*. The site remains were very similar to those noted at Chachara (Mk-3–43) and the site was tested as a comparison with the inland site. A series of test pits (Fig. 3.32) were excavated in an area of the site where concentrations of pottery, shellfish and fire-cracked stone were present on the surface. Post-depositional processes have affected the condition of the recovered pottery and shellfish and also appear to have eliminated any bone remains. The stratigraphy was very shallow, a maximum depth of 40cm only, and only a single layer was distinguishable. A total of five square metres was excavated.

Layer 1(0–30cm bd) was the only layer identified at the site. It consisted of a black (10YR 2/1) topsoil with occasional limestone pebbles/cobbles and basalt cobbles. The buildup of the matrix appeared to be largely the result of slopewash interspersed with midden debris. Midden remains were sparse. The layer sat directly on top of the basal limestone rock. Its thickness ranged from 6–8cm in TP 7 to 20cm in TP 1, 30cm in TP 4 and 4a and to 30cm in TP 5. All test pits, except TP 7 which was completely sterile, returned worn shellfish and pottery remains.

The Nuas site appears to be a single short-term late settlement site associated with the adjacent *nasara*. The site revealed similar layout and ceramic remains to the inland site of

Chachara. A radiocarbon determination on a shell from the upper levels of the cultural layer (TP 4, 0–15cm bd) returned a date of 570±50 BP (ANU-10530) 298–0 BP. This is somewhat later than the date from Chachara and suggests that this site type and the associated ceramics have some time depth.

Discussion of Malekula Sites

The stratigraphy of the Malekulan cave sites was relatively consistent, generally representing intermittent occupation with ash lenses from cooking fires, fire-cracked basalt and or coral cobbles, shellfish, bone and relatively sparse artifactual material. Ceramics were consistently found at the base of the cultural stratigraphy in all the cave sites and less frequently throughout all layers. In a number of coastal cave sites calcareous-tempered sherds were recovered from layers associated with initial human occupation of the area. Three of the cave sites returned a high concentration of faunal material. This is attributed to them having periodically been owl roost

Figure 3.31 Fiowl testpit section and plan

caves. The excavated open sites included an initial settlement site dating to c. 2700–2500 BP at Malua Bay that was associated with calcareous-tempered plainware ceramics, and two much later sites (post 500 BP) namely Nuas and Chachara, that were associated with ceremonial activity areas and extensive collections of the later style ceramics. The excavated sites have revealed only the two extremities of the ceramic sequence on the island. The recovered ceramic remains are presented in detail in Chapter 7 along with further evidence which suggests that a central section of the sequence does exist but that it was largely absent from the excavations carried out in Northwest Malekula.

The earliest settlement on the Northwest appears to have been concentrated on the coast around sheltered bays and reliable water sources. It is not until some time later that permanent settlement shifts inland and to more marginal areas of the coast. There is also some indication across a number of the sites that after initial settlement there is a period of abandonment or less intensive occupation of the area. More intensive occupation occurs again from c.1000 BP or later. There was no evidence, in any of the excavated sites, of pre-Lapita settlement in this area of the island.

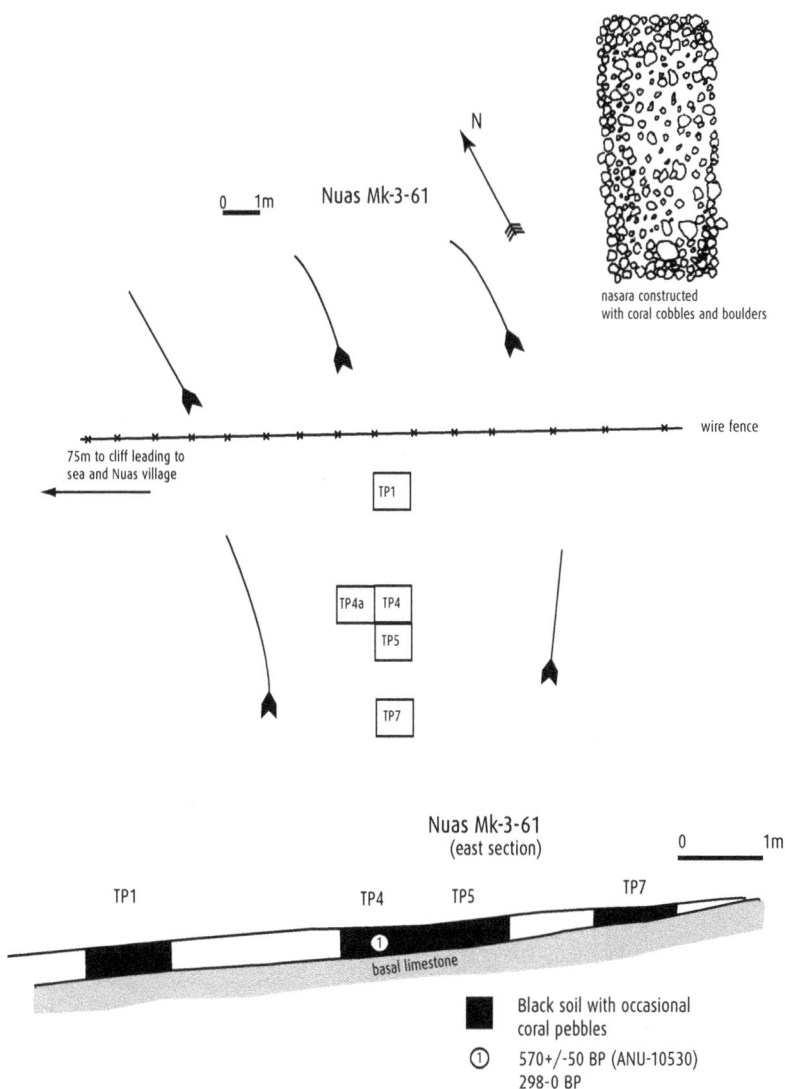

Summary

The stratigraphic and cultural sequences gleaned from the excavations on Erromango, Efate and Malekula have greatly expanded our knowledge of the early history of Vanuatu. Settlement pattern across the archipelago is becoming increasingly clear. It seems safe to conclude that the pattern of initial occupation in Vanuatu would fit a situation where Lapita colonists and their immediate successors were moving into an empty landscape and were thus able to set up settlements in the prime locations for habitation and canoe access which would have facilitated maximum utilisation of the marine and other faunal resources (e.g. Ponamla and Ifo on Erromango, Arapus, Mangaasi and Erueti on Efate, Malua Bay on Malekula and its small offshore islands and much of coastal Malo, Aore). This settlement pattern has been further confirmed on the small islands of northeast Malekula where four Lapita sites have been located (Bedford 2003). This phase of settlement was extremely fluid and at a number of sites initial occupation appeared to be relatively short-lived being followed by a period of abandonment or certainly less intensive occupation (e.g. Ponamla on Erromango, Malua Bay and coastal cave sites on Malekula). This may be due to a combination of factors one of which is that these sites were located on the more environmentally fragile leeward sides of the respective islands. There is certainly evidence that populations moved on to other areas of the same island or indeed to other islands. Other environmental factors may also have influenced the suitability of long-term settlement. The site of Arapus and Mangaasi display lengthy cultural sequences and certainly the tephra-rich soils found on the west coast of Efate would have been an attractive feature which would have facilitated intensive horticulture over a lengthy time period.

The archaeological record also suggests that it is only later, as populations grew and people spread out from these prime locations, that the interior areas, certainly of the larger islands became part of the settlement system (e.g. Erromango cave sites, and inland Malekula).

The more recent investigations have filled a number of archaeological blanks and enables a reassessment of earlier research and at the same time provides a platform from which to further address a number of the research questions detailed in the introduction. Already it has been established that pre-Lapita occupation of the archipelago seems very unlikely and a tentative settlement pattern has begun to emerge. Having outlined the excavated sites we move to the analysis and presentation of the recovered midden remains.

4

Ceramic analytical methodology

'We consider that typologies are tools made for a purpose, and as long as they can be shown to work for that purpose they require no more abstract justification than a crowbar' (Adams and Adams 1991:8).

This chapter outlines the methodology used for the analysis of the recovered ceramics from the excavations in Vanuatu. Firstly, a description of the series of attributes which were selected in order to characterise the ceramics is presented. Having established the analytical parameters for the recovered ceramics the results are then outlined in detail in subsequent chapters starting with ceramic remains from Erromango (Chapter 5), then Efate (Chapter 6) and finally Malekula (Chapter 7). Chapter 8 summarises the proposed ceramic chronologies for the different islands of Vanuatu and moves onto a discussion of intra and inter-archipelagic comparisons.

There is now a formidable list of sites from throughout the Pacific where large quantities of ceramics have been recovered. Some have been published in great detail, many others in much less detail and some in very little detail at all. Dating from Gifford's initial work in Fiji (Gifford 1951) a lengthy tradition of ceramic analyses has developed along with established methodologies. A number of different approaches have been used and, not surprisingly, have been specifically designed and adapted for specific collections and research questions.

The methodology used in the analysis of the Vanuatu ceramics concurs with the aims of Irwin in his study of the Mailu ceramics, in that it 'sets itself the task of being a practical exercise in classification rather than one of justification and semantics' (Irwin 1985:100). The analytical methods were designed specifically to describe the excavated ceramic materials including vessel form, decorative technique, motif form and fabric composition, as a first step in addressing the research issues outlined in Chapter 1. The ceramic data was utilised in the establishment of ceramic chronologies and cultural sequences for a number of different islands in Vanuatu which were then subsequently used in a comparison with ceramics from other archipelagoes to test current theories regarding cultural change and interaction.

Rather than attempting to construct an all new analytical procedure, this study borrows heavily from the already established methodologies used and modified by previous researchers who have focused on the ceramics from the Southwest Pacific, namely Irwin (1985), Specht (1969), Summerhayes (1996) and Wahome (1999), who in turn, along with most other archaeologists dealing with ceramics in the Pacific, have been at some point influenced by Shepard (1963). More recent comprehensive texts which have also proved useful include Arnold (1985) and Rice (1987).

Archaeological sites in the Pacific where ceramics are recovered are often shallow, have been heavily disturbed or even totally lack stratigraphy as in the case of ceramic deposits on reefs. Sherds can be small and worn and provide limited information regarding vessel form or decoration. This is not so debilitating when dealing with dentate stamped Lapita ceramics. The decoration is so distinctive and it has been the focus of such detailed research that even very poorly preserved sherds can provide useful information. However with the much more varied and unknown corpus of decoration and form associated with post-Lapita ceramics the analysis can be severely curtailed. Several of the sites excavated as part of this research have been singularly spectacular for their depth of stratigraphy and preservation of ceramics of the immediate post-Lapita period and hence have led to an emphasis on form and decoration. Combined with the illustrations of Garanger (1972) of the ceramics he recovered from Efate and the Shepherds they now provide one of the most detailed chronological sequences of post-Lapita ceramics from the Southwest Pacific.

The application of a single set of variables to large collections of ceramics from a number of different sites and different islands, often with greatly varying chronologies, was initially found to be somewhat problematical and cumbersome. However if meaningful comparisons were to be made it was an essential requirement. Bearing this in mind, some effort was made to avoid the over-elaboration of certain shape attributes (Irwin 1985:105) and the potential minutiae of detail that can be recorded from single sherds. If it is accepted that some variance in particular attributes can be found on a single pot (see for example Ambrose 2002:64) or between potters producing the same pots then the inclusion of too much fine detail may unnecessarily complicate the ceramic picture. This (mild) 'lumping' approach has both its adherents and detractors and is one of the components in the extensive associated literature involving the 'typological debate' (see Adams and Adams 1991:265). The approach developed during the period of analysis has been subjected to numerous changes and refinement and has ultimately provided a suitable framework for the analysis of the Vanuatu ceramics.

Outlined below are the set of discrete and continuous variables which were selected that primarily relate to vessel form, decoration and fabric. The attributes were assigned a numeric code for entry into a database which facilitated further manipulation. All of the sherds recovered from the excavations were included in the analysis.

Ceramic Attributes

1) **Class of sherd** (see Fig. 4.1 for vessels with named parts). These included 1) rim 2) base 3) body 4) carination 5) handle 6) rim and body 7) spout.

2) **Weight**. Each sherd was individually weighed and recorded in grams. Although often regarded as a somewhat unnecessary practice it can in some cases provide a further useful quantification of ceramic collections.

3) **Sherd thickness**. The procedure for measuring rim thickness followed that used by Summerhayes (1996:79) who in turn had followed Irwin (1985:107) and Specht (1969:78). Two measurements, A and B were recorded. Measurement A was taken on the rim itself and B on the body immediately below it. Some variation to the rule was required with measurement A when dealing with very different rim forms. The outline originally devised by Specht (1969:79) was followed here, namely A was taken at the point of maximum thickness on divergent or thickened rims or close to the lip on parallel or convergent rims.

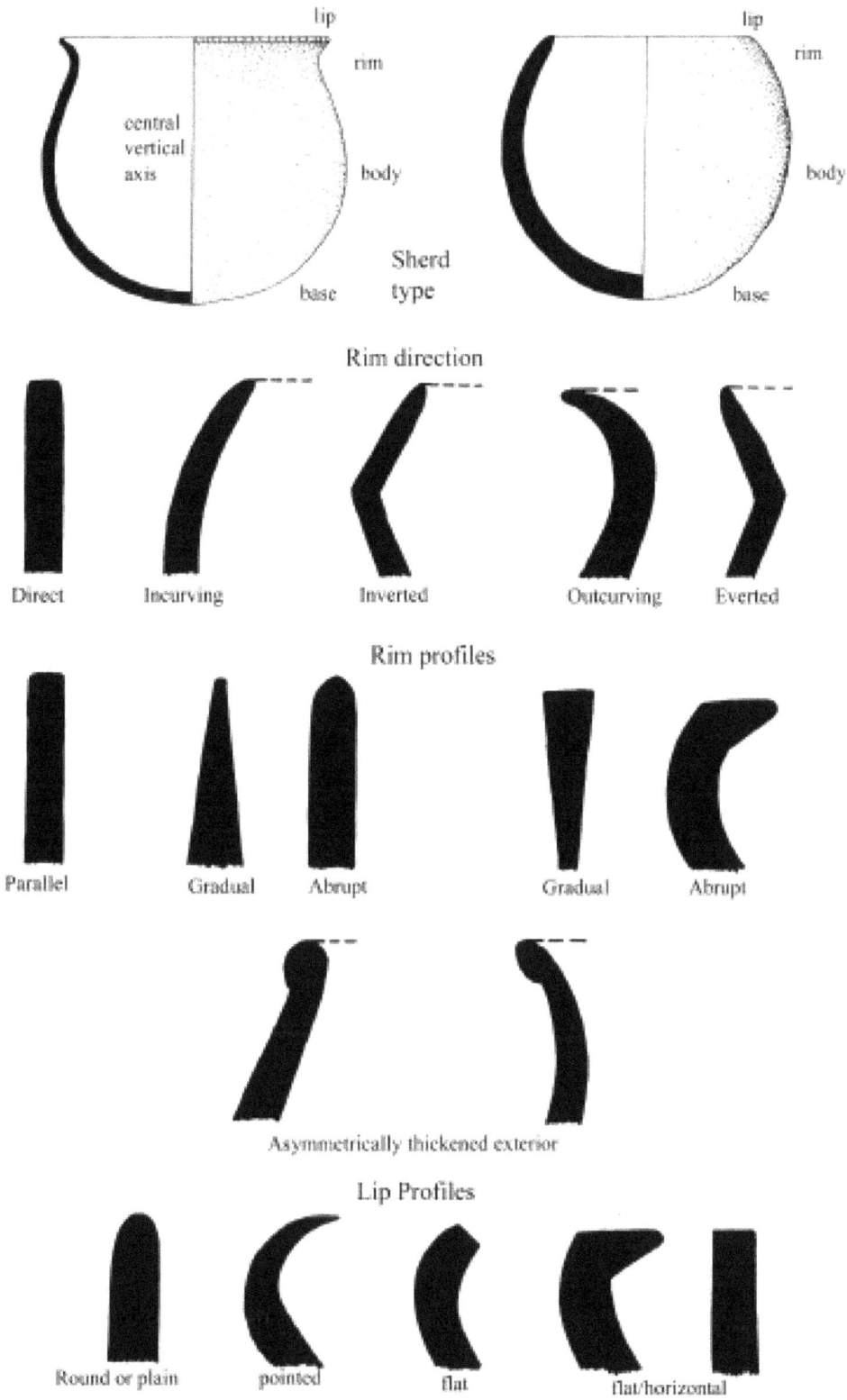

Figure 4.1 Sherd type and attributes of the rim and lip.

The measurement of body sherds was somewhat more straight forward. With sherds which displayed no discernible difference in thickness, only one measurement has been recorded. This is the case for the majority of body sherds as they are generally not large (no more than 4–8cm^2) and do not display great variability over such an area. Two measurements were taken on sherds which showed marked variation in thickness, namely the thickest and thinnest points. The further grouping of these measurements followed Wahome (1999), who indicated that thickness measurements could be helpful in terms of chronological differentiation and were more useful if grouped as such, 1–4mm, 4–8, 8–12, 12–16, 16–20 and so on.

4) **Fabric Analysis.** Basic analyses regarding fabric composition and structure were carried out on the excavated sherds. The main objectives with this aspect of the analysis were as follows: 1) to establish whether the ceramics were produced at or near the sites from which they were excavated or as is often the case pinpointed more generally to a particular island; 2) To identify any evidence for trade and exchange across single islands and or between islands and 3) whether any change of fabric composition could be identified through time. This information combined with details of vessel form and decoration provides a more comprehensive characterisation of the ceramic collections than any of these attributes alone (Arnold 1985:237).

The analysis of ceramic fabrics encompasses a number of analytical and interpretive procedures which have been increasingly refined over the last 40 years (Wilson 1978; Summerhayes 1987, 1997). Healthy debate over the appropriateness and merits of various techniques and methodologies continues amongst those involved in this specific area of research (Ambrose 1992; Hunt 1988, 1993; Summerhayes 1996). The emphasis that is placed on fabric analysis can vary greatly from one research project to another depending on the state of the collections and the research questions. The analyses carried out on the excavated ceramics from Vanuatu were designed simply to address the above objectives.

The primary focus of the fabric analysis concentrated upon the identification of non-plastic inclusions through petrographic examination. A more subjective description of the texture of a sherds' cross section was also included. This enabled the possible differentiation of sherds which, although they may have contained very similar non-plastic inclusions, utilised a different clay source or involved differing manufacturing techniques and so on. These various combinations of non-plastic inclusions and texture were assigned numeric codes which enabled further characterisation of a particular ceramic collection. The combined experience and expertise of both William Dickinson and Glenn Summerhayes provided vital guidance during this phase of the analysis.

Dickinson, while visiting Canberra in 1996, was able to peruse the excavated ceramic collections from Erromango and Malekula macroscopically and selected a number of sherds which he identified as representing the range of non-plastic inclusions. These selected sherds were then looked at microscopically to define further the identifications of the non-plastic inclusions (termed tempers by Dickinson). The various reports completed by Dickinson (see Dickinson 1998 for an outline of methodology), along with his detailed discussions of the geology of relevant islands, are included in Appendix 3. Further macroscopic analysis was carried out by Summerhayes on the ceramics from the 1997 excavations at the Mangaasi site on Efate and again a limited number were thin-sectioned and sent to Dickinson for petrographic analysis.

5) **Rim direction.** A total of five rim directions were defined (see Fig. 4.1). These categories can be seen as somewhat all-encompassing but proved effective for the analysis of the Vanuatu materials. Two long established and interconnected methods were employed for defining the direction of the rims. The orientation of the rim is established from an imaginary perpendicular central axis running down the centre of the original pot (Poulsen 1987:29)

combined with the placing of a horizontal surface on the lip which enables finer calibration of the rim angle in relation to the central axis (Shepard 1963:253).

   1) Direct: rims of this orientation do not curve towards the interior or the exterior of the vessel. As outlined by Summerhayes (1996:78) it can often be difficult to be completely certain about the identification of these rims if only smaller sherds are available and that the assigning of vessel forms is often not possible. This warning can be extended to any rim sherd with insufficient preserved detail.

   2) Incurving: In this case the rim inclines towards the interior of the vessel, with a convex profile.

   3) Inverted: Again the rim is inclined to the interior of the vessel but has a sharp angle on the exterior surface which forms a corner point.

   4) Outcurving: This category of rim is inclined to the exterior of the vessel, again with a relatively convex profile.

   5) Everted: Related to the above category in that the rim is inclined to the exterior of the vessel but includes a sharp angle on the interior of the vessel at the point of inflection.

6)   **Rim profile** (see Fig. 4.1). The rim profile relates to the relationship between the inner and outer walls of the rim as they proceed to the lip. Again well established conventions regarding the definition of rim profile were employed (Kirch and Rosendahl 1973; Irwin 1985; Poulsen 1987; Summerhayes 1996). Six profiles were identified.

   1) parallel: little perceptible change in rim profile towards the lip
   2) convergent gradual: rim profile thins towards the lip
   3) convergent abrupt: rim profile dramatically thins towards the lip
   4) divergent gradual: rim profile thickens towards the lip
   5) divergent abrupt: rim profile dramatically thickens towards the lip
   6) asymmetrically thickened exterior: related to the above divergent categories but distinct in that the form of thickening relates to either the folding over of the rim, the addition of extra clay or the rim being squeezed (Summerhayes 1996:79).

7)   **Lip profile** (see Fig. 4.1). The lip is defined as the top edge of a vessel i.e.; the meeting point of the inner and outer walls of the rim (Specht 1969; Summerhayes 1996:70)

   1) plain or rounded lip: a smooth semi-circular (convex) profile
   2) pointed: the apex of the lip is defined by a sharp point
   3) flat: lip surface that is relatively straight between the inner and outer walls of the rim
   4) flat horizontal: same as above but lip surface lies on a horizontal plain

8)   **Rim radius.** This attribute was only measured on rim/lip sherds that retained some degree of curvature which further enabled the diameter of the pot to be calculated. The lip sherds generally needed to have at least 4-5cms of the radius to be regarded as being able to provide a reliable measurement. The method of measurement simply involved matching the rim/lip sherd to a series of graduated concentric circles.

9)   **Vessel form** (see Figs 4.2–4.6). Three broad (designed as such to avoid over-classification) vessel form categories are outlined along with a number of sub-categories which when combined with other attributes such as rim and lip form give a very detailed picture of a particular vessel. The basic terminology and definition of vessel form have been adapted from Summerhayes (1996) and Wahome (1999). Note that the vessel form classification is restricted to excavated ceramics only and is not designed to be an all-encompassing Vanuatu catalogue. This is particularly relevant when focusing on the later ceramic material from Malekula. An attempt has been made to record as comprehensively as possible the full range of Malekula ceramics including previously surface collected and illustrated materials, but they have not been incorporated into this classification. That awaits further fieldwork, research and much refinement.

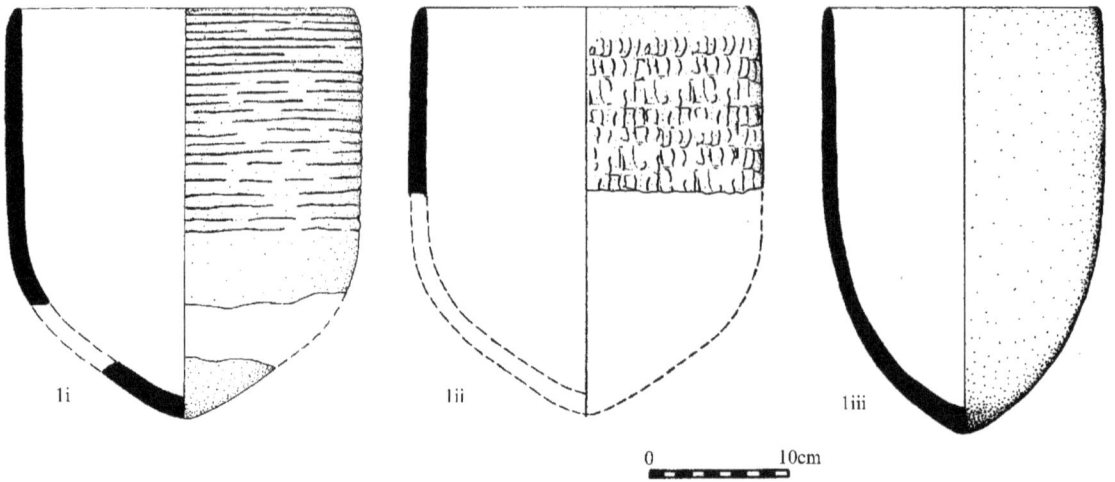

Figure 4.2 Vessel form, unrestricted vessels 1i to 1iii.

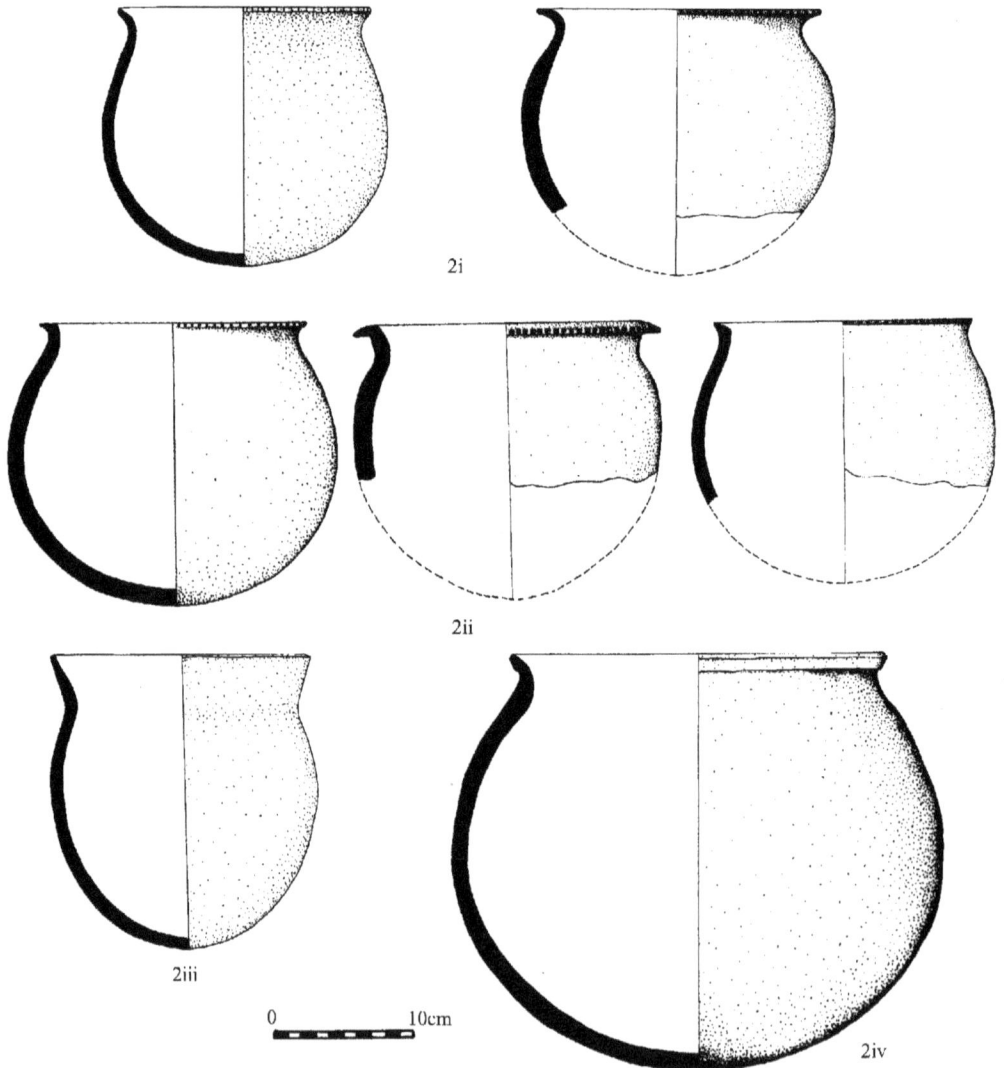

Figure 4.3 Vessel form, restricted vessels 2i to 2iv.

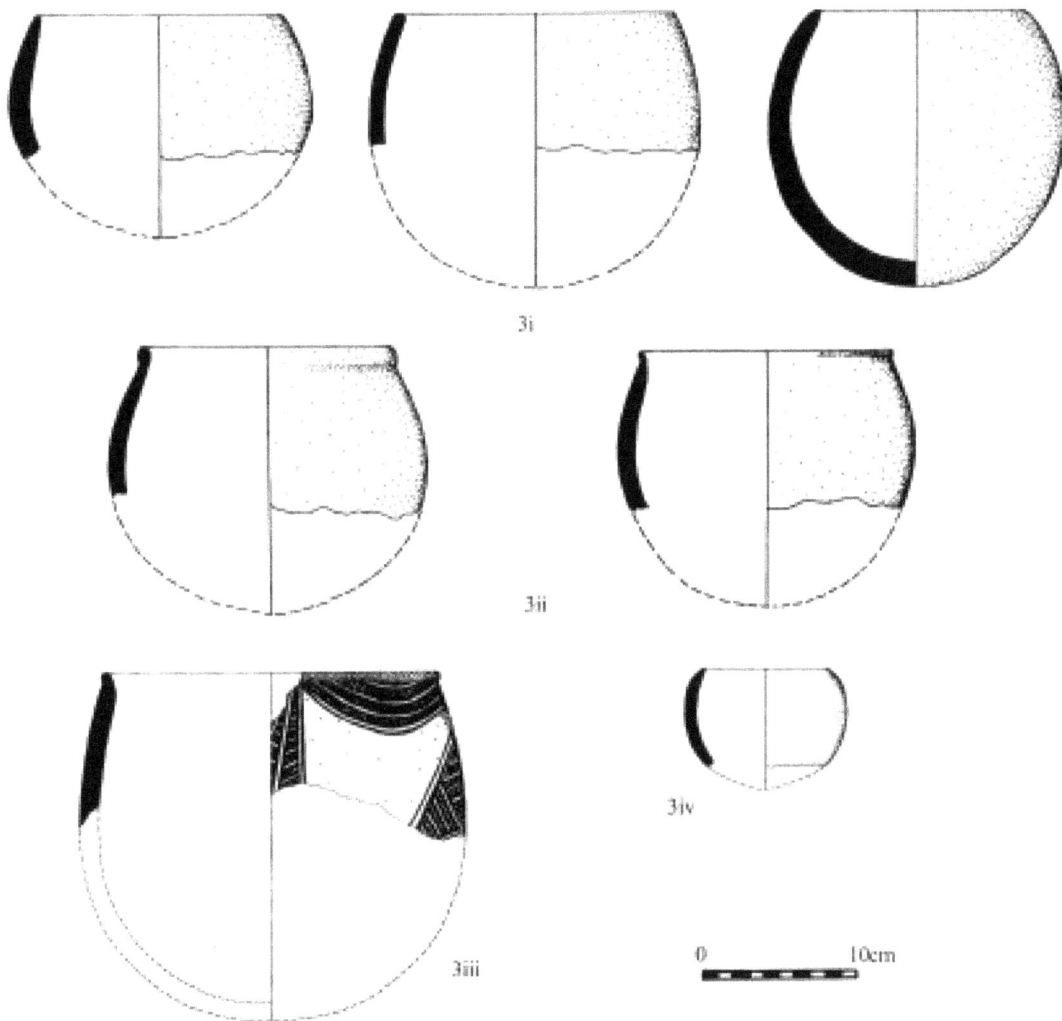

Figure 4.4 Vessel form, restricted vessels 3i to 3iv.

a) unrestricted vessels (Fig. 4.2)
1) open pot, tubular body, vertical walls and pointed base. Appears to be the immediate predecessor of the ceremonial *naamboi* (see below) which are often referred to as 'bullet shaped' pots. These pot forms are restricted to Malekula and other islands of northern Vanuatu only.
 i) ribbed only. Coil made pots with a rim diameter of between 24–26cm. Height approximately 14–15cm. Although the interior of the pot is generally smoothed the upper exterior part of the pot is not which gives it a ribbed appearance.
 ii) ribbed with 'scale' effect. Similar to above but exposed coils (ribs) have been pressed with a finger or thumb to create a 'scaled' effect.
 iii) non-ribbed, always decorated. Again coil made but the exterior of the pots has also been smoothed to facilitate decoration. Slightly smaller rim diameter of 20–22cm and height of again c. 14–15cm. Decorated with a wide variety of motifs utilising largely incision or punctation.

b) restricted vessels
2) globular pot with restricted neck (Fig. 4.3)

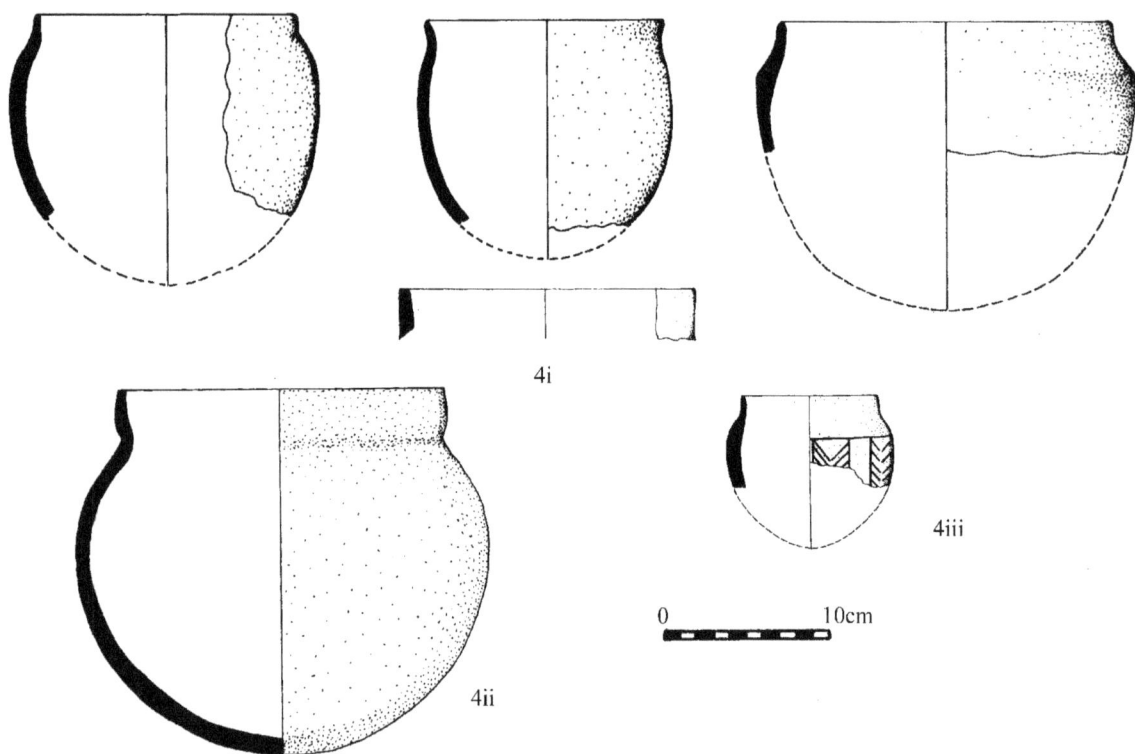

Figure 4.5 Vessel form, restricted vessels 4i to 4iii.

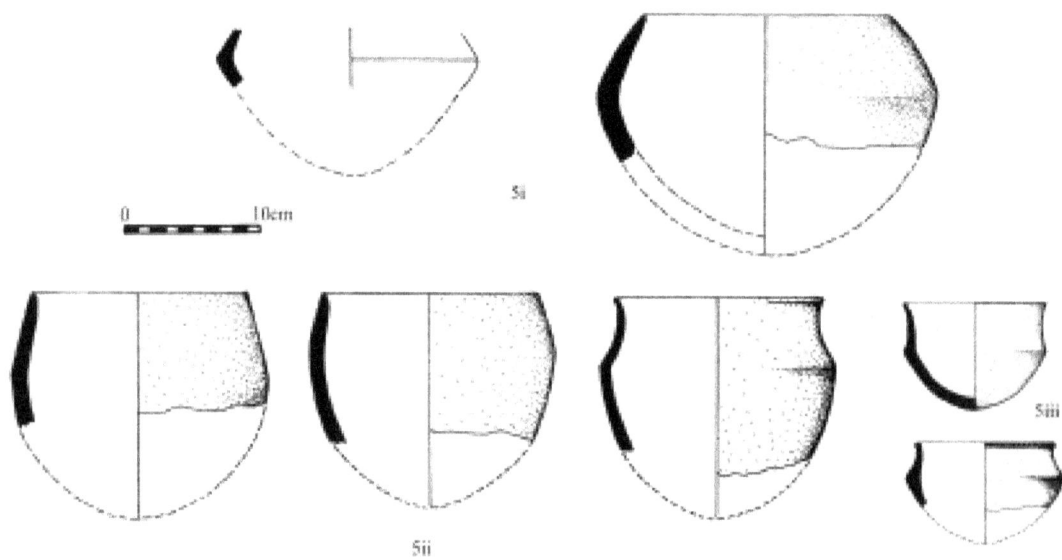

Figure 4.6 Vessel form, carinated vessels, 5i to 5iii.

> i) globular pot with restricted neck and outcurving rim. Somewhat of an ubiquitous form but generally associated with earliest ceramic production on all three of the studied islands, although a variant appears during the last phase of ceramic production on Efate.
>
> ii) globular pot with restricted neck and outcurving rim with flat expanded (often horizontal) lip. Although very similar to the above the distinctive rim/lip form sets it apart and it characterises a distinct phase of the ceramic chronology of Efate and the other central islands.
>
> iii) globular pot with restricted neck and outcurving, sharply everted rim. This pot form is associated with the earliest ceramic production on Malekula but was also recovered from Erromango and a variant appears during the latter part of the Efate sequence.
>
> iv) very large plain globular pot with externally thickened outcurving rim. A rim diameter of 24–26cm and height of c. 28cm. Features such as highly smoothed interior and exterior surfaces and the wide body curvature enable this pot form to be identified from body sherds only. This pot form is restricted to Malekula.

3) globular pot with incurving rim orientated towards the interior of the vessel (Fig. 4.4)

> i) symmetrically globular pot with rim orientated towards the interior of the vessel. Rim tends to be convergent gradual. This pot form dominates in the later part of the ceramic sequence of Efate and to a lesser extent on Erromango.
>
> ii) globular pot with an externally thickened rim which is orientated towards the interior of the vessel. Similar to 3i but merits a separate category as it represents a significant phase of the Efate and Erromango sequences. It appears to represent a transitional phase of vessel form, sandwiched between outcurving (2i) and incurving rim vessels (3i).
>
> iii) very large globular pot with externally thickened rim and distinctive incised and excised decoration. Both a rim diameter and height of c. 22cm. This pot form is restricted to Malekula.
>
> iv) globular cup. Similar to 3i but meriting a separate category due to the substantial size variance. To date only found on Efate.

4) globular pots with direct rims (Fig. 4.5)

> i) globular pot with direct rim. A relatively rare pot form with variants found on both Efate and Erromango.
>
> ii) globular pot with collared rim. Similar to above but distinctive thickened rim. To date only found on Malekula and associated with early period ceramics.
>
> iii) globular cup with direct rim. Again similar to 4i but due to substantial size variance merits a separate category. To date only found on Efate.

C) Other

5) carinated vessels (Fig. 4.6)

> i) sharply carinated vessels. The exterior carination is sharp and angular.
>
> ii) soft carination. The exterior carination is markedly less angular than 5i. To date only found on Efate.
>
> iii) carinated cup. Similar to 5i but due to substantial size variance given separate category. To date only found on Efate.

Decoration. This attribute is one of the most useful and often utilised in the study of ceramics in determining levels of communal interaction and transformation over time and therefore necessarily requires a more detailed analysis. Three distinct variables are utilised here to characterise the decorated ceramics, namely 10) location of decoration,

11) technique and 12) type of decoration or motif.

10)   **Location of decoration.** This is the more easily identified and defined category of the three related to decoration. The six areas of decoration are outlined below.

    1) lip

    2) inside lip: this refers to inner edge of horizontal lip

    3) outside lip: this refers to outer edge of horizontal lip

    4) flat area of horizontal lip

    5) rim: only noted if decoration is restricted to the rim rather than a decoration which includes rim and body. If the latter is the case rim and body are noted.

    6) body

11)   **Technique.** A total of five general techniques were identified with a larger number of sub-techniques. All are fairly self explanatory. These categories are again relatively broad but are specifically designed to describe technique and have avoided shifting into the terrain of decorative motifs which are discussed further below.

    a)   Incision

    1)   linear

    2)   geometric

    3)   gash

    4)   curvilinear

    5)   comb incised

    6)   miscellaneous (sherds too small to define any of above)

    b)   Impression

    1)   punctation: stick or similar tool impression

    2)   dentate stamping

    3)   fingernail: further divided into three techniques

    i)   fingernail impression: fingernail pressed into clay, perpendicular to the surface creating fine crescent-like designs.

    ii)   fingernail pinch: use of thumb and finger to create pinched clay effect.

    iii)   fingernail gouge: fingernail pressed into clay at an angle to the surface which creates a gouge effect.

    c)   applied relief

    1)   plain bands

    2)   notched bands

    3)   nubbins

    d)   notching (mostly on the lip)

    e)   excision

    1)   carving

    2)   perforation

12)   **Motif type.** For over 30 years the decorative elements found on Lapita dentate-stamped and incised ceramics have been intensively studied (Anson 1983; Donovan 1973; Mead *et al*. 1975; Sharp 1988; Siorat 1990). The Lapita material lends itself to detailed systematic motif analysis with its complex geometric nature and frequent occurrence of identical or near identical decorative elements (Anson 1983:16). Despite this fact, however, agreement as to the ideal methodology for the analysis of the Lapita design system remains somewhat disputed and unresolved territory (Anson 1990; Green 1990; Specht 1977).

Post-Lapita ceramics have received nothing like the focus and interest that have been directed towards Lapita ceramics. This is due to a number of factors not least that post-Lapita material is far less morphologically sophisticated and less decoratively structured (Golson 1992:165) than its predecessor, simply adding to the difficulty of the task of defining motifs. Often ceramics recovered from sites throughout the Pacific are very fragmentary, degraded and display very limited areas of much larger motifs. Hence the decoration is necessarily described in terms of technique rather than motif. Partly then, the nature of the collections are to blame for the lack of differentiation of decoration. These aspects are also directly related to the simple fact that few sites containing representative collections of post-Lapita material have been excavated and studied. Comprehensive studies of ceramics of this period are largely restricted to those carried out by Specht (1969) and Garanger (1972). Until much greater effort is focused on sites that are directly post-Lapita and a much larger corpus of representative decorated sherds are recovered and published, efforts to extricate ourselves from the all-encompassing 'incised and applied relief' label will be further impeded.

Some attempt to define a number of motifs from the Vanuatu ceramics was seen as vital, an attempt to move beyond the 'rather limited range of technological traditions for producing decoration on pottery' and 'simple technological categorisation' (Green 1990:33), to further differentiate the excavated material. The motif classification established here aspires to something much less than 'paradise', rather it is an initial attempt at an unsystematic basic level, similar to that carried out by Specht (1969) on the Buka material, to define 'distinctive' motifs. Similar difficulties as experienced by those working on Lapita ceramics were encountered. This included the difficulty of distinguishing and defining the significance of variation and similarity between various motifs and whether they should be assigned separate motif numbers. The universality of a number of decorative techniques and sub-motifs tends to limit their usefulness in terms of delineation and can unnecessarily crowd the classification. There were a number of sherds where the decoration was either too fragmentary and/or ubiquitous to be assigned a separate motif number. These sherds are however illustrated. The methods of categorisation and definition of motifs are many and varied (Rice 1987:244–272) and can often prove to be somewhat contentious (e.g. Lapita above). Validation of method can be less of a concern, if it is accepted that motif definition is often subjective, as is the case with this research, and influenced by the ceramic material under study.

Due to the large quantity of sherds studied and the fact that they originated from a number of archaeological sites from different islands identified motifs were separated by island (E-Erromango, Ef-Efate, M-Malekula) and arbitrarily assigned serial numeric codes. Following Specht (1969:83) the motif code numbers have no significance in terms of techniques, content or chronology. All the motifs feature amongst the illustrations of the ceramics from the various sites (Erromango Figs 5.1–5.16; Efate Figs 6.1–6.23; Malekula Figs 7.1–7.23) and are described and listed individually in Appendix 4. The Lapita dentate stamped sherds recovered from the excavations were assigned, where possible, motif numbers from Anson's inventory (1983). No vessel forms could be established from the Lapita dentate-stamped sherds.

5

Ceramics from Erromango

'It is essential as well that we not assume that Lapita exchange, in any region or province, was static over the temporal period that the distinctive Lapita ceramic complex can be recognised … in any instance we must be prepared to deal with processes of regionalisation, localisation and specialisation' (Green and Kirch 1997:20).

Ponamla ceramics

A total of 8419 sherds (plain body, base, rim, decorated and carinated) were recovered from the excavations at Ponamla (Table 5.1). The vast majority of the sherds (7387) were from the areal excavation named Area A (Fig. 3.1). Of those sherds 80% were plain body sherds, leaving 1442 sherds that were classified as being diagnostic; i.e. rims (1071 [75%]), decorated sherds (371) and one carinated sherd. A total of 121 clay wasters were also recovered. The ceramics recovered from Ponamla show a high degree of homogeneity both in terms of fabric, form and decoration, reflecting the relatively short-term nature of the occupation.

The most striking feature is a shift from essentially plainware at the lowest levels of the site to increasingly decorated ware in the upper levels of the site. A test pit program carried out across the site was designed to determine the extent of the site and if there existed any temporal or spatial variance (see Chapter 3). The ceramics recovered from the test pits were similar both in form (Figs 5.1, 5.2), decoration and fabric to those recovered from Area A and no temporal or spatial variation was indicated. Only sherds from Area A are analysed in any detail here.

Fabric
Three basic fabric types, all relatively similar, were identified amongst the excavated materials and are described below. The detailed petrographic analyses carried out by Dickinson are reproduced in Appendix 3 (see WRD 147).

Fabric 1) This fabric accounts for the vast majority of the sherds. It is characterised by pyroxene-rich, plagioclase-poor mineral inclusions, which are moderately sorted, subangular to subrounded

Table 5.1 Ponamla, Area A and Test pits. Excavated ceramic sample

	plain body	base	rim	carin	deco. %	cal. temper	wasters	weight (gm)
Area A								
Layer 1	2825	22	533	–	290 (7.9)	–	59	27485.22
Layer 2	1033	12	190	1	52 (4.1)	–	16	10586.56
Layer 3a	817	19	169	–	24 (2.3)	3	25	12839.02
Layer 3b	1034	9	125	–	3	5	19	8991.92
Layer 4	173	–	54	–	1	10	2	1474.61
Total	5882	62	1071	1	371	18	121	49537.51
%	(79.6)		(14.5)		(5.1)			
TP 5.1	7	–	3	–	–	–	–	33.9
5.2	13	–	2	–	2(11.7)	–	–	78.9
5.3	27	–	11	–	10(20.8)	–	–	216.7
5.4	428	–	59	–	41 (7.7)	–	5	2769.4
5.5	3	–	–	–	–	–	–	13.94
6.0	150	–	15	–	11 (6.2)	–	1	1437.2
7.0	167	–	29	–	13 (6.2)	–	–	1601.99
8.0	36	–	5	–	1 (2.3)	–	–	219.7
9.0	–	–	–	–	– –	–	–	
Total	831	–	124	–	78		6	6387.61
Grand total %	6713	62	1195	1	448	18	127	67764.91
	(79.73)		(14.19)		(5.3)			

aggregates and inferred to be of stream origin as the calcareous grains present appear to be derived from reworked limeclasts rather than reef detritus found amongst beach sands. The clay content of the fabric dominated with mineral inclusions making up approximately 20% of the density.

Fabric 2) Similar to above but with additional calcareous inclusions derived from reef detritus. A total of 18 sherds only exhibited this fabric type and all were restricted to the lowest levels of the site. The only vessel form (2iii) associated with this fabric is that with an everted rim (Fig. 5.1n).

Fabric 3) This fabric is again pyroxene-rich, plagioclase-poor as are the two above but the inclusions are very much more finely sorted and most likely of beach origin. The fabric has a grainy texture with less clay content. This fabric is only associated with several sherds, including a single carinated Lapita dentate-stamped sherd (Fig. 5.10i) and a direct parallel rim (Fig. 5.1b).

A tentative explanation for this fabric variance is outlined below. It may well have been that very soon after arrival a suitable clay source with a percentage of naturally occurring temper (Fabric 1) was identified close to Ponamla. Initially calcareous temper (Fabric 2) may have been deliberately added following previous practices elsewhere but that very quickly the clay source alone (Fabric 1), after some experimentation, was found to be suitable for pot manufacture without the addition of the calcareous temper. Fabric 3 appears to represent the addition of non-calcareous temper to the local clay source for the production of specialised vessels. A similar scenario has been proposed by Kirch in relation to ceramics both from Nuiatoputapu and Futuna (Kirch 1988b:155). However as also noted by Kirch the validity of the argument can only be further tested with detailed sampling and analysis of the clays in the area.

Plain sherds

The vast number of recovered sherds were plain body sherds with a significantly smaller percentage of basal sherds (Table 5.2). The sherds were weighed (grams), the thickness was then recorded and then they were assigned to a particular fabric type (see below). The total numbers of sherds per layer is largely insignificant as both layer thickness and total excavated areas of the

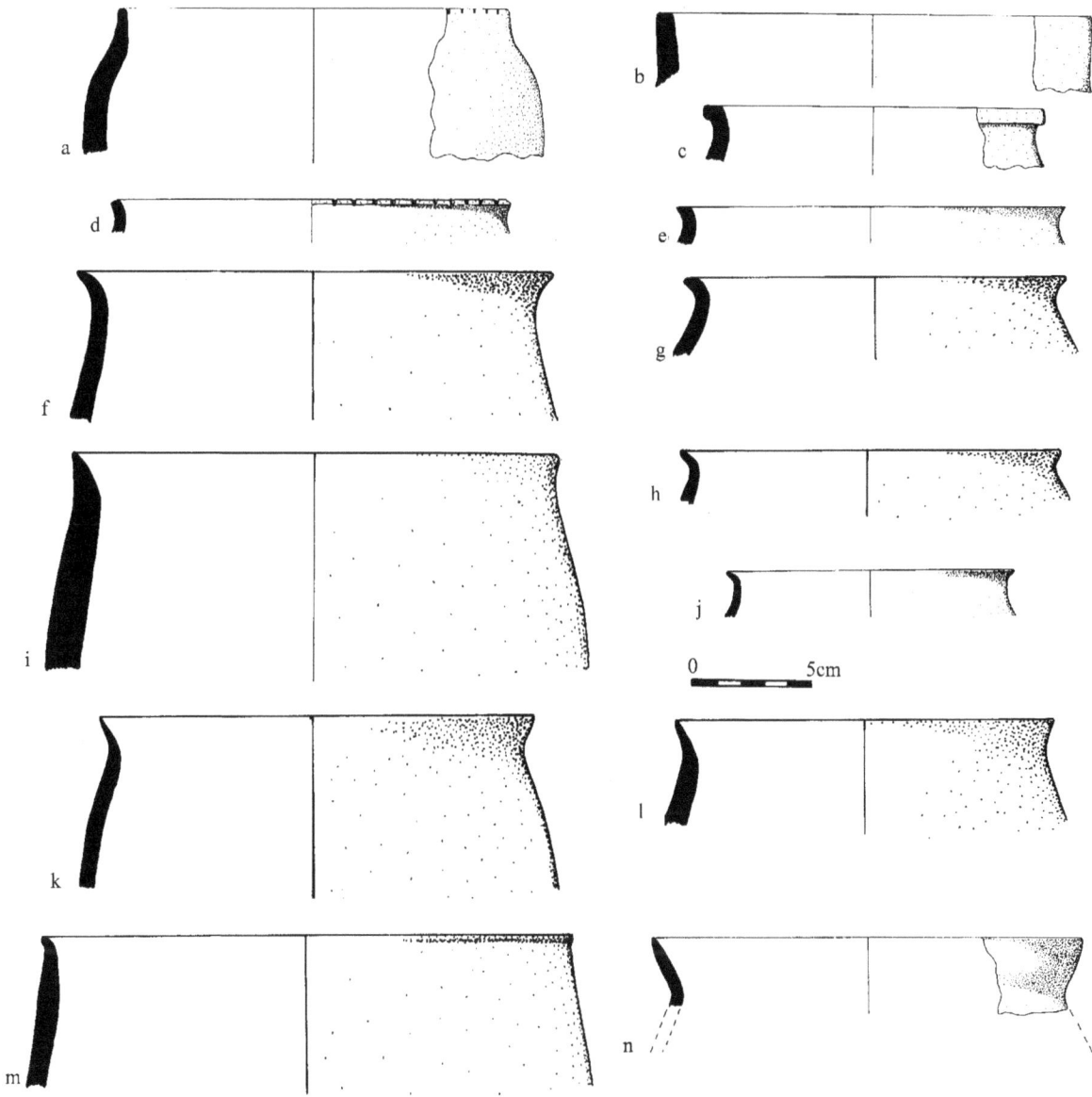

Figure 5.1 Ponamla vessel forms. a. vessel form 4i with incised lip; b. vessel form 4i with flat horizontal lip; c. vessel form 2ii with flat horizontal expanded lip; d. vessel form 2i with flat incised lip; e-l. vessel form 2i; m. vessel form 3ii; n. vessel form 2iii.

layers varied considerably. The majority of the sherds measured between 4-8 (27%) and 8-12 (67%) mm with much fewer falling in the 12–16 (5.3%) mm division, largely made up of basal sherds. There was no discernible change of thickness through time. The globular nature of the pots made the definitive identification of basal sherds somewhat problematic and was generally possible only for larger sherds. Only a handful of sherds fell outside the above noted measurements. A total of 11 sherds were identified as possessing a calcareous temper (Fabric 2) and they were concentrated in the lower levels of the site.

Those sherds which had not been exposed to weathering exhibited a number of common features. The colour ranged from a dusky red (10R 3/2-3/4) to a dark red (10R 3/6) no doubt largely due to the fact that the exterior of the pots was coated with a reddish slip. External surfaces on many of the body sherds, and more specifically in the area of the rim/body confluence, showed evidence of wiping in the form of fine striations.

Figure 5.2 Ponamla vessel forms. All vessel form 2i. c. Incised lip; f. externally expanded lip.

Table 5.2 Ponamla Area A plain sherds

	Layer 1	Layer 2	Layer 3a	Layer 3b	Layer 4	Total	%
body	2825	1033	817	1034	173	5882	
base	22	12	19	9	–	62	
1-4mm	1	1	1	–	–	3	
4-8mm	705	337	286	204	71	1603	26.9
8-12mm	1928	729	718	524	100	3999	67.2
12-16mm	142	69	61	46	2	320	5.3
16+	3	8	6	2	–	19	
weight	20923.66	8734.23	11294.85	7386.05	1198.72	49537.51	
cal. temper	–	–	2	3	6	11	

Rim and lip form

Consistent with the all-encompassing globular outcurving pot form are the rims which, apart from 13 are all (1058) outcurving. Four rims are direct, eight everted and one is incurving. The dividing of rim profile into four categories may have, in this case, been somewhat unnecessary. Much of the perceived variation may in fact relate to the 'sweep of the potters fingers' (Shepard 1963:246 quoted in Irwin 1985:105 but see also Ambrose 2002) particularly with those rims that are outcurving. Rim profile is only significantly different here with the parallel direct rims (4) with flat lip (Fig. 5.1b) and the one divergent abrupt rim with a flat horizontal lip (Fig. 5.1c). Incising on the lip was rare with only ten cases being recorded.

Table 5.3 Ponamla Area A. Rim and lip forms

	Layer 1	Layer 2	Layer 3a	Layer 3b	Layer 4	Total
No	533	190	169	125	54	1071
Rim direction						
direct	-	-	1	2	1	4
outcurving	531	189	166	119	53	1058
everted	1	1	2	4	-	8
incurving	1	-	-	-	-	1
Rim profile						
convergent gradual	353	150	122	87	42	754
parallel	158	27	43	30	10	268
externally thickened	21	13	4	8	2	48
divergent abrupt	1	-	-	-	-	1
Lip form						
plain	463	170	153	113	39	958
point	49	20	16	10	14	109
flat/horiz.	1	-	-	1	-	3
flat	-	-	-	1	1	1
deco	7	-	1	1	1	10
Cal. temper	-	-	1	2	4	7

Vessel form

The collection is characterised by globular outcurving rim vessels, namely form 2i (Figs 5.1, 5.2). The only variation amongst this vessel form was largely restricted to orifice diameter. Of those rims (107) that gave some indication of orifice diameter 96 or 90 % were between 18-20cm, 6 of 16cm, 3 of 22cm, along with single examples of 12 and 24cm.

The very limited variation that is seen in vessel form comes largely from the lowest levels of the site. Variation tends to be related to rim form with a total of 13 rim sherds suggesting some digression. Two different direct rim vessels (form 4i) were recorded (Figs 5.1a and 5.1b) along with several examples of everted rims (form 2iii) (Fig. 5.1n). Three single examples of form 2i variants were also recorded (Fig. 5.1c, d and e). One incurving rim sherd (Fig. 5.1m) was also identified and seems to be rare example of a form 3ii vessel. A carinated dentate-stamped sherd indicated that one example of a further vessel form (form 5i) was also present at the site.

Decoration

The dominant decorative technique employed at Ponamla was fingernail impression which is in turn divided into three categories, impressed, pinch and gouge (Table 5.4). The three categories of fingernail decoration taken together make up 85% of the entire number of decorated sherds (Figs 5.3–5.7). Fingernail pinch dominates, making up 52% of the decorated material followed by impression with 30% and finally gouge with only 3%. Incised sherds make up a much smaller component of the collection, namely 11.5%. It is predominately simple linear, often parallel incision (11.3%) (Fig. 5.8d–h and 5.9). More complex single examples of incised motifs are also represented (Fig. 5.8a, b). The remaining decorated materials (3.5%) comprise often single examples of combined decoration (Fig. 5.10c–h), and incision of the lip (10 examples). Examples of combined decoration generally consist of fingernail impression and incision. The paucity of decorated sherds recovered from the lower levels of the site is dramatically illustrated in Table 5.4. As already mentioned a single carinated dentate-stamped sherd (Fig. 5.10i) was also found in Area A. Recovered from Layer 2 it was in secondary deposition. One other dentate-stamped sherd (Fig. 5.10j) was also recovered from Ponamla, again in secondary deposition in the top layer of TP 5.5.

Table 5.4 Ponamla Area A. Summary of decorative techniques and location

	Layer 1	Layer 2	Layer 3a	Layer 3b	Layer 4	Total	%
Deco technique							
Dentate	–	1	–	–		1	
Incision							
linear	26	7	6	2	1	42	(11.3)
curvilinear	2	–	1	–	–	3	
geometric	3	–	1	–	–	4	
complex	4	–	–	–	–	4	
Fingernail							
impression	95	11	3	1	1	111	(30)
pinch	147	34	11	–	–	192	(51.75)
gouge	11	–	–	–	–	11	(2.9)
Punctate	2	–	–	–	–	2	
Combination of above	6	–	2	–	–	8	
Total	296	53	24	3	2	371	
Location							
body	282	53	21	2	1	359	
rim	1	–	2	–	–	3	
lip	7	–	1	1	1	10	

Motifs

A total of 35 separate non-dentate stamped motifs were identified. These are listed and described in Appendix 4. Twenty-four of the motifs relate to variations in fingernail decoration, again highlighting its predominance. Many of the motifs are represented by only single or several examples and in some cases are clearly components of much larger motifs or designs. Examples of motifs that appeared to dominate at the site included E-motif 1 (21) (Fig. 5.3a–c), E-motif 4 (35) (Fig. 5.4a,b), E-motif 17 (14) (Fig. 5.5a) E-motif 21 (30) (Fig. 5.6f–l) and E-motif 24 (24) (Fig. 5.7b–e). A greater number of decorated sherds were too fragmentary to be assigned to a particular a motif. This category was dominated by examples of fingernail pinch (45), fingernail impression (6) and miscellaneous linear incision (5). Lip modification was restricted to incision (10) (E-Lip motif 1). Only one of the dentate-stamped motifs could be assigned a motif classification according to Anson, namely M187 (Fig. 5.10i). The other dentate stamped sherd displayed paired parallel straight lines (Fig. 5.10j). Two incised motifs (E-motif 25 [Fig. 5.8a] and 28 [Fig. 5.8d]) also showed some close affinity with Lapita motifs (Anson 175, 318 and 369 respectively).

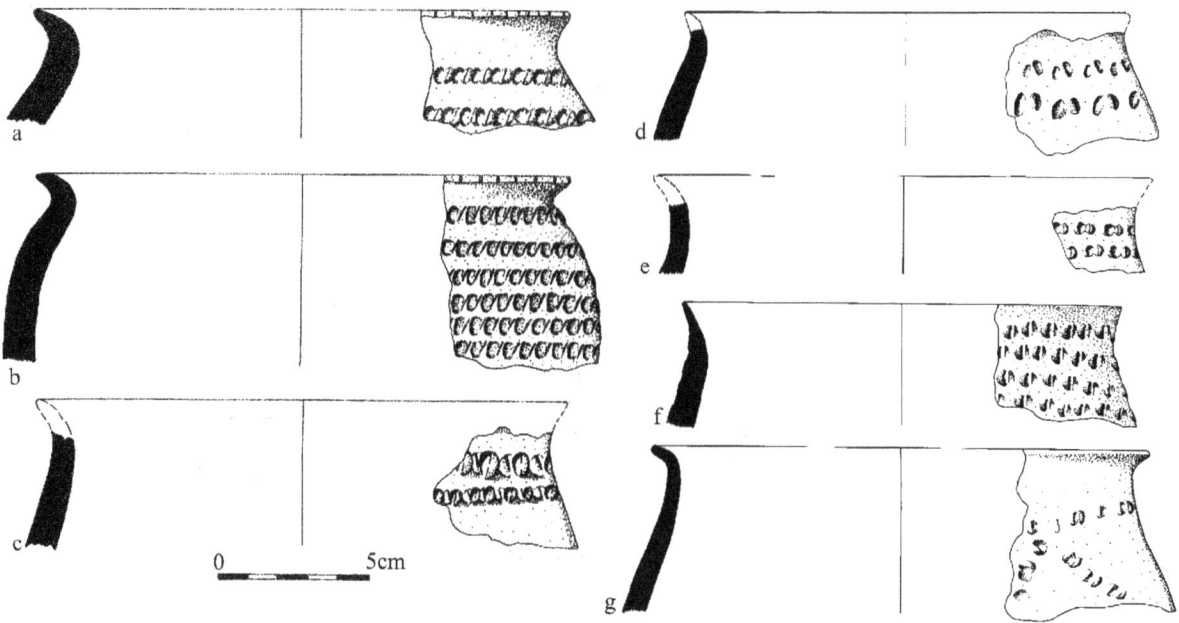

Figure 5.3 Ponamla vessel form and decoration. All vessel form 2i. a–c. (E-motif 1) plus incised lip (a and b) (E-lip motif 1); d–f. E-motif 2; g. E-motif 3.

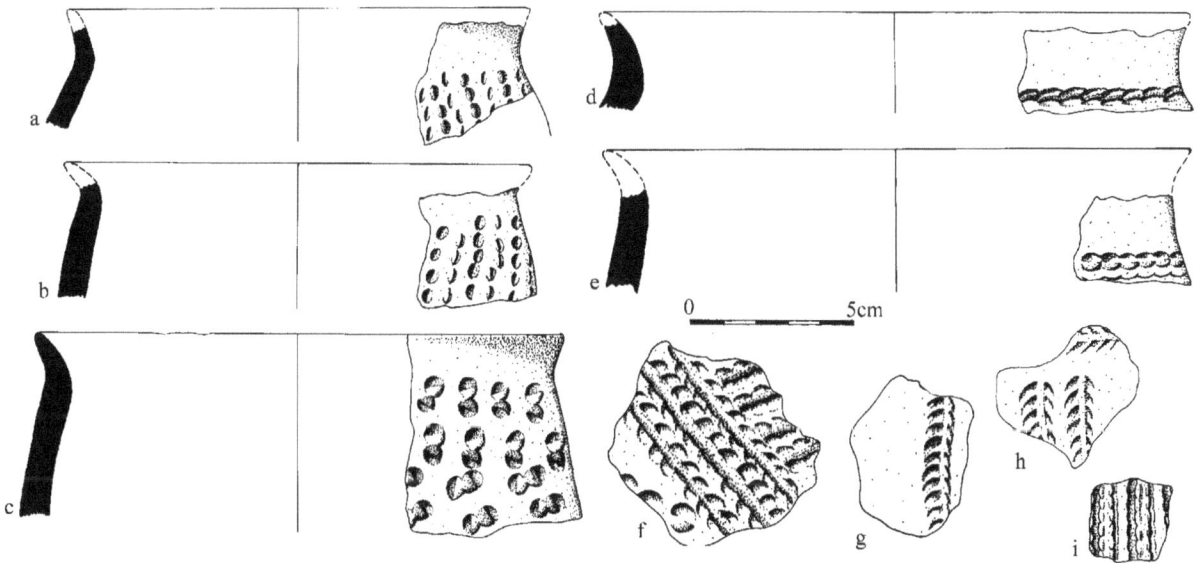

Figure 5.4 Ponamla vessel form and decoration. All vessel form 2i. a and b. E-motif 4 c. E-motif 5; d and e. E-motif 6; f. E-motif 7; g and h. E-motif 8; i. E-motif 9.

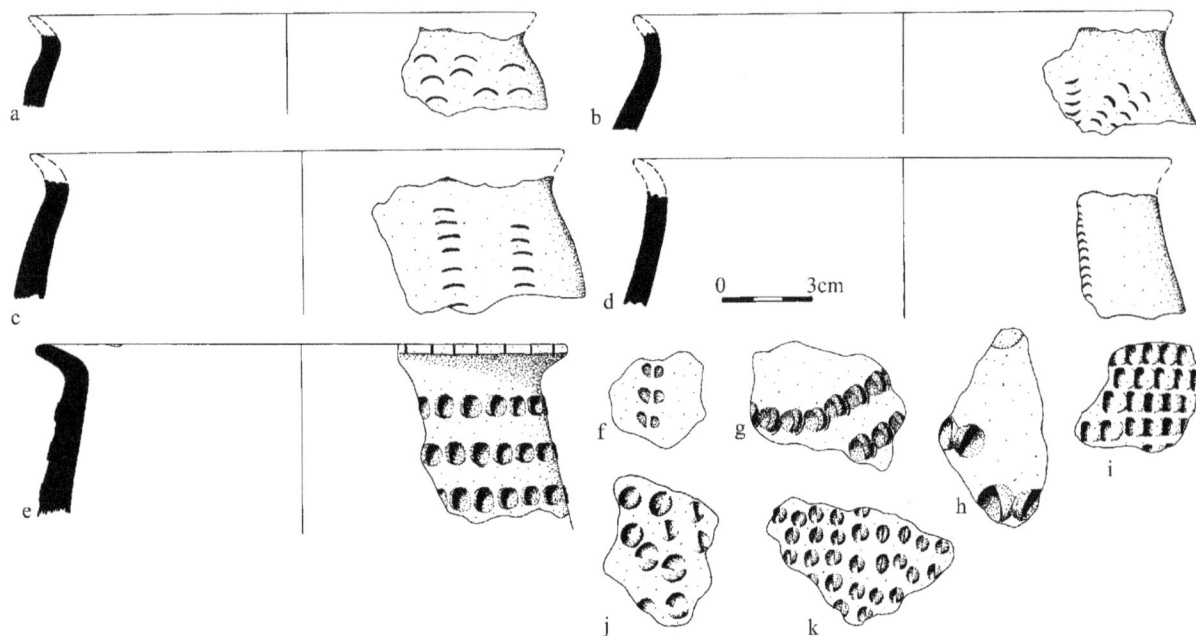

Figure 5.5 Ponamla vessel form and decoration. Vessel form 2i. a. E-motif 17; b. E-motif 18; c. E-motif 19; d. E-motif 20; e. E-motif 16; f. E-motif 10; g. E-motif 11; h. E-motif 12; i. E-motif 13; j. E-motif 14; k. E-motif 15.

Figure 5.6 Ponamla decorated sherds. a. E-motif 17; b. E-motif 21; c and d. E-motif 18; e. E-motif 19; f–l. E-motif 21; m. E-motif 22; n. E-motif 23.

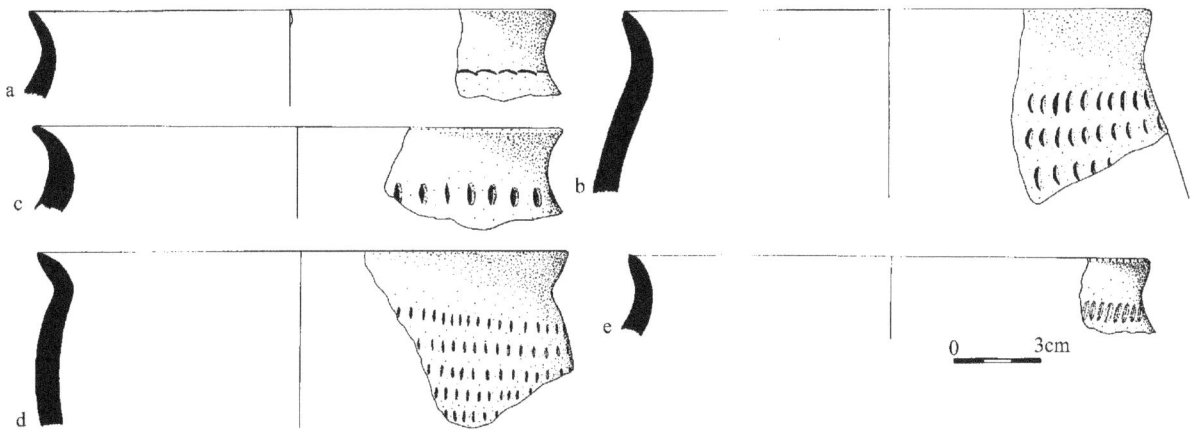

Figure 5.7 Ponamla vessel form and decoration. All vessel form 2i. a E-motif 23; b–e. (E-motif 24).

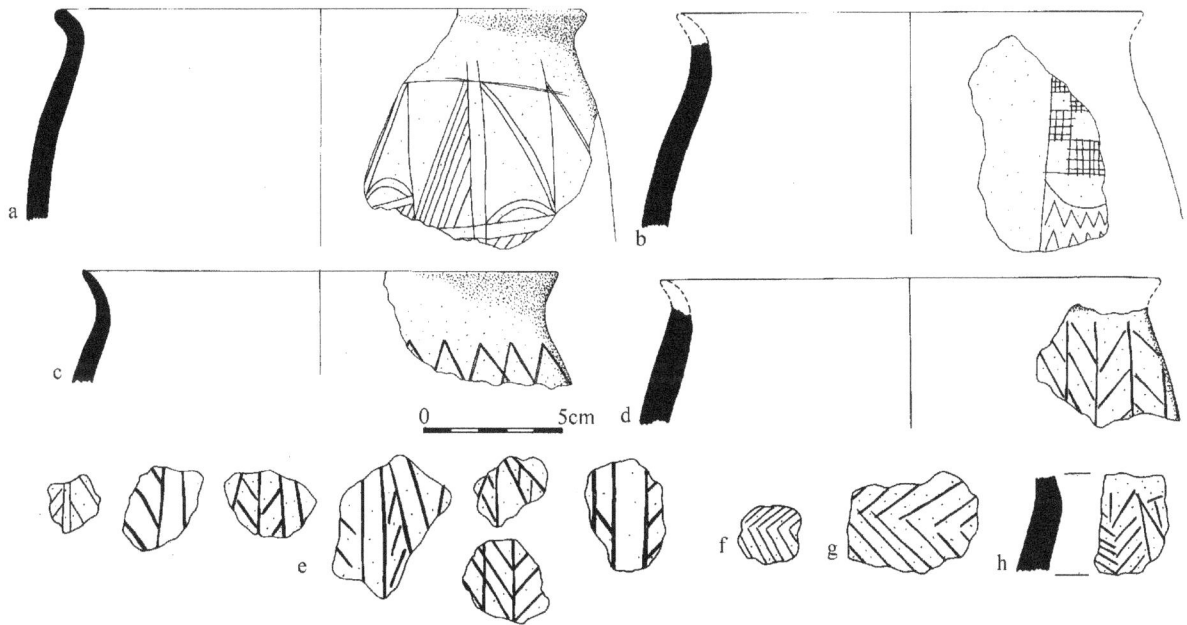

Figure 5.8 Ponamla vessel form and decoration. Vessel form 2i. a. E-motif 25; b. E-motif 26; c. E-motif 27; d. E-motif 28; e. E-motif 28; f–h. E-motif 29.

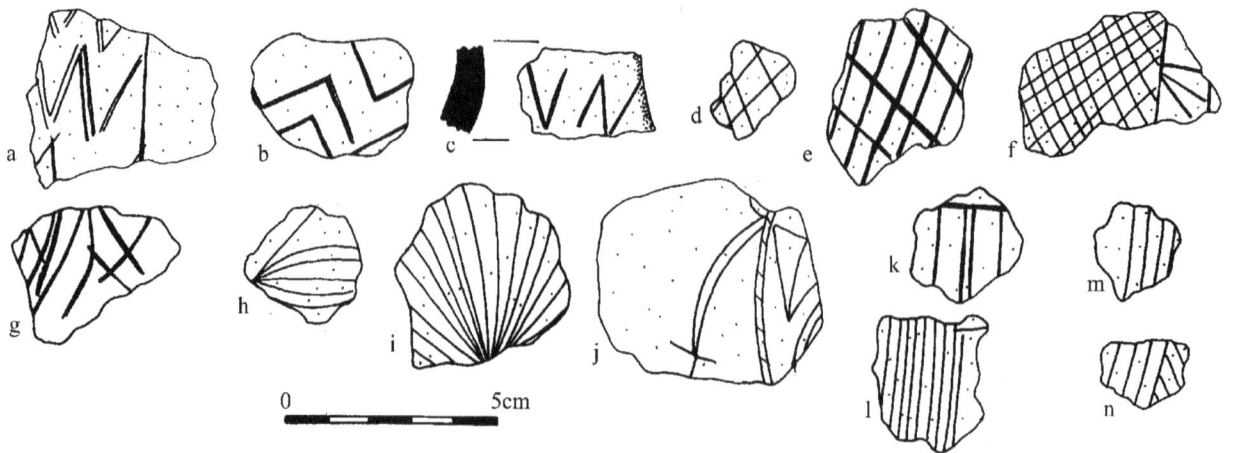

Figure 5.9 Ponamla decorated sherds. a–c. E-motif 30; d-f. cross-hatch incision; g–i. curvilinear incision; j. E-motif 31; k-n. parallel linear incision.

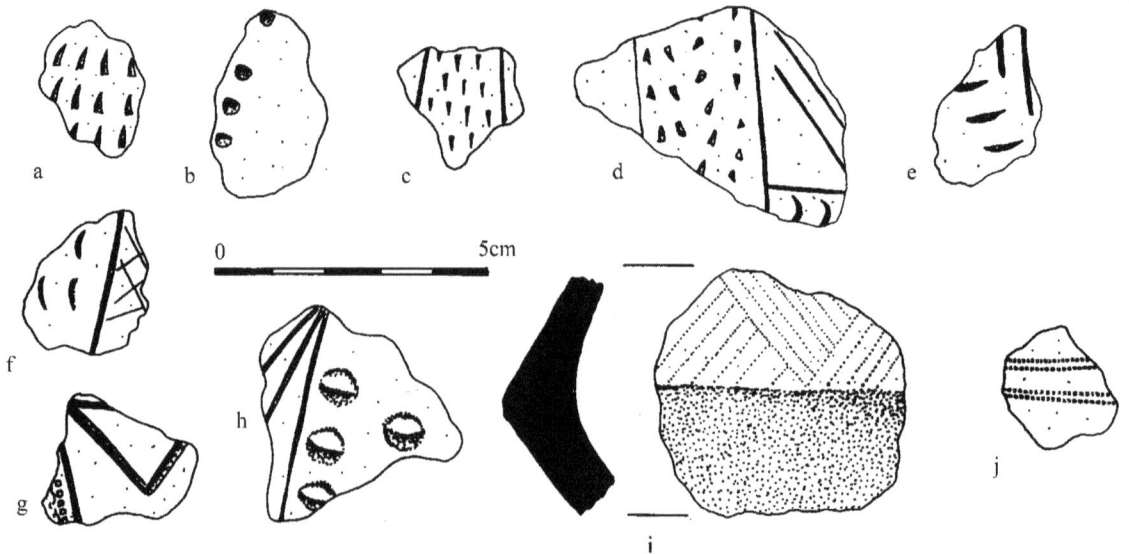

Figure 5.10 Ponamla decorated sherds. a-b. punctate; c. E-motif 32; d. E-motif 33; e. fingernail impression and linear incision; f. E-motif 35; g. incision plus punctate; h. incision plus fingernail gouge, E-motif 35; i. Carinated sherd with Lapita dentate stamped motif (Anson 187) j. Parallel double lines of dentate stamping.

Ponamla Summary

The ceramics at Ponamla can be broadly summarised as follows:

- a homogeneous assemblage in terms of vessel form, fabric and decoration
- the site dated from c. 2800BP when plainware vessels (form 2i), made at the site, were completely dominant. The assemblage largely post-dated initial Lapita settlement of the island and represented the perseverance of the domestic cooking vessel.
- at c. 2600 BP a multitude of motifs (35) began to appear on the same vessel form. The large number of contemporary motifs might be indicative of a number of potters who through this method of identification were expressing delineation between household units. Decoration is dominated by fingernail impression. Although only two dentate stamped sherds were recorded at the site, later incised motifs (particularly E-motif 25 and 28) demonstrate a close generic connection to Lapita motifs.
- Very soon after 2500 BP the site was abandoned.

Ifo ceramics

The ceramics recovered from the Ifo site spanned a considerably longer time period than those from Ponamla and were much less concentrated largely due to the nature of their deposition (see Chapter 3). Details of the previous excavation at the site by Spriggs and analysis of the excavated materials by Wickler (1985) have been outlined. Firstly, it must be said that the initial excavations at Ifo were only preliminary in nature and that Wickler also incorporated ceramics recovered from other sites on Erromango into his analysis. Definitive conclusions regarding the form, decoration and other aspects of the Ifo ceramics were hampered by the small size of the sample (286) where Wickler was obliged to include even what he described as ceramic 'crumbs' to boost the numbers. The analysis focused largely on sherd fabric which demonstrated both intra and inter-site similarities and variability. Interpreting the implications of this variance was however limited by the nature of the sample. One of the stated aims, that of determining the relationship between Lapita and Mangaasi ceramic traditions remained elusive. The two traditions were identified on Erromango; Lapita by a single dentate-stamped sherd in secondary deposition and the latter by a much larger sample of what was described as a 'regional variant' of Mangaasi (Spriggs and Wickler 1989:82) characterised by fingernail decoration. Although the evidence was somewhat inconclusive, as noted earlier it was this site along with a number of others that Spriggs (1984) argued could be identified as a 'transitional site', suggestive of gradual cultural change from Lapita to Mangaasi-related cultures. Clearly with larger scale excavations this question could be further addressed. The sherds from the 1983 excavations were examined but not included in the analysis below as they provided little added information due to their size, condition and the small overall sample.

A much larger number of sherds were retrieved in 1996 from relatively undisturbed deposits which included a larger sample of *in situ* Lapita ceramics. When this material is combined with that from Ponamla a much more complete ceramic sequence for Erromango can be demonstrated. A total of 902 sherds were recovered from all the excavations at Ifo in 1996. This included 645 plain body sherds, 113 rims and 128 decorated body sherds (see Table 5.5). The greatest concentration of sherds (659 75%) which encompassed the first one thousand years of occupation at Ifo came from Trenches B, C and D, a total excavated area of 16m^2 (Fig. 3.4). Further test pitting across the site, as at Ponamla, enabled site areas to be established and to determine any spatial or temporal variance. The results of the test pitting are outlined in Appendix 2. Only the sherds from Trenches B, C and D are discussed in any detail here.

Fabric

Again the central analytical tool used for delineating the fabrics was petrography. The full petrographic report for selected Ifo sherds (WRD 147) is reproduced in Appendix 3.

Fabric 1) This fabric is characterised by plagioclase and pyroxene-rich mineral inclusions which are moderately to finely sorted, subangular to subrounded aggregates, contained within a silty clay. This is by far the most common fabric and is associated with the ceramics produced during the latter part of the sequence. Calcareous inclusions were also noted but their finely sorted nature are suggestive of a weathered limestone origin. Sherds associated with this fabric were not red-slipped and the sherd surfaces were generally softer and less well preserved suggesting that originally they were less highly fired.

Fabric 2) This second fabric contains fine grained, pyroxene-rich plagioclase poor mineral inclusions within a fine silty clay. Calcareous inclusions were not in evidence. This fabric would seem to have been highly fired due to the well preserved nature of the sherds and when the surface has not been degraded the remains of a red-slip is in evidence.

Table 5.5 Ifo excavated ceramic sample

Trench B,C,D	plain body	base	rim	body deco	deco	fabric1	f2	f3	f4	weight
Layer 1	262	6	63	75	105	364	22	13	7	8005.1
Layer 2	104	8	11	12	16	89	14	22	9	1228.1
Layer 3	42	–	2	5	5	16	10	23	4	548.2
Layer 4	58	–	4	7	9	3	8	55	3	544.2
Total	**466**	**14**	**80**	**99**	**135**	**472**	**54**	**113**	**23**	**10325.6**
%	70%		12%	15%	20%	71%	8%	17%	3%	
TP2	10	–	1	–	1	–	–	11	–	19.7
TP3	32	1	8	4	9	15	8	13	10	473.3
TP4	5	–	–	–	–	5	–	–	–	15.6
TP5	11	–	–	1	1	14	–	–	–	89
TP6	6	–	1	1	–	5	3	–	–	57.2
Area A	64	–	5	6	7	57	4	13	1	876
TP8	2	–	–	–	–	1	–	1	–	9.9
TP9	13	–	5	–	–	16	1	–	1	130.3
TP10	8	–	–	–	–	7	1	–	–	91.1
TP11	4	–	–	–	–	4	–	–	–	18.8
TP12	64	–	9	14	15	45	10	26	7	714.3
TP13	25	–	4	3	4	28	–	3	–	51.9
TP14	–	–	–	–	–	–	–	–	–	–
TP15	3	–	–	–	–	7	–	–	–	38.4
Total	**155**	**1**	**33**	**29**	**37**	**204**	**27**	**67**	**19**	**2585.5**
Grand total	**645**	**15**	**113**	**128**	**172**	**676**	**81**	**180**	**42**	**12911.1**
%	70%		13%	14.5%	19.5%	74.5%	9%	20%	5%	100%

Fabric 3) This is a largely calcareous-tempered fabric with plagioclase and pyroxene rich mineral inclusions. The calcareous temper is very distinctive and tends to dominate the matrix suggesting it has been intentionally added. The fabric has a grainy friable texture with less clay content. It is largely associated with the sherds recovered from the lower levels of the site. The vast majority are plain sherds but those that are decorated are either dentate stamped or incised sherds with a red slip.

Fabric 4) Similar to above but calcareous temper is less dominant. On initial inspection this fabric was thought to be distinctive but may in fact simply reflect some degradation of the calcareous temper over time or its decreasing use.

Variation in fabric, although limited, can be identified from the Ifo sherds. All show some similarity in their makeup and seem most likely to have been manufactured using local materials, but some change in technique over time is indicated. The greatest variance is a shift from the predominantly calcareous-tempered wares to non-calcareous tempered wares which over time are less highly fired and possibly made from clay that was naturally tempered as was argued for in the case of the Ponamla ceramics. The deliberate addition of a calcareous temper appears to be associated with the manufacture of specialised vessels, i.e. ceremonial vessels, which very quickly became obsolete.

Brief mention must also be made of a single sherd recovered from this site in 1983 which has been identified as having an anomalous suite of mineral inclusions (Spriggs and Wickler 1989:82). It would appear to be a calcareous-tempered sherd with mineral inclusions exotic to Vanuatu. Dickinson pointed to its likely origin as being New Caledonia. No further evidence for this fabric type was recovered in 1996.

Plain sherds

A total of 480 plain sherds were excavated from Trenches B, C and D which made up 73% of the sherd total. Again similar difficulty was encountered when attempting to differentiate between body and basal sherds from globular pots. Only 14 basal sherds were positively identified. There are perceptible changes in the assemblage that occur over time. Overall the sherds are thinner in the lower levels, concentrated between the 4–8 and 8–12mm divisions. This changes in the upper levels with an increasing number of sherds appearing in the 12–16 and the 16-20mm divisions. This corresponds with a change in fabric over time with Fabric 3 dominating in the earlier layers and being superseded by Fabric 1 in the upper layers. The significantly smaller numbers of sherds made up of Fabric 2 and 3 spread throughout the stratigraphy limit any meaningful commentary.

Table 5.6. Ifo plain sherds from Trenches B,C,D

	Layer 1	Layer 2	Layer 3	Layer 4	Totals
No	268	112	42	58	480
weight	5881.1	1228.1	443.2	478.2	8030.6
4-8mm	39 (14.5%)	20 (18%)	7 (16%)	11 (19%)	77
8-12mm	158 ((59%)	67 (60%)	32 (76%)	45 (77%)	302
12-16mm	57 (21%)	15 (13%)	2 (4.5%)	2 (3.5%)	76
16-20mm	13 (6 base) (5%)	10 (8 base) (9%)	1	—	24
Fabric 1	236	78	14	3	331
2	12	9	7	8	36
3	13	19	18	45	95
4	7	6	4	2	19

Rim/lip form

A total of 80 rims were excavated from Trenches B, C and D, 78% of which came from Layer 1. This disproportion unfortunately limits meaningful comparison between the layers. The rim sherds of Layer 1 are dominated (66%) by incurving rims, often with thickened exteriors and all of Fabric 1. This same rim form also accounts for 52% of the entire rim collection. Within Layer 1 these rims are associated with lesser numbers of outcurving (12[19%]) and direct (9[14%]) rims. Layer 2 sees a dramatic drop in both numbers of rims and examples of incurving rims along with a diversification of fabric. Although the sample size is substantially reduced it appears that direct and outcurving rims were more dominant in Layer 2. The only rims that were recovered from Layers 3 and 4 were either direct or outcurving and all were of Fabric 3.

Vessel form

The sherds from the lower levels of the site associated with dentate-stamped and incised decoration were of such a fragmentary nature that little information could be ascertained regarding vessel form. Only three examples of a single vessel form were positively identified from Layer 4, namely vessel form 2i (Fig. 5.11i), made up of Fabrics 3 and 4. All were undecorated plainware. From Layer 3 a globular pot with direct rim (form 4i) (Fig. 5.12a), and two with wide horizontal rims (form 2ii) constructed of Fabrics 2 and 3 (Fig. 5.11g, h) was all that could be identified. These were again plainware vessels. The incurving globular vessel, 3i (Figs 5.11b–d), which is predominately decorated, appears in Layer 2 and along with form 3ii completely overwhelms any other vessel form in Layer 1. From Layer 2 a total of seven vessels comprising two forms were identified, namely form 2i (6) and 3i (1). As with all the information gleaned from the ceramic remains the results are skewed by the much greater quantity recovered from Layer 1. Four vessel forms were identified, namely forms 2i, 3i, 3ii and 4i. Vessel forms 3i (21) and 3ii (24) are totally dominant with 45 sherds indicating this vessel form. All were constructed using Fabric 1 and except for two examples, all are decorated. The thirteen examples of form 2i were made up of Fabrics 1(7), 2(2) and 4(4). Globular vessels with direct rims (form 4i) accounted for 15 of the identified vessels in Layer 1. All were made from Fabric 1.

Table 5.7. Ifo Trenches B, C, D, Rim and lip form

	L.1	F1	F2	F3	F4	L.2	F1	F2	F3	F4	L3	F1	F2	F3	F4	L4	F1	F2	F3	F4
No	63					11					2					4				
Rim direction																				
direct	9	8	–	–	1	4	2	–	1	1	2	–	–	2	–	1	–	–	1	–
outcurving	12	10	1	–	1	4	–	2	–	2		–	–	–	–	3	–	–	2	1
incurving	42	42	–	–	–	2	1	1	–	–		–	–	–	–		–	–	–	–
Rim Profile																				
conv. grad	21	21	–	–	–	3	1	2	–	–		–	–	–	–	2	–	–	1	1
parallel	7	6	–	–	1	3	–	1	–	2	1	–	–	1	–	1	–	–	1	–
ext. thick	22	22	–	–	–	2	1	–	–	1		–	–	–	–		–	–	–	–
div. abrupt	4	3	–	–	1	2	–	–	1	1		–	–	–	–	1	–	–	1	–
div. grad.	8	8	–	–	–	1	1	–	–	–	1	–	–	1	–		–	–	–	–
Lip form																				
plain	43	42	–	–	1	10	3	3	1	3	1	–	–	1	–	1	–	–	–	1
pointed	2	1	–	–	–		–	–	–	–		–	–	–	–		–	–	–	–
flat horiz.	10	9	–	–	1	1	–	–	1	–	1	–	–	1	–	2	–	–	1	–
flat	6	6	–	–	–		–	–	–	–		–	–	–	–		–	–	–	–
deco	8	8	–	–	–	1	–	–	1	–		–	–	–	–	1	–	–	1	–

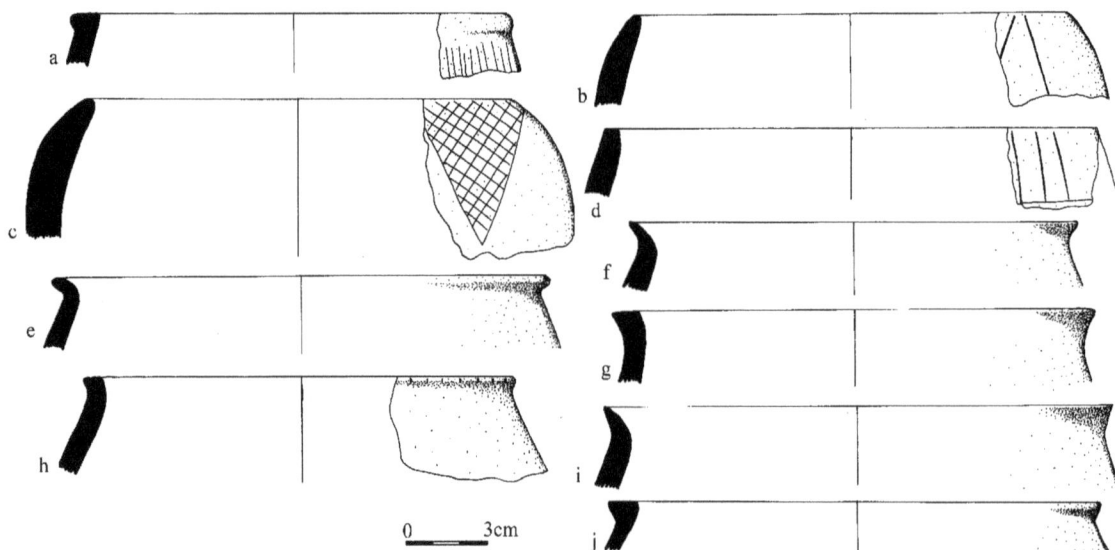

Figure 5.11 Ifo vessel form and decoration. a. vessel form 3ii (fabric 1) plus E-motif 41; b. vessel form 3i (fabric 1) with linear motif; c. vessel form (fabric 1) 3i plus E-motif 42; d. vessel form 3i (fabric 1) with linear incision; e and f. vessel form 2i (fabric 1); g. vessel form 2ii (fabric 3); h. vessel form 2ii (fabric 2) with incised lip; i. vessel form 2i (fabric 3); j. vessel form 2ii (fabric 1).

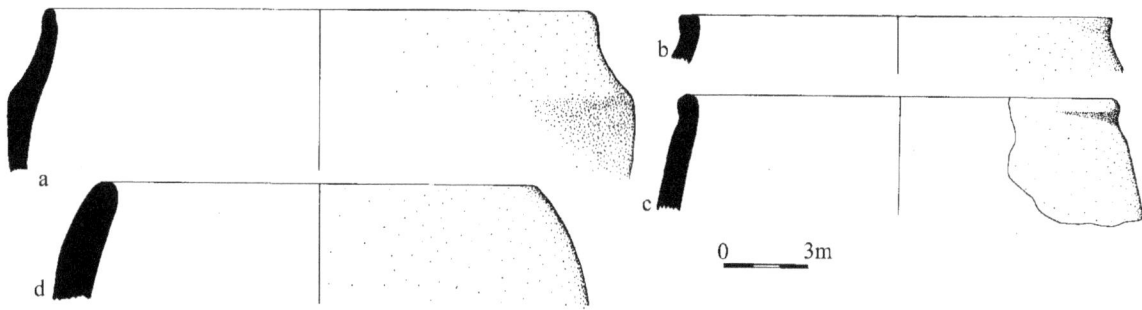

Figure 5.12 Ifo vessel form. a and b. vessel form 4i; c. vessel form 3ii; d. vessel form 3i.

Decoration

The decorated sherds from Ifo are completely dominated by motifs produced with the fingernail (77%) (Figs 5.13–5.16). Fingernail pinch is the most common (64%) followed by fingernail impression (12.5%) and a single example of fingernail gouge. The decoration is overwhelmingly associated with Fabric 1 and Layer 1. Six fingernail decorated sherds are also associated with Fabric 3 from Layers 2 and 3. Incision is a minor component largely restricted to Layer 1 with only two examples from Layer 4. A single example of a sherd decorated with punctate was also recorded (Fig. 5.13o). Decoration which combined a number of techniques was largely restricted to a motif made up of incision and fingernail impression (Fig. 5.14a).

Although hindered by the small sample size from the lower layers, several observations can also be made regarding changing decoration through time (see Table 5.8). Dentate-stamped ceramics made up only 8 % of total sample but represent almost 100% of the decorated sherds from Layer 4 (Fig. 5.17a–g). A minor component of incision, and impressed circles (Fig. 5.17h–l) was also present. All of these sherds are associated with Fabric 3. Smaller numbers of decorated sherds were recovered from Layer 3 (5) but they are again dentate (3) along with the first appearance of fingernail pinching (2). All five sherds were of Fabric 3. Greater numbers of fingernail decoration appear only in Layer 2 along with the predominance of Fabric 1. A single dentate-stamped sherd in secondary deposition was also recovered from this layer (Fig. 5.17g). Finally there is the explosion of fingernail decoration accounting for 83% of the decoration on sherds from Layer 1.

Decoration of the lip was rare (11 examples mostly from Layer 1). Notching (E-lip motif 2) was recorded in three cases along with six examples of either fingernail pinch (E-lip motif 4) or impression (E-lip motif 3). One of these gave a crenellated appearance to the rim (E-lip motif 5) (Fig. 5.14i). Two sherds had dentate decoration on the lip (Fig. 5.17f, g). Decoration was most frequent on the body of the pots. The relatively high number of rims where decoration was recorded is largely due to the fact that many of the rims are associated with incurving pots where there is some difficulty differentiating between the rim and body. Decoration as a percentage of the total from each layer also increased dramatically, hovering between 11-12% from Layer 4 through to Layer 2 and then jumping to 30% in Layer 1.

Table 5.8 Ifo Trenches B, C, D. Summary of decorative technique and location

	L1	F1	F2	F3	F4	L2	F1	F2	F3	F4	L3	F1	F2	F3	F4	L4	F1	F2	F3	F4	TOTAL (%)
Deco tech.																					
Dentate	–	–	–	–	–	1	–	–	1	–	3	–	–	3	–	7	–	–	7	–	11(8)
Incision																					
Linear	7	7	–	–	–		–	–	–	–		–	–	–	–	1	–	–	1	–	8(6)
curvilinear	–	–	–	–	–		–	–	–	–		–	–	–	–		–	–	–	–	–
geometric complex	1	1	–	–	–		–	–	–	–		–	–	–	–	1	–	–	1	–	2
Fingernail																					
impression	17	14	3	–	–		–	–	–	–		–	–	–	–		–	–	–	–	17(12.5)
pinch	70	67	3	–	–	14	10	–	4	–	2	–	–	2	–		–	–	–	–	86(64)
gouge	1	1	–	–	–	1	–	1	–	–		–	–	–	–		–	–	–	–	1
Punctate	1	–	1	–	–		–	–	–	–		–	–	–	–		–	–	–	–	1
Combination of above	6	6	–	–	–		–	–	–	–		–	–	–	–		–	–	–	–	6 (4)
Notching	3	3	–	–	–		–	–	–	–		–	–	–	–	1	–	–	1	–	4
Total	105					16					5					9					135
Location																					
body	75	68	7	–	–	15	10	1	4	–	5	–	–	5	–	7	–	–	7	–	102
rim	30	30	–	–	–	1	–	–	1	–		–	–	–	–	2	–	–	2	–	33
lip	8	8	–	–	–	1	–	–	1	–		–	–	–	–	1	–	–	1	–	10

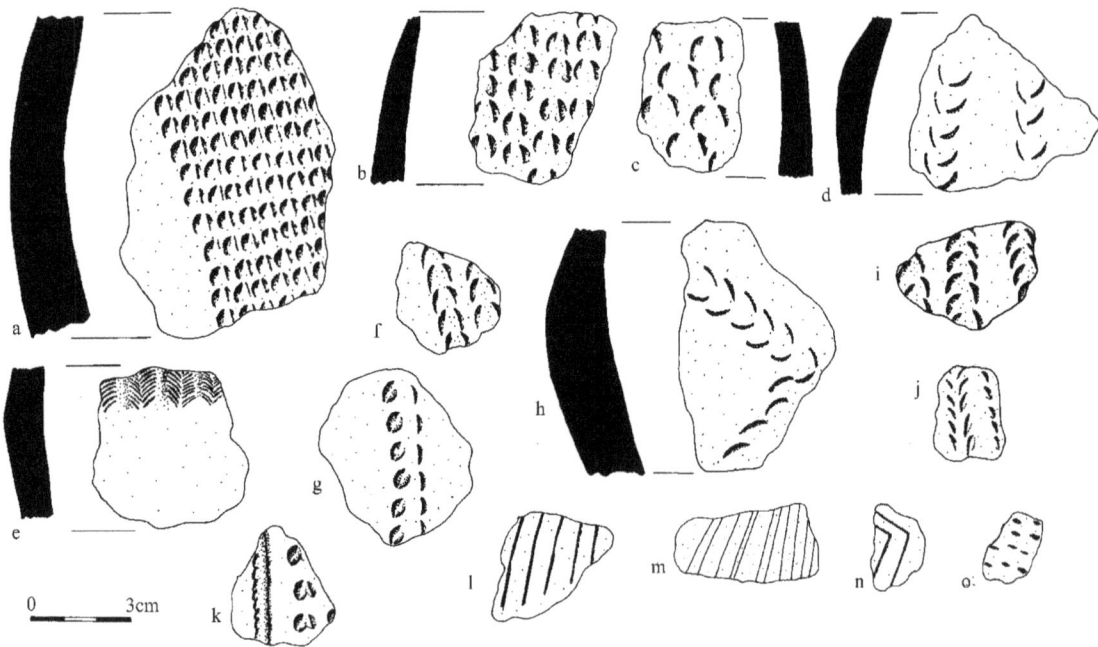

Figure 5.13 Ifo decorated sherds. a-i, fabric 1. a-c. E-motif 4; d. E-motif 8; e. E-motif 37; f. E-motif 8; g. E-motif 8; i. E-motif 8; j. E-motif 8 (fabric 3); k. E-motif 39 (fabric 2); l-m. parallel linear incision (fabric 1); n. geometric parallel incision (fabric 2); o. punctate (fabric 2).

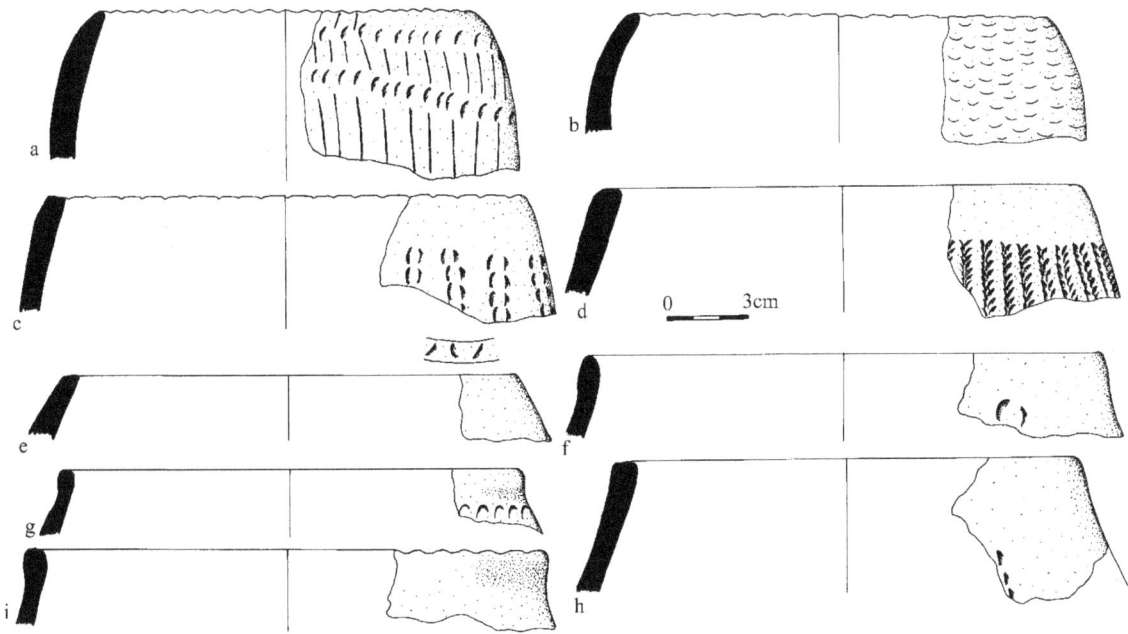

Figure 5.14 Ifo vessel form and decoration (all fabric 1 except h which is fabric 3). a. vessel form 3i plus E-motif 40 with notched lip; b. vessel form 3i plus E-motif 17 with notched lip; c. vessel form 3i plus E-motif 38 with notched lip; d. vessel form 3i plus E-motif 36; e. vessel form 3i with fingernail pinching on the lip; f. vessel form 3ii plus fingernail pinch; g. vessel form 3ii plus E-motif 23; h. vessel form 4i with fingernail pinch; i. vessel form 4i with deep fingernail pinch on the lip creating crenellations.

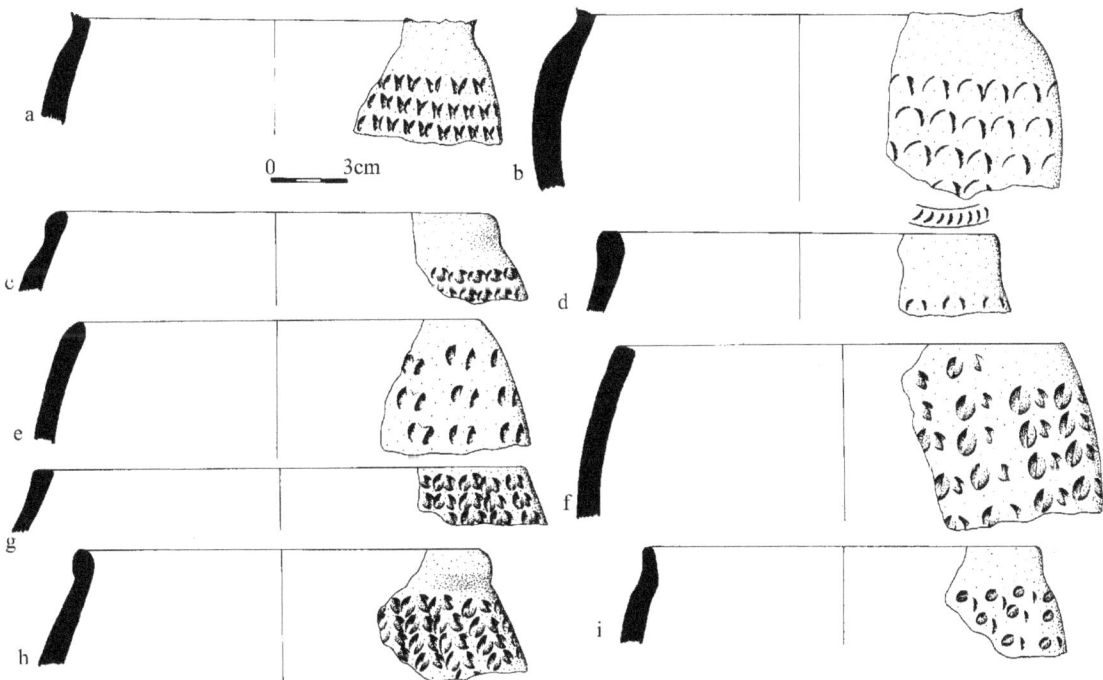

Figure 5.15 Ifo vessel form and decoration (all fabric 1). a and b vessel form 2i plus E-motif 1; c vessel form 3ii plus E-motif 1; d vessel form 3ii with E-motif 2 plus fingernail impression on the lip (E-lip motif 3); e–g. vessel form 3i plus E-motif 4; h. vessel form 3ii plus E-motif 4; i. vessel form 4i plus E-motif 4.

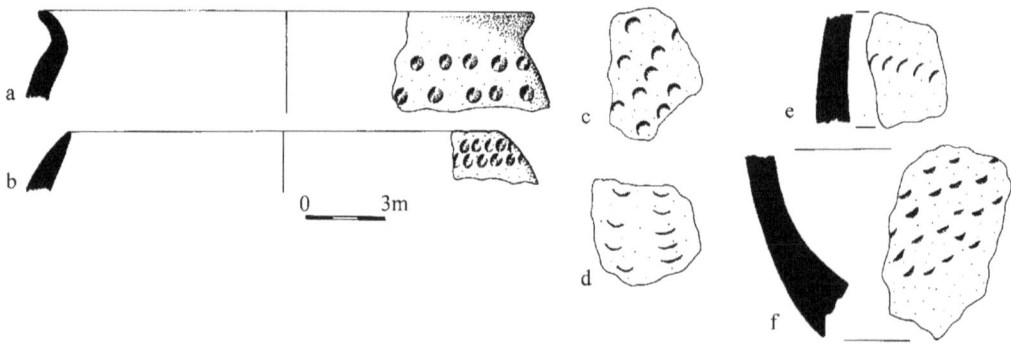

Figure 5.16 Ifo vessel form and decoration. a. vessel form 2i (fabric 2) plus E-motif 15; b. vessel form 3i (fabric 1) plus E-motif 15; c. E-motif 17 (fabric 1); d–e. E-motif 19 (fabric 1); f. E-motif 17 (fabric 1).

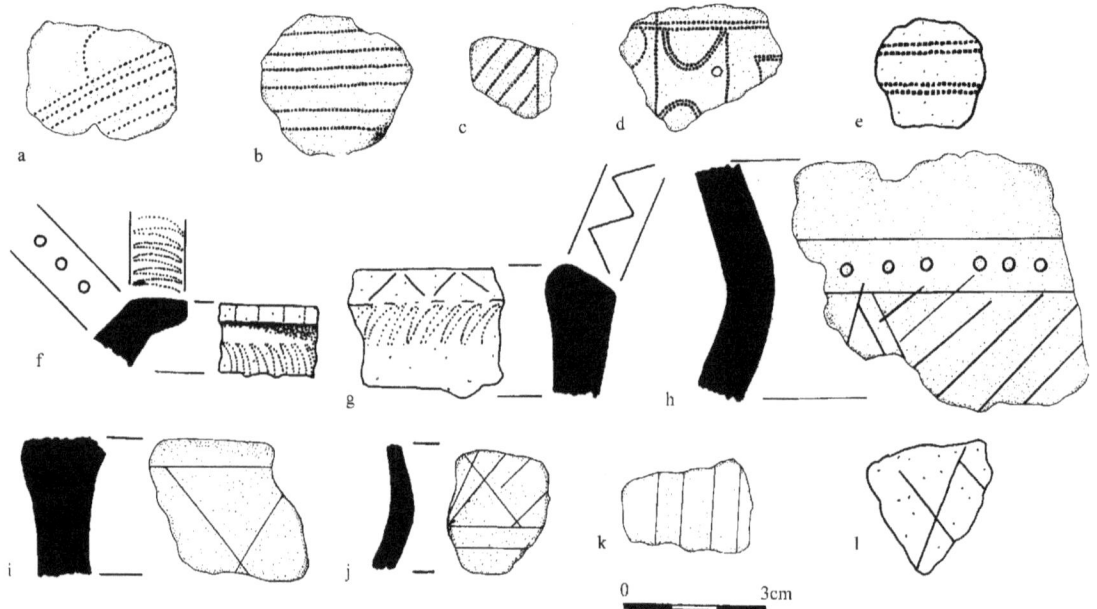

Figure 5.17 Ifo dentate-stamped and incised sherds. a–g. dentate-stamped sherds (a. TrD, L.4; b. TrD, L.4; c. TrD, L.4; d. TrD, L.4; e. TrD, L.4; f. dentate-stamped and impressed circles (TrD, L.4); g. TrC, L.1; h–l. incised sherds (h. TrD, L.3; i. TP2, L.3; j. TrB, L.4; k. TP13, L.4; l. TP3, L.4).

Motifs

A total of 15 separate motifs (non dentate-stamped) were identified from the ceramics at the Ifo site (see Appendix 4). They are again, like Ponamla, dominated by motifs comprised of fingernail decoration (12). The most frequently recorded were E-motif 1 (10) (Fig. 5.15a–c), E-motif 4 (11) (Fig. 5.15e–i), E-motif 8 (11) (Fig. 5.13d, f–g) and E-motif 37 (5) (Fig. 5.13e). Eight of the motifs are shared with those from Ponamla and all are associated with fingernail decoration. Those sherds which were too fragmentary to assign to a motif category were dominated by fingernail pinch (45), followed by lesser quantities of fingernail impression (6) and simple linear incision (5). The seven distinct motifs from Ifo comprise of four associated with fingernail decoration (E-M36–39) (see Figs 5.13e, k and 5.14c, d), one with fingernail impression and incision (E-M40) (Fig. 5.14a) and others with parallel linear incision and bordered incised cross hatch (E-M41–42) (Fig. 5.11a, c). Lip modification was more varied than Ponamla with four categories represented (E-Lip motif 2-5).

Eleven dentate-stamped sherds were recovered in association with a number of rectilinear and geometrically incised sherds (Fig. 5.17). Many of these sherds were too fragmentary to be assigned to any particular motif. Four sherds only could be assigned a motif number using Anson's classification system. They included 431 (Fig. 5.17d), 254 in combination with 417 (Fig. 5.17f), 254 (Fig. 5.17g) and finally 417 (Fig. 5.17h) in combination with incision.

Ifo and Erromango Summary

The recovered sherds from Ifo provide a fragmentary record of the entire ceramic sequence from Erromango and can be broadly summarised as follows:

- Lapita dentate and incised ceramics were associated with the initial settlement of Ifo (and Erromango) some 3000 years ago (see Chapter 3). The ceramics were produced on Erromango and certainly as indicated by the recovered wasters from Ponamla, they were produced at the site. The comprise a clay that was initially calcareously tempered. A red slip was also in evidence.

- Dentate decoration and calcareous tempering were short-lived (several hundred years at the most) and were replaced by fingernail impressed decoration and a more restricted range of vessel form. The appearance of generally very different motif forms and the dominance of a single decorative technique is indicative of changing social dynamics and the role that the decorated ceramics plays within it. The changes may possibly be signifying increased territoriality, development of group identification and or increased proletarianisation of the population (Yen quoted in Kirch 1988b). There is a perceptible change in vessel form over time from globular outcurving rim vessels (2i) to globular incurving rim vessels (3i and 3ii). Fabric composition and manufacturing techniques also changed over time. Attention was less focused on tempering the clay, there was no sign of red slipping and firing temperatures may also have been lower. Sherds became thicker over time.

- More frequent decoration is common in the latter part of the sequence. It was totally dominated by fingernail decoration. Distinctive more crudely made, thick-walled, decorated, incurving rim vessels appeared at the very end of the sequence. Ceramic production and use ceased around 2000 BP.

- By combining the ceramic remains from Ifo with those from Ponamla we are able to establish a detailed picture of ceramic form, decoration and chronology which was in evidence during the initial 1000 years of human settlement on the island. At Ponamla the ceramic sequence is seen in fine detail in the over two metres of cultural stratigraphy which accumulated over a period of 300 years. At Ifo, a similar depth of material has accumulated over a longer period of time, up to 1000 years. Components of the full sequence were in evidence but somewhat compressed. The Ponamla site enabled further refinement of the Ifo sequence.

At Ifo, *in situ* Lapita ceramic remains were recovered from the lowest cultural layers of the site. Vessels were both decorated and plain. The decorated vessels disappeared from the sequence to be replaced by a plainware phase which was dramatically illustrated at Ponamla from its lowest cultural layers dating from c. 2800 BP. Fingernail and incised decoration appears on identical vessel forms at around 2600 BP. This was again best seen in the remains from Ponamla. After 2600 BP a change in vessel form saw a tendency towards incurving rims with fingernail decoration being retained as the dominant decorative technique. Eight motifs were shared between the two sites and all were associated with fingernail decoration. The ceramic remains add further strength to the argument of a basic cultural continuity in Vanuatu between Lapita and the cultures which followed (Spriggs 1984).

6

Efate ceramics

When a day's work on the excavation is over and dusk falls, it is time for conversation in the village, dialogue between the researcher and his hosts, a very precious time for learning, understanding and evaluating. So it was that one evening at Natapao (Lelepa Island), I inquired where Mangaasi and the farea Serelepa were, wondering to myself whether this place had ever really existed. It was right in front of us, on the coast of Efate (Garanger 1996:69).

Introduction

The two sites of Mangaasi and Erueti along with their associated ceramic remains, which were excavated by Garanger (1972, 1982) on the island of Efate, have been briefly summarised in Chapter 2. The fact that the ceramics from these two sites are central to an understanding of the more recent results obtained from the Mangaasi site requires that they are further outlined in detail below.

Mangaasi

The establishment of the Mangaasi ceramic sequence and the definition of its distinctive decoration and form was achieved by Garanger at the culmination of an intensive period of survey and excavation on the islands of Efate and the Shepherds (Garanger 1966, 1971, 1972, 1982). An initial report signalled the discovery of the distinctive ceramics, which were frequently decorated with incision and applied bands, from the Shepherds and offshore islands of Efate (primarily Lelepa), (Garanger 1966:76). Tentative connections were made with recovered ceramics from Fiji and New Caledonia. Further surveys on the west coast of Efate recovered a total of 13,206 sherds, over 40% of which were decorated with incisions and or applied relief. These sherds provided the basis for a detailed description of the decoration associated with the tradition where a total of 12 major motifs with a series of sub-motifs (32 motifs in all) were defined (Garanger 1972:47; Figs 87-96). The ceramic tradition was ultimately named Mangaasi (1971, 1972) after the site located on the west coast of Efate where Garanger recovered a further 17,000 sherds. The justification for this was that the earliest date associated with the tradition was recovered from Mangaasi along with an abundant and varied collection of ceramics. The site was also seen as a possible central place for the manufacture of pottery (Garanger 1971:54).

Although it was argued that Mangaasi ceramics displayed aspects of conservatism over a 2000 year period an Early and Late Mangaasi-style was identified. Early Mangaasi was characterised by discontinuous applied relief, pinched bands and handles, while Late Mangaasi could be defined by numerous incised motifs and the persistence of applied notched bands (Garanger 1971:54). Vessel form, that is globular incurving pots, remained consistent throughout the sequence apart from the presence of small bowls or cups being restricted to the earliest levels (Garanger 1972:109). Despite this seemingly well defined ceramic chronology and associated characterisation of the tradition, Garanger highlighted a number of anomalies. These anomalies, along with a number of others, were to be further highlighted in some detail by both Ward (1979, 1989) and Spriggs (1984, 1997).

Garanger noted the disturbed nature of the stratigraphy at the Mangaasi site made it difficult to accurately define the ceramic chronology (Garanger 1972:46, 133). Disturbance was most dramatically demonstrated by the recovery, from a number of different horizons, of 52 sherds belonging to a single pot (Garanger 1972:48). There was also some difficulty dating the transition from the Early and Late traditions, a period when discontinuous applied relief, pinched bands and handles began to disappear (Garanger 1972:133). In fact Garanger went further and managed to demonstrate that the actual division of the sequence into an Early and Late Mangaasi could be regarded as somewhat tenuous. When outlining the distribution of motifs throughout the site it was pointed out that Motifs 1–6 were homogeneously distributed throughout the layers, an aspect which certainly supported the concept of some conservatism of design through time. Applied pinched bands, applied discontinuous relief, along with handles were found only in the two lower layers (Garanger 1972:53; Fig.110). It is these latter motifs and handles, not present in Layer 1, which enabled Garanger to define an Early and Late Mangaasi. He did point out (also commented on by Ward 1989:160) however that these motifs were far from characteristic of the different layers and that if considered in relation to the total percentage of motifs in each layer their contribution to the totals was almost imperceptible (Garanger 1972:53; Fig.111).

Other aspects of chronology also proved to be somewhat problematic. The earliest dates for the site of around 2600 BP were generally accepted at the time despite the acknowledged disturbed nature of the stratigraphy. More confusing was the termination date for ceramic production which was variously stated as ceasing around 1700 AD (Garanger 1972:58) or around the end of the sixteenth century (Garanger 1972:127) (see Ward 1989 for more detailed discussion).

Although the globular incurving vessel form was said to change little for over 2000 years (Garanger 1972:109), Garanger did note that a number of rim forms associated with Erueti-style pots (all less than 2% of total per layer) were recovered from the Mangaasi site (Garanger 1972:55). It was also noted more generally that in the late Efate and late Tongoa levels, flared rims were more common and the lips were more varied (Garanger 1972:109). No further comment or explanation was made regarding these two points. However a survey of the illustrated sherds confirms that a distinct outcurving rim vessel is associated with the vast majority of rim sherds which exhibit continuous notched applied relief decoration (although rims displaying this decoration were not well represented at Mangaasi). A number of incised vessels also appear to have outcurving rims (see Garanger 1972: Figs 78, 95).

Erueti

The recovered ceramics from the Erueti site were quite distinct from those recovered from Mangaasi, although small numbers of sherds of both traditions were recovered from the respective sites. The ceramics from Erueti were dominated by plain sherds belonging to globular outcurving rimmed vessels often with a distinctive wide flat lip. These flat lips were often notched on the outer edge and were occasionally decorated on the flat lip itself. Decoration on the body of the pots was largely restricted to incision. A number of carinated sherds were recovered as were a number of dentate stamped sherds (Garanger 1972:27). The Erueti-style ceramics were seen as being part of the Lapitoid tradition (Garanger 1972:29). As already noted, Garanger concluded that due to the fact that a number of Early Mangaasi-style sherds were recovered from the lowest levels of the Erueti site, a site which post-dated the Mangaasi site, the Mangaasi tradition pre-dated the Erueti (Lapitoid) tradition in Vanuatu and that the two traditions were un-related (Garanger 1971:61).

None of the conclusions to which Garanger arrived, regarding the ceramic remains from his excavations in Vanuatu, could be seen as particularly unjustified or suspect at the time. In fact the archaeological programme there can still be regarded, over 30 years later, as one of the more ambitious and successful projects ever carried out in the Pacific. Despite the fact that recent research has led to the substantial modification of earlier conclusions and associated theories related to the ceramic remains from Vanuatu, it was Garanger's initial research which first identified ceramic variation in Vanuatu and it is only due to fact that the excavations and recovered artefacts were published in such detail that some reassessment of the results has been possible (Ward 1989:154). Literary-based reassessments do, however, have their limitations and further questions regarding Garanger's original assertions began to be raised following the series of excavations on a number of other islands throughout Vanuatu (Bedford *et al.* 1998). It was with these issues in mind that the re-excavation of the Mangaasi site commenced in 1996.

Mangaasi 1996–1999

The ceramic remains from the test pits excavated between 1996-1999 at Mangaasi have enabled the establishment of a basic characterisation and chronological framework for the ceramics from the site. The analysis presented here is not designed to be a detailed inventory of all possible vessel forms and motif permutations or combinations found on Efate and the other central islands of Vanuatu. Rather it is a re-assessment of the ceramic sequence, to provide a general framework into which additional motifs, decorative techniques and distinctive vessel forms can be placed. For the finer characterisation and definition of the ceramic sequence, which will be presented in Chapter 8, both the earlier results of Garanger (1972, 1982) along with aspects of the recovered ceramics from Arapus (Bedford and Spriggs 2000; Spriggs and Bedford 2001) have been incorporated.

The ceramic materials analysed here were all recovered during the excavations of 1996-1999 at the Mangaasi site. The excavation strategy (test pit grid) and the complicated vertical and horizontal stratigraphy of the site had the potential to unnecessarily complicate the presentation of the results. In order to simplify the comparison of test pits and the presentation of the ceramic assemblage a number of points need to be emphasised. As noted in Chapter 3 *in situ* Erueti-style ceramics were encountered in TPs 9, 12, 10, 4 and 5 while the *in situ* Mangaasi-style ceramics were largely restricted to test pits closer to the sea, namely TPs 1, 2, 15, 16 and 10 (although somewhat complicated by mixing caused by a tidal wave deposit). It is the remains from these above test pits that has enabled the establishment of the ceramic sequence from the site. Therefore it is the ceramic remains from these test pits and in particular those from the identified *in situ* cultural horizons which are highlighted below. Seven of the excavated test pits (TPs 6, 7, 8, 11, 13, 14 and 18) lay largely outside the area of concentrated settlement, although the upper layers of TPs 7 and 11 did return the most concentrated samples of Late Mangaasi-style ceramics.

All the recovered ceramics comprise a broadly similar fabric (see below) both in terms of texture and mineral inclusions. The complete ceramic sample was analysed and a comprehensive breakdown of the ceramic sample including by test pit and layer is presented in full detail in

Bedford 2000b (Appendix 5: Tables 5.2–5.87). The sample is summarised here with selected tables only being included to highlight specific aspects of the assemblage (Tables 6.2–6.21). All vessel forms and associated decoration are illustrated (Figs 6.1–6.23) and a description of design motifs is listed in Appendix 4.

Mangaasi ceramic sample 1996–1999

A total of 5811 sherds were recovered from the 18 test pits which were excavated over the four field seasons at the site (Table 6.1). Of those, 4362 (75%) were plain body or basal sherds, 664 (11%) were decorated body sherds, 742 (13%) were rim and rim/body sherds, which were both decorated and non-decorated, and two were carinated sherds. A total of 13 handles and 26 wasters were recovered. A single possible spout was also identified.

Table 6.1 Mangaasi excavated ceramic sample

TPS	PLAIN SHERDS	DECO BODY	RIM SHERD	DECO*	LIP DECO.	CARIN.	HANDLE	WASTER	WEIGHT (gm)	TOTAL
1/15	372	102	49	134 (25%)	12	–	1	1	4147.89	525
17	285	68	35	89 (23%)	18	–	3	1	5885.2	392
10	304	100	77	134 (27%)	23	–	2	2	6914.51	485
12	477	19	52	31 (5.2%)	47	–	2	7	8915	557
9	1467	63	330	99 (5%)	293	1	2	8	32448.8	1871
2	285	54	42	78 (20%)	18	1	–	1	5813.33	383
3	150	16	22	20 (10.5%)	15	–	1	–	2351.49	189
4	202	24	35	34 (13%)	22	–	1	2	4494.8	263
5	155	22	26	26 (12%)	17	–	–	3	2077.67	207
6	35	19	5	22 (36%)	2	–	–	–	511.7	59
7	58	58	15	64 (49%)	7	–	–	–	1970.1	131
8	51	24	4	28 (35%)	–	–	–	–	794.63	79
11	78	35	8	42 (35%)	6	–	–	–	1089.56	121
13	36	14	5	20 (36%)	2	–	–	–	1496.5	55
14	97	8	16	15 (12%)	6	–	–	–	1734.8	121
16	288	31	16	40 (12%)	7	–	1	–	4438.7	338
18	22	7	5	12 (31%)	1	–	–	1	583.5	35
Total	4362	664	742	863	496	2	13	26	89,660.58	5811

* lip decoration not included

This broad outline disguises somewhat the variation found across the site particularly the ratio of plain to decorated sherds (see Table 6.1). Although the recovered ceramics displayed considerable variation in vessel form, decorative technique and design motifs over time, evidence from a range of characteristics point to the evolutionary nature of the sequence. The methodology used in the analysis of the Mangaasi ceramics is outlined in Chapter 4.

Fabric

A selection of sherds from the sites originally excavated by Garanger was analysed petrographically by both William Dickinson and Con Key (Garanger 1972:110–112). The mineral inclusions found in the Mangaasi sherds were dominated by angular to subangular plagioclase feldspar grains, subordinate rock fragments composed of pale brown volcanic glass plus minor amounts of clinopyroxene and opaque iron oxides. The moderately to poorly sorted texture of the inclusions implied derivation from stream sands. Especially characteristic of the Mangaasi sherds are the pale brown pumiceous glassy grains.

Two sherds recovered during the excavations in 1997 from former foreshore deposits below the depth of excavation achieved by Garanger were sent to Dickinson for petrographic analysis (see Appendix 3, WRD-138). It was confirmed that the two sherds resembled typical indigenous Efate sherds.

Both Dickinson and Key along with Warden (Garanger 1972:112–113) concluded that it seemed most likely that the examined sherds from Efate and the Shepherd Islands had been produced locally and that it was doubtful they had a single origin. Numerous wasters were recovered from the excavations of 1996–1999 at Mangaasi as they were by Garanger (1972:51), providing further evidence of local pottery production. The ceramics recovered from Mangaasi then, appear to have been produced on site from local materials and show no significant change in their mineral composition over time.

Plain sherds

A total of 4362 plain sherds were recovered from the excavations of 1996–1999 which represented 75% of total number of sherds. However relative numbers of plain sherds varied greatly from test pit to test pit and even within single test pits, inversely reflecting the percentage of decorated sherds. Test pits in which Erueti cultural horizons were identified displayed greater percentages of plain sherds (e.g. TP9 (78%) (Table 6.2) and TP12 (86%) (Table 6.3) than those test pits containing Mangaasi cultural horizons (TP1/15 (70%) (Table 6.4) and TP 17 (72%) (Table 6.5). This variation is clearer if percentages of decorated sherds (decorated body and rim sherds) per test pit are emphasised (e.g. TP 9 (5%), TP 12 (5.2%); TP1/15 (25%), TP 17 (23%) (see Tables 6.17, 6.20, 6.15, 6.21). Changing frequencies of plain sherds within single test pits was most clearly demonstrated in those containing Erueti cultural horizons (e.g. TP 9, L9d (87%)-L9a (64%); TP 12 L9c (82%)-L9a (73%) where the percentages of plain sherds decreased over time.

The globular nature of the vessel forms limited the potential for differentiating between body and basal plain sherds. Definitive identification of basal sherds (71) was only possible with large sherds or was assumed from sherd thickness. There appeared to be little perceptible change in the thickness of the plain body sherds through time or across the site (see Tables 6.2-6.5). The thickness of the vast majority of all the plain sherds fell between either 8-12 or 12-16mm, with much lesser percentages measuring 4–8 or 16–20mm. No sherds measured less than 4mm in thickness while a total of 52 measured between 20–24mm.

Rim/lip and vessel form

A number of distinctive rim and lip forms have facilitated the characterisation of the vessel forms represented at Mangaasi (see Tables 6.6–6.14). Erueti-style vessels are characterised by a number of distinct forms whose frequencies change over time. The earliest layers of the site associated with Erueti-style ceramics are dominated by vessel form 2ii (Fig. 6.1b–f) with occasionally decorated horizontal wide flat lips (Fig. 6.2).

TABLE 6.2. MANGAASI TP 9 SUMMARY OF PLAIN SHERDS

	LAYER 1/2I	LAYER 9A	LAYER 9B/8A	LAYER 9C	LAYER 8B/9D	TOTAL
body	60	81	197	659	424	1421
base	3	7	13	18	5	46
carination	–	–	–	–	1	1
wasters	1	–	–	6	1	8
1-4mm	–	–	–	–	–	–
4-8	5	8 (9%)	13 (6%)	45 (6.6%)	43 (10%)	114
8-12	33 (52%)	30 (34%)	87 (41%)	298 (44%)	201 (47%)	649
12-16	20 (31.7%)	41 (46%)	91 (43%)	240 (35%)	154 (36%)	546
16-20	4	7 (8%)	15 (7%)	67 (10%)	28 (6.5%)	121
20-24	1	2	4	27	3	37
weight	844.5gm	1083.4	3506.2	13461.4	5992.2	24887.7

TABLE 6.3. MANGAASI TP 12 SUMMARY OF PLAIN SHERDS

	LAYER 1/2I	LAYER 9A	LAYER 9B	LAYER 8/9C	LAYER 11	TOTAL
body	41	38	67	57	271	474
base	–	1	–	1	1	3
carination	–	–	–	–	–	–
wasters	–	5	–	2	–	7
1-4mm	–	–	–	–	–	–
4-8	1	3	12	4	47	66
8-12	26 (63%)	16 (41%)	30 (45%)	34 (57%)	112 (41%)	218
12-16	11 (27%)	18 (46%)	22 (33%)	18 (31%)	88 (32%)	157
16-20	3	2	3	1	24	33
20-24	1	–	–	1	–	2
weight	347.3	1018.8	1785.6	911.4	3027.5	7090.6

TABLE 6.4. MANGAASI TP 1/15 SUMMARY OF PLAIN SHERDS

	LAYER 1/2I	LAYER 3A	LAYER 4A/4B	LAYER 3B	LAYER 11	TOTAL
body	23	94	61	144	48	370
base	–	–	2	–	–	2
carination	–	–	–	–	–	–
wasters	–	–	–	–	1	1
1-4mm	–	–	–	–	–	–
4-8	5	11	4	7	2	29
8-12	15 (65%)	50 (53%)	24 (38%)	81 (56%)	24 (50%)	194
12-16	3 (12%)	27 (29%)	30 (48%)	52 (36%)	19 (40%)	131
16-20	–	6	3	4	3	16
20-24	–	–	2	–	–	2
weight	214.0	1012.4	1621.2	1858.8	627.4	5333.8

Table 6.5. Mangaasi TP 17 Summary of plain sherds

	LAYER 1/2I	LAYER 3A-3C	LAYER 4A-4D	LAYER 3D	LAYER 5III-11	TOTAL
body	25	144	75	13	23	280
base	–	–	4	–	1	5
carination	–	–	–	–	–	–
wasters	–	1	–	–	–	1
1-4mm	–	–	–	–	–	–
4-8	1	–	–	–	–	1
8-12	21 (84%)	63 (44%)	27 (34%)	8	16 (70%)	135
12-16	3	68 (47%)	41 (52%)	4	5	121
16-20	–	13	11	1	3	28
20-24	–	–	–	–	–	–
weight	206.3	2104.7	1506.9	237.3	202.8	4258.2

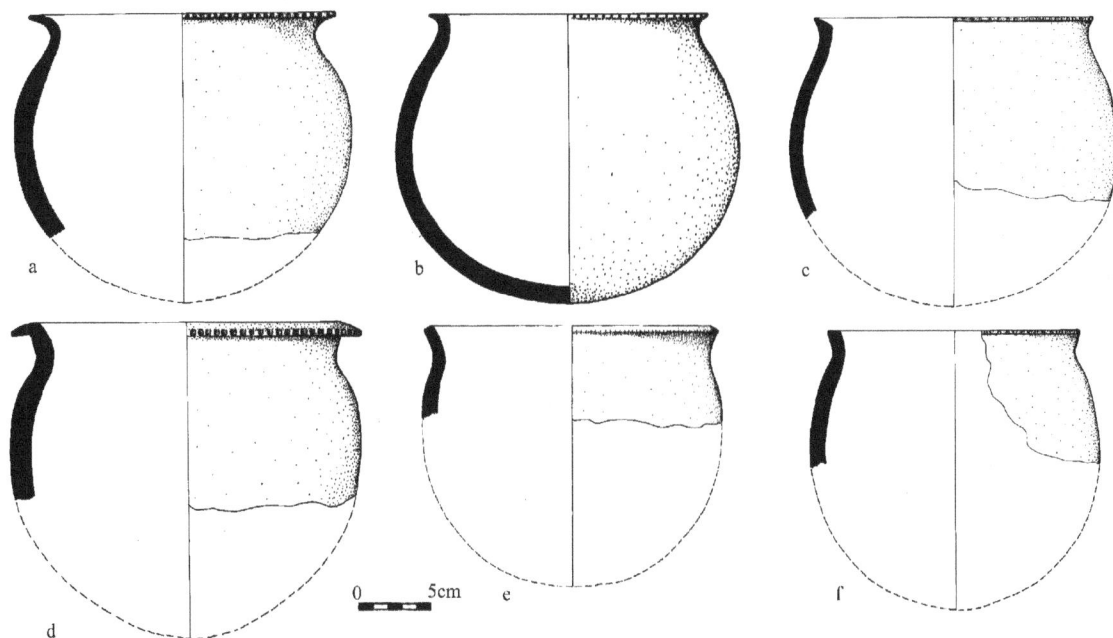

Figure 6.1 Arapus (a) and Erueti-style plainware vessels. a. vessel form 2i with notched lip (Ef-lip motif 1); b. vessel form 2ii; c-e. vessel form 2ii variants.

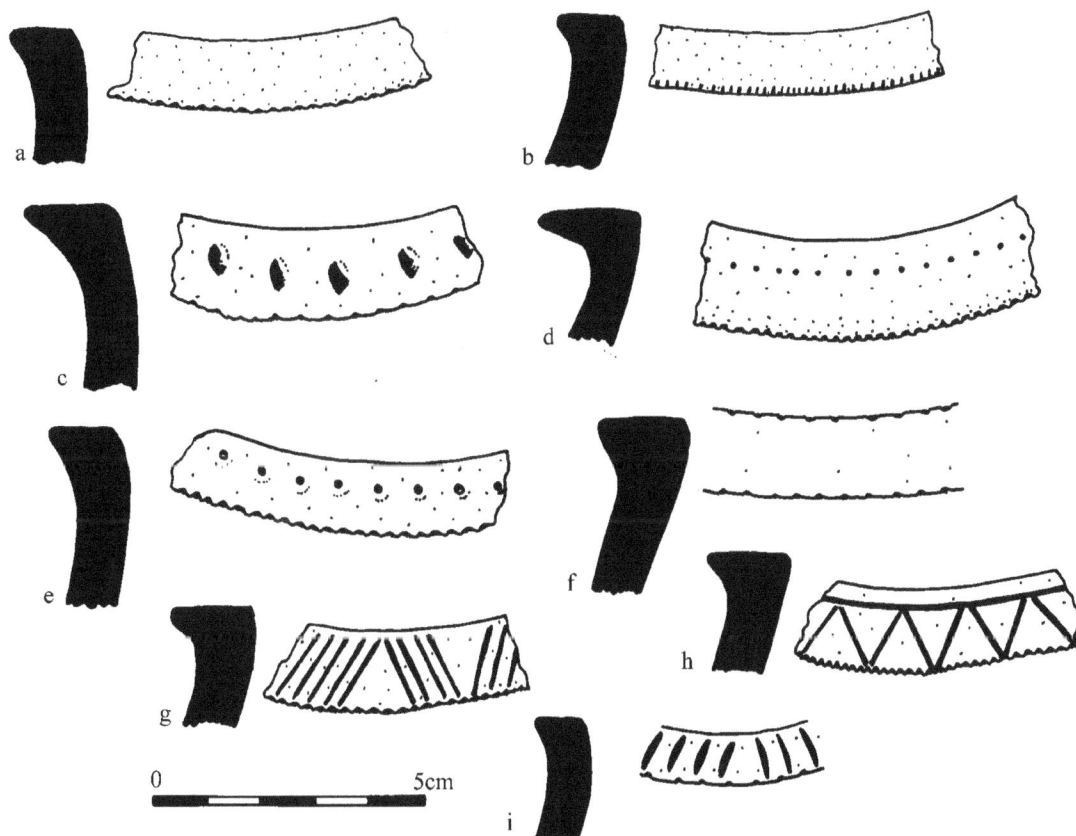

Figure 6.2 Erueti-style vessels, lip decoration. a. notching on the lip (Ef-lip motif 1); b. incised lip (Ef-lip motif 2); c-e. exterior notching plus punctate (Ef-lip motif 3); f. notching on interior and exterior of lip (Ef-lip motif 4); g. oblique parallel incision (Ef-motif 5); h. incised zigzag (Ef-lip motif 6); i. parallel notching (Ef-lip motif 7).

Table 6.6. Mangaasi TP 1/15 Rim/lip and vessel form

	LAYER 1/2I	LAYER 3A	LAYER 4A/4B	LAYER 3B	LAYER 11	TOTAL
Rim direction						
direct	-	-	-	1	-	1
outcurving	-	-	-	1	3	4
everted	-	-	2	-	-	2
incurving	3	10	11	18	-	42
Rim profile						
converg. gradual	1	8	12	16	1	38
converg. abrupt	-	-	-	-	-	-
parallel	-	-	-	1	1	2
ext. exp.	2	2	1	2	-	7
divergent abrupt	-	-	-	1	1	2
divergent gradual	-	-	-	-	-	-
Lip form						
plain	3	10	12	16	1	32
point	-	-	-	-	-	-
flat/horiz.	-	-	-	1	1	2
flat	-	-	1	3	1	5
Vessel form						
2i	-	-	-	1	1	2
2ii	-	-	-	-	2	2
2iii	-	-	2	-	-	2
3i	1	2	10	16	-	29
3ii	2	8	1	2	-	13
4i	-	-	-	1	-	1
Total	3	10	13	20	3	49

Table 6.7 Mangaasi TP 2. Rim/lip and vessel form

	LAYER 1/2I/2II	LAYER 3A	LAYER 4	LAYER 3B	LAYER 11	TOTAL
Rim direction						
direct	-	-	-	-	-	-
outcurving	-	12	8	-	4	24
everted	-	-	-	-	-	-
incurving	5	13	-	-	-	18
Rim profile						
converg. gradual	4	11	1	-	-	16
converg. abrupt	-	3	-	-	-	3
parallel	-	1	-	-	-	1
ext. exp.	1	3	-	-	-	4
divergent abrupt	-	7	7	-	4	18
divergent gradual	-	-	-	-	-	-
Lip form						
plain	5	15	1	-	-	21
point	-	-	-	-	-	-
flat/horiz.	-	8	7	-	4	19
flat	-	2	-	-	-	2
Vessel form						
2i	-	2	1	-	-	3
2ii	-	8	7	-	4	19
3i	4	11	-	-	-	15
3ii	1	3	-	-	-	4
3iv	-	1	-	-	-	1
Total	5	25	8	-	4	42

Table 6.8 Mangaasi TP 3. Rim/lip and vessel forms

	LAYER 1/2I	LAYER 2II	LAYER 4	LAYER 3	LAYER 11A-C	TOTAL
Rim direction						
direct	-	-	-	-	1	1
outcurving	2	-	1	-	10	13
everted	-	1	-	-	-	1
incurving	3	1	2	-	1	7
Rim profile						
converg. gradual	3	2	-	-	3	8
converg. abrupt	-	-	-	-	-	-
parallel	1	-	-	-	-	1
ext. exp.	-	1	2	-	1	4
divergent abrupt	1	-	1	-	8	10
divergent gradual	-	-	-	-	-	-
Lip form						
plain	4	3	2	-	3	12
point	-	-	-	-	-	-
flat/horiz.	1	-	-	-	7	8
flat	-	-	1	-	2	3
Vessel form						
2i	1	1	-	-	2	4
2ii	1	-	1	-	8	10
3i	3	1	-	-	1	5
3ii	-	1	1	-	-	2
4i	-	-	-	-	1	1
Total	5	3	2	-	12	22

Table 6.9 Mangaasi TP 4. Rim/lip and vessel form

	LAYER 2I	LAYER 5I	LAYER 5II	LAYER 9	LAYER 11	TOTAL
Rim direction						
direct	-	-	-	-	-	-
outcurving	2	2	1	15	2	22
everted	-	-	-	-	-	-
incurving	1	8	1	3	-	13
Rim profile						
converg. gradual	1	7	1	8	1	18
converg. abrupt	-	-	-	-	-	-
parallel	-	-	-	-	-	-
ext. exp.	-	1	-	-	-	1
divergent abrupt	2	2	1	10	1	16
divergent gradual	-	-	-	-	-	-
Lip form						
plain	1	8	1	8	1	19
point	-	-	-	-	-	-
flat/horiz.	2	2	1	8	1	14
flat	-	-	-	2	-	2
Vessel form						
2i	-	-	-	5	1	6
2ii	2	2	1	10	1	16
3i	1	7	1	2	-	11
3ii	-	1	-	-	-	1
3iv	-	-	-	1	-	1
Total	3	10	2	18	2	35

Table 6.10 Mangaasi TP 5 rim/lip and vessel form

	LAYER 1/2I/2II	LAYER 5I	LAYER 9	LAYER 11	TOTAL
Rim direction					
direct	-	-	-	-	-
outcurving	1	-	16	-	17
everted	-	-	-	-	-
incurving	3	6	-	-	9
Rim profile					
converg. gradual	2	5	1	-	8
converg. abrupt	-	-	-	-	-
parallel	-	-	-	-	-
ext. exp.	1	1	-	-	2
divergent abrupt	1	-	15	-	16
divergent gradual	-	-	-	-	-
Lip form					
plain	3	6	1	-	10
point	-	-	-	-	-
flat/horiz.	1	-	13	-	14
flat	-	-	2	-	2
Vessel form					
2i	-	-	1	-	1
2ii	1	-	15	-	16
3i	1	5	-	-	6
3ii	1	1	-	-	2
3iv	1	-	-	-	1
Total	4	6	16	-	26

Table 6.11. Mangaasi TP 9 rim/lip and vessel form

	LAYER 1/2I	LAYER 9A	LAYER 9B/8A	LAYER 9C	LAYER 8B/9D	TOTAL
Rim direction						
direct	-	4	6	15	13	38
outcurving	-	8	35	10	86	139
everted	-	-	-	-	-	
incurving	14	14	7	10	7	52
Rim profile						
converg. gradual	7	10	15	26	20	78
converg. abrupt	1	-	-	2	2	5
parallel	1	2	4	14	12	33
ext. exp.	5	8	7	5	5	30
divergent abrupt	-	6	23	84	67	180
divergent gradual	-	-	-	4	7	11
Lip form						
plain	11	18	16	32	25	102
point	-	-	2	5	3	10
flat/horiz.	3	3	23	57	43	129
flat	-	5	8	40	35	88
Vessel form						
2i	-	2	8	17	20	47
2ii	-	6	29	98	72	205
3i	8	7	1	3	4?	23
3ii	5	7	3	3	-	18
3iv	1	1	-	1	-	3
4i	-	3	5	12	9	29
5i	-	-	-	-	1	1
5ii	-	-	3	1	-	4
Total	14	26	49	135	106	330

Table 6.12. Mangaasi TP 10 rim/lip and vessel form

	LAYER 1/2I	LAYER 3A	LAYER 9A	LAYER 8/9B/11	TOTAL
Rim direction					
direct	-	-	2	-	2
outcurving	-	3	5	13	21
everted	1	1	-	-	2
incurving	6	24	21	1	52
Rim profile					
converg. gradual	7	25	21	5	58
converg. abrupt	-	-	-	-	-
parallel	-	1	1	-	2
ext. exp.	-	-	1	-	-
divergent abrupt	-	2	5	9	16
divergent gradual	-	-	-	-	-
Lip form					
plain	7	26	21	5	52
point	-	-	-	-	-
flat/horiz.	-	1	6	8	15
flat	-	1	1	1	3
Vessel form					
2i	-	1	1	4	6
2ii	-	2	4	9	15
2iii	1	1	-	-	2
3i	6	24	20	1	51
3ii	-	-	1	-	1
4i	-	-	2	-	2
Total	7	28	28	14	77

Table 6.13. Mangaasi TP 12 rim/lip and vessel form

	LAYER 1/2I	LAYER 9A	LAYER 9B	LAYER 8/9C	LAYER 11	TOTAL
Rim direction						
direct	-	2	-	-	-	2
outcurving	4	3	5	6	15	33
everted	-	-	-	-	-	-
incurving	4	5	6	2	-	16
Rim profile						
converg. gradual	2	4	1	1	6	13
converg. abrupt	-	-	-	-	-	-
parallel	-	1	-	-	-	1
ext. exp.	2	3	4	2	-	11
divergent abrupt	4	2	6	5	9	26
divergent gradual	-	-	-	-	-	-
Lip form						
plain	3	7	4	3	6	20
point	-	-	-	-	-	-
flat/horiz.	4	1	6	5	9	25
flat	1	2	1	-	-	4
Vessel form						
2i	-	-	-	1	6	7
2ii	4	3	6	5	9	27
3i	2	1	1	-	-	4
3ii	2	3	2	2	-	9
3iv	-	1	-	-	-	1
4i	-	2	-	-	-	2
5ii	-	-	2?	-	-	2?
Total	8	10	11	8	15	

Table 6.14. Summary of rim diameters for the various vessel forms

vessel form	10cm	12cm	14cm	16cm	18cm	20cm	22cm	24cm
2i	-	-	-	2	7	2	-	-
2i (late)	-	-	-	7	1	1	-	-
2ii	-	1	2	19	50	22	7	-
2iii	-	-	-	-	-	1	-	-
3i	-	1	3	29	16	7	5	4
3ii	-	-	1	15	4	2	-	-
3iv	6	-	-	-	-	-	-	-
4i	1	-	2	4	6	-	3	-
5i	-	-	-	2	-	-	-	-
5ii	-	-	2	1	2	-	-	-
5iv	1	-	-	-	-	-	-	-

This was most dramatically demonstrated in TP 9, Layers 9c/9d (Table 6.11) where these predominantly plain vessels made up 70% of the identified vessel forms. Although other test pits where Erueti cultural horizons were recorded returned much smaller samples, the dominance of vessel form 2ii can also be seen (e.g. TP 12, Layer 8/9c (60%) (Table 6.13); TP 10, Layer 8/9/11 (64%) (Table 6.12); TP 4, Layer 9 (55%) (Table 6.9); TP 5, Layer 9 (93%) (Table 6.10). The rim diameter of this vessel form ranged between 12-22cm but the greatest number (90%) measured between 16-20cm (Table 6.14). Vessel form 2ii was also frequently recorded within the former foreshore deposits in test pits closer to the sea (e.g. TP 1/15, Layer 11 (66%) (Table 6.6); TP 2, Layer 11 (100%) (Table 6.7); TP 3, Layer 11 (66%) (Table 6.8)).

Often found in association with vessel form 2ii (but always in lesser quantity), particularly in the earliest Erueti cultural horizons, was vessel form 2i (Fig. 6.1a) (e.g. TP 9, Layer 9c/d (15%); TP 12, Layer 8/9c (40%); TP 10, Layer 8/9/11 (28%); TP 4, Layer 9 (28%)). As with vessel form 2ii, vessel form 2i was also recorded amongst the former foreshore deposits. The more recent excavations at Arapus (Bedford and Spriggs 2000; Spriggs and Bedford 2001) have revealed that vessel form 2i actually pre-dates vessel form 2ii and can be shown to represent an earlier ceramic phase interpreted as the cooking component of a Lapita assemblage. This is further discussed in Chapter 8. After a short time period of perhaps only one hundred years or quite possibly much less, vessel form 2i disappeared from the sequence, being replaced by vessel form 2ii. Vessel form 2i did make a reappearance but it was amongst the much later Mangaasi cultural horizons (see below). In these later contexts it was always associated with decorated sherds that can now be identified as representing the last phase of ceramic production on Efate. Vessel form 2i then, was associated with both the very earliest phase and the last phase of the ceramic sequence. The rim diameters of these vessels associated with both the early and late phases ranged between 16–20cm (Table 6.36).

Other vessel forms represented (often single examples) amongst the earliest Erueti cultural horizons, included vessel forms 3i (Fig. 6.3b), 3ii (Figs 6.3g, 6.4a, c), 4i (Fig. 6.3d), 5i, 5ii (Fig. 6.5a, b) and 5iii. All of these vessel forms were decorated and may have been associated with ceremonial activity. This suggestion was more clearly evidenced at the Arapus site where more intact examples of vessel forms 5i (Fig. 6.6c) and 5ii (Fig. 6.6a, b) were recovered amongst predominantly plain ware vessels (2ii). The angle of the carination became less pronounced over time to the point where these vessels appear to have transformed into vessel form 3i. Small cups represented by vessel forms 3iv (Fig. 6.7a) and 5iii (Fig. 6.7c, g) were recovered from early and late Erueti cultural horizons. These vessel forms were more often undecorated although several decorated examples were recovered from Mangaasi (Fig. 6.7d) and Arapus (Fig. 6.7i, j). These vessel forms which appeared in the sequence several hundred years after initial settlement in Vanuatu are also found in others sites further east where kava connections have been made (Green and Davidson 1974 Vol. 2:129). Kava it has been argued was first domesticated in Vanuatu at some time soon after first settlement (Lebot *et al.* 1997:23). The rim diameter of these small cup-like vessels was consistently around 10cm (see Table 6.14).

Remains of handles were also recovered from Erueti cultural horizons (Fig. 6.8 a-c) but none remained attached to any vessels. The undecorated handles were all tubular in form with an oval cross-section (Garanger's loop handles). A single sherd from the Arapus site does demonstrate (Fig. 6.8j) that handles (although very rare) were associated with the earliest Erueti-style ceramics.

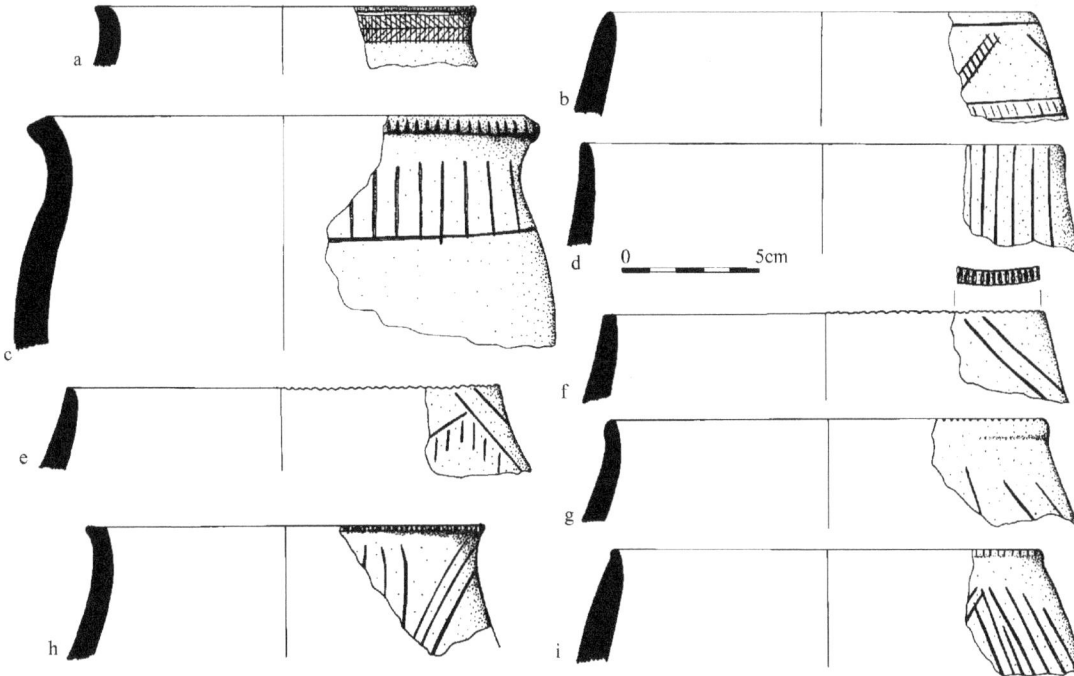

Figure 6.3   Erueti-style vessel form and decoration. a. vessel form 2ii and Ef-motif 23; b. vessel form 3i and Ef-motif 32; c. vessel form 2ii and Ef-motif 28; d. vessel form 4i? and Ef-motif 28; e. vessel form 5ii plus Ef-motif 30; f. vessel form 5ii; g. vessel form 3ii; h. vessel form 2ii; i. vessel form 3ii and Ef-motif 31.

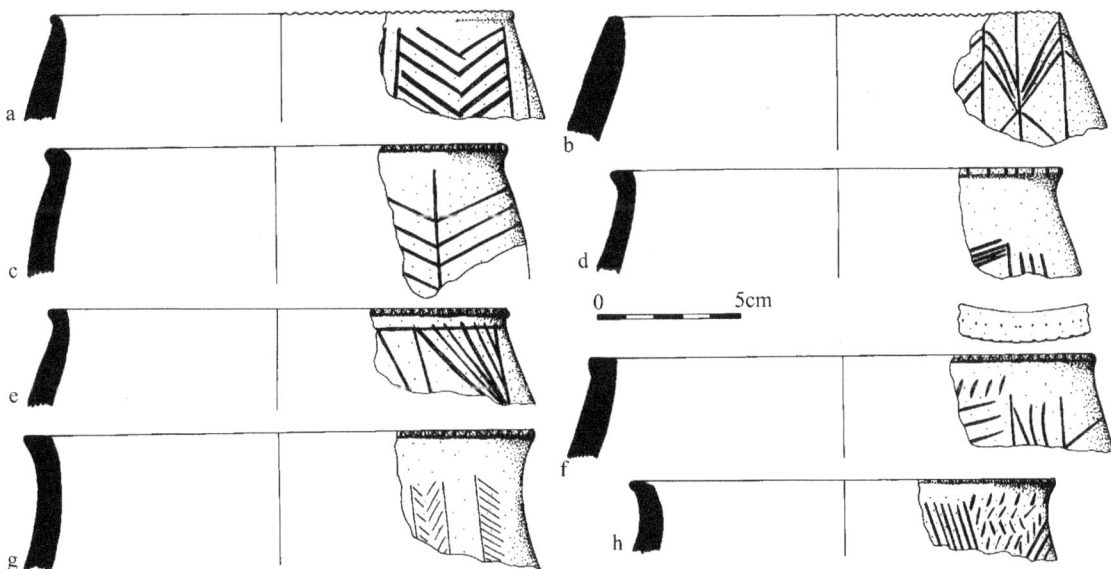

Figure 6.4   Erueti-style vessel form and decoration. a. vessel form 3ii plus Ef-motif 11; b. vessel form 3i and Ef-motif 8; c. vessel form 3ii plus incised motif; d-f. vessel form 2ii plus incised motif; g. vessel form 2ii and Ef-motif 39; h. vessel form 2ii and Ef-motif 51.

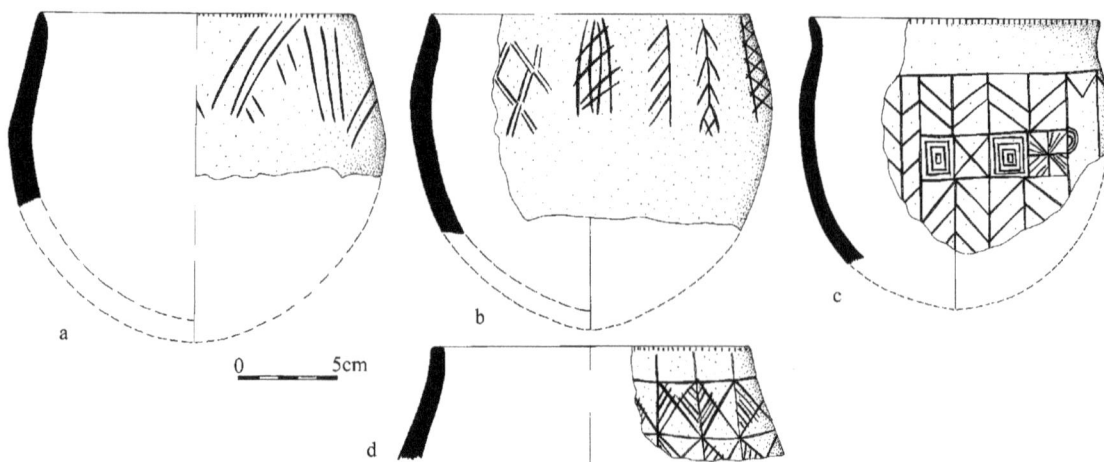

Figure 6.5   Late Erueti-style vessel form and decoration. a. vessel form 5ii and Ef-motif 30; b. vessel form 5ii and Ef-motif 19; c. vessel form 4i and Ef-motif 20; d. vessel form 4i and Ef-motif 24.

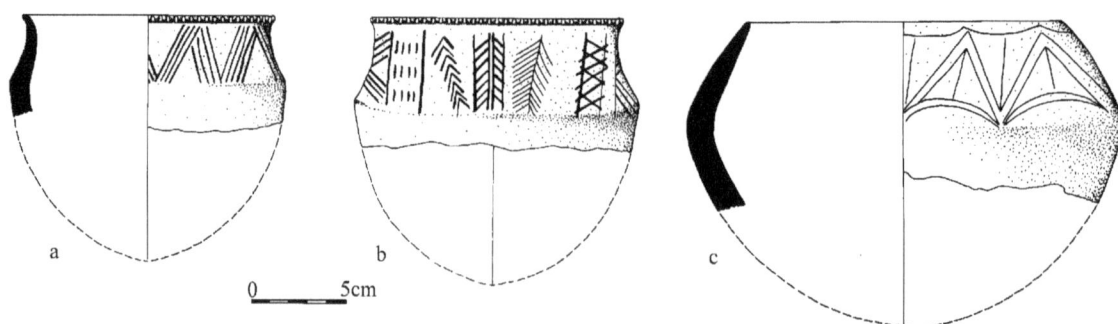

Figure 6.6   Early Erueti-style vessel form and decoration. a. vessel form 5ii and Ef-motif 31; b. vessel form 5ii and Ef-motif 21; c. vessel form 5i and Ef-motif 27.

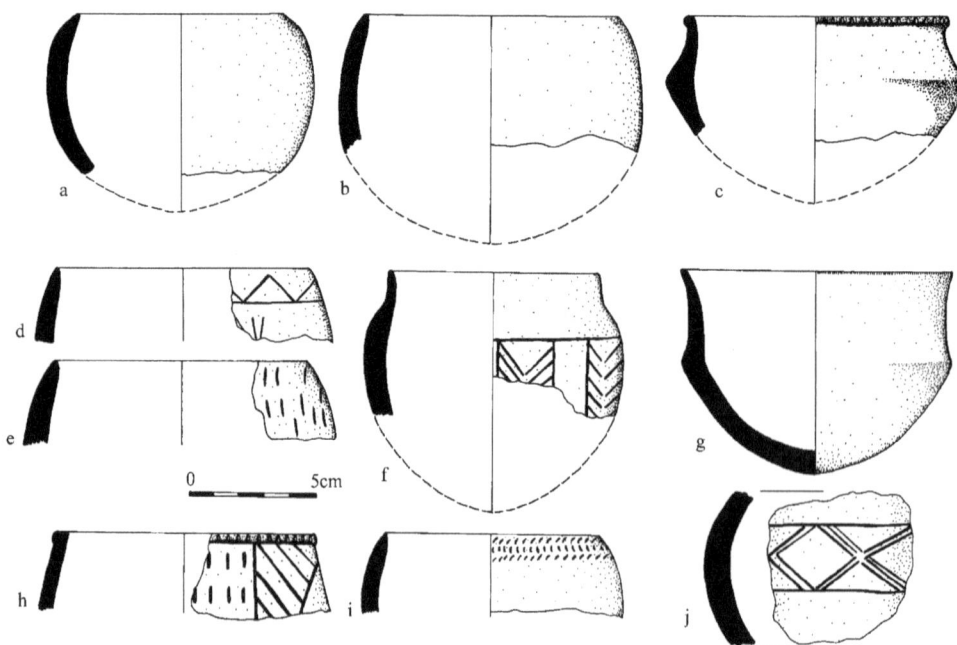

Figure 6.7   Erueti (a-h) and Mangaasi-style (i) cups. a-b. vessel form 3iv; c. vessel form 5iii; d. vessel form 3iv and Ef-motif 25; e. vessel form 3iv and Ef-motif 14; f. vessel form 4iii and Ef-motif 6; g. vessel form 5iii; h. vessel form 5iii? and Ef-motif 16; i. vessel form 3iii and Ef-motif 56; j. vessel form 4iii? and Ef-motif 26.

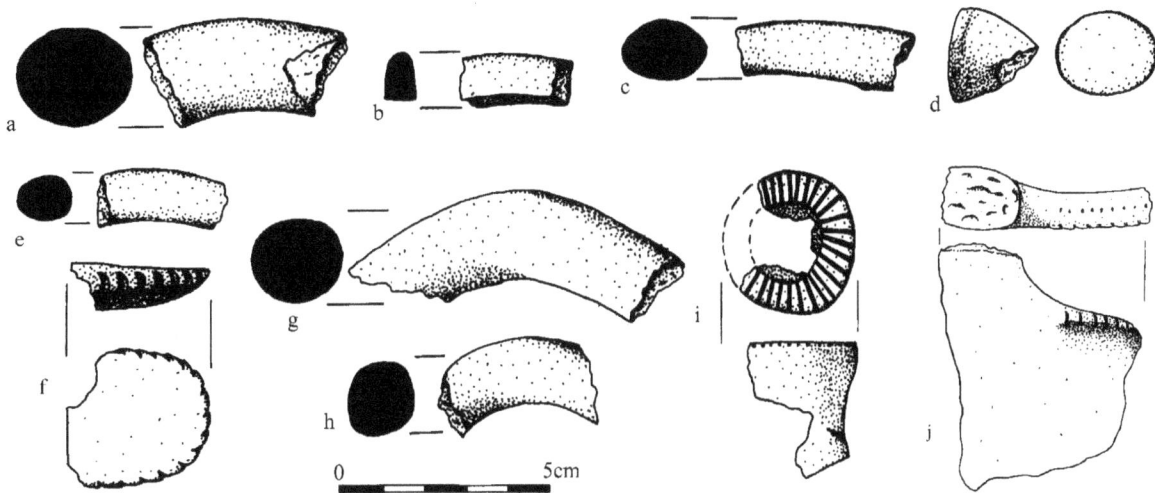

Figure 6.8 Assorted Erueti and Mangaasi-style handles and spout? a–c. loop handle (a. TP12, L.9a; b. TP9, L.9b; c. TP9, L.9b); d. knob (TP10, L.3b); e. loop handle (TP1, L.4a); f. handle (ear) (TP10, L.3b); g-h. loop handle (g. TP17, L.3b; h. TP17, L.3b); i. spout (TP10, L.3b); handle attachment (TP28, 150-170cm bd).

Early Erueti-style vessel forms can be summarised as follows. They were predominately outcurving rim vessels dominated by vessel form 2ii with wide flat lips along with much lesser quantities of vessel form 2i. Handles, of a single form, were present but rare. Other vessel forms which were identified but also rare, included 3i, 3ii, 3iv, 4i, 5i, 5ii and 5iii.

In the later Erueti cultural horizons a change in vessel forms is clearly demonstrated (e.g. TP 9, TP 12) (Tables 6.11 and 6.12). Vessel forms 2i and 2ii became much less frequent or disappeared to be replaced by increasing numbers of incurving rim vessels (3i and 3ii). Carinated vessels also appeared to become increasingly globular. Occasional examples of 3iv were also recorded.

The trend towards a predominance of incurving globular vessels was further emphasised in the Mangaasi cultural horizons. Vessel form 3i was completely dominant with generally lesser percentages of form 3ii (e.g. TP 1/15, Layer 3a/3b (95%); TP 17, Layer 3 (92%); TP 10, Layer 3a (86%)). Although not recovered from *in situ* deposits, but able to be inferred through motif associations were small numbers of vessel form 3iv (Fig. 6.7e). The rim diameters of vessel form 3i ranged between 12-24cm but the vast majority (80%) measured between 16-20cm. Vessel form 3ii tended to have a smaller rim diameter range (14-20cm) but again the majority measured between 16-20cm (see Table 6.14).

A limited number of handles were recovered from the Mangaasi cultural horizons. They included a notched horizontal ear (Fig. 6.8f) (Garanger 1972:55) and a knob-like handle (Fig. 6.8d) both of which appear to have been fitted to the rim of incurving vessels (3i). Loop handles (Fig. 6.8e, g, h) were recovered from test pits consisting of Mangaasi cultural horizons but from mixed or potentially mixed deposits which suggests the possibility that this form of handle was associated with Erueti-style rather than Mangaasi-style vessels. Of the wide range of handles illustrated by Garanger that were still attached to vessels, all appeared to be associated with vessel form 3i. A single possible spout (Fig. 6.8i) was also recovered from the Mangaasi cultural horizon but no further indication of vessel form was apparent.

Late Mangaasi-style vessel forms, although not well represented in the recent excavations (as opposed to those of Garanger) were clearly associated with vessel form 2i and 2iii. These late vessel forms were best represented in TPs 7 and 11 (e.g. TP7, L.5i; TP11, L.2i). No handles appeared to be associated with these vessel forms.

A basic transformation over time of vessel form can be demonstrated at the Mangaasi site. The early Erueti-style vessels are characterised by a predominance of outcurving rim vessels, primarily form 2ii with lesser quantities of 2i. Handles of a single form were present but very rare. Other vessel forms were present (3i, 3ii, 3iv, 4i, 5i, 5ii and 5iii) but were again rare. A greater variety of vessel forms was noted with the early Erueti-style ceramics. Over time incurving rim vessels (3i and 3ii) became more frequent, coinciding with a decrease in the number of outcurving rimmed vessels. Sharply carinated vessels (5i) became increasingly rounded (5ii) and ultimately disappeared from the record. Occasional examples of small cups (3iv) were found throughout the Erueti cultural horizons. Mangaasi-style vessels are initially characterised by globular incurving rimmed vessels (3i). A variety of handle forms are associated with these vessels but were relatively rare. Vessel form 3iv was also associated with the Mangaasi-style ceramics. Late Mangaasi-style vessels, associated with the last phase of ceramic use on Efate, appeared to return to an outcurving rim form (2i and 2iii) bereft of handles.

Decoration

Excluding decoration of the lip, a total of 863 (15%) decorated sherds were recovered from the Mangaasi site (Tables 6.15–6.21). As noted above in the discussion of plain sherds, frequencies of decorated sherds varied greatly across the site and within single test pits (Tables 6.15–6.21). Decoration was less frequent in the early Erueti cultural horizons (e.g. TP9, L9d (1.4%)(Table 6.16); TP12, L. 8/9c (0%)(Table 6.16)) but increased over time (e.g. TP9, L.9a (17%); TP12, L.9a (15%)). Decorated sherds from Mangaasi cultural horizons always comprised more than 20% (e.g. TP1/15, L3a (24.6%) (Table 6.17)) and often up to 30% (e.g. TP17, L.3a-c (30%)(Table 6.18)) of the ceramic sample. Lip decoration, which consisted almost exclusively of notching, showed an inverse trend with the great majority of lips from the early Erueti cultural horizons exhibiting decoration (e.g. TP9, L.9d (91.5%)), a slightly reduced percentage can be seen in later Erueti horizons (e.g. TP9, L.9a (88%)) leading to a dramatic decrease in the incidence of notching recorded in the Mangaasi cultural horizons (e.g. TP10, L.3a (10%)(Table 6.19)). Other varieties of lip decoration, which were comparatively rare, were restricted to the Erueti-style vessel form 2ii with horizontal wide flat lips (Fig. 6.2).

Table 6.15 Mangaasi TP 9 Summary of decorative techniques and location

DECO TECHNIQUE	LAYER 1/2I	LAYER 9A	LAYER 9B/8A	LAYER 9C	LAYER 8B/9D	TOTAL
Incision						
linear	17	12	18	9	9	65
geometric	6	8	6	6	3	29
curvilinear	-	-	-	2	-	2
gash	-	1	6	2	2	11
complex	-	-	1	-	-	1
Impression						
punctate	1	1	2 (flat lip)	7 (flat lip)		11
Applied relief						
notched bands	-	-	-	-	-	-
plain bands	-	-	-	-	-	-
nubbins	-	-	-	-	-	-
Excision						
perforation	-	-	-	-	1	1
notching	5	21	46	107	92	271
Total	29	43	79	133	107	391
Location	-	-	-	-	-	-
lip ext.	1	9	34	94	75	209
lip int.	-	-	-	-	-	-
flat area of lip	1	-	4	7	2	14
rim	6	4	3	3	5	21
body	20	19	17	9	8	73
lip	7	14	11	18	21	65

Table 6.16 Mangaasi TP 12 Summary of decorative techniques and location

DECO TECHNIQUE	LAYER 1/2I	LAYER 9A	LAYER 9B	LAYER 8/9C	LAYER 11	TOTAL
Incision						
linear	7	5	1	-	2	15
geometric	3	3	7	-	-	13
curvilinear	-	-	-	-	-	-
gash	-	3	-	-	-	3
complex	-	1	-	-	-	1
Impression						
punctate	-	1	-	1 (flat lip)	-	2
Applied relief						
notched bands	1	-	-	-	-	1
plain bands	-	-	-	-	-	-
nubbins	-	-	-	-	-	-
Excision						
perforation	-	-	-	-	-	-
notching	5	7	10	8	15	45
Total	16	20	18	9	17	80
Location						
lip ext.	5	3	8	5	9	16
lip int.	-	-	-	-	-	-
flat area of lip	-	-	-	1	-	1
rim	3	4	3	-	-	10
body	11	8	9	-	2	30
lip	1	5	2	3	6	17

Table 6.17. Mangaasi TP 1/15 Summary of decorative techniques and location

DECO TECHNIQUE	LAYER 1/2I	LAYER 3A	LAYER 4A/4B	LAYER 3B	LAYER 11	TOTAL
Incision						
linear	4	18	19	36	3	80
geometric	2	9	15	16	-	42
curvilinear	-	-	-	-	-	-
gash	3	8	7	17	2	37
complex	-	-	1	-	-	1
Impression						
punctate	-	1	1	5	-	7
Applied relief						
notched bands	1	-	3	-	-	4
plain bands	-	2	-	1	-	3
nubbins	-	1	-	-	-	1
Excision						
perforation	-	-	-	-	-	-
notching	2	2	1	3	2	10
Total						
Location						
lip ext.	-	-	-	1	2	3
lip int.	-	-	-	-	-	-
flat area of lip	-	-	-	1	-	1
rim	1	9	10	12	-	32
body	7	22	28	40	4	101
lip	2	2	1	3	-	8

Table 6.18. Mangaasi TP17 Summary of decorative technique and location

DECO TECHNIQUE	LAYER 1/2I	LAYER 3A-3C	LAYER 4A-4D	LAYER 5III-11	TOTAL
Incision					
linear	8	26	1	1	36
geometric	1	19	4	1	25
curvilinear	-	-	-	-	-
gash	3	8	-	1	12
complex	-	-	-	-	-
Impression					
punctate	1	4	-	-	5
Applied relief					
notched bands	-	-	-	-	-
plain bands	-	1	-	-	1
nubbins	1	1	-	-	2
Excision					
perforation	-	-	-	-	-
notching		3	10	5	18
Total					
Location					
lip ext.	-	1	4	-	5
flat area of lip	-	-	-	-	-
rim	6	10	4	1	21
body	11	51	4	2	68
lip	-	2	6	5	13

Table 6.19 Mangaasi TP 10 Summary of decorative techniques and location

DECO TECHNIQUE	LAYER 1/2I	LAYER 3A	LAYER 9A	LAYER 8/9B/11	TOTAL
Incision					
linear	1	33	42	12	88
geometric	5	11	17	2	35
curvilinear	-	-	-	-	-
gash	2	18	20	-	40
complex	-	-	2	-	2
Impression					
punctate	1	1	9	1	12
Applied relief					
notched bands	-	3	2	-	5
plain bands	-	-	-	-	-
nubbins	-	3	2	-	5
Excision					
perforation	-	-	-	-	
notching	-	2	7	7	16
Total					
Location					
lip ext.	-	2	6	4	12
lip int.	-	-	-	1	1
flat area of lip	-	-	1	1	2
rim	4	14	14	1	33
body	5	49	65	10	129
lip	-	-	1	3	4

Erueti-style decoration consisted exclusively of incised motifs (see Figs 6.3–6.13) apart from notching on the lip and much rarer examples of punctation found only on the horizontal surface of wide flat lips (vessel form 2ii). There is no applied relief of any kind. Two examples of perforation were recorded (Fig. 6.12j, k). Both linear and geometric incision dominated, followed by much fewer examples of incised gashes. These three varieties of incision are found both separately and in association. Curvilinear incision was recorded but was rare. Incision associated with Erueti-style decoration tends generally to be thicker and heavier than later Mangaasi-style decoration.

Although Mangaasi-style decoration included virtually the full spectrum of techniques apart from dentate stamping there was great variation through the sequence. Initially Mangaasi-style decoration was predominately represented by incision (linear, geometric and gashes) and to a lesser extent punctation. These techniques appeared initially to be utilised separately rather in association (Figs 6.14, 6.15, 6.16, 6.17). Lip decoration completely fell from favour and disappeared from the decorative repertoire.

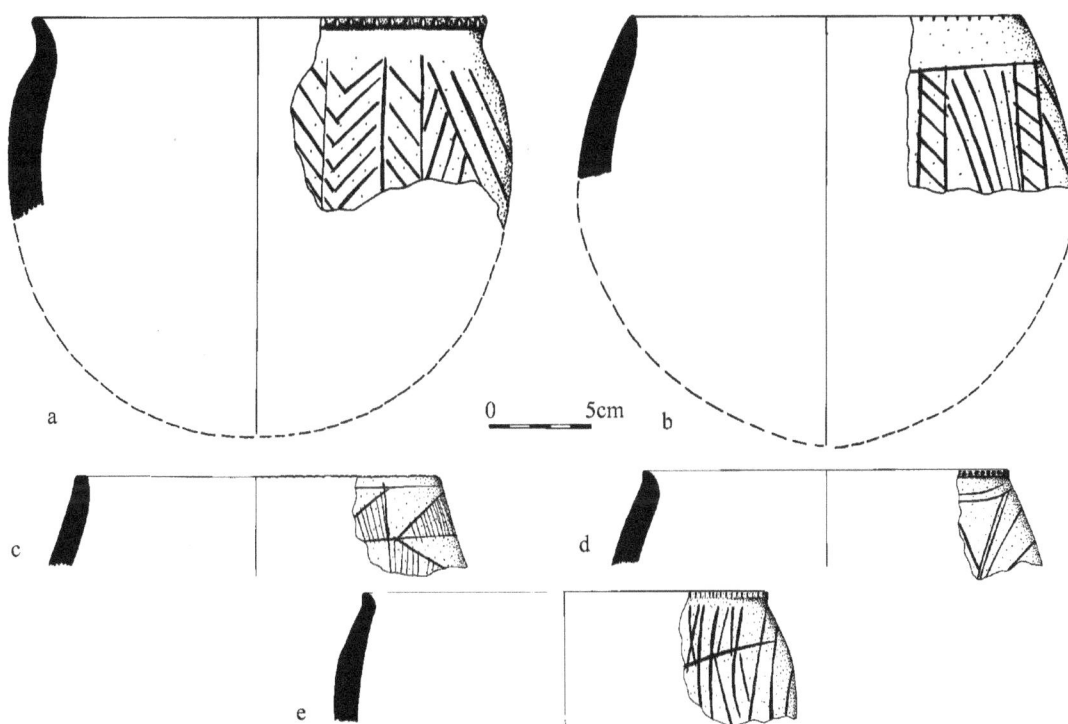

Figure 6.9 Erueti-style vessel form and decoration. a. vessel form 2i and Ef-motif 2; b. vessel form 3i and Ef-motif 12; c. vessel form 4i and Ef-motif 65; d-e. vessel form 3ii and linear incision.

Figure 6.10 Erueti-style incised decoration. a. Ef-motif 1; b–c. Ef-motif 9; d. Ef-motif 9; e. Ef-motif 4; f. Ef-motif 10; g. Ef-motif 3; h. Ef-motif 9; i. Ef-motif 9; j. Ef-motif 8, k. Ef-motif 3.

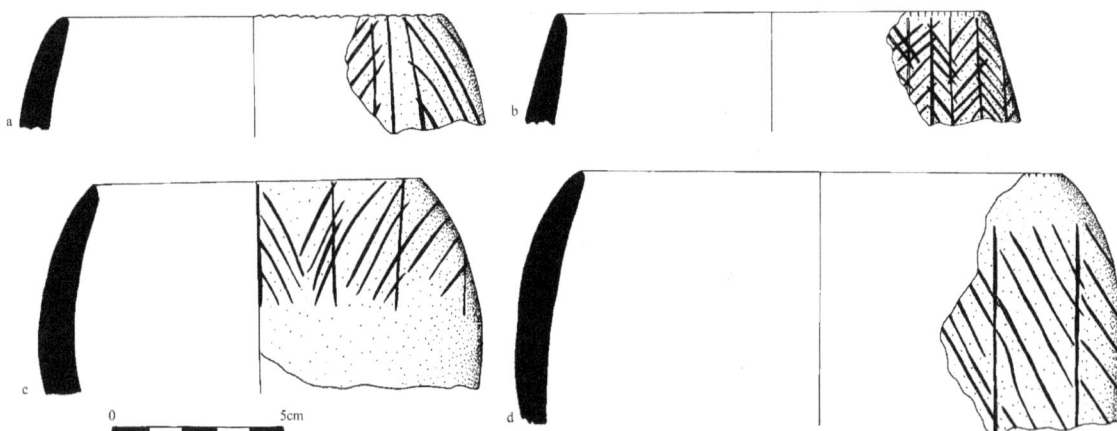

Figure 6.11 Late Erueti-style vessel form and decoration. a-d. vessel form 3i (a. Ef-motif 5; b. Ef-motif 1; c. Ef-motif 5; d. Ef-motif 3).

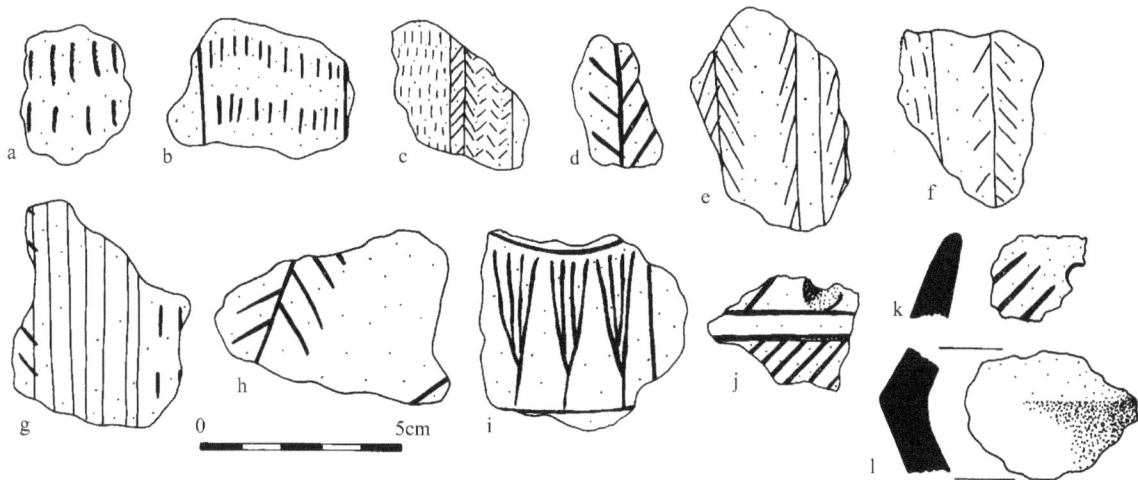

Figure 6.12 Miscellaneous Erueti-style decoration. a. Ef-motif 14; b. Ef-motif 18; c. Ef-motif 13; d. Ef-motif 38; e. Ef-motif 36; f. Ef-motif 37; g. linear incision plus gashes; h. Ef-motif 38; i. Ef-motif 43; j. geometric incision plus perforation; k. vessel form 3i, parallel incision plus perforation; l. vessel form 5i.

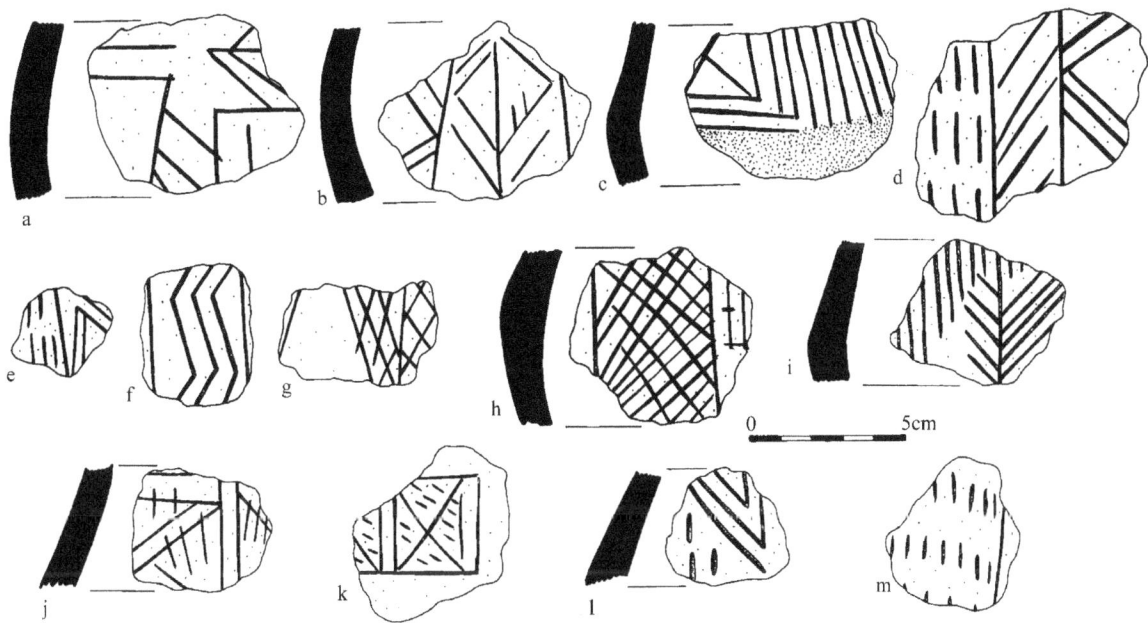

Figure 6.13 Erueti-style incised decoration. a. Ef-motif 34; b. Ef-motif 33; c. Ef-motif 29; d. Ef-motif 15; e. linear incision plus gashes; f. Ef-motif 7; g-h. incised crosshatch; i. linear incision; j. geometric incision plus gashes; k. Ef-motif 35; l. geometric incision plus gashes; m. Ef-motif 18.

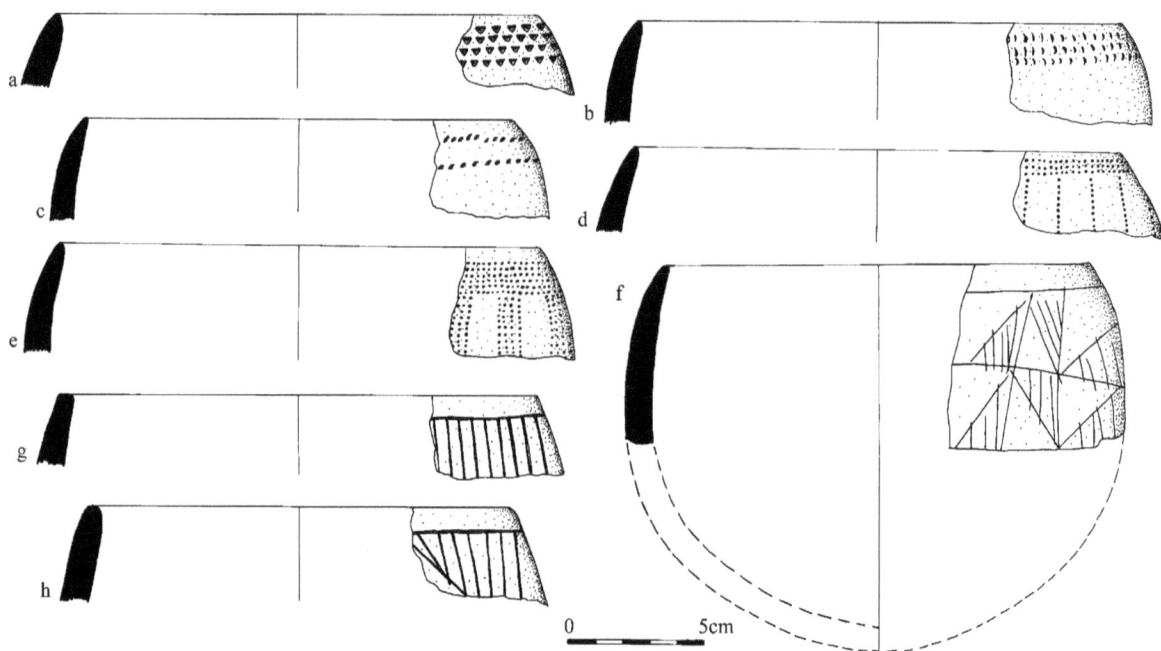

Figure 6.14 Mangaasi-style vessel form and decoration. a-f. all vessel form 3i with punctate (a-e) and incised decoration (f-h) (a-c. Ef-motif 44; d. Ef-motif 49; e. Ef-motif 48; f-h. Ef-motif 65).

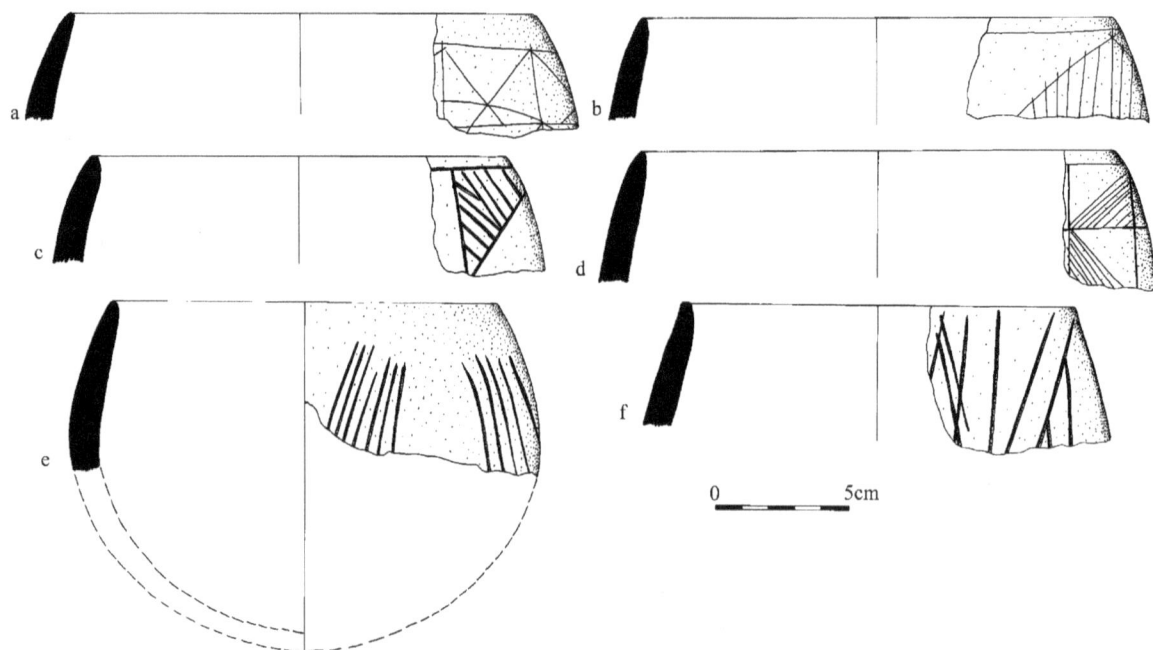

Figure 6.15 Mangaasi-style vessel form and decoration. All vessel form 3i with incised decoration (a. Ef-motif 63; b-c. Ef-motif 65; d. Ef-motif 24; e. Ef-motif 31; f. parallel incision.

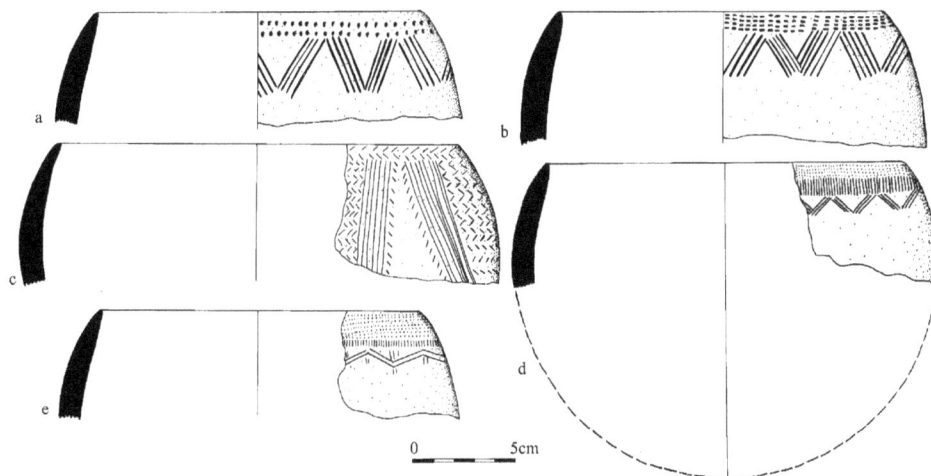

Figure 6.16 Mangaasi-style vessel form and decoration. All vessel form 3i with linear incision, punctation and gashes (a-b. Ef-motif 45; c. Ef-motif 50; d. Ef-motif 47; e. Ef-motif 46).

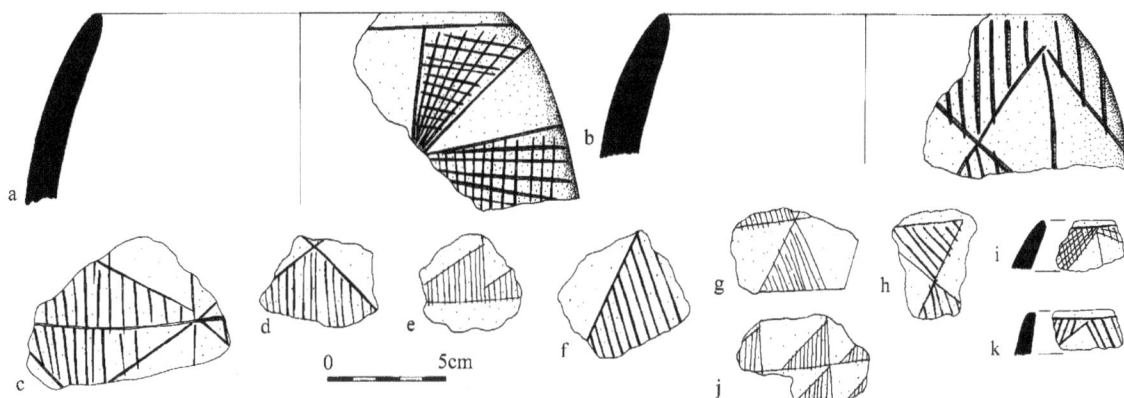

Figure 6.17   Mangaasi-style vessel form and decoration. a. vessel form 3i and Ef-motif 66; b. vessel form 3i and Ef-motif 64; c-d. Ef-motif 68; e-h. Ef-motif 65; i. vessel form 3i and Ef-motif 57; j. Ef-motif 65; k. vessel form 3i and Ef-motif 65.

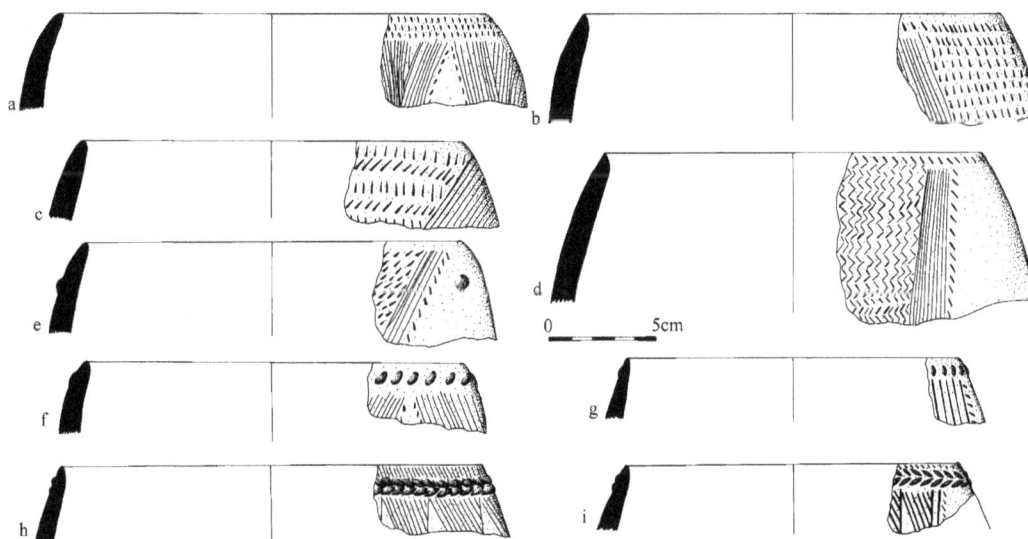

Figure 6.18   Mangaasi-style vessel form and decoration. All vessel form 3i decorated with incision, gashes and applied relief. A. Ef-motif 57; b. Ef-motif 52; c. Ef-motif 51; d. Ef-motif 50; e. Ef-motif 53; f. Ef-motif 59; g. Ef-motif 59; h-i. Ef-motif 58.

Table 6.20 Mangaasi TP 11 Summary of decorative technique and location

DECO TECHNIQUE	LAYER 1	LAYER 2I	LAYER 1I	TOTAL
Incision		-		
linear	1	12	7	20
geometric	-	8	6	14
curvilinear	-	-	-	-
gash	-	4	2	6
complex	-	-	-	-
Impression				
punctate	-	8 (1 flat lip)	3	11
Applied relief				
notched bands	-	5	1	6
plain bands	-	-	-	-
nubbins	-	-	-	-
Excision				
perforation	-	-	-	-
notching	1	1	-	2
Total	1	38	19	58
Location				
lip ext.	-	1	-	1
lip int.	-	-	-	-
flat area of lip	-	1	-	1
rim	1	5	-	6
body	1	21	15	37
lip	1	-	-	1

Table 6.21 Mangaasi TP 7 Summary of decorative technique and location

DECO TECHNIQUE	LAYER 1/2I/2III	LAYER 5I	LAYER 5II	LAYER 1I	TOTAL
Incision					
linear	-	24	1	-	25
geometric	-	21	5	-	26
curvilinear	-	2	-	-	2
gash	-	1	-	-	1
complex	-	1	-	-	1
Impression					
punctate	-	15	2	-	17
Applied relief					
notched bands	-	14	1	-	15
plain bands	-	-	-	-	-
nubbins	-	-	-	-	-
Excision					
perforation	-	-	-	-	-
notching	-	1	8	-	9
Total	-	79	17	-	
Location					
lip ext.	-	1	6	-	7
lip int.	-	-	-	-	-
flat area of lip	-	-	-	-	-
rim	-	5	1	-	6
body	-	52	6	-	58
lip	-	-	2	-	2

Handles may also have been associated with this phase of decoration and were certainly present soon after. Over time the motifs became increasingly complex and the various techniques and designs were found in combination (Figs 6.18, 6.19, 6.20). Discontinuous applied relief appeared to be associated with already established motifs that can be associated with earlier generic designs suggesting that this form of decoration may represent a later elaboration (Fig. 6.18f–i). Plain continuous applied bands (i.e. not notched) also appeared to be associated with this phase, although this form of decoration was not well represented in the recent excavations. The reconstructed vessel illustrated by Garanger (1972:Fig.139) which features plain continuous applied bands also incorporates a number of other decorative techniques (punctation and geometric incision) which along with the design motif provides further evidence of the transitional nature of the sequence, in this case from Early to Late Mangaasi.

Later Mangaasi-style decoration was again characterised by a multitude of decorative techniques and combinations. Examples of this decoration were retrieved principally from test pits closer to the sea (TP11, Table 6.20) or in the upper most layers of test pits where the deposits were somewhat dispersed (TP7, Table 6.21).

A distinctive decorative marker is continuous notched applied bands (Fig. 6.21g–n, 6.22a–k) which are found in association with a multitude of other decorative techniques including incision, comprising both linear and geometric motifs (infilled grids and fine cross-hatch). Parallel linear incision infilled with rows of punctation (Fig. 6.23a, d) and or incised ladder-like motifs are also frequently in evidence. This last phase of the sequence is best represented by ceramics recovered by Garanger from his excavations and surface collections. Neither discontinuous applied relief or handles appear to be associated with this phase of the ceramic sequence.

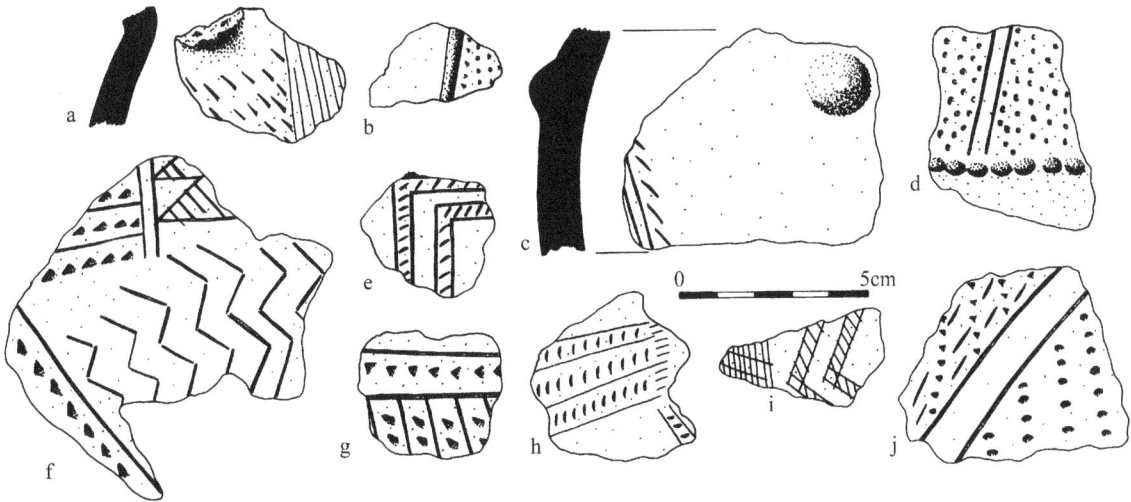

Figure 6.19   Mangaasi-style vessel form and decoration. a. vessel form 3i and Ef-motif 52; b. Ef-motif 62?; c. nubbin plus Ef-motif 53; d. Ef-motif 61; e. Ef-motif 85; f. Ef-motif 84; g-h. Ef-motif 83; i. ladder-like incision; j. linear incision, punctation and gashes.

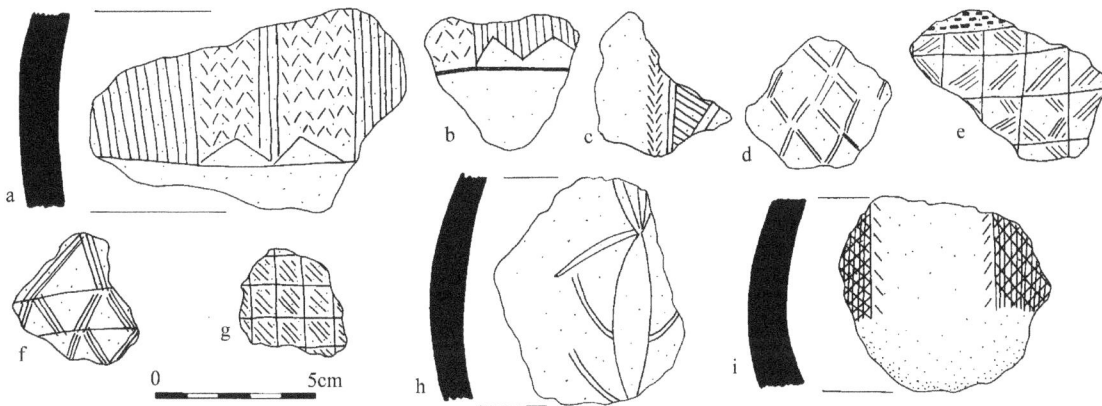

Figure 6.20   Mangaasi-style decoration. a. Ef-motif 76; b. Ef-motif 76; c. Ef-motif 40; d. Ef-motif 73; e. Ef-motif 72; f. Ef-motif 74; g. Ef-motif 71; h. Ef-motif 77; i. Ef-motif 75.

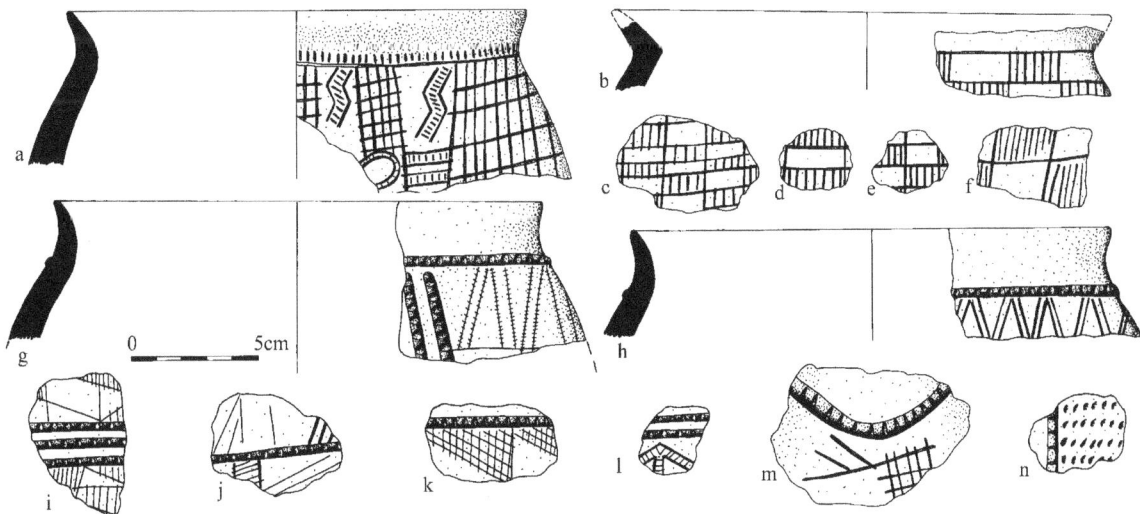

Figure 6.21 Late Mangaasi-style vessel form and decoration. All vessel form 2i with incised, punctate and applied decoration. a. Ef-motif 22; b. Ef-motif 69; c-f. Ef-motif 69; g. Ef-motif 93; h. Ef-motif 78; i. Ef-motif 82; j. notched applied band plus incision; k. Ef-motif 80; l. Ef-motif 79; m. Ef-motif 94; n. Ef-motif 81.

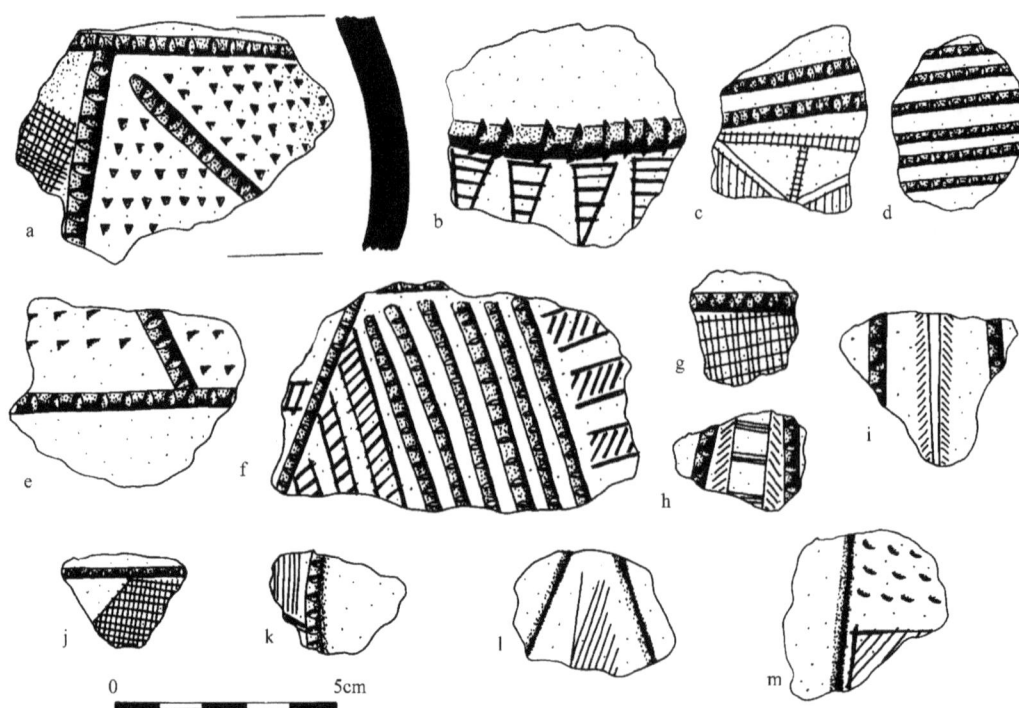

Figure 6.22 Late Mangaasi-style applied relief, incised and punctate decoration. a. Ef-motif 86; b. Ef-motif 87; c. Ef-motif 88; d. parallel rows of applied relief; e. Ef-motif 86; f. Ef-motif 89; g. Ef-motif 90; h. Ef-motif 91; i. Ef-motif 92; j. Ef-motif 90; k. applied relief and linear incision; l-m. Ef-motif 62?

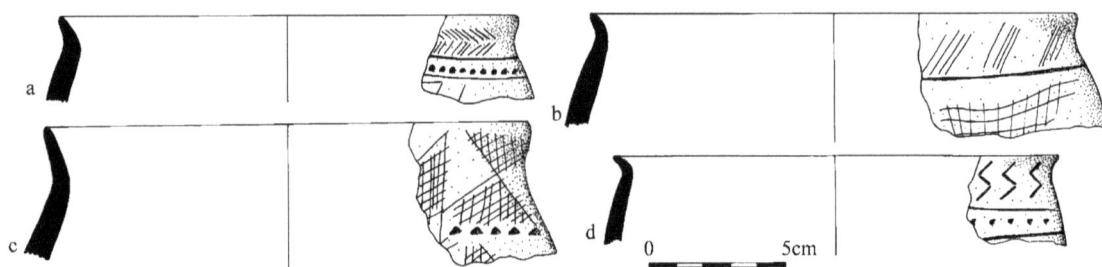

Figure 6.23 Late Mangaasi-style vessel form and decoration. All vessel form 2i. a. Ef-motif 95; b. Ef-motif 96; c. Ef-motif 97; d. Ef-motif 98.

Motifs

A total of 95 motifs (see Appendix 4 and Figs 6.2–6.23) were identified from the ceramic remains recovered from both the Mangaasi (1996–1999) and Arapus sites. The majority of the decorated sherds (468 [55%]) could not be assigned a motif due to their fragmentary nature and a number of the identified motifs may only be parts of much larger designs. Sherds that could not be assigned a particular motif were dominated by linear (278[59%]) and geometrically incised (132[28%]) sherds, a reflection of the dominance of these techniques in the composition of the motifs. Much lesser numbers of decorated sherds displaying incised gashes (16[3.5%]), punctation (11[2%]), notched applied bands (29[6%] could also not be assigned to motifs. Despite these limitations attempts to establish an initial inventory of motifs has proved useful. Most of the identified motifs were represented by only single or several examples and in a number of cases were clearly components of much larger designs. The excavations have greatly increased the repertoire and finer detail of motifs associated with Erueti-style pottery. Motifs associated with Mangaasi-style pottery were also able to be identified with some frequency but were not present in large numbers, particularly those associated with the Late Mangaasi-style. This is further highlighted when perusing Garanger's illustrations (Garanger 1972) where a much greater sample of the Late Mangaasi-style can be found, albeit in many cases too fragmentary to define complete motifs. Crucial to the identification of these often complex motifs which comprise a multitude of decorative techniques are collections of large sherds. In only a few cases were these present either from the recent excavations or Garanger's original investigations. Further research focusing on Mangaasi-style ceramics is required to establish a more complete inventory of motifs.

Erueti-style motifs are dominated by those produced by incision both linear and geometric (Figs 6.3–6.13). They are generally composed of a single decorative technique although there are a number of rare exceptions that are more complex (e.g. Ef-motifs 19 [Fig. 6.5b], 20 [Fig. 6.5c] and 21 [Fig. 6.6b]). Several of the Erueti-style motifs have clear generic connections to incised Lapita motifs (e.g. Ef-motifs 20 [Fig. 6.5c, d] [see Anson M 156 and 369); 27 [Fig. 6.6c] [Anson 175]; 31 [Fig. 6.6a] [Anson 158, 159, 297]) further confirming the 'Lapitoid' associations originally suggested by Golson (1971) and Garanger (1972:29). Those Erueti-style motifs which dominated included Ef-motif 1 (Fig. 6.11b) (11), 9 (Fig. 6.10b, c, g) (11), 18 (Fig. 6.12b) (10), 30 (Fig. 6.5a) (16) and 31 (Fig. 6.6a) (54). A number of motifs and/or elements of motifs can be identified throughout the ceramic sequence i.e. Erueti to Mangaasi. Some continued in evidence throughout the chronological sequence in a largely unmodified form such as Ef-motif 14 (Fig. 6.12a), 24 [Fig. 6.5d], 31 [Fig. 6.6a] and 38 [Fig. 6.12d, h]), or became combined into increasingly complex motifs (e.g. Ef-motifs 50–55 [Figs 6.18a–e] are clearly derived from a combination of Ef-motifs 13 [Fig. 6.12c], 14 [Fig. 6.12a], 31 [Fig. 6.6a] and 44 [Fig. 6.14a–c]). Other motifs can be shown to have completely transformed over time. Punctation, for example, was initially only found on the lip of Erueti-style vessels [Ef-lip motif 3], but later transferred to the vessel rim [Ef-motif 44] before it became only a component of increasingly complex motifs such as Ef-motif 62. The example of Ef-motif 62 illustrated by Garanger (1972:Fig. 139), decorated with plain continuous applied bands combined with other decoration, has already been mentioned as clearly displaying a transitional form of motif.

As noted, discontinuous applied relief initially appeared only with already established motifs (e.g. Ef-motif 53[Fig.6.18e] and 59[6.18f]), but over time applied relief in general became more elaborate and an increasingly dominant motif component in the form of notched applied bands (eg. Ef-motifs 86–89 [Figs 6.22a–f]). Other Late Mangaasi-style motifs often found in association with notched applied bands included Ef-motif 83 (Fig. 6.19f) and incised and punctate designs such as Ef-motifs 91–94 (Figs 6.23a–d). More frequent motifs associated with Mangaasi-style pottery included Ef-motif 44 (17), 50 (55), 57 (57), 69 (10) and 83 (26).

Lip modification was largely restricted to Erueti-style ceramics at Mangaasi. Notching (Ef-lip motif 1) as opposed to incision of the lip completely dominated this category of motif. The only other lip decorations identified at the site, although in relatively small numbers, were restricted to the wide flat lips associated with vessel form 2ii (Fig. 6.2). Most frequent of these was punctation (Ef-lip motif 3) on the lip (11), followed by rare examples of notching both on the interior and exterior of the lip (Fig. 6.2f) and three different incised motifs (Fig. 6.2 g–i).

Summary

- Petrographic analysis of a chronological range of sherds indicated an Efate origin for the mineral inclusions, and clay wasters were recovered across the site. The pottery appears to have been made at the site from local materials. The composition of the fabric of the sherds from the Mangaasi site remained largely consistent throughout the sequence.

- Erueti-style ceramics were associated with the initial human occupation of the Mangaasi site some 2800 years ago. They are characterised by a variety of vessel forms dominated by outcurving rim, plainware cooking vessels which were almost always notched on the lip. A small percentage of the vessels were decorated, principally with incised motifs, and may have represented the ceremonial component of the assemblage. Handles were present but rare. Change over time can be demonstrated within the Erueti cultural horizons. After 2500 BP globular incurving rim vessels became more predominant and decoration, still largely restricted to incision, became more frequent. Notching on the lip continued as a regular decorative feature.

- By 2200 BP vessel form was completely dominated by globular incurving rimmed pots which were almost always decorated. There was both an increasing variety of motifs and techniques of decoration in evidence. Handles were associated with this vessel form but were relatively rare. Both discontinuous applied relief and plain bands also appeared around this date. Decoration of the lip however, abruptly disappeared from the record. It is the appearance of this new suite of distinctive motifs in association with the overwhelming predominance of a single vessel form which warrants the labelling of this phase of the ceramic sequence as Mangaasi.

- Change in the Mangaasi-style ceramics over time was also in evidence although less clearly defined due to the smaller sample recovered and the disturbed nature of the deposits. Late Mangaasi can be best characterised, although not exclusively, by notched applied bands which are always found in association with a host of other decorative techniques. Handles did not seem to be present. Vessel form appeared to return to one with an outcurving rim. On present evidence ceramic production and use on Efate ceased sometime around 1200 BP.

- A change in the delineation of vessel function was also seen through the sequence. It is argued that cooking vessels and ceremonial vessels at least early on in the sequence can be seen to be represented by the plain and decorated vessels respectively, as has been argued in the case of Lapita assemblages. This changed over time as almost all vessels became decorated. This was due no doubt to a number of factors. It may have been that several generations after Lapita arrival we may be seeing the gradual proletarianisation of the population occurring (Yen quoted in Kirch 1988b:245), and at the same time an increasing localised identity may have been developing which might have then encouraged an increasing proliferation of decoration.

- A developmental transformation from Erueti to Mangaasi-style ceramics (much modified from Garanger's original sequence) can be identified both in terms of vessel form, fabric and decoration.

7

Malekula ceramics

Singsing blong Lapita

Long yia 1996 Septemba 26,

Taem ia we mifala i stap work long Malua Bay

Boss blong mifala Steward Berward i kam long Kantri Ostralia,

Em nao i boss blong mi long wok blong mi (ripet)

Taem ia mifala i kantinu wok long eria ia,

Mekem se mifala i faenem wan pis Lapita,

Jimmyson William i faenem wan pis Lapita ia,

Em nao i filwoker long Not Wes eria (ripet)

Ale behin mifala i glad blong composem song ia,

Emi wan histri blong Malekula,

Blong talem long yu spirit blong Bubu i kam bak.

*(song composed by the SPR Band on 5/10/96 at Albalak village to
celebrate the recovery of a dentate-stamped sherd from Malua Bay)*

Introduction

As outlined earlier a total of fifteen cave sites (eleven of which are discussed here) and four open sites were excavated on the northwest coast of Malekula (Fig. 3.16) in line with a research strategy designed to answer a number of specific questions. All site stratigraphy and associated dates are outlined in detail in Chapter 3.

Although results gleaned from the excavations and the later analyses went some way in addressing the archaeological *terra incognita* that was Malekula, substantial gaps in the archaeological sequence do remain and in many respects this research can be regarded as only very preliminary. Now more fully understood are the two ends of the cultural sequence on Malekula. One is specifically related to the timing and associated cultural remains of the initial settlement of Malekula and the other with the last 500 years prior to European contact. It was not possible to define the central portion of the ceramic sequence due to two inter-related factors: the sparse remains that were recovered from most of the caves and the likelihood that this area of Malekula experienced a period of abandonment or less intensive settlement after initial colonisation. Many of the excavated cave sites returned only a handful of plain sherds from which only limited information could be extracted (see Table 7.1). Despite this fact certain patterns could be identified, particularly when related to the more abundant materials excavated from the open sites of Malua Bay, Chachara, Nuas and the cave site of Navaprah. The central focus of this current research on Malekula then is the characterisation of the early and late ceramic remains for which a robust sample was recovered within stratified contexts. The methodology used for the characterisation of the ceramic remains has been outlined in Chapter 4. Of the total number of excavated sherds (1710) 90% were plain body sherds which somewhat limited the analysis, although several vessels could be identified through their distinctive form or fabric. Very distinctive body sherds included calcareous-tempered sherds which could be attributed to one of the three known early vessel forms (2i, 2iii and 4ii) (Figs 7.1a, 7.2, 7.3), ribbed sherds (where the coils of clay used in the manufacture of

Table 7.1 Malekula ceramic sample (excavated sites only). Layers are site specific

SITES	PLAIN SHERDS											RIM	DECO	FABRIC		WEIGHT	TOTAL	WIDTH IN MM (PLAIN SHERDS)						
	Sur	L1	L2	L3	L4	L5	L6	L7	L8	rib	total			F1	F2	gm		1-4	4-8	8-12	12-16	16-20	20-24
Malua Bay Mk-3-55	–	128	76	332	133	–	–	–	–	2	669	45	20	459	260	4834.4	719	37	396	197	36	3	
Chachara Mk-3-43	–	497	–	–	–	–	–	–	–	216	497	33	56	–	629	20439.04	629			399	124	7	
Nuas Mk-3-62	–	85	–	–	–	–	–	–	–	36	85	3	9	–	96	1388.5	96		2	40	38	2	
Navaprah Mk-3-47	–	–	18	9	66	–	–	–	–	15	86	7	5	67	26	1027.8	93	5	45	21	14	1	
Woplamplam Mk-3-26	–	6	5	2	–	–	–	–	–	2	13	–	–	4	9	204.1	13			8	5		
Bartnator Mk-3-32	–	1	–	–	–	1	–	–	–	–	2	–	–	–	–	25	2				1		
Waal Mk-3-39	–	2	2	–	–	–	–	–	–	–	4	2	1	5	2	172.3	7		1	2	1		
Malua One Mk-3-40	–	–	–	–	–	–	4	3	3	–	10	–	–	4	6	380.9	10			3			
Woapraf Mk-3-41	–	–	–	–	1	1	2	5	–	–	9	3	5	10	2	75	12	1	4	3	2		
Yalo Sth Mk-3-48	–	–	2	7	8	–	–	–	–	–	17	–	1	8	10	470.6	18		7	3	6		
Wambraf Mk-3-56	–	13	2	3	–	–	–	–	–	4	18	1	1	–	20	603.5	20			2	14	2	
Navapule A Mk-3-58	–	8	3	–	–	–	–	–	–	3	11	–	–	–	11	181.5	11			9	2		
Navepule B Mk-3-59	21	–	–	–	–	–	–	–	–	3	21	3	5	–	28	901.6				5	14	2	
Navepule C Mk-3-60	–	2	3	–	–	–	–	–	–	–	5	1	–	–	6	108.7	6		1	2	1	1	
Fiowl Mk-3-61	–	32	11	17	6	–	–	–	–	1	66	3	–	2	67	765.3	69		13	21	19	13	1
Totals											1513						1708						

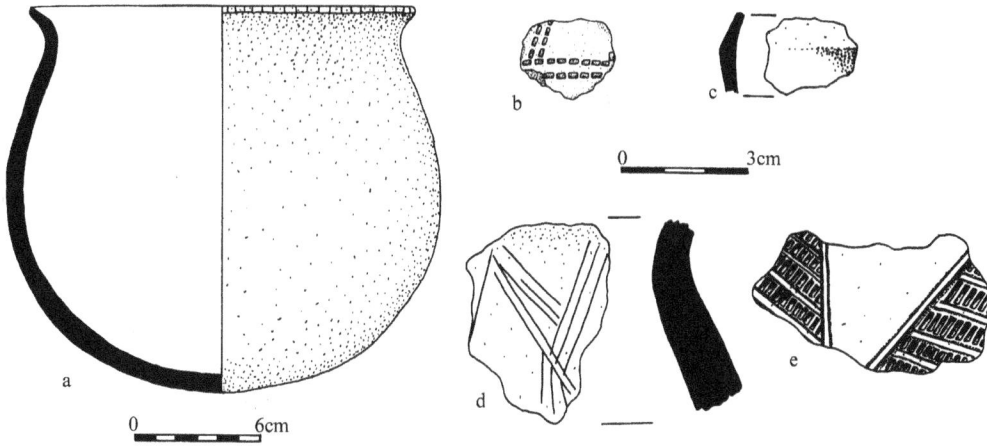

Figure 7.1 Malekula vessel form and decoration. Malua Bay School ceramics. a. vessel form 2i; b. dentate-stamped Lapita sherd (TP9, L.3); c. carinated sherd (TP9, L.3); d. incised sherd; e. incised and excised decoration (M-motif 45).

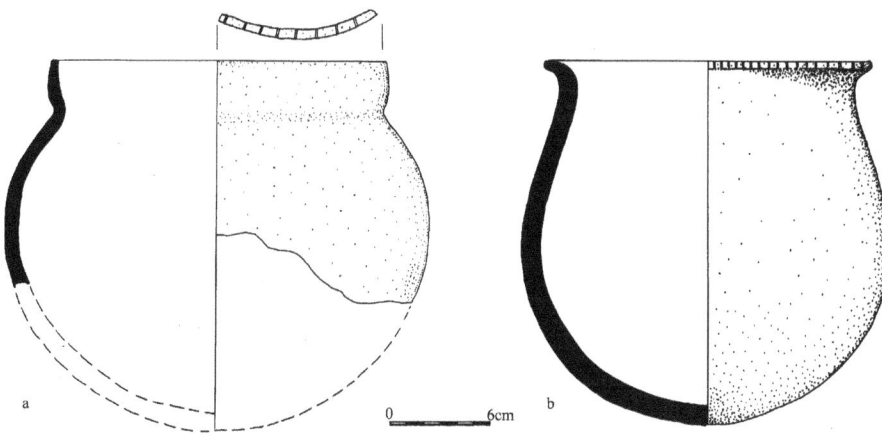

Figure 7.2 Malekula vessel form. a. vessel form 4ii with incised lip; b. vessel form 2i.

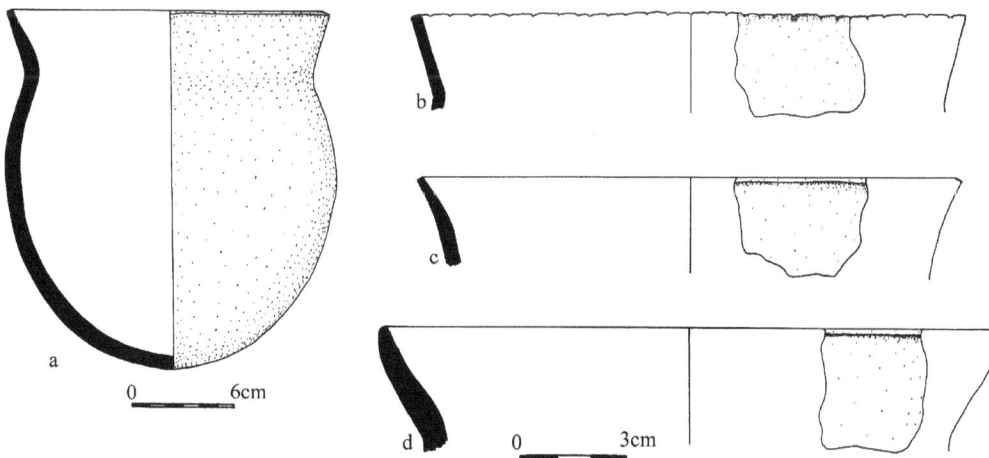

Figure 7.3 Malekula vessel form. a. vessel form 2iii; vessel form 2iii with notched rim; c-d. plain everted rims.

the pots have not been smoothed) of vessel form 1i (Fig. 7.4a), thick smoothed (interior and exterior) sherds with a very wide curvature which related to vessel form 2iv (Fig. 7.5a). Ninety percent of the recovered sherds came from the four sites of Malua Bay, Chachara, Nuas and Navaprah.

Surface collected sherds and those held in museum collections hinted at a much richer and diverse range of vessel form and decoration but very few of these were recovered from excavated contexts. Although a substantial section of the ceramic sequence remains elusive the above-mentioned sherds are illustrated to provide a tantalising glimpse of the diversity in form and decoration which as yet cannot be placed within a chronological context (Figs 7.6–7.13). The presentation of this surface collected material is not included within the analytical structure that was utilised for the excavated materials but an attempt was made to establish any variation in vessel form or decoration. None of the surface collected sherds were calcareously-tempered. Also briefly discussed are those ceramic vessels known as *Naamboi*, thought to be largely restricted to the south of Malekula, that have been used in various ceremonial capacities from at least the early ethnographic period up until today. The geographical distribution of vessel forms and motifs from both the excavated and surface collected sherds is included later in this chapter along with a discussion of associated implications.

Fabric

Not surprisingly over a 3000 year period some variation in fabric composition is to be expected especially on an island the size of Malekula. Definition of fabrics was established through a combined macroscopic and petrographic analysis of the sherds. Petrographic analysis was undertaken by William Dickinson. Dickinson had previously studied 20 sherds from various locations around Malekula that had been collected by Richard and Elizabeth Shutler. The sherds selected by the Shutlers were identified as being indigenous to Malekula but the mineral inclusions demonstrated great variability which precluded the identification of specific areas of ceramic manufacture. A further 12 sherds were selected from the recently excavated and surface collected materials and sent to Dickinson (see Appendix 3 for full reports WRD-155 and 180). These sherds included two from the earliest excavated deposits on the northwest coast along with a further ten being selected on the basis of them being related to a particular pot form or decoration.

Dickinson's reaction (pers. comm.) in describing them as a 'morass' (petrographically speaking), further confirmed both the widespread nature of ceramic manufacture on Malekula and that similar pot forms and associated decoration were being made throughout the island, rather than being restricted to a limited number of manufacturing centres. All the sherds, bar one (see below), were indigenous to Malekula. The most distinctive difference that could be pinpointed was that the early sherds were calcareously-tempered and the later sherds were not. For the purposes of the analyses below the fabrics are divided into these two broad categories.

Fabric 1) Two sherds were selected from the earliest levels of the Malua Bay and Navaprah sites. Both contained calcareous temper along with volcanic sand rich in felsitic volcanic rock fragments characteristic of other Malekula mineral inclusions which are moderately to well sorted, fine-to-medium-grained sands composed largely of subrounded grains, probably of beach derivation. This fabric characterises the thus far earliest excavated ceramics from Malekula and is associated only with the first few hundred years of settlement.

Fabric 2) This all-encompassing fabric is characterised by having an eclectic group of non-calcareous mineral inclusions that can be said to be indigenous to Malekula. The volcanic rock fragments are again dominated by felsitic grains which are typically moderately sorted, subangular to subrounded volcanic sands, which probably represent stream alluvium derived from the island's interior. Attempting to pinpoint specific areas of manufacture was seen as an unnecessary (particularly as the petrographic results pointed towards the pottery being village specific) and potentially inconclusive exercise (Dickinson pers. comm.), beyond the scope of this research.

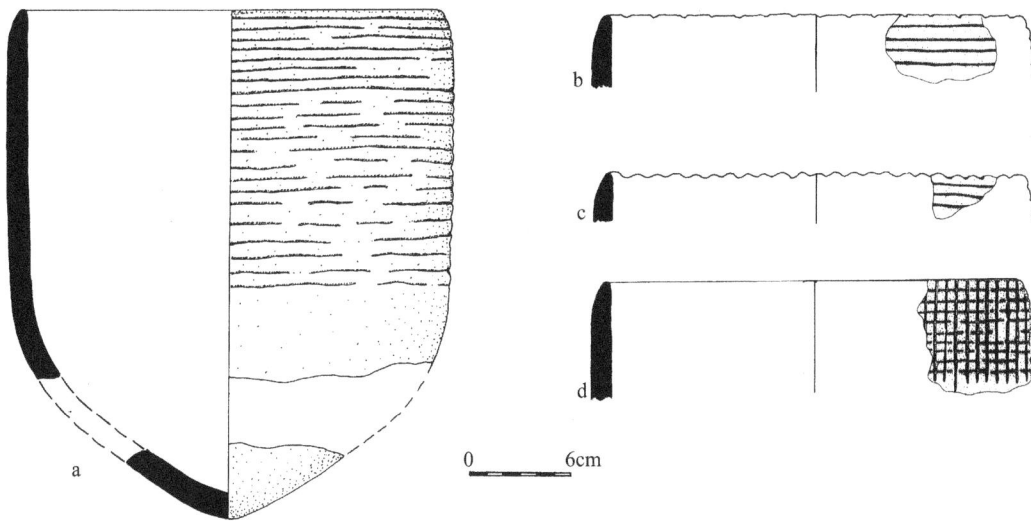

Figure 7.4   Malekula vessel form and decoration. Vessel form 1i. a. vessel form 1i with ribbed exterior (M-motif 1); b. vessel form 1i with incised rim; c. vessel form 1i with notched rim; d. vessel form 1i with vertical incision (M-motif 3).

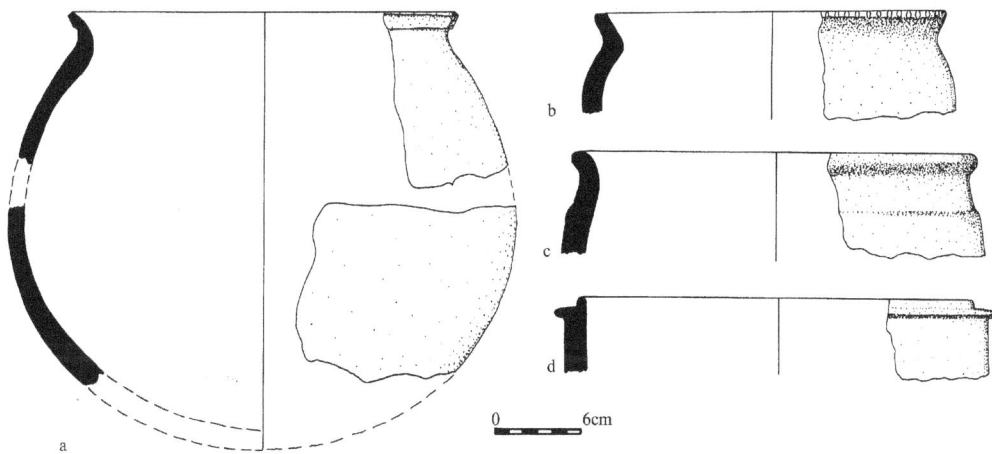

Figure 7.5 Malekula vessel form. Miscellaneous excavated and surface collected sherds. a. vessel form 2iv; b–d. miscellaneous surface collected vessel forms.

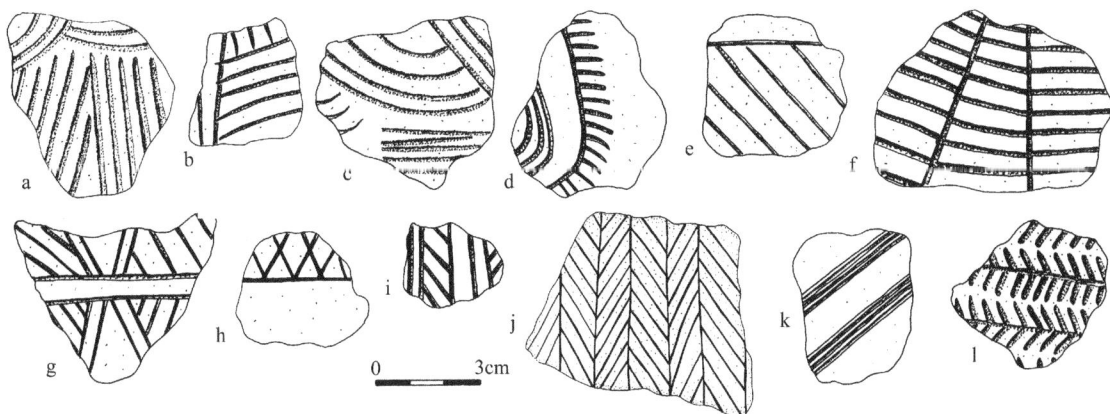

Figure 7.6 Malekula decorated sherds. a–c. M-motif 20; d. M-motif 21; e. M-motif 22; f. M-motif 23; g. M-motif 24; h. M-motif 25; i–j. M-motif 26; k. parallel linear incision, M-motif 27; l. M-motif 28.

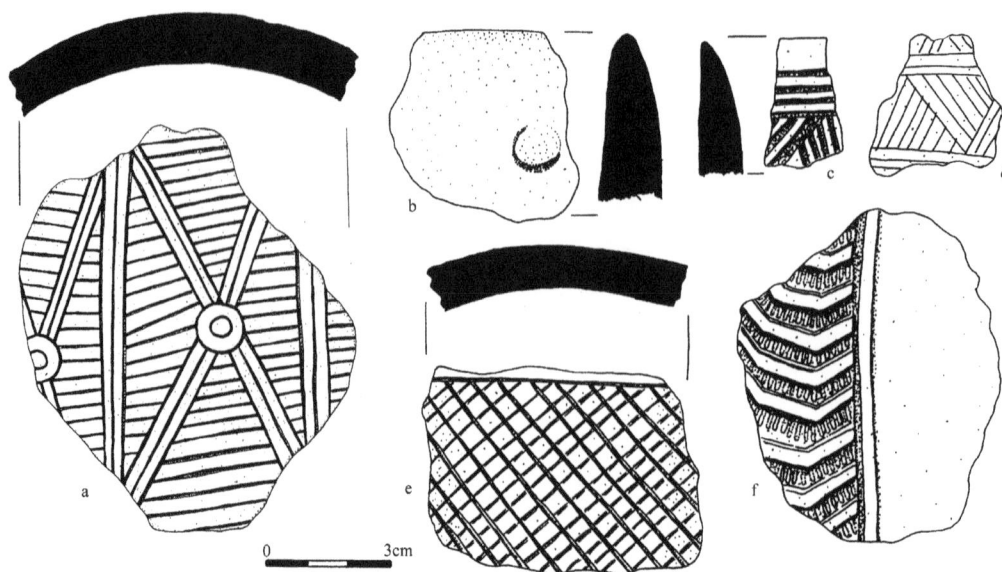

Figure 7.7 Malekula vessel form and decoration. All vessel form 1iii. a. M-motif 29; b. fingernail gouge; c-d. M-motif 4; e. M-motif 25; f. M-motif 30.

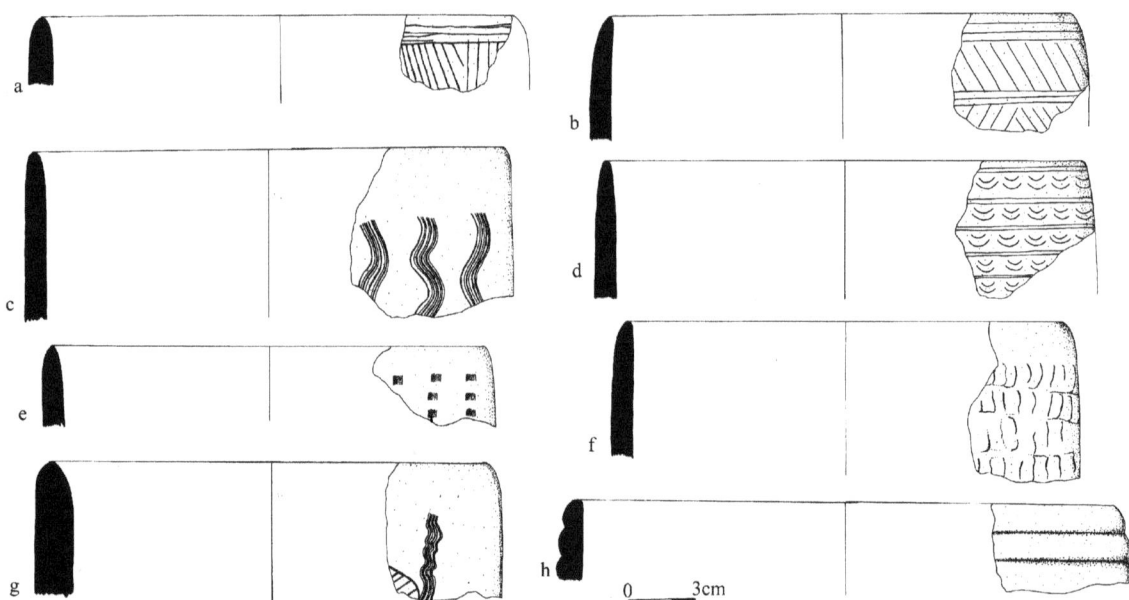

Figure 7.8 Malekula vessel form and decoration. a–b. vessel form 1iii (M-motif 4); c. vessel form 1iii (M-motif 31); d. vessel form 1iii (M-motif 32); e. vessel form 1iii (M-motif 33); f. vessel form 1ii (M-motif 2); g. vessel form 1iii (M-motif 34); h. vessel form 1i (M-motif 35).

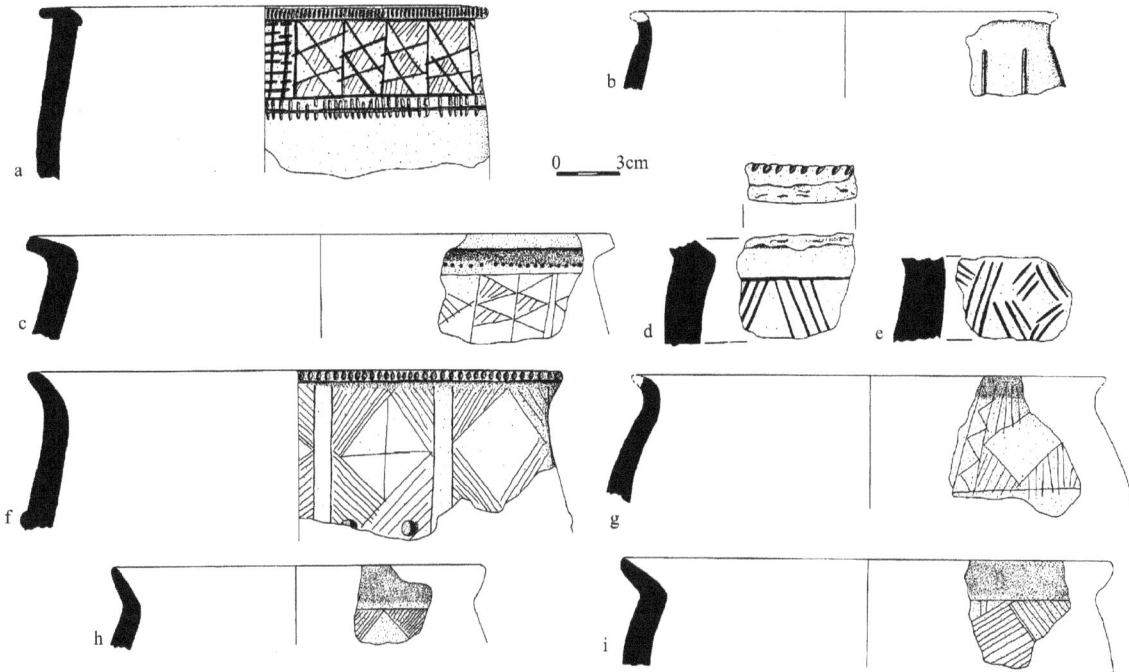

Figure 7.9 Malekula vessel form and decoration. Surface collected sherds. a. M-motif 36; b. M-motif 37; c. M-motif 38; d. M-motif 39; e. M-motif 40; f. M-motif 41; g. M-motif 42; h. M-motif 43; i. M-motif 44.

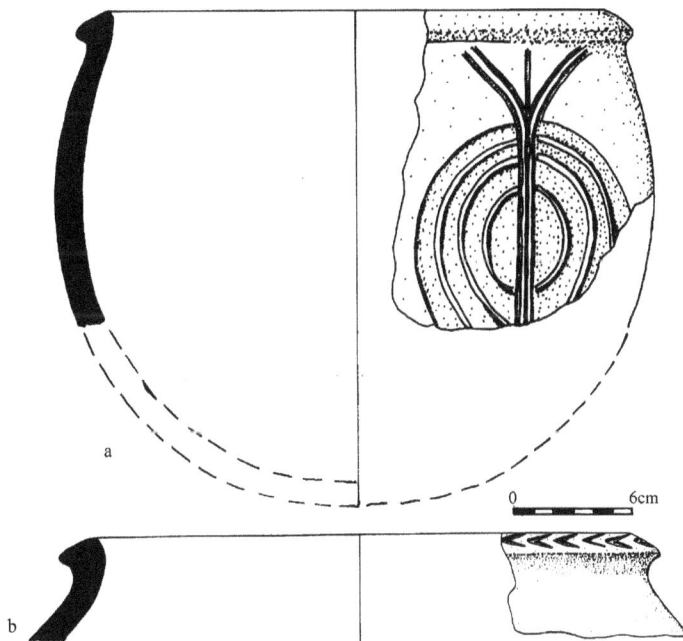

Figure 7.10 Malekula vessel form and decoration. Surface collected sherds. a. M-motif 47; b. excised lip.

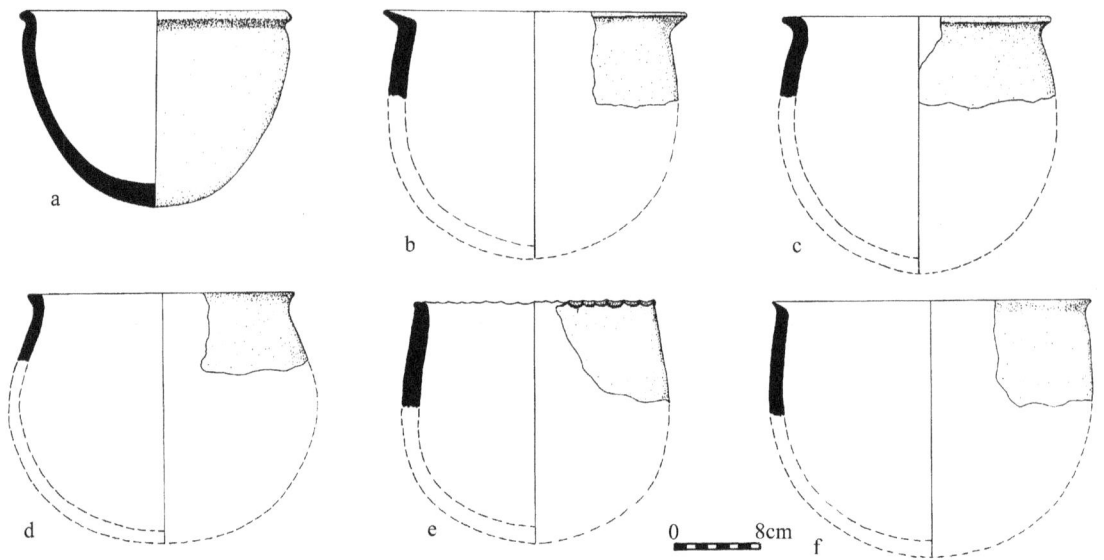

Figure 7.11 Malekula vessel form. Miscellaneous surface collected sherds. a. unrestricted vessel with groove below rim; b–d. globular vessels with restricted necks; e. unrestricted vessel with crenellated lip; f. unrestricted vessel with direct rim.

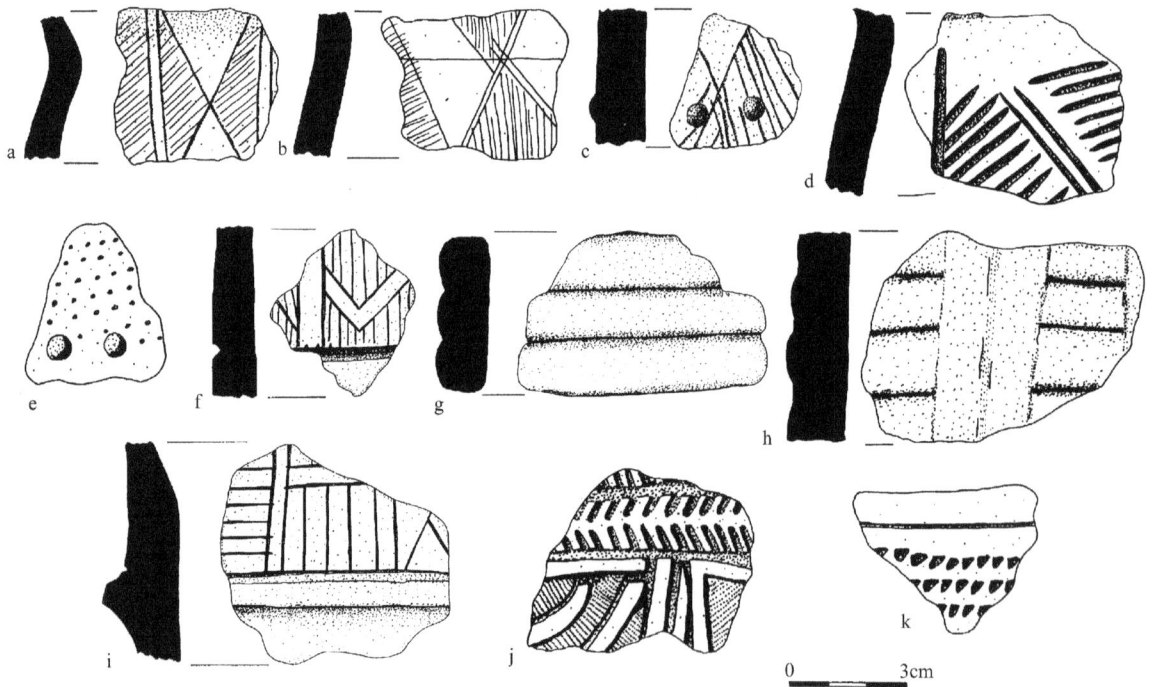

Figure 7.12 Malekula surface collected decorated sherds. a-b. M-motif 43; c. M-motif 41; d. M-motif 29; e. M-motif 48; f. M-motif 49; g-h. M-motif 35; i. M-motif 50; j. M-motif 51; k. M-motif 52.

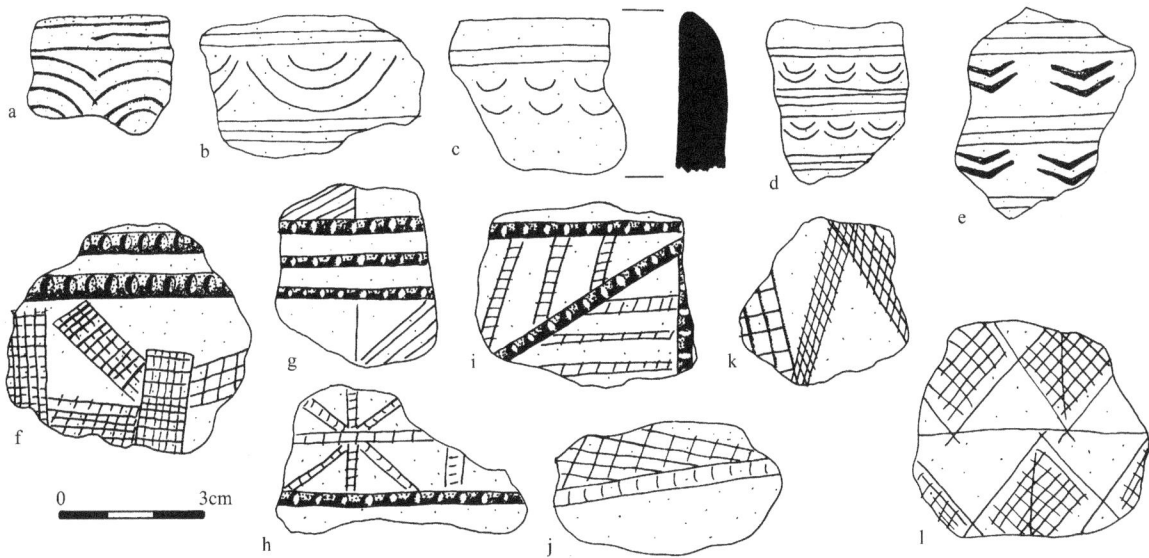

Figure 7.13 Malekula surface collected decorated sherds. a. M-motif 5; b-d. M-motif 32; e. M-motif 53; f. M-motif 54; g. M-motif 55; h. M-motif 56; i. M-motif 57; j. M-motif 58; k. M-motif 59; l. M-motif 60.

Malua Bay School Mk-3-55

A total of 719 sherds were recovered from the 11 test pits excavated at the Malua Bay School site, 669 (93%) of which were plain sherds, 45 (6%) were rims, 4 decorated body sherds and one was a carinated sherd (Tables 7.2 and 7.3; Fig. 7.1c). Several ribbed sherds (i.e. unmodified surface of coil constructed vessels) were also noted. The Malua Bay School ceramic assemblage related to two distinct occupations. Those recovered from the lower layers were associated with a short-term colonising Lapita settlement. The sherds were excavated from within a matrix of concentrated cultural debris which had been buried beneath a tephra (see Chapter 3). Sherds recovered from the upper layers were more dispersed and had been subjected to vigorous post-depositional processes.

The 669 plain sherds can be divided into two fabric types (Table 7.4). Sherds from the lower layers comprise exclusively of Fabric 1 while those from the upper layers are predominately Fabric 2. It seems likely that any Fabric 1 sherds recovered from the upper layers (1 and 2) are the result of displacement. The change in fabric composition over time is accompanied by increasing thickness (Table 7.4). The table below (7.3) shows the distribution of plain sherds recovered from the various test pits. The presence or absence of Fabric 1 sherds facilitates the delineation of the early settlement site which can be seen in the lower layers of TPs 3, 4, 7, 8, 9 and 10 (Fig. 3.30).

Table 7.2 Malua Bay School ceramic sample

	plain	ribbed	carinated	rim	deco body	deco
Layer 1	128	2		12	1	2
Layer 2	76			2	2	2
Layer 3	332		1	23	1	9
Layer 4	133			8		7
Total	669	2	1	45	4	20

Table 7.3 Malua Bay School. Distribution of plain sherds

	Layer 1	Layer 2	Layer 3	Layer 4	Total	weight (gm)
TP 1	4	3	–	–	7	127.3
TP 2	5	8	–	–	13	166.1
TP 3	3	–	22	–	25	161.9
TP 4	9	22	110	8	149	708
TP 5	9	–	–	–	9	93.9
TP 6	9	–	–	–	9	71.3
TP 7	11	–	41	10	62	381.5
TP 8	29	14	38	20	101	832.2
TP 9	45	29	114	81	269	1848.4
TP 10	4	–	7	14	25	87.2
TP 11	–	–	–	–	–	–
Total	128	76	332	133	669	4477.8

Table 7.4 Malua Bay School. Plain sherds, fabric and thickness

	Layer 1	Layer 2	Layer 3	Layer 4	Total
Fabric 1	5	32	260	127	424
Fabric 2	123	44	72	6	245
Total	128	76	332	133	669
1–4mm	–	–	28	9	37
4–8mm	41	22	235	98	396
8–12mm	68	40	65	24	197
12–16mm	19	11	4	2	36
16–20mm	–	3	–	–	3

Rim, lip and vessel form

Three vessel forms could be positively identified from the excavated sherds. Two of those forms, 2i and 2iii, were related to the earliest occupation of the area and were composed of Fabric 1. Small (Fig. 7.1a) and large (Fig. 7.2b) variants of form 2i were present in the assemblage. All rims associated with these vessel forms are outcurving. The only variation that was noted was in lip profile where everted rims tended to have flat lips while outcurving rims tended to be plain. Incision on the lip was common. In the upper layers two ribbed sherds indicated the presence of vessel form 1i. More tentative is the suggested presence of vessel form 3iii, indicated only by a distinctively decorated body sherd (Fig. 7.1e).

Other rim sherds, all Fabric 2 and all from the uppermost layers, were too fragmentary to be assigned to particular vessel forms although all appeared to be outcurving and seem likely to be associated with a group of large plain globular vessels (Fig. 7.11) which have been reconstructed from large surface collected sherds.

Decoration

Any decoration was restricted largely to incision on the lip (16 examples). A single dentate-stamped Lapita sherd (Fig. 7.1b) was also recovered but was too fragmentary to be assigned a motif number. The only other decorated sherds, one incised (Fig. 7.1d) and one incised and impressed sherd (Fig. 7.1e) were for similar reasons unable to be assigned motif numbers although the latter sherd may be part of M-motif 45.

Chachara Mk-3-43

This site is associated with a relatively late short-term occupation located in the high inland area of the Northwest (Fig. 3.16). Cultural remains can be seen scattered on the ground surface amongst and between two ceremonial structures. The ceramic remains recovered from a series of test pits were sparse and indicated that the cultural deposits were shallow and ephemeral. The largest corpus of excavated ceramics were recovered from a mound feature adjacent to one of the stone platforms and is presented in detail here. The sherds appeared to be restricted to two vessel forms with limited variation in decoration. The limited number of sherds (28) recovered from the other test pits showed no variation in form or decoration. Sherds were also surface collected from the site and they further confirmed the homogeneous nature of the ceramics. Several of the surface collected decorated sherds are illustrated and added to the corpus of motifs. Due to the fact that there did not appear to be any stratigraphic variation across the site, the excavated assemblage was grouped as a single unit.

All of the pottery comprised Fabric 2, and was restricted to two vessel forms, namely 1i and 1iii. A total of 629 sherds were excavated from the mound feature and the test pit. Of those, 601 (95%) were recovered from the mound feature. The sherds were mainly plain body sherds (497) either ribbed (216) or smoothed (281) with a smaller component of rim sherds (33, 32 of which were ribbed) and decorated sherds (56).

Vessel form 1i with its distinctive ribbed exterior (Fig. 7.4a) dominated the assemblage. The upper half of the body has a ribbed appearance where, as noted, the coils of clay used in the manufacture of the pots have not been smoothed while the lower half and base of the pots tended to be smoothed. Decoration of these pot forms is limited to very occasional examples of notching on the lip (Fig. 7.4b,c) and more occasional examples of parallel vertical incision (Fig. 7.4d). The rims are of course direct and the lips are generally plain. Rim diameter ranges from 24–26cm.

In the case of vessel form 1iii the whole body has been smoothed to facilitate decoration. The vessels tend to be smaller than 1i with a rim diameter of between 20–22cm. Decorative techniques are dominated by incision, both linear and curvilinear and occasionally in association with punctation (see Figs 7.14–7.17). Punctation occurs in a limited number of cases as a single decorative technique (Fig. 7.15) as does fingernail pinching. Of those decorated sherds that could be assigned to a vessel form all appeared to be associated with vessel form 1iii, apart from the single example of vertical parallel incision (Fig. 7.4d) found on vessel form 1i. A number of distinctive decorative motifs tended to dominate the assemblage, particularly M-motif 4, 16 and 18. It is suggested here and further elaborated on below that these two quite distinct vessels (1i and 1iii) served two distinct and separate functions, one (1i) domestic and the other (1iii) ceremonial.

Nuas Mk-3-62

The site of Nuas was also associated with late period ceramics. It appeared to represent a short-term occupation centred near a ceremonial structure located near the coast. The ceramics are very similar to those recovered from Chachara in terms of vessel form and decoration. A total of 96 sherds were either surface collected (9) or excavated (87) at Nuas (Table 7.1). All comprised Fabric 2.

The sherd assemblage was dominated by plain body sherds (85 [88%]), almost fifty percent (36) of which were ribbed on the exterior indicating a predominance of both coil production and vessel form 1i. One rim of this vessel form was notched. Vessel form 1iii was also in evidence as indicated by a number of the larger decorated sherds. Decoration was largely restricted to either incision (7) (Fig. 7.6a–c), both linear and curvilinear, or punctation (2) (Fig. 7.8e). The decorated

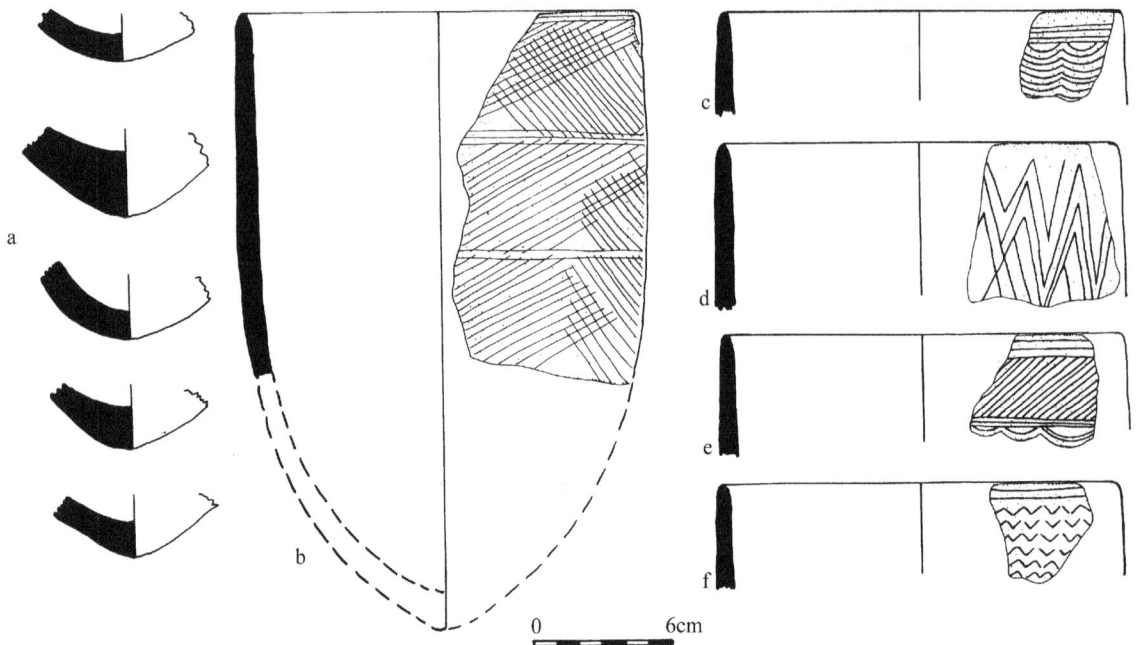

Figure 7.14 Malekula vessel form and decoration. Vessel form 1iii. a. assorted basal sherds; b. M-motif 4; c. M-motif 5; d. M-motif 6; e. M-motif 7; f. M-motif 8.

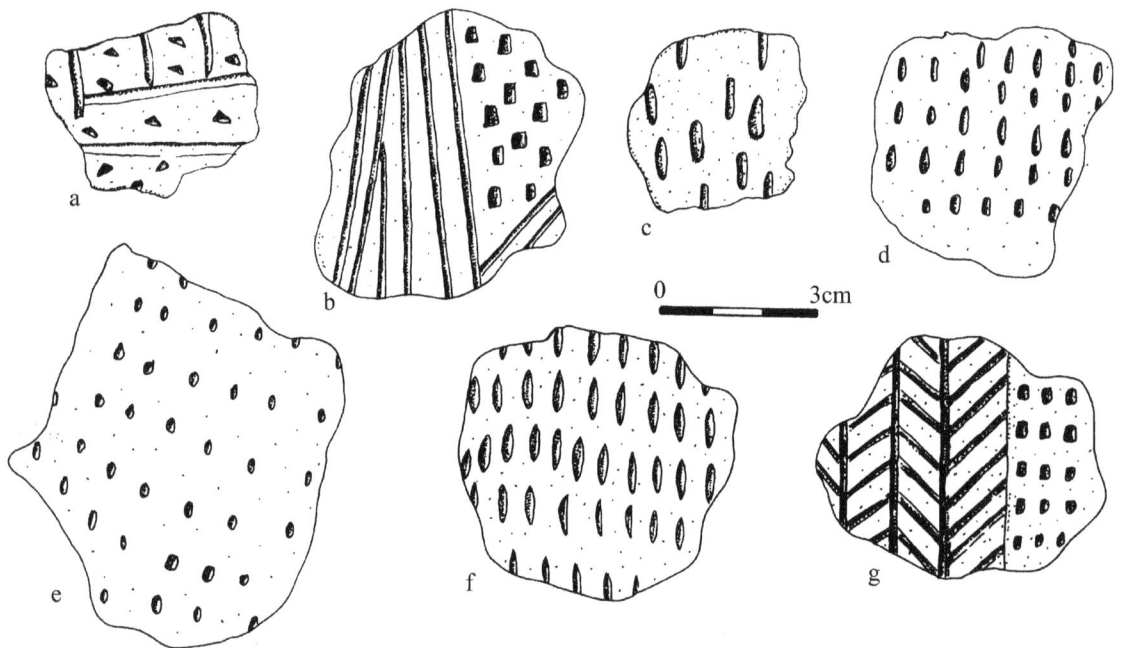

Figure 7.15 Malekula decorated sherds. a. M-motif 9; b. M-motif 10; c-f. punctate; g. M-motif 11.

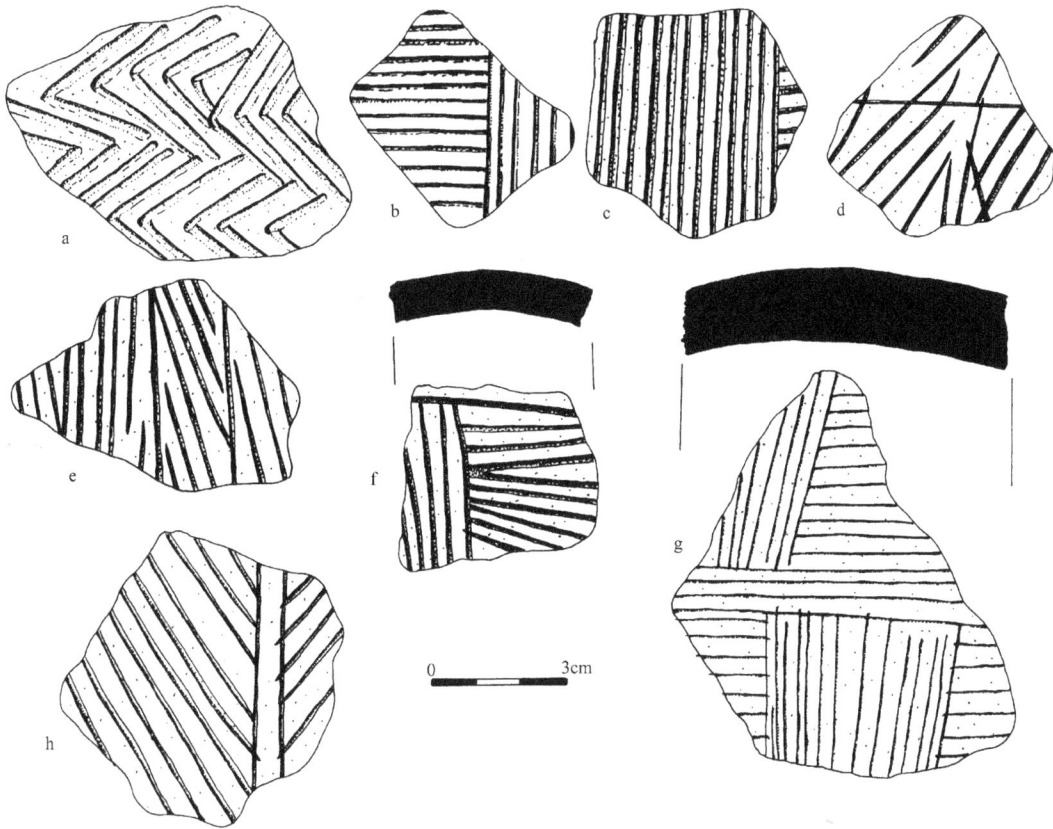

Figure 7.16 Malekula vessel form and decoration. All vessel form 1iii. a. M-motif 12; b–c. M-motif 13; d–e. linear incision; f–g. M-motif 14; h. M-motif 15.

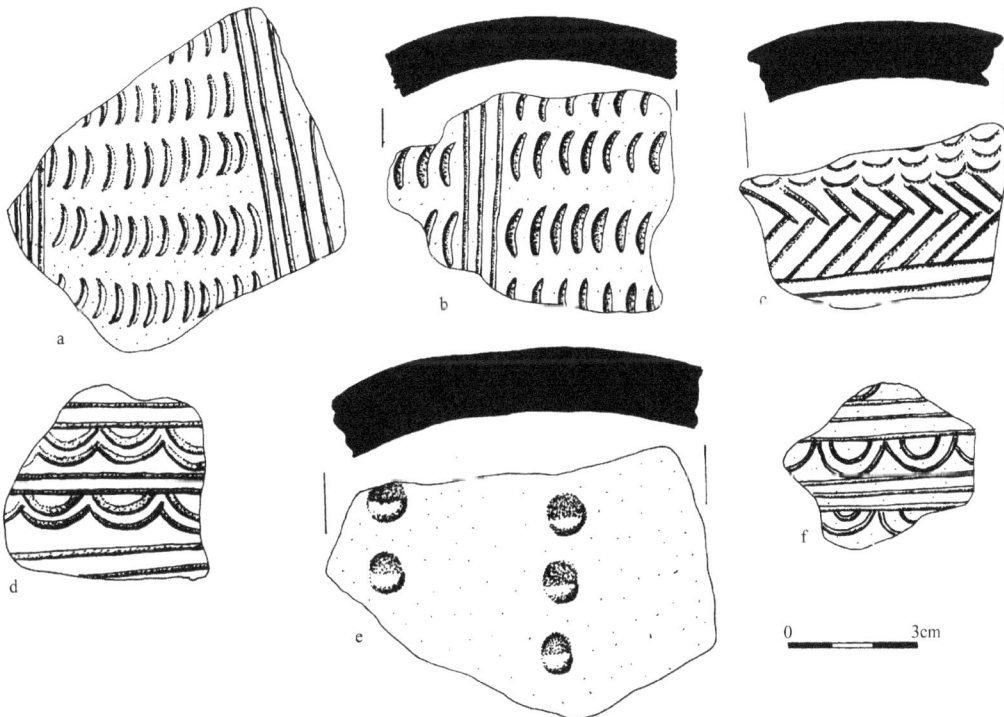

Figure 7.17 Malekula vessel form and decoration. All vessel form 1iii. a–b. M-motif 16; c. M-motif 17; d-f. M-motif 18; e. M-motif 19.

sherds were generally too fragmentary to be assigned separate motif numbers. Comparison with the decorated sherds from Chachara indicate a number of shared motifs. Several of the motifs recorded at Nuas have also been recorded from outside the Northwest area.

Navaprah Mk-3-47

The recovered ceramics from Navaprah were dominated by calcareous-tempered sherds (Fabric 1) recovered from Layer 4 which were associated with the initial use of the cave dating to c. 2700–2500 BP. The number of sherds recovered from other layers was very few, particularly those with diagnostic detail. Rims and decorated sherds were only recovered from Layer 4. This has limited both the prospects for positive identifications of vessel form and a chronological comparison between layers. Despite these factors the sherds from Navaprah have confirmed some crucial details of the ceramic sequence from the Northwest where several recurrent trends, discussed below, can be identified.

Table 7.5 Navaprah ceramic sample

| | plain sherds | | | | | rib | rim | F1 | F2 | wght | Total |
	1-4mm	4-8	8-12	12-16	16-20						
L.1	-	-	-	-	-	-	-	-	-	-	-
L.2	-	1	8	8	1	11	-	-	18	425.1	18
L.3	-	2	3	4	-	4	-	1	8	211.5	9
L.4	5	42	10	2	-	-	7	59	-	391.2	66
Total	5	45	21	14	1	15	7	67	26	1027.8	93

A total of 93 sherds were recovered from excavations at Navaprah of which 86 (92%) were plain body sherds and 7 (8%) were rim sherds. As can be seen in Table 7.5 the majority of the sherds were recovered from Layer 4 and all comprised Fabric 1. The larger rim sherds enabled two vessel forms to be identified. They were globular pots with restricted orifices and outcurving rims (2i and 2iii). Some variation in rim form and sherd thickness enabled the further delineation of vessel form which led to the definition of three distinct vessels. Two types of vessel form 2i were present, one a finer, smaller (rim d=16cm) pot (Fig. 7.1a) and the other a more robust and larger (rim d=20cm) version (Fig. 7.2b). Both featured notching on the lip. The other vessel form represented at Navaprah (2iii) was indicated by two rim/body sherds neither of which showed any notching on the lip. No other decoration was recorded from any of the sherds at Navaprah.

Although only plain body sherds (all Fabric 2) were recovered from the other layers of the cave distinctive manufacturing techniques and pot form enabled further vessel forms to be identified. Ribbed sherds appear in Layer 3 (4) and increased in number in Layer 2 (11). These very distinctive sherds are associated with pot form 1i. The other distinctive body sherds (4) that were identified are those from the very large thick walled globular pot (2iv). As noted sherds of this vessel form are smoothed both on their interior and exterior surfaces. This feature along with the distinctive wide body curvature which can be identified on larger body sherds facilitated the identification of this vessel form. The provenance of the Layer 3 sherds are in some doubt as they were retrieved from the large oven feature and may include material mixed from the layers above and below.

The ceramic sequence at Navaprah, along with a series of radiocarbon dates indicates that the cave has had early and late periods of intensive short-term usage, punctuated by a long periods of much less intensive use or even abandonment. It is a sequence that recurs at a number of sites in northwest Malekula. The earliest ceramics are calcareously-tempered and are predominately plainware vessels. The later ceramics, dating to the last 1000 years, are fewer in number and are dominated by the ribbed vessel form (1i). No ceramics were retrieved from Layer 1 which post-dates 500 BP.

Other excavated sites

The sparse number of sherds recovered from the other excavated sites, most of which were plain, rather limited their usefulness. They do however conform to a general pattern which was highlighted at the sites where more abundant ceramics were recovered. In caves where evidence associated with the earliest human use of the caves was present, calcareously-tempered (Fabric 1) sherds were recovered from the lowest cultural layers. This contrasted with the upper layers and in cave sites which were utilised at a later period where the sherds tend to be thicker and comprised a much more eclectic fabric (Fabric 2). The stratigraphy of the caves also confirmed what appears to be a gap in the archaeological and settlement record, possibly indicating a period of abandonment or at least a period of less intensive use. The ceramic remains from these sites are presented in Table 7.1 and are discussed below by site.

Woplamplam Mk-3-26
Thirteen sherds were retrieved from the test pit excavated at Mk-3-26. All were plain sherds composed of both Fabric 1 (4) restricted to the lowest stratigraphic layers of the site and Fabric 2 (9) restricted to the upper layers. Two ribbed sherds can be ascribed to vessel form 1i (Fig. 7.4a) while Fabric 1 sherds are most likely to represent fragments of one or more of the early vessel forms (2i, 2iii, 4ii) identified from the sites above (Figs 7.1a, 7.2, 7.3).

Waal Mk-3-39
A total of seven sherds were recovered from the 1.5 by 1m test pit excavated at the entrance of this cave. Three of the sherds were diagnostic as to form or decoration. On the surface of the cave floor a single decorated sherd (Fabric 2) with applied relief and incised cross-hatch (M-motif 54) (Fig. 7.13f) was recorded and from the lowest levels of the stratigraphy calcareous-tempered sherds (Fabric 1) including two rims were retrieved. Both rims were notched and can be ascribed to vessel form 2i and 4ii (Fig. 7.2a).

Malua One Mk-3-40
The sherds from Mk-3-40 comprised both Fabric 1 and 2. A total of 10 plain sherds in all were recovered, the thinner calcareously-tempered (Fabric 1) sherds from the lower levels of the site, while the thicker sherds associated with Fabric 2 were restricted to the upper layers. The Fabric 1 sherds seem most likely to be associated with one of the vessel forms already identified from other sites and several of the Fabric 2 sherds are clearly associated with vessel form 2iv.

Woapraf Mk-3-41
The twelve sherds retrieved from excavation at Mk-3-41 included nine plain body sherds and three rims. Sherds from the lowest layers of the test pit were all calcareously-tempered (Fabric 1) and the rim forms enabled the identification of vessel forms 2i and 4ii. Sherds recovered from the upper layers were all Fabric 2 and were associated with vessel form 2iv. No decoration was recorded apart from notching of the lip on the earlier vessel forms.

Yalo South Mk-3-48
Despite the fact that a total of 17 sherds were excavated from Mk-3-48 all but one were plain sherds and only minimal information could be gleaned. Seven sherds with calcareous temper (Fabric 1) were retrieved from the lowest levels of the test pit. All were plain body sherds. The nine other plain body sherds, all recovered from later layers, were substantially thicker and composed of Fabric 2. Several of these sherds could be assigned to vessel form 2iv due to their distinctive interior and exterior finish and wide curvature. A single body sherd with excised decoration (M-motif 46) (Fig. 7.18c) was excavated from the upper most layer of the test pit and resembles that seen on vessel form 3iii (Fig. 7.18a).

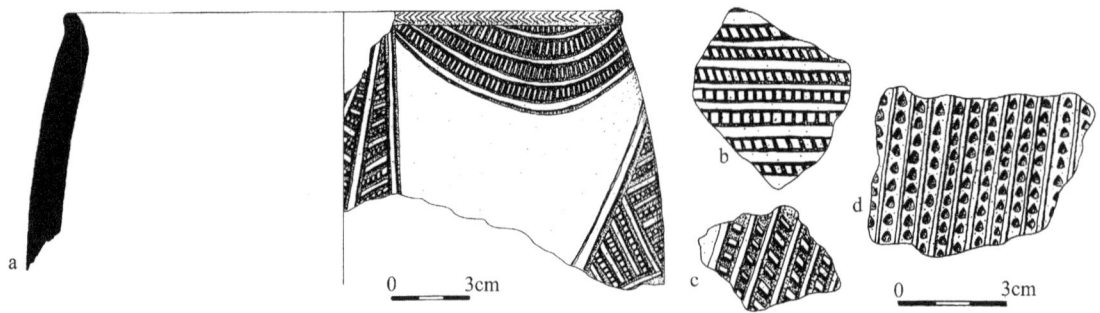

Figure 7.18 Malekula vessel form and decoration. Vessel form 3iii. a. M-motif 45; b-d. M-motif 46.

Wambraf Mk-3-56

Twenty sherds were recovered from this rockshelter site, 18 of which were plain body sherds, along with one rim and one decorated sherd. They were spread thinly throughout the stratigraphy. All of the sherds were Fabric 2. Four ribbed sherds indicated the presence of vessel form 1i. Several other distinctive plain sherds hinted at the presence of vessel form 3iii. The one decorated sherd displayed simple linear incision.

Navepule A Mk-3-58

Eleven sherds made up the total sample recovered from the test pit excavated at Mk-3-58. All were plain sherds composed of Fabric 2 and were dispersed throughout all excavated layers. Three sherds displayed the distinctive ribbed exterior associated with vessel form 1i, while several others can be attributed to vessel form 3iii. No Fabric 1 sherds were in evidence at this inland site.

Navepule B Mk-3-59

A promising surface scatter of 21 sherds collected from the interior of this cave proved to be somewhat misleading. The stratigraphy of the test pit was largely sterile with only a single decorated sherd (Fig. 7.18b) being recovered from the uppermost layer. It showed close resemblance to a large surface collected sherd (Fig. 7.18a). All the sherds were composed of Fabric 2. A number of basal sherds and distinctive ribbed sherds again pointed to the presence of vessel form 1i. Vessel form 1iii was also represented by a single large rim sherd decorated (Fig. 7.8c) with distinctive wavy incision, M-motif 31. Three decorated sherds displaying M-motifs 21, 45 and 46 (unable to be ascribed a vessel form) completed the ceramic collection from Mk-3-59.

Navepule C Mk-3-60

A total of only 6 sherds, 5 plain body sherds and 1 rim were excavated from Mk-3-60. The sparse ceramic remains were scattered throughout the stratigraphy. All were associated with Fabric 2. Little information regarding form could be gleaned from the fragmentary remains.

Fiowl Mk-3-61

The bulk of the 69 sherds recovered from Mk-3-61, were worn plain body sherds (66) that had been subjected to a variety of post-depositional processes. Sixty-seven of the sherds could be assigned to Fabric 2 while only two sherds of Fabric 1 were present. One ribbed sherd from the upper layers of the site could be ascribed to vessel form 1i. One rim sherd of note (Fig. 7.8f) is that associated with vessel form 1ii. This was the only example recovered from excavated contexts on the northwest coast of Malekula but has been frequently recorded in other areas of Malekula and further afield.

Surface collected ceramics

Surface collected ceramics include both those recovered during the 1995 and 1996 field seasons along with a varied assortment held at the Vanuatu National Museum and a number of whole pots held at the Sydney Museum. Malekulan ceramic remains held by the Vanuatu National Museum include numerous sherds that have been collected by the VCHSS since its inception in 1990, an original type collection supplied by the Shutlers and other oddities such as sherds collected by John Hedrick and Barry Weightman in 1973 during the construction of the Norsup airport. A number of *Naamboi* from a variety of sources are also housed at the Museum. Sherds from other northern islands (apart from Santo) are few and far between but again the valuable work of the VCHSS in collecting surface sherds during surveys has provided some useful information. The archaeological excavations and survey work of Hedrick (nd) on Malo and Galipaud (1996c, 1998a, 1998b) on Malo, Santo and the Torres Islands have enabled further and more detailed inter-island comparisons to be carried out.

The surface collected material has been essential in a number of cases in further defining the excavated ceramics and instrumental in establishing an initial overview of the geographic distribution of vessel forms and decoration both across Malekula and other northern islands (see Tables 7.6, 7.7, and 7.8). A full listing of motifs recorded from Malekula is found in Appendix 4.

Table 7.6. Distribution of excavated Malekula vessel forms

✔ present   ? present but vessel form unconfirmed

Vessel form Site/Islands	1i(fig 7.4a)	1ii(fig7.8f)	1iii(fig 7.14)	2i(fig 7.1a)	2i(fig 7.3b)	2iii(fig 6.3)	2iv(fig 7.5a)	3iii(fig 7.18a)	4ii (fig 7.2a)
Excavated Malekula									
Malua Bay	✔			✔	✔	✔		✔	
Chachara	✔		✔						
Nuas	✔		✔						
Navaprah	✔			✔	✔	✔	✔		
Woplamplam				?	?	?		✔	?
Waal						✔			✔
Malua One				?	?	?	✔		?
Woapraf				?	?	✔	✔		✔
Yalo Sth	✔			?	?	?	✔	✔	?
Wambraf	✔						✔		
Navapule A	✔						✔		
Navapule B	✔		✔					✔	
Fiowl	✔	✔		?	?	?			?
Malekula Surface									
Vao	✔	✔	✔						
South West Bay	✔	✔	✔					✔	
Wala	✔	✔	✔						
Tenmaru	✔		✔						
Norsup/Lakatoro	✔	✔	✔						
Uripiv	✔	✔	✔						
Espeigle Bay	✔	✔	✔						
Albalak	✔								
Metkhun	✔								
Pitar	✔		✔						
Other Islands									
Malo	✔	✔	✔	✔					
Santo	✔	✔							
Pentecost	✔	✔	✔						
Ambae	✔								
Maewo			✔						

Table 7.7. Distribution of Malekulan vessel forms from surface and museum collections (multiple examples only)

Vessel form (figure) Malekula	7.19–7.23 naamboi	7.9f	7.10	7.11a	7.11f	7.11d	7.11e	7.11b	7.11a
Malua Bay			✔	✔	✔	✔	✔	✔	✔
Tenmiel		✔	✔	✔	✔			✔	
Lakatoro		✔	✔						
Norsup			✔	✔		✔			
Eldru	✔		✔						
Wala			✔						
Vao						✔			
Uripiv						✔			
Pitar			✔						✔
Metkhun							✔		✔
SW Bay	✔	✔							
Lambumbu	✔								
Brenwe	✔								
Melpmes	✔								
Laravet	✔								
Lembelag	✔								
Toman	✔								

Table 7.8. Distribution of design motifs from excavated and surface collected sherds (multiple examples only)

MOTIF SITE/ISLAND	1	2	3	4	5	12	18	19	20	21	25	26	29	31	32	33	35	41	43	45	46
Malekula																					
Malua Bay	✔										✔	✔								✔	✔
Chachara	✔		✔	✔	✔	✔	✔	✔													
Nuas	✔								✔							✔					
Yalo Sth																				✔	✔
Navepule B	✔								✔					✔						✔	✔
Espeigle Bay	✔	✔				✔					✔	✔									
Norsup	✔	✔						✔													
Eldru											✔							✔			
Lakatoro	✔	✔																✔	✔		
Metkhun	✔			✔		✔										✔			✔		
Albalak	✔		✔							✔											
Tenmiel	✔											✔						✔	✔	✔	
Tenmaru	✔		✔																		
Vao	✔	✔																			
Wala	✔	✔	✔	✔											✔	✔					
Uripiv	✔	✔																			✔
SW Bay	✔												✔					✔	✔		
Other Islands																					
Santo	✔		✔																		
Pentecost	✔	✔	✔				✔								✔						
Malo	✔	✔						✔	✔	✔						✔					
Ambae	✔																				
Maewo							✔														

Both the excavated and surface collected ceramics indicate that the cultural homogeneity which existed during the earliest phase of Malekulan colonisation is repeated for some period over at least the last 500 years leading up until European contact. Similar vessel forms and or decorative motifs dating to this period are found from the north to the south and from the coast to the centre of the island. The ubiquitous vessel forms comprise forms 1i, 1ii and 1iii, while at least 21 motifs were recorded from more than one location on Malekula (Table 7.8). This homogeneity of vessel forms (1i, 1ii, 1iii) along with a number of motifs (M-1, 2, 3, 18, 19, 20, 32, 33) can also be extended to all of the islands north and west of Malekula and south of the Banks Islands (Santo, Malo, Pentecost, Ambae Maewo), again indicating regular contact and interaction amongst the inhabitants of these islands (see Tables 7.6 and 7.8). This was initially hinted at by Hedrick (nd) who recognised the similarity of a single motif (M-33) from Norsup (Malekula) with that of material recovered from Malo. Galipaud (1996a:123) in a more wide ranging survey further outlined a number of similar vessel forms and decoration that were found across the islands of Malekula, Santo, Pentecost and Ambae.

This region which has been previously described as an 'Oceanic Mediterranean' (Bonnemaison 1996:208) shares a roughly similar political structure based around a chiefly system that is consolidated through a hierarchy of grades where men of power come to the fore through ritual and economic tests (which contrasts significantly with the chiefly system of the centre and south of Vanuatu which is based on titles rather than achievements (Bonnemaison 1996:200)). Archaeological evidence of this system is manifested in the remains of the *nasara* which are associated with various grade-taking ceremonies which are an integral component of the political system. The earliest date thus far associated with these *nasara* and the distinctive late ceramics as described above, is a radiocarbon determination of c. 550 BP from the inland site of Chachara. Clearly these sites and the political structures associated with them have a pedigree of at least 500 years on Malekula, as do the associated ceramics.

What explanations then might be suggested for the appearance in the north of Vanuatu of a very distinctive ceramic style at around at least 500 BP or earlier which at this stage of research appears to have no apparent antecedents and which also appears to have coincided with the development of a divergence in the nature of the political structures of the north with those from the centre and south of Vanuatu? Green (1997, 1999) has more recently speculated that influence from Near Oceania in the form of sustained contact or even migrations of non-Austronesian speakers might well have contributed to substantial change in this part of Remote Oceania during the last 1000 years. Certainly there are clear links which can be identified in the oral traditions, ethnographic observations, archaeological and biological data (Hagelburg 2001). There are similarities to be found in pot form, manufacturing and decorative techniques and motifs between northern Vanuatu and areas to the northwest (Specht 1972). It also seems unlikely that such specialised cultural practises as headbinding and the production of full-circle pig tusks (Speiser 1996) found both in northern Vanuatu and southern New Britain have developed independently. Influence from the west seems like an increasingly likely scenario and certainly oral traditions in the north of Vanuatu talk of the grade taking ceremonies or *nimanggi* as being a more recent innovation (D. Tryon pers. comm.). Further elucidation, however, of the cultural transformations which occurred in the region will only be forthcoming through further intensive archaeological research both in Vanuatu and areas to the west specifically targeting this period.

Appearance of *Naamboi*

Some breakdown in the above pattern of ceramic similarity across the 'Oceanic Mediterranean' is seen in the last several hundred years when either distinctive vessel forms and or decoration began to appear on a number of islands or ceramic production ceased altogether. This disruption is characterised on Malekula by the appearance of the very distinctive thick walled bullet-shaped pots and cylinders (Figs 7.19–7.23) known as *Naamboi*, noted in some detail by Layard (1928) and

Deacon (1934) which have not to date been recorded on other islands. These vessels appear to signal the last phase of ceramic production on Malekula and although they were recorded in the early twentieth century as being integral in a number of ceremonies, there was no evidence of their manufacture which appeared to have ceased. Oral traditions ascribed them to having been produced by a mythical white-skinned people known as the *Ambat* (Layard 1928:210).

According to Layard who visited South West Bay for a week in May 1915 the pottery was kept in sacred places and occasionally used in various ceremonies relating to the increase of the fertility of pigs (Layard 1928:210). In a detailed article Layard (1928) described and illustrated the various vessel forms and decoration encountered in southern Malekula (see Figs 7.19–7.23). Deacon (1934) who spent a much longer period in southern Malekula, but who also ventured to other areas of the island, provided further details. It seems the *Naamboi* were somewhat more widespread (further confirmed by museum collections and a single sherd from the northwest (see Table 7.7)) and used in a number of ceremonial activities relating to either increased productivity/fertility, burials or control of the weather. When not being used in ceremonies they were kept in specific sacred areas (Deacon 1934:597). Deacon noted the presence of the pottery in the South West Bay area, both on the coast and inland. Toman Island, off the south west coast, was recognised as being a particularly important centre for the fertility ceremony in which these pots were involved. The pots were also recorded in the central west coast area (Lambumbu) where their function was again ceremonial but related to the proliferation of breadfruit (sospan blong breadfruit or *Ambwi nembet*). Sacred stones were placed inside the pots (Deacon 1934:597). At Lembelag they were associated with 'hurricane magic' and control of rainfall (Deacon 1934:671) and in the South West Bay area they were recorded as being variously placed within or on top of burials (Deacon 1934:649).

These *Naamboi* are generally tubular in form with a conical base hence attracting the generic term 'bullet' pots. Large and small versions have been recorded along with a multitude of decoration. The earliest detailed descriptions of form and decoration were those given by Layard

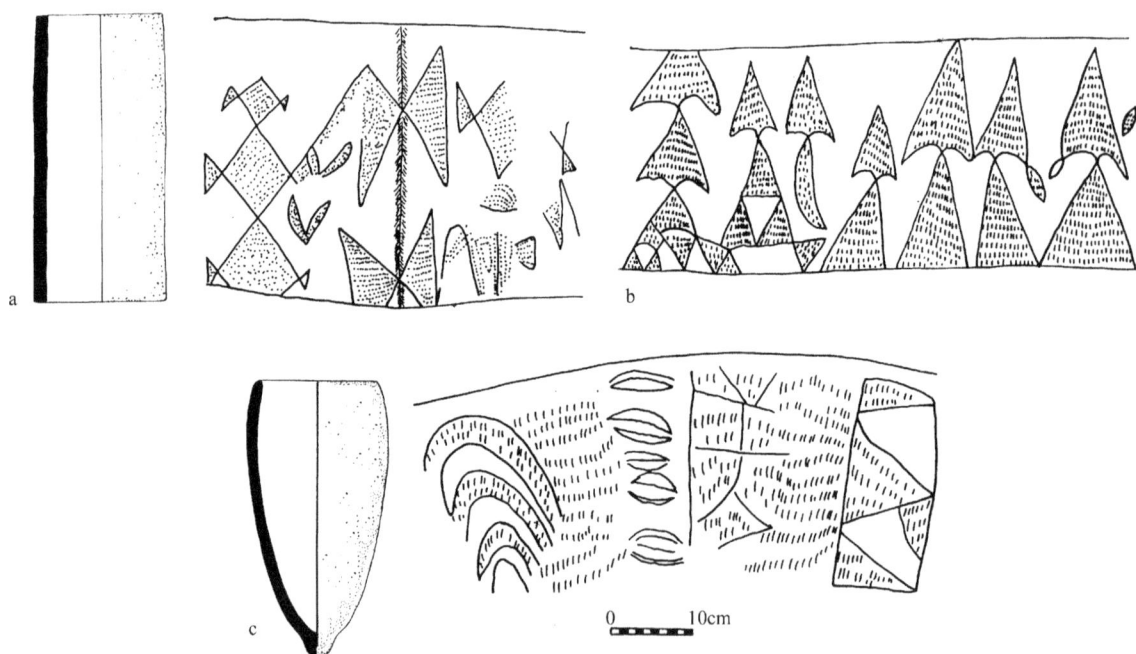

Figure 7.19 Malekula vessel form and decoration. Naamboi recorded by Layard (1928). a-b. decorated cylinders; c. decorated pot.

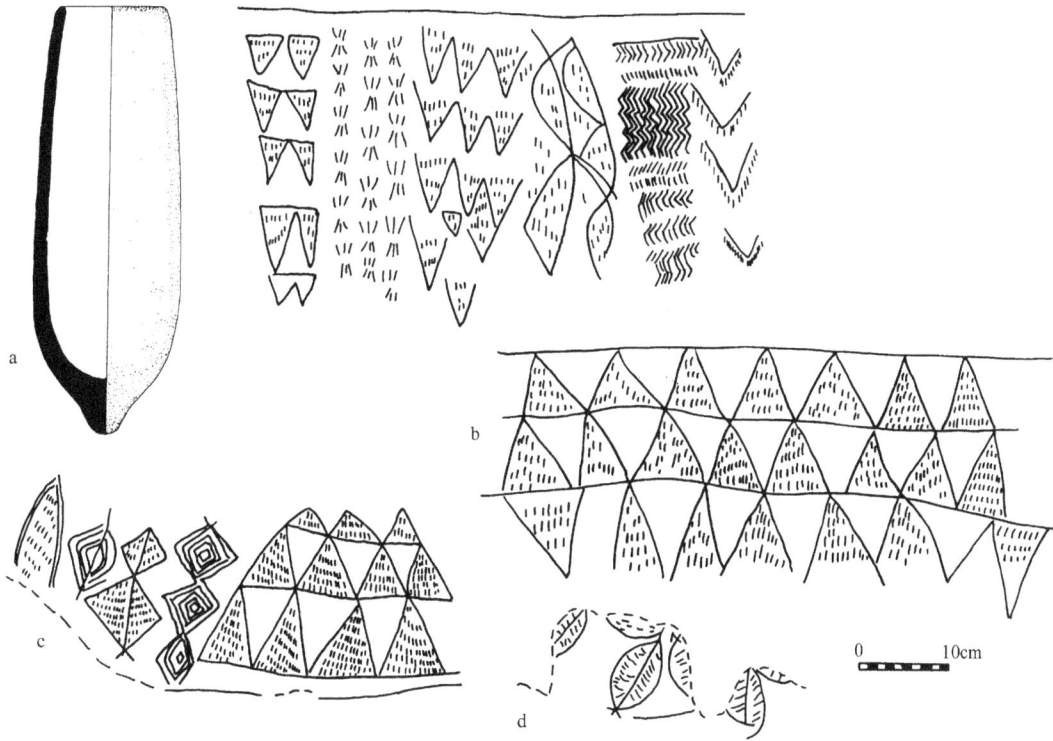

Figure 7.20 Malekula vessel form and decoration. Naamboi recorded by Layard (1928). a. Naamboi vessel; b–d. Naamboi vessel fragments.

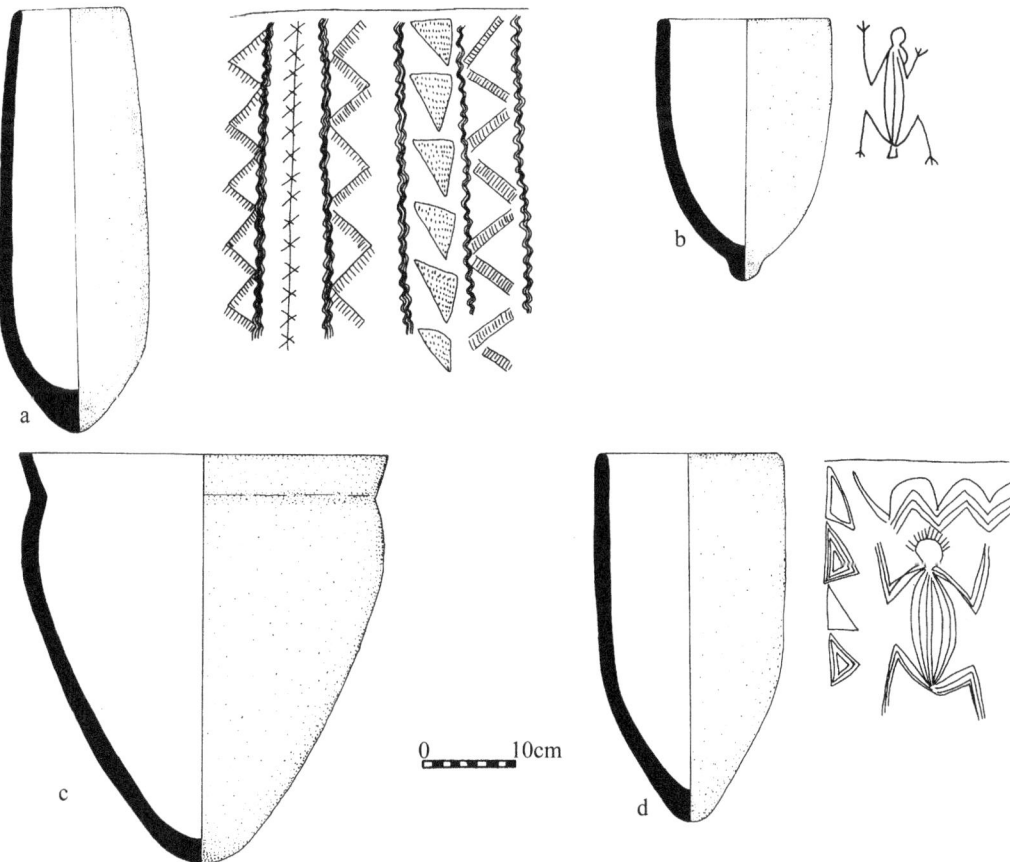

Figure 7.21 Malekula vessel form and decoration. Miscellaneous Naamboi. a. Naamboi, Vanuatu Cultural Centre; b. Naamboi recorded by Layard (1928); c. anomalous vessel form recorded by Layard (1928); d. Naamboi recorded by Deacon (1934).

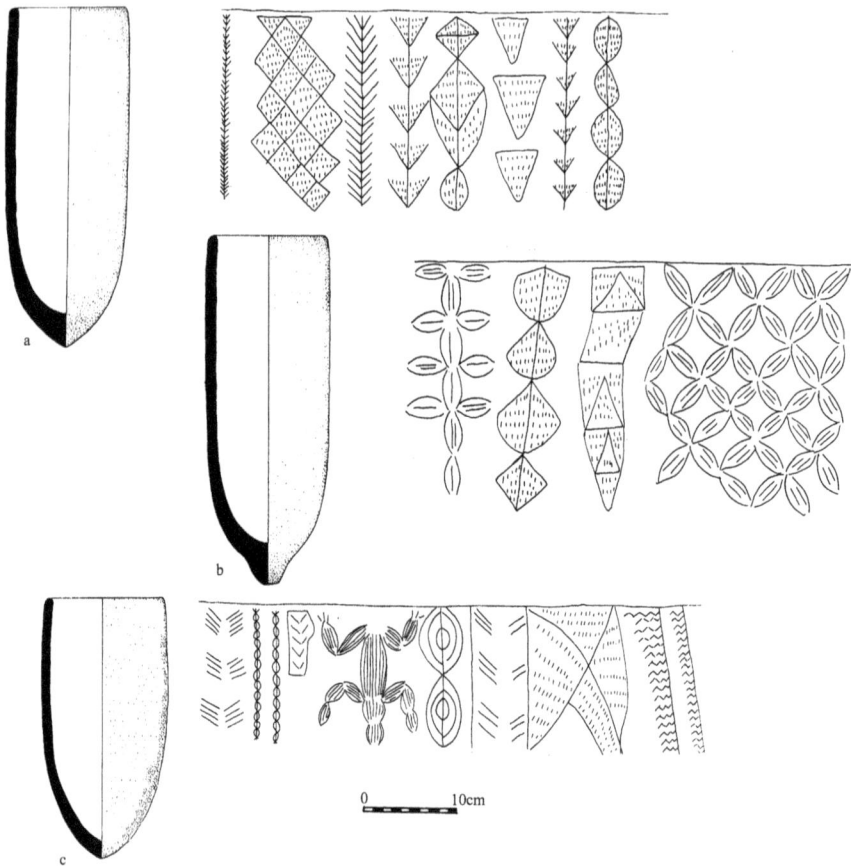

Figure 7.22 Malekula vessel form and decoration. Naamboi held by the Australian Museum. a. Laravat village, west coast; b–c. Southwest Bay.

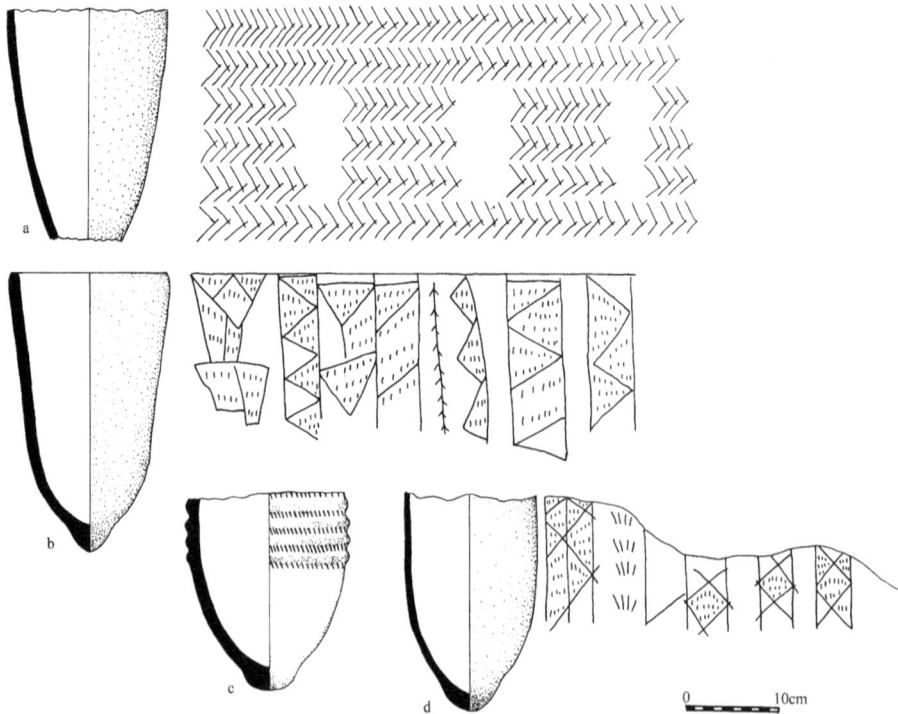

Figure 7.23 Malekula vessel form and decoration. Naamboi held by the Vanuatu Cultural Centre. a. north west area; b. Malekula; c. anomalous ribbed Naamboi; d. Southwest Bay.

(1928) and Deacon (1934). More recent research by Cayrol (1992) has included a detailed description of examples from both the Musée de l'Homme and the Vanuatu National Museum. As a component of this current research a further three pots have been recorded from the Australian Museum (Sydney) along with the five held at the Vanuatu National Museum. Included within the category of *Naamboi* are the ceramic cylinders (Fig. 7.19a, b) recorded by Layard (1928) which are still used as ceremonial objects in the South West Bay area today. One large globular everted rim vessel (Fig. 7.21c) recorded by Layard remains somewhat anomalous and hence is not categorised as *Naamboi*. The *Naamboi* are coil built, with smoothed interior and exterior surfaces facilitating decoration. Generally the decoration on these vessels consists of varied complex geometric motifs created by incision and punctation. A number of anthropomorphic figures have also been recorded (Fig. 7.21b, d). One single example (Fig. 7.23c) appears to accentuate the coils creating a ribbed effect with added punctation in the hollow between the ribs.

Little research has focused on these vessel forms since the early ethnographic work apart from the recent thesis by Cayrol (1992) which relied on the ethnographic record both in Vanuatu and other parts of the Pacific to explore in detail the functional status of the *Naamboi*. The vessel form clearly appears, at least on typological grounds, to have developed from the earlier decorated bullet-shaped pots (vessel form 1iii) found throughout Malekula. The walls and the base of the *Naamboi* show a dramatic increase in thickness compared to the earlier vessel forms and their cumbersome size and form render them increasingly less suitable for cooking purposes. A somewhat speculative scenario which may have led to the development of the *Naamboi* is that the earlier ribbed and decorated pots represented cooking and ceremonial vessels respectively. If cooking in pots was for whatever reason abandoned it would have signalled the end of the ribbed form leaving only the decorated vessels from which the highly ceremonial *Naamboi* developed. This would have had the effect of radically changing the dynamics of pottery production on the island as it would have become increasingly specialised with much smaller quantities being produced. The difficulty with this argument is of course explaining the abandonment of cooking in pots, an often discussed Pacific-wide phenomenon, that remains somewhat of a mystery (Green and Davidson 1974; Kaeppler 1973; Leach 1982; Le Moine 1987; Marshall 1985). It would seem to have happened very recently on Malekula and potentially with further archaeological research in regions where it has been so recently abandoned, some greater understanding of the complexities involved is likely to be reached.

The multiple inter-related factors which led to the decline and finally the complete loss of pottery making on the island seem to coincide with a shrinking sphere of interaction with other islands in the north and the development of increasing territoriality across Malekula itself. The increased contacts with Europeans from the early nineteenth century, with introduced diseases and the later labour recruitment drives, would have caused dramatic disruptive effects on the social and cultural systems. Certainly from at least the late nineteenth century significant population decline was being commented on throughout Vanuatu (Harrisson 1937; McArthur 1967; Spriggs 1997:255–260). Although the causes leading to the collapse of ceramic production on the island remain somewhat elusive it is clear that one effect would have been to elevate the increasingly rare surviving vessels to sacred status. The ethnographic record relating to ceramic usage collected by Layard and Deacon, when there was no evidence of ceramic production on Malekula, appears to post-date this transformation to tabu status. Even today, in the south of Malekula, the remnant sacred vessels or parts of vessels are still used in various ceremonial capacities.

Speiser (1996[1923]:230) and later Huffman (1996:182-194) both detail the long established trade routes that existed amongst the northern islands of Vanuatu including one for pots which originated on the west coast of Santo then on to Malo to Vao and then down the east coast of Malekula. The archaeological evidence for the presence of Santo pots on Malekula is thus far lacking although this is quite possibly due a lack of research in particular areas. However a single surface collected sherd from Tenmaru is very similar to the late Santo material both in terms of form and decoration (Fig. 7.9b) and the results of a petrographic analysis carried out by Dickinson (Appendix 3 WRD-184) certainly indicates Santo affiliations.

Summary

- Ninety per cent of the excavated ceramics were recovered from the four sites of Malua Bay School, Chachara, Nuas and Navaprah. These sites enabled both the establishment of the two ends of the ceramic sequence on Malekula and facilitated the analysis of the more sparse remains from the other excavated sites.

- Ceramics first appear on Malekula some 3000 years ago although the sites excavated on the Northwest coast largely post-date the earliest Lapita dentate-stamped phase. The earliest recovered ceramics appear to represent a plainware phase dating to around 2700–2500 BP. A limited number of vessel forms were identified all of which had been calcareously-tempered and made on Malekula. A single dentate-stamped sherd was found in association with the plainware ceramics at the Malua Bay School site.

- Although the ceramic sequence post-dating this initial settlement phase is far from complete the variety of decorative motifs and vessel forms demonstrated amongst the surface collected sherds indicates that the manufacture and use of ceramics did continue at least in some areas of Malekula for some considerable period. It is too early at this stage of the research to determine whether the ceramic sequence was continuous through to the later traditions or not.

- The last phase of ceramic production on Malekula is characterised by the bullet shaped, coil made pots that appear to have no apparent antecedents. These appear in the sequence no earlier and possibly sometime later than 1000 BP. The vessels are both decorated and plain, most likely associated with cooking and ceremonial activities respectively. Similar vessel form and design motifs have been identified both across Malekula and other islands of northern Vanuatu indicative of some form of cultural homogeneity. The ceramics are found in association with *nasara*, one of the archaeological signatures indicative of the distinctive northern political systems. The homogeneity of ceramics survived at least until the last several hundred years when ceramics either died out on some northern islands or became island specific.

- Recently surface collected sherds and those held in museum collections demonstrate a much more diversified ceramic sequence than that which was revealed from the excavations. Although highly speculative at this point there is the hint of a possibility of a lengthy continuous sequence, albeit unevenly distributed.

- The final phase and demise of ceramics on Malekula is recorded in the ethnographic record and associated with *Naamboi* which appear to have evolved from earlier forms found across Malekula.

- Ceramics recovered from Malekula both from the earliest and latest phases of the sequence are made on Malekula most probably throughout the island rather than at specific manufacturing centres.

8

Vanuatu Ceramic Sequences and Inter-regional Comparisons

'For Tikopia and the Polynesian outliers, it is clear that we cannot construct such elegantly simple models of culture change (as in East Polynesia), for we have had in operation the complexity wrought by a virtual stream of cultural arrivals and external contacts' (Kirch and Yen 1982:346).

The recovered ceramics from the various excavations on the three different islands have enabled the establishment of a series of chronological sequences. These are outlined in detail below by island (see also Figs 8.1–8.15). General summaries defining the broad characteristics of the various identified phases are emphasised here along with chronological information. Finer detail regarding the recovered ceramics by site is found in relevant chapters. Inter-island Vanuatu comparisons are made and these are then followed up with more general comparisons of other sites in the Southwest Pacific which have either been claimed to have links with Erueti or Mangaasi-style ceramics (Ambrose 1991; Kennedy 1982; Kirch 1983; Kirch and Yen 1982; Sand 1995; Specht 1969; White and Downie 1980) or are sites which have been cited as displaying elements of an Applied and Incised Relief tradition argued as having some affinity across the Southwest Pacific (Spriggs 1984, 1990a, 1993, 1997, 2000; Wahome 1997, 1999).

Erromango

The ceramics associated with the initial settlement of Erromango are characterised by Lapita dentate and incised material (Fig. 5.17) which was predominately calcareously tempered. The largest sample was recovered from the earliest layers of the Ifo site. The limited number of decorated sherds and or rims has however restricted definition of vessel form and design motifs. Those motifs that could be identified and correlated to Anson's classification system included M

187, 254, 254 plus 417 and 417 in combination with incision (Figs 5.10i; 5.17). Plainware cooking vessels (form 2i) were also identified as being a component of this phase of the ceramic sequence. The phase was short-lived, appearing around 3000 BP and dropping out by c. 2800 BP.

From 2800 BP a distinct plainware phase has been identified. Vessel form was largely restricted to outcurving rim vessels (form 2i) (Fig. 8.1) with decoration being restricted to only very occasional notching on the lip. This vessel form and phase of the sequence was most dramatically illustrated at Ponamla, the site after which this phase is named. This again appears to have been a somewhat short-lived phase being transformed by the appearance of a multitude of largely fingernail incised motifs from c. 2600 BP.

The greatly increased number of decorated sherds and wide variety of motifs initially appeared on identical vessel forms to those characteristic of the Ponamla Phase (form 2i) (Figs 8.2, 8.3). Calcareous tempering had, by this stage, been dropped from the manufacturing process. Plainware vessels continued as a component of the ceramic repertoire. Notching on the lip remained a rare decorative technique as it had been in the earlier Ponamla Phase. A total of 35 separate design motifs were identified, 24 of which were composed of fingernail decoration. Much rarer were incised sherds and associated identifiable design motifs, but two in particular demonstrate close affinity to Lapita motifs (E-motif 25 and E-motif 28 [Anson motifs 175, 318 and 369]). E-motif 28 also featured amongst the early post-Lapita motifs recovered from Efate (Ef-motif 1). Although this phase of the sequence was most clearly identified at the Ponamla site, it is proposed that it be called the Early Ifo Phase, after the site at which these decorative motifs, which characterise the post dentate-stamped Lapita ceramics on Erromango, were first identified (Spriggs 1984). The Early Ifo Phase then, initially appeared at around 2600 BP and continued for 200 years or so, transforming into what can be identified as a distinctive Late Ifo Phase from c. 2400 BP.

From 2400 BP some simplification in vessel form can be seen with outcurving rims being replaced by incurving rims (forms 3i and 3ii) on vessels which tended to have thicker walls and were possibly less highly fired. Decoration became increasingly common during this phase but was still dominated by design motifs created with fingernail impression (Fig. 8.4). Some continuation in motif form can be demonstrated as eight of the fifteen identified motifs belonging to the Late Ifo Phase were also found amongst the Early Ifo Phase ceramics. Not able to be clearly demonstrated chronologically at the site but inferred from vessel form and design motifs are a number of vessels which appear to signal the end of the ceramic sequence on Erromango at no later than 2000 BP. They are again thicker walled vessels that display a number of distinct motifs, often in association with notching on the lip (Fig. 8.4h, m, n).

Four ceramic phases have been proposed for the 1000 year sequence identified on the island of Erromango, namely Lapita (c. 3000–c. 2800 BP), Ponamla (c. 2800–c. 2600 BP) and Early (c. 2600–c. 2400 BP) and Late Ifo (c. 2400–c. 2000 BP) Phases (Figs 5.17; 8.1–8.4). The ceramics were made at or near the sites from which they were recovered and they demonstrate the continuous evolutionary nature of the sequence on the island. Although it can be seen from the presence of the Lapita dentate-stamped ceramics along with a number of immediately post dentate-stamped Lapita incised designs that initial settlement of the island was part of the widespread colonisation of the region by members of the Lapita Cultural Complex, whatever connections this may have initially represented appear to have been somewhat short-lived. Very dramatically demonstrated is the rather sudden development of a distinctly Erromangan ceramic style displaying quite distinct and unique design motifs which have not been found at other sites in Vanuatu.

Although the ceramics from Erromango were initially described as a regional variant of Mangaasi (Spriggs and Wickler 1989) and later as having clear parallels with Mangaasi (Bedford 1999), the further detailed analysis of the Erromango sherds, along with the re-excavation of the Mangaasi site, have shown that these earlier assertions can no longer be supported. It was this revelation which has led to a reassessment of the validity of the widely accepted concept of an

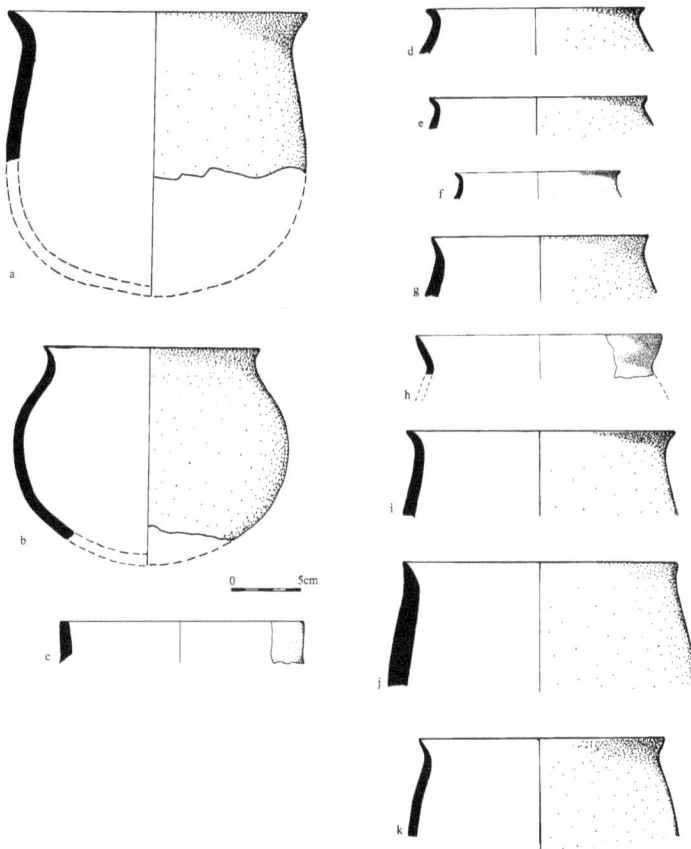

Figure 8.1 Ponamla Ware c. 2800-2600 BP. Vessel forms 2i (a,b, e–h, j–l) and 2iii (h).

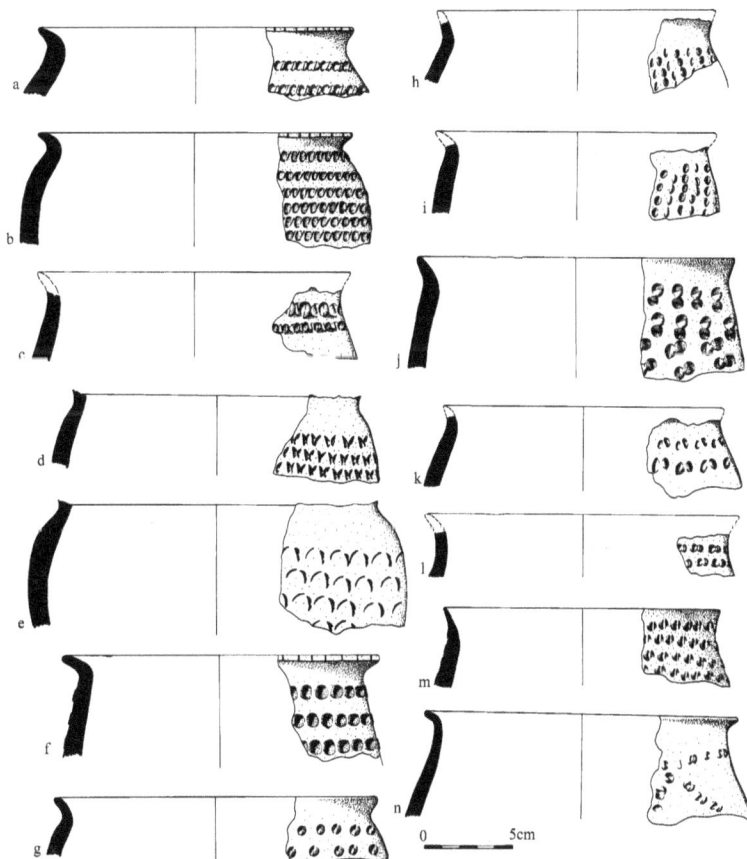

Figure 8.2 Early Ifo Ware c. 2600–2400 BP. All vessel form 2i displaying a variety of motifs. a–e. E-motif 1; f. E-motif 16; g E-motif 15; h–i. E-motif 4; j. E-motif 5; k-m. E-motif 2; n. E-motif 3.

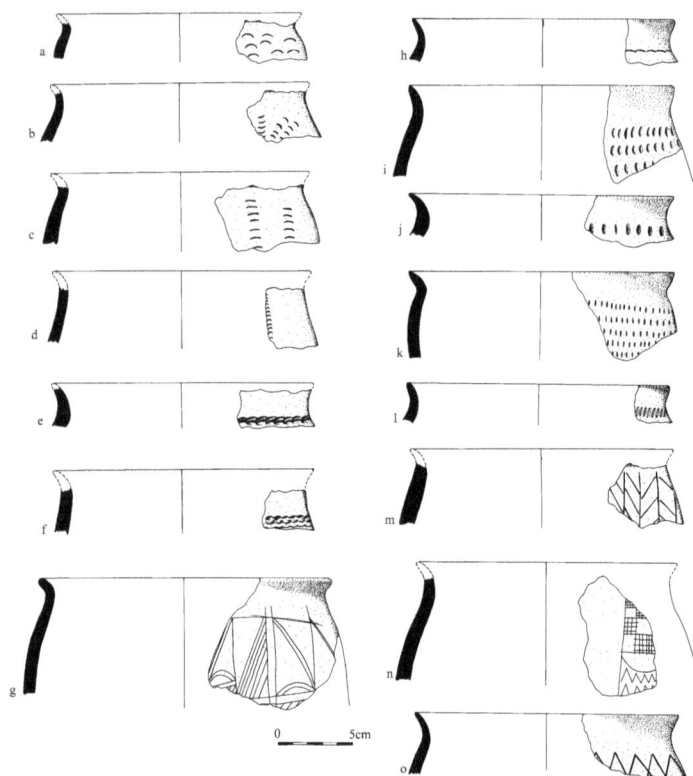

Figure 8.3 Early Ifo Ware c. 2600–2400 BP. All vessel form 2i displaying a variety of motifs. a. E-motif 17; b. E-motif 18; c. E-motif 19; d. E-motif 20; e. E-motif 6; f. E-motif 7; g. E-motif 25; h. E-motif 23; i-l. E-motif 24; m. E-motif 28; n. E-motif 26; o. E-motif 27.

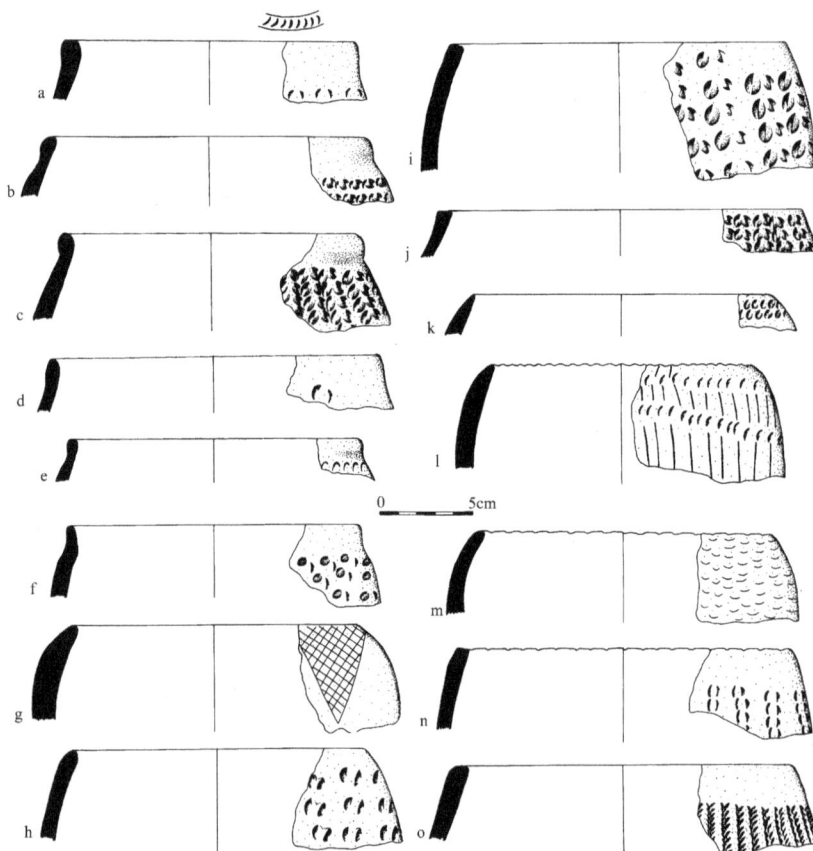

Figure 8.4 Late Ifo Ware c.2400–2000 BP. a. vessel form 3ii, E-motif 2; b. vessel form 3ii, E-motif 1; c. vessel form 3ii, E-motif 4; d. vessel form 3ii; e. vessel form 3ii, E-motif 23, f. vessel form 4i, E-motif 4; g. vessel form 3i, E-motif 42; h-j. vessel form 3i, E-motif 4; k. vessel form 3i, E-motif 15, l. vessel form 3i, E-motif 40; m. vessel form 3i, E-motif 17; n. vessel form 3i, E-motif 38; o. vessel form 3i, E-motif 36.

Incised and Applied Relief tradition, that could be identified across the Southwest Pacific, and which underwent some form of synchronous change indicating continuing levels of interaction and communication (Spriggs 1984, 1997, 2000; Wahome 1997, 1999). This issue is further expanded on below, following an assessment of a number of ceramic assemblages which have been cited as examples supporting the scenario of a widespread post-Lapita ceramic tradition.

Efate

Excavation at the Arapus site enabled the identification of a distinct vessel form which can be seen as being associated with the Lapita settlement of Efate but representing the cooking component of the assemblage only (Bedford and Spriggs 2000; Spriggs and Bedford 2001). This phase of the ceramic sequence has been named Arapus after one of the ancestral household areas where it was first recognised. Initial interpretations of the Arapus site and ceramic remains were that they either indicated that there were distinct activity areas at the site and that the location of the ceremonial and/or high status dentate-stamped pottery had yet to be located (Bedford and Spriggs 2000, Spriggs and Bedford 2001) or that the site may have lacked any such ceremonial or status association and may represent what Yen (quoted in Kirch 1988b:245) originally described as the 'proletarianisation' of the Lapita Cultural Complex. Although further fine dating of the site and detailed publication is forthcoming, continued research at the site (2001–2003) and the discovery of a Lapita site on the south coast of Efate with elaborate vessel forms and western-style dentate decoration (Bedford *et al.* 2004) now suggest that while the Arapus site is indeed associated with first settlement on the west coast of Efate it may in fact post-date initial settlement of the island and the Lapita dentate-stamping phase by at least some 100 years if not longer.

The appearance of the Arapus Phase has thus far has been dated to c. 2900 BP. The pottery was characterised by globular pots with outcurving rims, frequently displaying notching on the lip (Fig. 8.5). The phase appears to have been short-lived, perhaps in the order of only 50–100 years or so. Over time the outcurving rims became increasingly horizontal, developing into the distinctive wide flat lips which characterise the Early Erueti Phase plainware vessels (Fig. 8.6).

The Erueti Phase of the ceramic sequence on Efate, originally identified by Garanger (1971) at the eponymous Erueti site is divided into two phases, Early and Late. The Early Erueti Phase appeared at around 2800 BP and is characterised by a variety of vessel forms, principally dominated by plain outcurving rim vessels (form 2ii) which are almost always notched on the lip (Fig. 8.6). A minor component of the assemblage represents continuity from the Arapus Phase in the form of vessel form 2i. Other vessel forms represented (although often by single examples only), included 3i (Fig. 8.7 g), 3ii (Fig. 8.7 e), 4i (Fig. 8.7 f), 5i (Fig. 8.8 c) and 5ii (Fig. 8.8 a, b). These vessel variants tended to be decorated and therefore may have represented the ceremonial component of the assemblage. A minor component of form 2ii vessels were also decorated (Fig. 8.7 a–d, l–o). Small cups, namely 3iv (Fig. 8.7 j, k), 4iii (Fig. 8.7q) and 5iv (Fig. 8.7 h, i), were also a component of the Early Erueti Phase.

Decoration on the Early Erueti Phase ceramics consisted exclusively of incised motifs and notching on the lip. Punctation was only identified on the horizontal surface of the wide flat lips of vessel form 2ii (Fig. 8.7 b). There was no applied relief associated with this phase of the sequence. Both linear and geometric incision dominated with very few examples of incised gashes. A number of quite complex and varied incised motifs were also identified (Fig. 8.8 b). Several vessel forms and motifs, such as Ef-M 20, 27, 31 (Figs 8.8 a, c, f, g) have been argued as having generic Lapita connections. Handles of a single form were present but rare (Fig. 5.17f).

By c. 2500 BP a perceptible change in the ceramic assemblage, both in terms of vessel form and decoration, can be identified. It is at this point a Late Erueti Phase is proposed. Vessel forms 2i and 2ii became much less frequent or disappeared to be replaced by vessel forms 3i and 3ii.

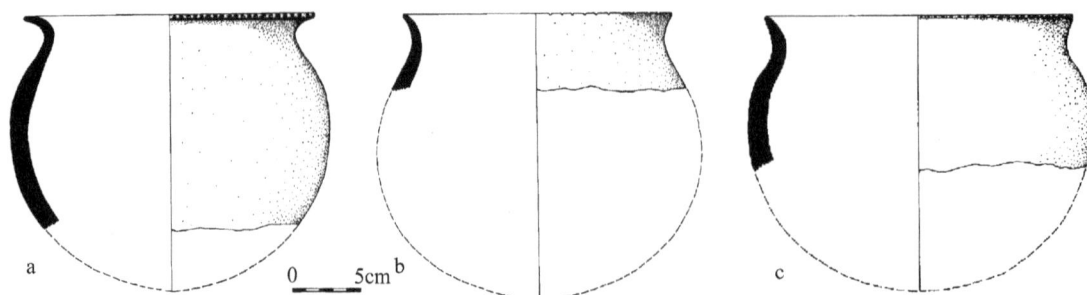

Figure 8.5 Arapus ware. All vessel form 2i with notched lip.

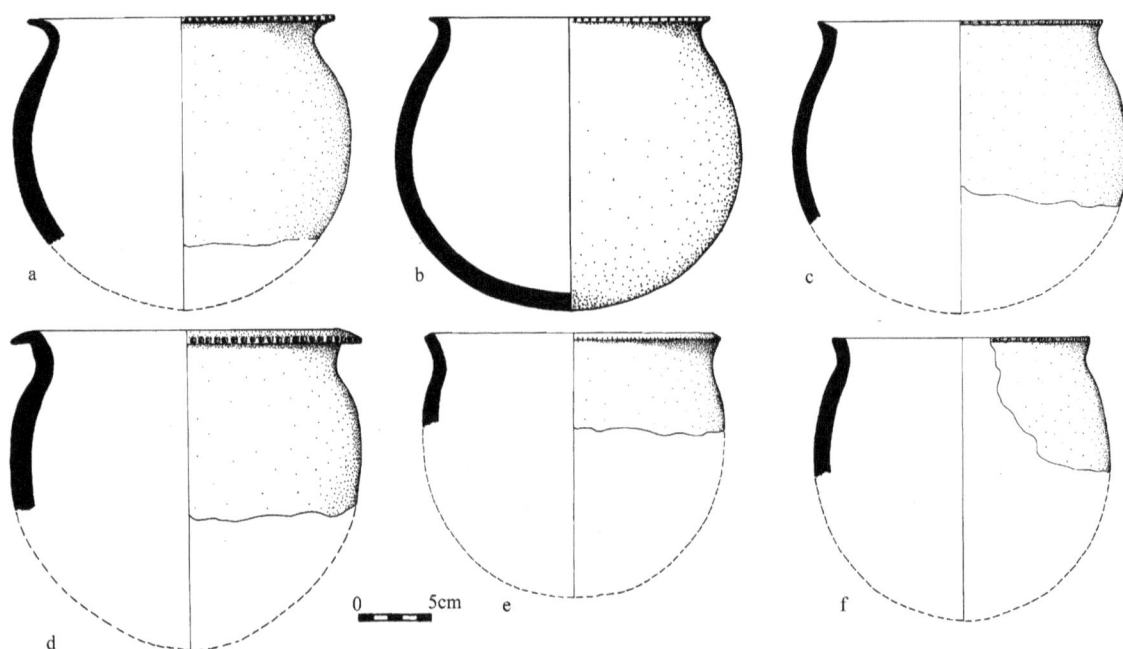

Figure 8.6 Early Erueti Ware (plain) c. 2800–2500 BP. Vessel form 2i with notched lip; b. vessel form 2ii with horizontal wide flat lip; c–d. vessel form 2ii variant with wide flat notched lip; e–f. vessel form 2ii variant with flat lip and parallel rim profile.

Carinated vessels also became increasingly globular (Fig. 8.8 d–e). Small cups in the form of 3iv were still present. Decoration, still largely restricted to incision (Fig. 8.9), became more frequent, suggesting that the proposed clear distinction between the plain utilitarian vessels and decorated ceremonial vessels was beginning to fade. Notching of the lip remained a regular decorative feature. Certain motifs identified in the Early Erueti Phase assemblages continued through into this Late Erueti Phase (i.e. Ef–M 14 [Fig. 6.12a], 24 [Fig. 8.8g] and 38 [Fig. 6.12d, h]) and a number could still be argued as having some affinity with Lapita motifs (i.e. Ef–M 20 [Fig. 8.8f], 24 [Fig. 8.8g] and 31 [Fig. 8.8a]).

From c. 2200 BP the appearance of a range of distinctive decorative techniques and motifs associated with the overwhelming predominance of a single vessel form justifies the further division of the sequence. It is from the beginning of this phase that elements associated with Garanger's original Mangaasi tradition began to appear, therefore it is proposed that this phase be labelled the Early Mangaasi Phase. Vessel form 3i was completely dominant throughout this phase

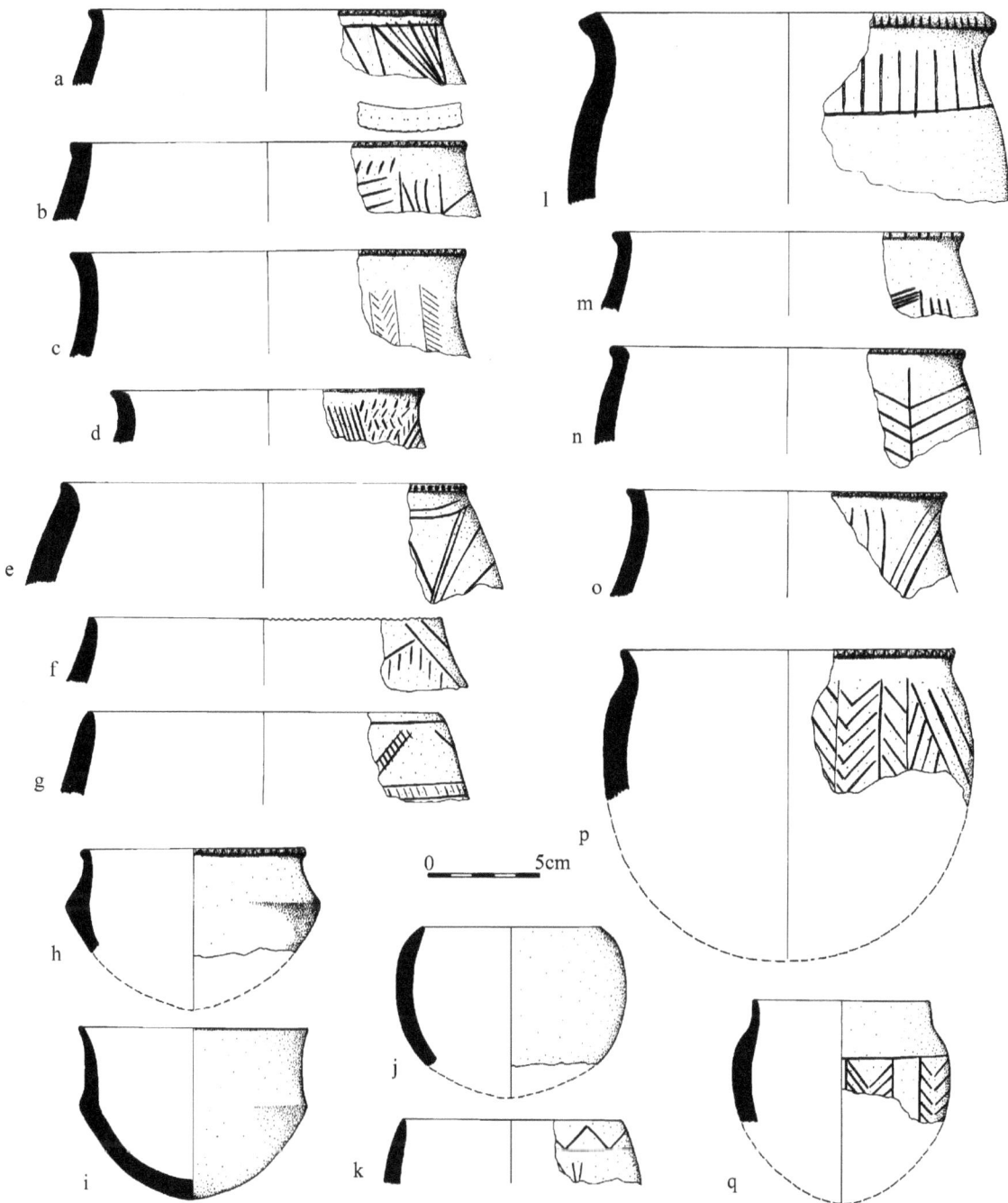

Figure 8.7 Early Erueti Ware c. 2800–2500 BP. a–d. vessel form 2ii with assorted decoration and notched lip (a, c, d Ef-motifs 1, 39, 51); e. vessel form 3ii with incised decoration and notched lip; f. vessel form 4i?, Ef-motif 30; g. vessel form 3i, Ef-motif 32; h-i. vessel forms 5iii, plain carinated cups with notched and incised lip; j-k. vessel form 3iv, plain and incised Ef-motif 25; l-o. vessel form 2ii with incised decoration; p. vessel form 4i, Ef-motif 2; q. vessel form 4iii, Ef-motif 6.

of the ceramic sequence. Vessel form 3ii was also recorded and it seems to represent continuity from the Late Erueti Phase. Small cups continued to be associated with this phase of the sequence (Fig. 8.10 l–m). Decorative techniques associated with the Early Mangaasi Phase cover a wide spectrum and there is great variation through the sequence and in motif structure. Initially it appears that decoration was predominately incision (linear, geometric and gashes) and to a lesser extent punctation, utilised separately rather than in combination (Fig. 8.10). Initially motifs may have been less complex. Lip notching which was a distinctive modal decorative attribute in the Erueti phases

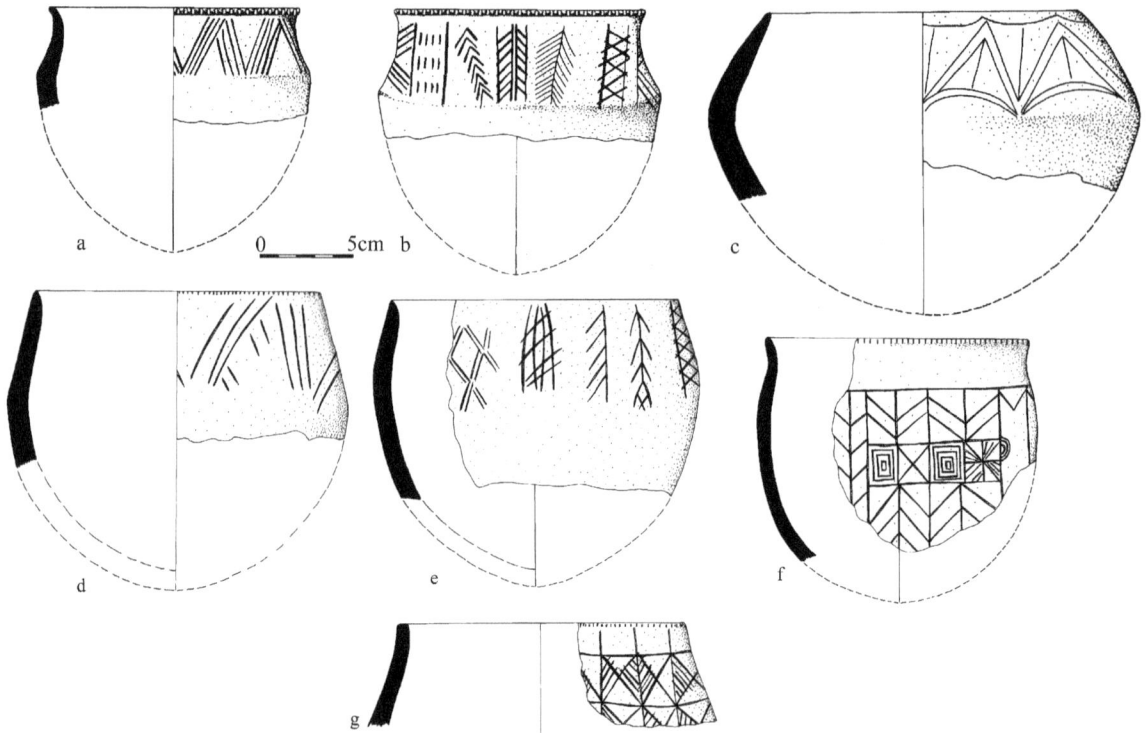

Figure 8.8 Early and Late Erueti Ware (decorated). a-c. Early Erueti ware c. 2800–2500 BP; a. vessel form 5ii, Ef-motif 31; b. vessel form 5ii, Ef-motif 29; c. vessel form 5i, Ef-motif 27; d–g. Late Erueti Ware c. 2500–2200 BP: d. vessel form 5ii, Ef-motif 30; e. vessel form 5ii, Ef-motif 19; f. vessel form 4i, Ef-motif 20; g. vessel form 4i, Ef-motif 24.

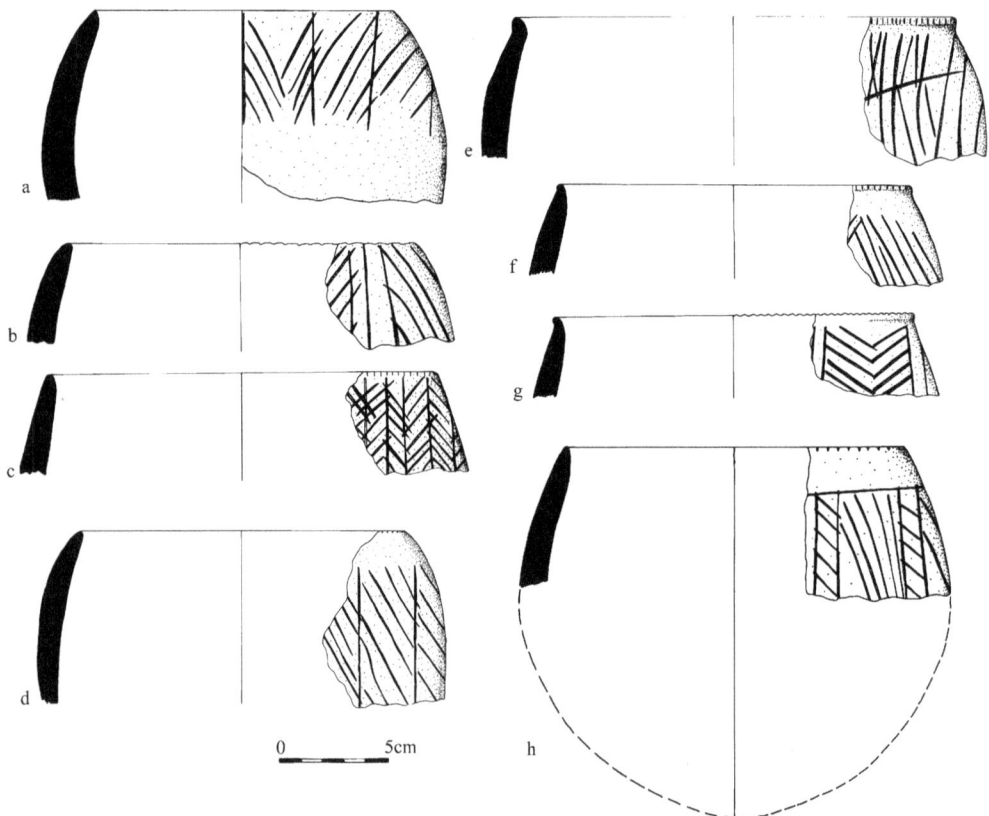

Figure 8.9 Late Erueti Ware c. 2500–2200 BP. a. vessel form 3i, Ef-motif 5; b. vessel form 3i, Ef-motif 5, plus notched lip; c. vessel form 3i, Ef-motif 1; d. vessel form 3i, Ef-motif 3; e. vessel form 3ii, plus notched lip; f. vessel form 3ii, Ef-motif 31; g. vessel form 3ii, Ef-motif 11; h. vessel form 3i, Ef-motif 12.

was no longer in evidence. Motifs became increasingly complex over time, with the various techniques and motifs being combined (Fig. 8.11). Discontinuous applied relief always appears to be associated with already established motifs that have generic connections to earlier forms (Garanger 1972:Figs 131, 132; Fig. 8.11 e–h) confirming that the discontinuous applied relief represents a slightly later expansion of the motif repertoire and that certain motif forms or design elements can be seen as providing continuity from earlier phases of the sequence. Continuous applied plain and pinched bands (Garanger 1972:Fig. 128; Fig. 8.11o) also make an appearance during this phase of the sequence although these techniques were relatively rare. The applied bands appear to replace the incised oblique parallel incision which formed a similar triangular motif. The applied triangle motifs are infilled by a number of other decorative techniques including punctation and geometric

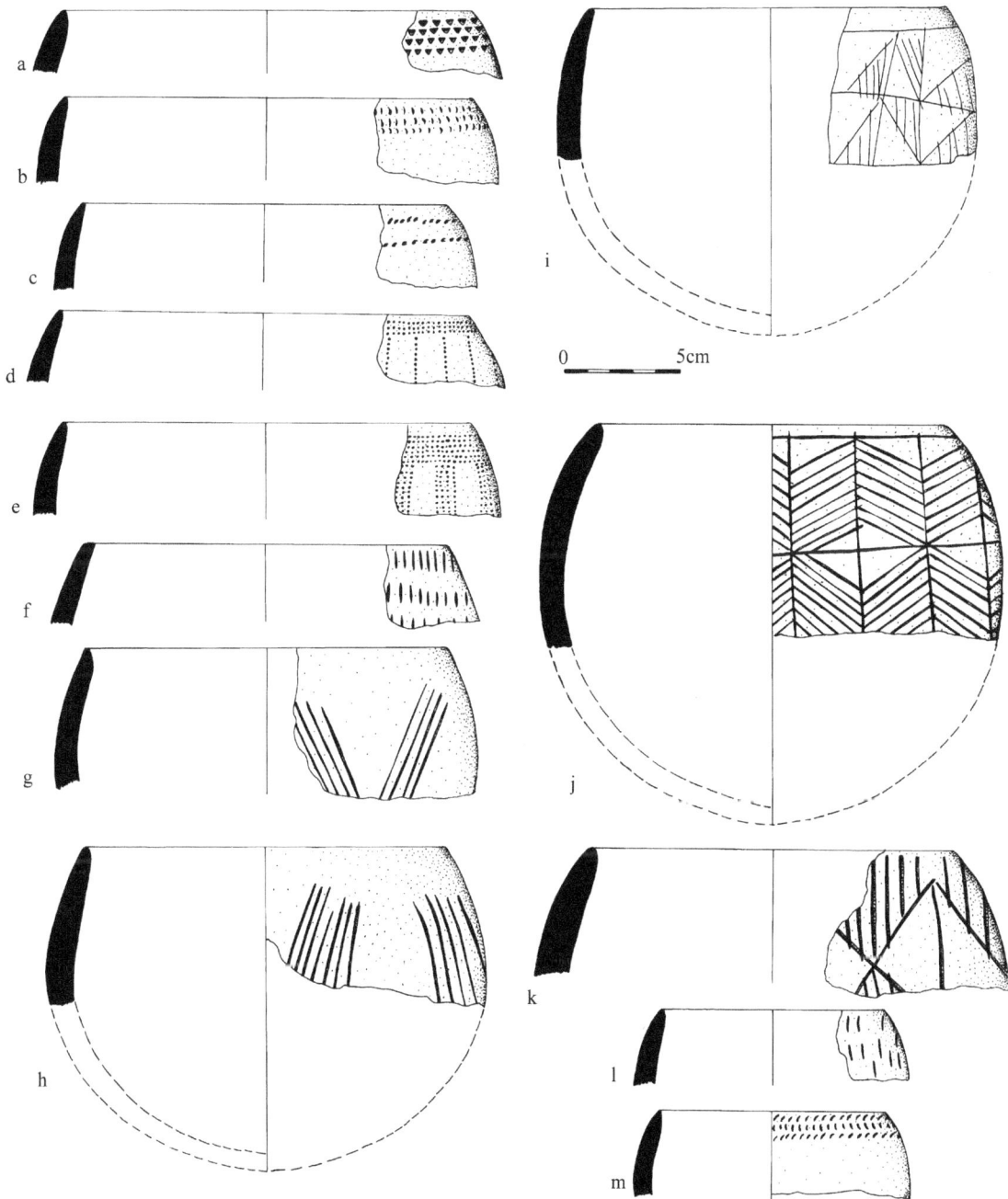

Figure 8.10   Early Mangaasi Ware c. 2200–1600 BP. a–k. vessel form 3i (a–c. Ef-motif 44; d. Ef-motif 49; e. Ef-motif 48; f. Ef-motif 14; g–h. Ef-motif 31; i. Ef-motif 65; j. Ef-motif 70; k. Ef-motif 64); l–m. vessel form 3iv (l. Ef-motif 14; m. Ef-motif 56).

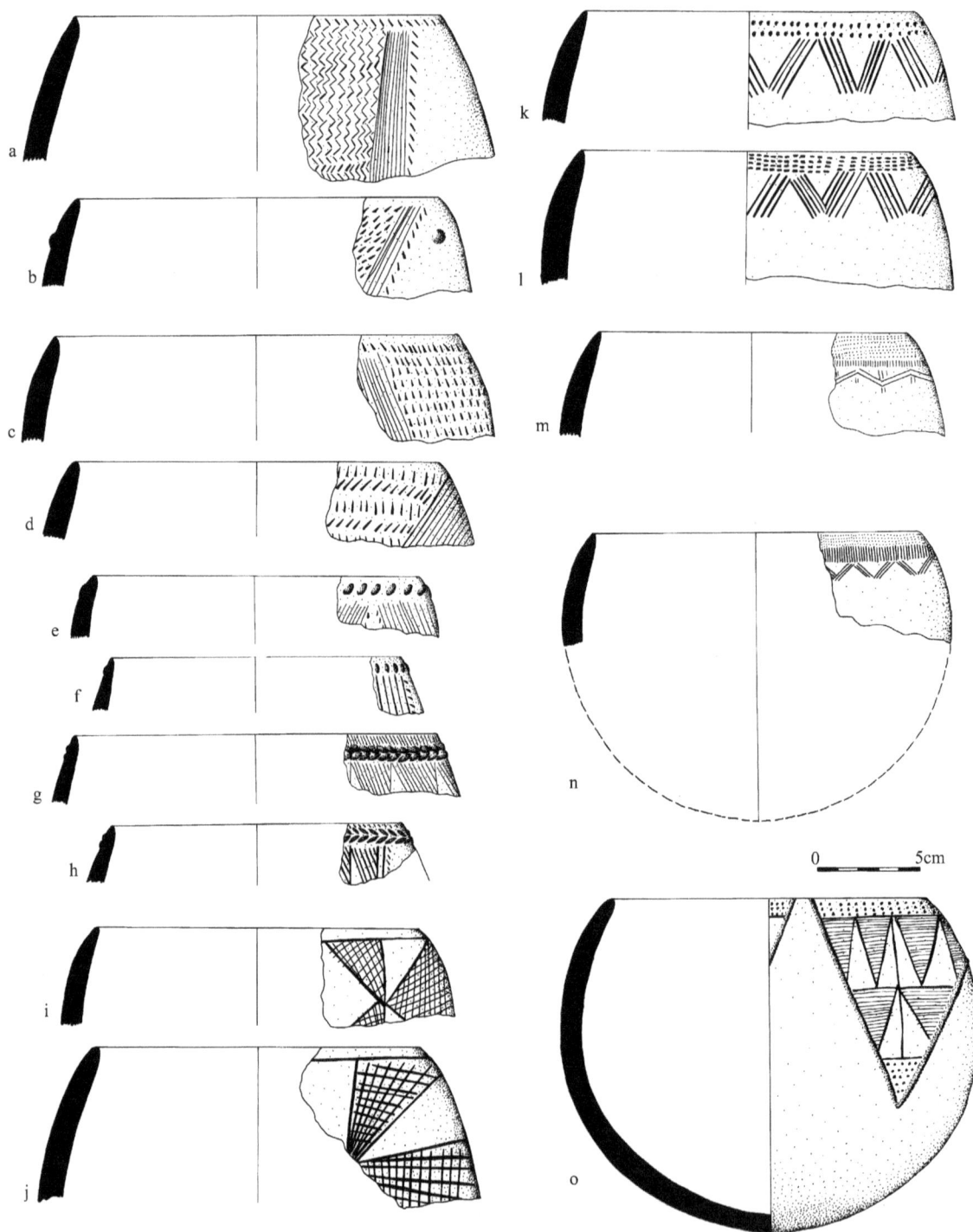

Figure 8.11 Early Mangaasi Ware c. 2200–1600 BP. All vessel form 3i. a. Ef-motif 50; b. Ef-motif 53; c. Ef-motif 52; d. Ef-motif 51; e–f. Ef-motif 59; g–h. Ef-motif 58; i–j. Ef-motif 66; k–l. Ef-motif 45; m. Ef-motif 46; n. Ef-motif 47; o. Ef-motif 62.

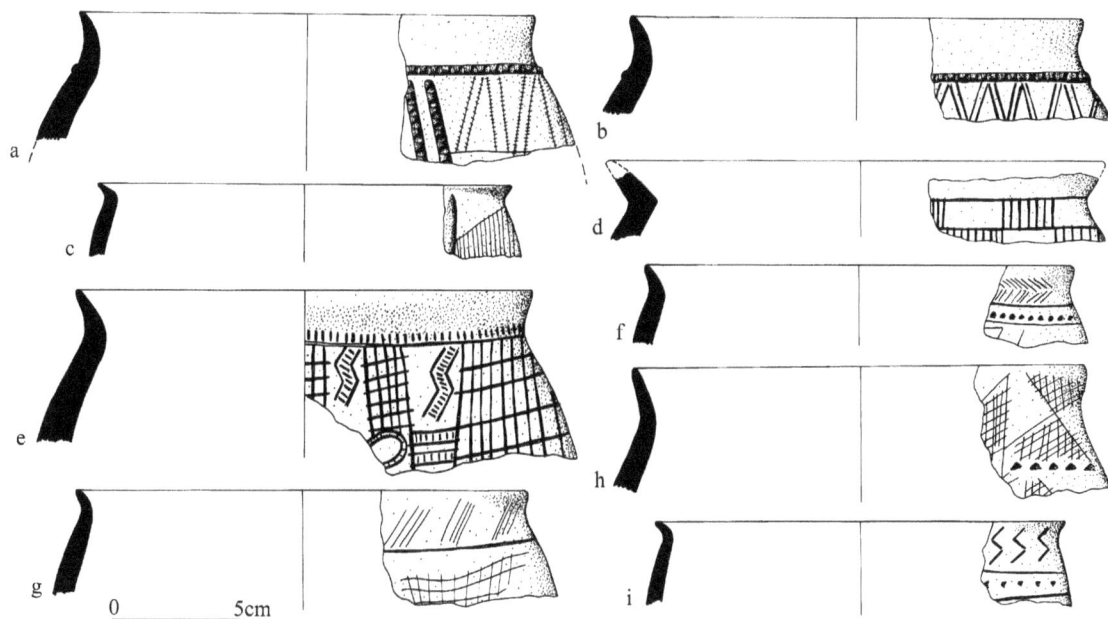

Figure 8.12 Late Mangaasi Ware c. 1600–1200 BP. a-c. vessel form 2i (a. Ef-motif 93; b. Ef-motif 78; c. applied relief and incision); d. vessel form 2iii (Ef-motif 69); e–i. vessel form 2i (e. Ef-motif 22; f. Ef-motif 95; g. Ef-motif 96; h. Ef-motif 97; i. Ef-motif 98).

incision. These motifs provide further evidence of the evolutionary and continuous nature of the sequence. Important to note is that these bands were not notched. This latter attribute became a distinctive feature only during the Late Mangaasi Phase.

Only a limited number of handles were recovered from the recent excavations at Mangaasi but they can be supplemented with the further range of forms illustrated by Garanger (1972:Fig.134). Of those that were attached to vessels all appeared to be associated with vessel form 3i and with a number of motifs that are exclusively associated with this phase of the sequence. As suggested in relation to discontinuous applied relief, handles appear to be associated with already established motifs, hinting that their appearance might be a slightly later addition to the Early Mangaasi Phase of the sequence. Vessels with handles were relatively rare which might be suggestive of them representing a ceremonial component of the assemblage.

The Late Mangaasi Phase is characterised by the reappearance of a variant of vessel form 2i (Fig. 8.12), which was first seen some 1500 years previously during the Arapus Phase. Lesser quantities of vessel form 2iii were also recorded. The vessels are decorated with a multitude of techniques in combination, although discontinuous applied relief and handles appear absent (Fig. 8.12). Notched applied bands were a modal attribute of this phase of the sequence. This last phase of the sequence is best represented by sherds recovered by Garanger from his excavations and surface collections. A whole host of motifs can be identified which were not recovered from the recent excavations at Mangaasi, but these can be, on the whole, reliably placed within the Late Mangaasi Phase due to a combination of the distinctiveness of the decorative techniques and associated vessel forms.

Much further refinement of the Mangaasi Phase of the sequence, particularly the Late Phase, is required both in terms of its chronology and characterisation. At this stage of the research it is emphasised that the date is circa 2200 BP for its appearance and circa 1200 BP for its termination at the Mangaasi site. The division at 1600 BP between Early and Late can also only be seen as an estimation. Certainly the Late Mangaasi Phase ceramics recovered from the recent excavations were restricted to levels post-dating 1600 BP. A predominance of modal attributes (notched applied relief and vessel form 2i) that are associated with the Late Mangaasi Phase ceramics were also found amongst Garanger's surface collected sherds which points to them being related to the last phase of ceramic production on the island.

The almost 2000 year ceramic sequence at Mangaasi/Arapus has been divided into five phases, namely Arapus (c.2900–c.2800 BP), Early Erueti (c.2800–c.2500 BP), Late Erueti (c.2500–c.2200 BP), Early Mangaasi (c. 2200–c.1600 BP) and finally Late Mangaasi (c.1600–c.1200 BP). It can be demonstrated through the sites of Erueti and particularly Teouma that the ceramic sequence on Efate first appears with the initial settlement of the island by Lapita colonists with the full suite of dentate-stamped vessels. On current data this appears to have occurred 100 years or so before the settlement of the Arapus site and is a relatively short-lived ceramic phase, thus explaining the absence of any dentate-stamped component at Arapus. The ceramic sequence on Efate was dynamic and changing throughout but can be largely shown to have followed a developmental trajectory, from Lapita through to the later phases. The pottery was made at or near the Mangaasi/Arapus site from local materials. Direct parallels with aspects of this sequence have been identified in the Shepherd Islands (Garanger 1972).

Unrelated but also recovered from Efate were the cord-marked sherds from the Mele Plain (Garanger 1972:Fig 33). These sherds have now been identified as being derived from the Jomon tradition of Japan (Dickinson *et al.* 1999). Their presence on Efate remains somewhat mysterious but the most parsimonious explanation would suggest that they have been deposited during the nineteenth or twentieth centuries.

Malekula

As outlined in Chapter 7, Malekula has potentially a 3000 year ceramic sequence. However the excavations were only able to characterise the two ends of that sequence. Although the results of research carried out between 1995 and 1997 on Malekula could not definitively demonstrate that Lapita was in evidence there around 3000 BP, it was argued that it could be assumed that this was the case (Bedford 2000b, 2001). Subsequent research on the small islands off the northeast coast have been able to confirm such a scenario (Bedford 2003). Research on the Northwest coast revealed the widespread presence of an essentially plainware ceramic phase dating from 2700 BP.

It appears that this largely plainware phase post-dates the initial Lapita occupation of Malekula. Three vessel forms have so far been identified, namely 2i, 2iii and 4ii (Fig. 8.13). Infrequent notching of the lip is the only regular decorative feature. A single dentate-stamped sherd was also found in association with the plainwares. It is proposed that this phase of the sequence on Malekula be named the Malua Phase after the site from which the earliest dates and largest sample were recovered. The phase appears to last for up to several hundred years, at least in this area of Malekula. The end point at c. 2500 BP is not well defined but all the sites from where the Malua plainwares were recovered were suggestive of short-term occupation.

Although the ceramic remains post-dating the Malua Phase are poorly represented, there are hints of some diversity in vessel form and decoration from the surface collected sherds from throughout the island. The sparse remains however enable only the most cursory of speculations. There are indications from the variety of surface sherds studied that the ceramic sequence continues after the Malua Phase (post 2500 BP) in some parts of the island although there was little excavated archaeological evidence of that to be found in the Northwest area.

A more comprehensive picture of the ceramic sequence of the last 1000 years, although far from complete, is beginning to emerge. Dating from at least 600 BP, distinctive coil-made 'bullet' shaped pots (Figs. 8.14, 8.15) appear in the archaeological record. They comprise two basic forms, one where the exterior of the vessel has not been smoothed which leaves a ribbed effect (form 1i) (Fig. 8.14a), and another similar pot where the outside surface has been smoothed to facilitate decoration, of which there is some variety both in technique and design motifs (Fig. 8.15). It has been noted that these same vessel forms and some of the decorative motifs are found across Malekula and a number of other northern islands of Vanuatu. It is proposed that this distinctive phase of the ceramic sequence be known as the Chachara Phase after the site from which a large sample of the distinctive pottery which defines this phase was recovered along with the thus far earliest associated date. Chachara Ware appears to have continued up to the time of European contact or shortly before c. 200 BP. Galipaud (1996a:4) has included the above ribbed vessel forms (form 1i) in a proposed Olpoï-style but this was a very tentative, chronologically unknown and somewhat ill-defined style related largely only to construction techniques.

Appearing to overlap somewhat and clearly developing from the earlier Chachara Ware are the *Naamboi* Wares unique to Malekula (Figs 7.19–7.23). Although detailed in the ethnographic period for their use in ceremonial activities there is no record of them being manufactured. It appears that the manufacture of these vessels dates from before European contact to a period when cooking in pots was no longer practised. At around the time of initial European contact the manufacture of the *Naamboi* appears to have ceased. *Naamboi* are dominated by the 'bullet' form with a variety of decorations. Ceramic tubes restricted to the south of Malekula are also included in the *Naamboi* Wares.

Three ceramic phases can now be identified in the archaeological record on Malekula. They are a Lapita Phase (Bedford 2003), a Malua Phase relating to an immediately post dentate-stamped period (c.2700–c.2500 BP) and the Chachara Phase tentatively dating from c. 600 BP to c. 200 BP. The ceramic record on Malekula is rich and varied and covers a considerable time depth, but its further definition must await future research.

Inter-island comparisons

Many of the islands of Vanuatu have received little archaeological attention, and on a number of islands i.e., Ambrym, Tanna and Aneityum, geomorphological considerations alone limit the archaeological prospects (Spriggs 1981; Spriggs and Wickler 1989). So any ceramic comparisons are largely restricted to assemblages recovered through the work of Galipaud in the Torres Islands (1998b), Ward (1979) in the Banks, Hedrick (nd) and Galipaud (1998a) on Malo, and Garanger (1972) in the Shepherds. Also of value have been the surface collections that have been carried out throughout the archipelago by the staff of the Vanuatu Cultural and Historic Sites Survey.

Dentate-stamped Lapita ceramics dating to c. 3000–2700 BP have now been identified on Malo, Aore, Tutuba, Malekula (Vao, Atchin, Wala, Uripiv), Efate (Erueti and Teouma), and Erromango (Ifo), but it can be argued that there is the potential that they will be found on most other islands of the archipelago (Bedford 2003; Bedford *et al.* 2004; Galipaud 1998a). This phase of the ceramic sequence associated with initial human settlement was in some cases short-lived, the dentate-stamped component drops out and the ceramic assemblages are then characterised by plainwares (e.g. Toga, Torres; Pakea, Banks Islands, Arapus, Efate; Malua Bay, Malekula) which in some instances also contain the occasional dentate-stamped vessel (e.g. Malua Ware, Malekula; Ponamla Ware, Erromango) or a minor incised component (e.g. Early Erueti Ware, Efate).

Soon after the appearance of this plainware phase across the archipelago, the ceramic sequence from Erromango began to follow its own largely independent trajectory, dropping out of

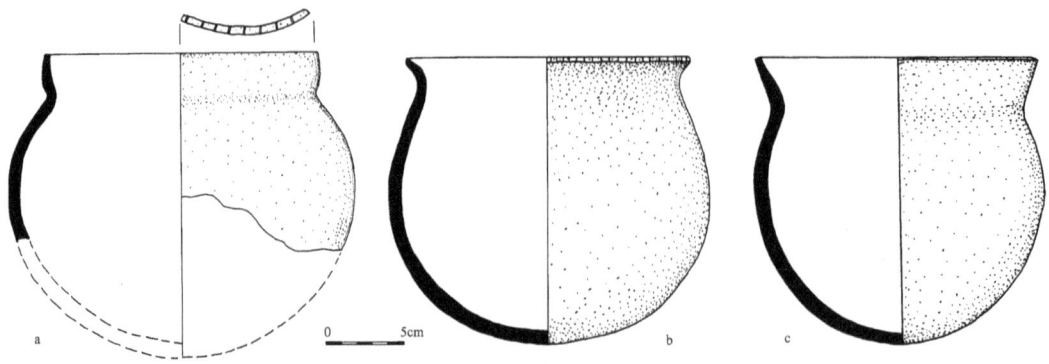

Figure 8.13 Malua Ware 2700–2500 BP. a. vessel form 4ii; b. vessel form 2i; c. vessel form 2iii.

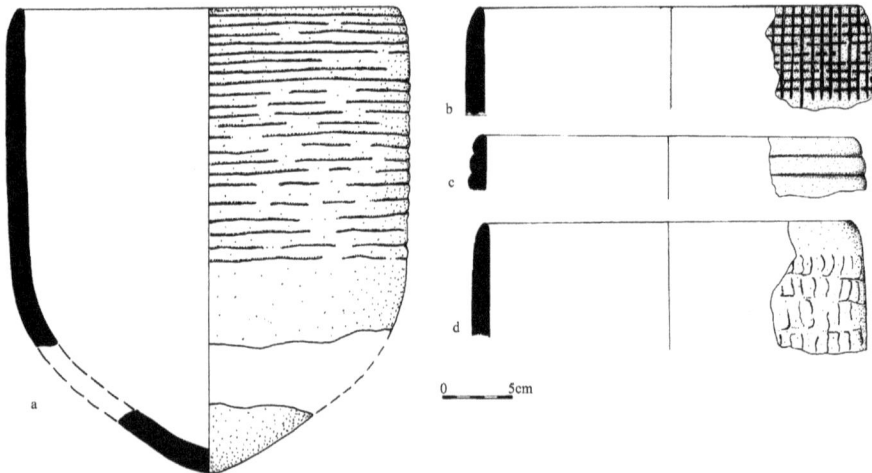

Figure 8.14 Chachara Ware c. 600–200 BP. a–c. vessel form 1i (a. M-motif 1; b. M-motif 3; c. M-motif 35); d. vessel form 1ii (M-motif 2).

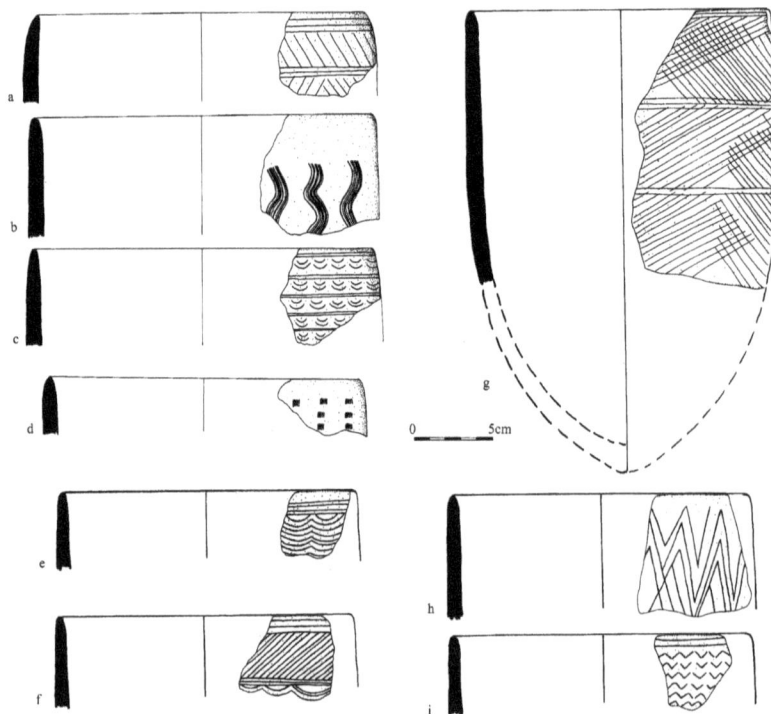

Figure 8.15 Chachara Ware c. 600–200 BP. a-i. vessel form 1iii (a. M-motif 4; b. M-motif 31; c. M-motif 32; d. M-motif 33; e. M-motif 5; f. M-motif 7; g. M-motif 4; h. M-motif 6; i. M-motif 8).

whatever interactive sphere sustained the ceramic homogeneity seen with Lapita. From 2600 BP quite distinctive decorative techniques and motifs define the Erromangan ceramic sequence, which continued for a further 600 years until the craft disappeared completely from the cultural repertoire. This same scenario might also be predicted for the islands further south such as Tanna and Aneityum.

To the north on Efate (occasionally visible from the north end of Erromango) an early Lapita colonising site has been identified at Teouma where preliminary observations of the dentate-stamped Lapita show clear parallels with ceramics further west (Bedford *et al.* 2004). A detailed 1600 year ceramic sequence post-dating Lapita has now been established for Efate. Strong parallels with this sequence have been identified, from both surface collections and excavations, on the offshore islands near Efate and in the Shepherd Group.

The assemblage from the Erueti site on Efate comprised predominantly Erueti Phase ceramics with minor components of both the earlier Lapita and later Mangaasi Phase ceramics. Erueti Ware has been noted on islands offshore from Efate, principally Lelepa (Garanger 1972:Fig. 79) but was also present on Emae in the Shepherds (MacLachlan 1939). Although the ceramic remains excavated by Garanger from Makura and Tongoa in the Shepherd Group (Garanger 1971, 1972) were said to be principally characterised by his Mangaasi style, there are indications that Erueti Ware was also present in the lowest layers of his excavations. The earliest dates from these excavations (c. 2540 BP on Makura) concur well with dates of c. 2500 BP for the Late Erueti Phase where wide flat lips were on the decrease but notching on the lip remained a modal attribute. From the lowest levels at Makura, Erueti Ware rims were in evidence (Garanger 1972:Fig. 220) and notching on the lip was identified as a persistent decorative feature (Garanger 1971:53). A similar picture can be seen on Tongoa where the earliest date for the ceramic remains (c.2460 BP) again fits well with the date for the Late Erueti Phase on Efate. The variety of rim forms in the lower levels at Tongoa sites was much greater than those recovered from the later levels or than at the Mangaasi site. Notching on the lip can again be seen to be as being relatively common, in the order of some 50% (Garanger 1972:53).

Pottery in the Shepherd Group does not appear in the archaeological record after 1000 BP which corresponds well with the revised sequence on Efate and dates associated with Mangaasi-style ceramics recovered earlier by the Shutlers (dating to c. 1000 BP) from Fila Island, located offshore from Efate (M.E. and R. Shutler 1965, 1968). Aknau Ware, a distinctive internally decorated pottery identified by Garanger (1972:96, 97), remains restricted to the Shepherds region although a single sherd was reported on Retoka, off Efate (1972:59). Aknau Ware was argued to have appeared around the end of the ceramic sequence in the Shepherds and like other ceramic traditions in the region did not continue after 1000 BP (Garanger 1972:82).

Excavations in the Banks Islands (Ward 1979) have also produced Erueti-like ceramics. Some wide flat rims, regular notching on the lip (40%) and carinated vessels all support this interpretation and would also fit with Ward's date of c. 2600 BP for the earliest phase of the site. Sherds which might be ascribed to the Mangaasi Phase as now outlined for Efate included only four sherds with applied relief, 28 incised sherds (although their size precludes definitive motif identification) and punctation. The small number of diagnostic sherds (168), however, limits somewhat the reliability of any conclusions. Ward's termination date of around 2000 BP for ceramics on the Banks Islands is looking more secure. Although not published in any detail Ward recovered sherds from a number of the other islands in the Banks, both through excavation and survey (Ward 1979:Appendix III 3-1- 3-13). The material was similar to that recovered from Pakea and further dates supported a c. 2000 BP termination date for ceramics in the Banks Islands in general (Ward pers. comm.). Their rarity on the ground surface (the Shutlers reported seeing none during their brief survey) further supports the argument of a relatively early date for the cessation of ceramic production in the Banks Islands. Certainly there is no evidence to date that suggests similar ceramics to those found on Malekula during the last 500 years were present.

There is little indication on any of the other islands of Vanuatu of Erueti-like ceramics except some vague parallels with a number of poorly provenanced sherds from Malo. However this disjunct distribution may be partly explained by a lack of archaeological research. It has at least been determined that Erueti Ware is not found further south than Efate.

Late and Early Mangaasi Ware, as has been demonstrated by Garanger and others, can be found throughout Efate and its offshore islands and up into the Shepherd Group. Its known distribution outside this area remains patchy at best and largely ill-defined. Again this can be primarily explained by a lack of archaeological research, but it can certainly be said that it does not feature in the Erromango cultural sequence, which by this date was non-ceramic. Ceramics it seems, on the basis of current research, disappeared from the archaeological record of the Shepherds and all islands further south to Efate by c. 1000 BP. Further south again ceramics disappeared some 1000 years earlier.

The ceramic remains from the other islands of northern Vanuatu are largely only known from surface collected material and generally associated with ethnographically recorded industries based on the west coast of Santo (Galipaud 1996c; Guiart 1958; MacLachlan 1939; Shutler 1968; Speiser 1996 [1923]). The *Naamboi* of Malekula have also received some detailed coverage (Cayrol 1992; Deacon 1934; Layard 1928). The recent archaeological research that has been carried out on Malekula has established that the ceramic record is rich and varied but requires much greater definition. The last 1000 years of the sequence is beginning to be revealed. It is characterised at least in the north of Malekula by 'bullet' shaped vessels either plain or decorated which have been named Chachara Ware and thus far have been dated from between c. 600–200 BP. From surface collections on other northern islands including Santo, Pentecost, Ambae, Malo and Maewo similar vessel forms and decoration have been identified, suggesting at least in the north of Vanuatu some form of cultural homogeneity existed up until the last few hundred years when ceramic production either ceased, as on most islands, or diverged on others, as was the case on Santo.

The known ceramic sequences from Vanuatu can thus be summarised as follows (see also Fig. 8.16). Lapita dentate and incised ceramics and plainware vessels were associated with initial settlement across the archipelago. Dentate-stamping was dropped as a decorative technique early on in the sequence to be replaced by largely plainware assemblages with a more restricted range of vessel forms. This is a Pacific-wide phenomenon (Kirch 1997:146–150) which suggests that whatever function the Lapita dentate-stamped vessels served, it was short-lived and very soon redundant. Ceramic assemblages that comprise predominately plainwares in Vanuatu are represented by Arapus, Early Erueti, Malua and Ponamla Wares. Less definitive but also hinted at are probably comparable ceramics recovered from Malo and the Banks and Torres Islands. Ceramic assemblages directly post-dating this plainware phase are really only known in any detail from Erromango, Efate and the Shepherds. The Erromango sequence began to follow its own trajectory from this point (Early to Late Ifo Ware), differentiating from the sequence found from at least Efate to the Shepherds (Early and Late Erueti to Early and Late Mangaasi). Ceramic production ceased on Erromango at c. 2000 BP and some 800–1000 years later on Efate and the Shepherds.

Ceramic production continued in the north of Vanuatu but at this stage of research it cannot be established whether or not it was a continuous sequence. Only on Malekula has any chronological detail of the late ceramic traditions been established, comprising of the distinctive 'bullet' shaped vessels both plain and decorated (Chachara Ware) dating from at least 600 BP to c. 200 BP. Direct parallels with surface collected sherds from other northern islands point to a period of cultural homogeneity. The last vestiges of the ceramic art to be found on Malekula are represented by the *Naamboi* which on form alone can be argued to have developed from the earlier Chachara Ware. By the time of regular European contact *Naamboi* were only recorded as being used

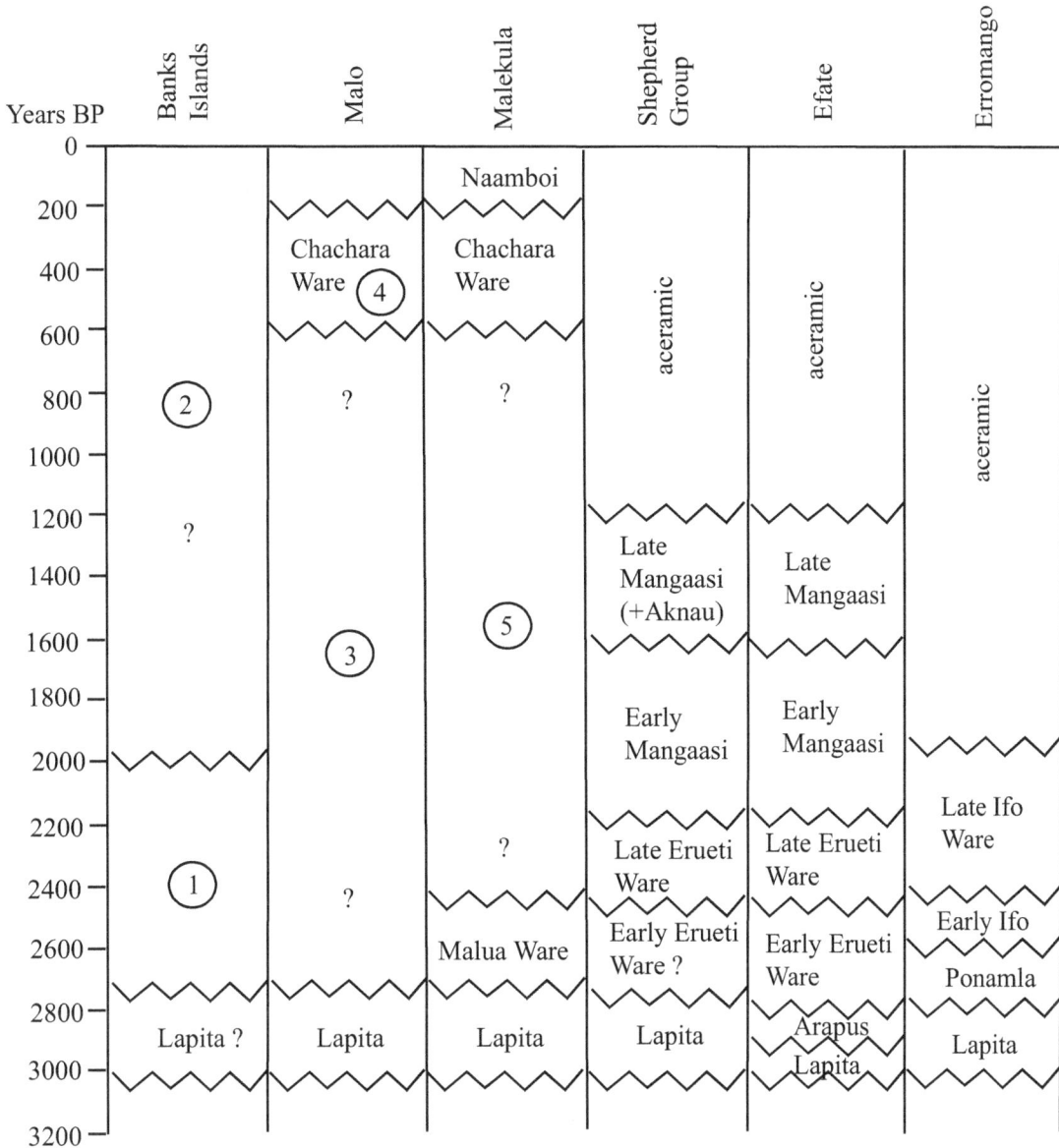

Figure 8.16 Ceramic sequences of Vanuatu.
1. The limited number of diagnostic sherds precludes finer definition; 2. Ward argued that ceramics disappeared from the Banks Islands around 2000 BP based on the results of Pakea only (Ward 1979:7-43); 3. The recovered ceramics from Malo have been overwhelmingly associated with Lapita with a general lack of evidence for any ceramics which parallel the sequences to the south; 4. The only other ceramics reported from Malo apart from Lapita sherds are those that resemble the late Chachara wares from Malekula; 5. Excavations on the northwest of Malekula failed to link the early and late ceramic traditions; 6. Lapita has been more recently found on the small islands of the north east coast of Malekula (Bedford 2003).

in a ceremonial capacity and there was no record of their manufacture. The *Naamboi* vessel form is thus far only known from Malekula. Pottery production continues in Vanuatu today but only on the west coast of Santo where it appears to show some generic connection to the distinctive late Santo vessel forms and decoration that are known only from surface collected materials.

Inter-regional comparisons

The more recent excavations in Vanuatu have provided detailed ceramic sequences at least from the islands of Efate and Erromango which enable a reassessment both of earlier results and conclusions from that archipelago and also from the broader region. It now seems appropriate to re-visit those sites where investigators originally claimed some sort of connection, in terms of homologous ceramic traits, to the Mangaasi ceramic tradition of Vanuatu as outlined by Garanger (1972).

Many of these sites were excavated over 20 years ago or more during a period which can be regarded as the pioneering exploratory phase of archaeological research in the Southwest Pacific. A number of clearly distinct ceramic styles began to be recognised from the very first scientific archaeological excavations carried out in the region (Gifford 1951; Gifford and Shutler 1956; Golson 1959) and these were later to be classified under the broad 'three great ceramic tradition' paradigm i.e. Lapita, Paddle Impressed and Incised and Applied Relief traditions (Golson 1968:10; Garanger 1972:122). It was within this all-encompassing paradigm that many of the subsequently excavated ceramics were initially arranged. Research focusing on Lapita sites predominated, as has been the case ever since, but the excavations of Specht on Buka (1969) and Garanger in Vanuatu (1972) had begun to reveal the rich variety and nature of the 'Incised and Applied Relief' tradition. These researchers identified similarities both between the respective traditions from Buka and Vanuatu and with other 'Incised and Applied Relief' traditions (see further detail of Buka comparisons below). Garanger noted parallels in the ceramics from both Fiji and New Caledonia with those of the Mangaasi tradition in Vanuatu (1972:124) but emphasised that research was only in the preliminary stages and that it was difficult to further determine their relationship (Garanger 1972:126). He tentatively suggested that the ultimate origins of the Mangaasi tradition lay much further to the west possibly in the Northeast of New Guinea where he noted parallels with Markham Valley pottery (Garanger 1972:124).

As a result of these pioneering excavations nascent theories regarding the widespread and interconnected nature of this Incised and Applied Relief tradition began to take hold (Frimigacci 1981; Golson 1972; Kennedy 1982; Kirch and Yen 1982; Spriggs 1984). These early assertions, despite having to rely heavily on the original results of Specht and Garanger, have only rarely been questioned (Bulmer 1999:570; Reeve 1989:55) and continue to have considerable influence right up to the present, with many of the earlier unmodified conclusions becoming incorporated into contemporary theory regarding colonisation, settlement and cultural transformation (Bedford 1999; Galipaud 1996a; Gorecki 1996; Gosden *et al.* 1989; Green 1997; Spriggs 1990a, 1993, 1997, 2000; Wahome 1997, 1999). In the most detailed and wide-ranging study of post-Lapita ceramics, and the first to utilise a range of statistical methods, Wahome (1997, 1999) has argued that the ceramic remains from Island Melanesia indicate continued contacts between various regions dating from the Lapita period, through to the post-Lapita period and right up to 800 BP, at which time increased regional diversification is evidenced (Wahome 1997:122).

Despite these claims, the Incised and Applied Relief tradition remains, as it did over thirty years ago, the least well known of the three traditions originally proposed by Golson (1968:10). Whilst Lapita now commands the attention of whole edited volumes and books and the Paddle impressed ware of at least New Caledonia has been interpreted as the domestic component of the Lapita assemblages, the Incised and Applied Relief tradition continues to be largely under-researched and the least well defined in terms of stylistic definition, its distribution in space and time and the human behaviour and cultural relationships it represented. Terminology has frequently been loosely applied. Categorising an assemblage as belonging to the Incised and Applied Relief tradition has often simply only enabled it to be differentiated from a dentate stamped or paddle-impressed assemblage. Assemblages are often stratigraphically mixed, poorly dated and/or the decorated sherds are small or in poor condition, making it difficult to identify motifs and/or vessel forms as opposed to simply decorative technique.

As previously argued (Bedford 2000a, 2000b), the Mangaasi ceramic tradition as defined by Garanger has been a somewhat convenient category over the last 30 years. With a general lack of detailed study or focus on post-Lapita style ceramics the Mangaasi tradition, spanning such a great time period and displaying a multitude of decorative techniques, has provided a structure into which almost all non-Lapita material could be placed. Major modification of this structure has now taken place and consequently this has affected the interpretation of a number of sites and their possible relationships vis-à-vis homologous ceramic chronologies and traits. A review of a number of these sites is outlined below, beginning with Vanuatu's nearest neighbours and then moving to progressively more distant locations.

Before venturing out into the various archipelagoes, some comment must be made on some aspects of the research and conclusions arrived at by Wahome (1999) regarding the Incised and Applied Relief tradition. Wahome argued that the post-Lapita ceramics he analysed indicated that there was some kind of watered down version of the earlier Lapita 'community of culture' throughout Melanesia which continued up until 800 BP. First it must be acknowledged that Wahome was dealing with a rather imprecise data set all of which had been recovered prior to 1995. Since that time an enormous amount of new data has been recovered, particularly in the case of Vanuatu, which has coincided with a period of major chronological reassessment of many ceramic sequences, including Lapita. The ceramic samples utilised by Wahome were often from sites that had been mixed, they were poorly dated, small in both sample and sherd size and in some cases poorly preserved. The majority of the ceramics used in his Melanesia-wide comparison were from published and unpublished sources rather than from the direct study of collections. Defined ceramic attributes therefore were in some cases, of necessity, very broad, particularly those associated with motifs (see for example Wahome 1999:Figs 3.14 and 3.28 for a demonstration of the wide variance found within single decorative categories). In other words a sound methodology was not matched by a suitable ceramic data set. Wahome himself lamented the quality of the available data admitting that they were too complex, poorly dated and often missing a large number of values, to fully support the argument of an inter-related ceramic tradition (Wahome 1999:177–178). Not being able to handle the vast majority of the Melanesia-wide assemblages included in his study Wahome was obliged to employ a different statistical approach to the Correspondence Analysis (Wahome 1999:96) that he had used on the Manus collections. In the Melanesia-wide study he utilised Proximity Analysis (Simple Matching and Jaccard's Indices) which is suitable for binary (presence or absence) category data (Wahome 1999:175), and a method that can accommodate less detailed broader attributes. However broad attributes can also produce a lumping effect which can be further accentuated if assemblages are mixed. Wahome was often not comparing well defined ceramic sequences but rather mixed assemblages that were poorly dated (e.g. Mangaasi and Erueti).

His study leaves us with a confusing picture where the ceramics assigned to the Incised and Applied Relief tradition appear to cluster into three groups i.e. 1) sites in Fiji and New Caledonia, 2) sites in New Ireland, Buka, Tikopia and Ifo on Erromango and 3) sites in Manus and Vanuatu (Wahome 1999:176–177). To highlight the all-encompassing nature of Wahome's attributes and the less than secure conclusions that were reached, further statistical tests were undertaken on his data set (Wahome 1999:Fig. 7.1). The same Proximity Analysis (Simple Matching and Jaccard's Indices) using SPSS Version 9.0 was applied to all of the sites as a single group rather than separating them into Post-Lapita and Late Prehistoric as Wahome had done prior to his statistical analysis (unfortunately the New Caledonian Nera and Fiji Modern attributes were not included in his list of selected attributes (Wahome 1999:Fig. 7.1)). What results is the grouping of a number of sites and assemblages which are known to have little association both in terms of chronology or homologous ceramic traits (Tables 8.1, 8.2 and Figs 8.17 and 8.18) which must cast doubt on the reliability of the results.

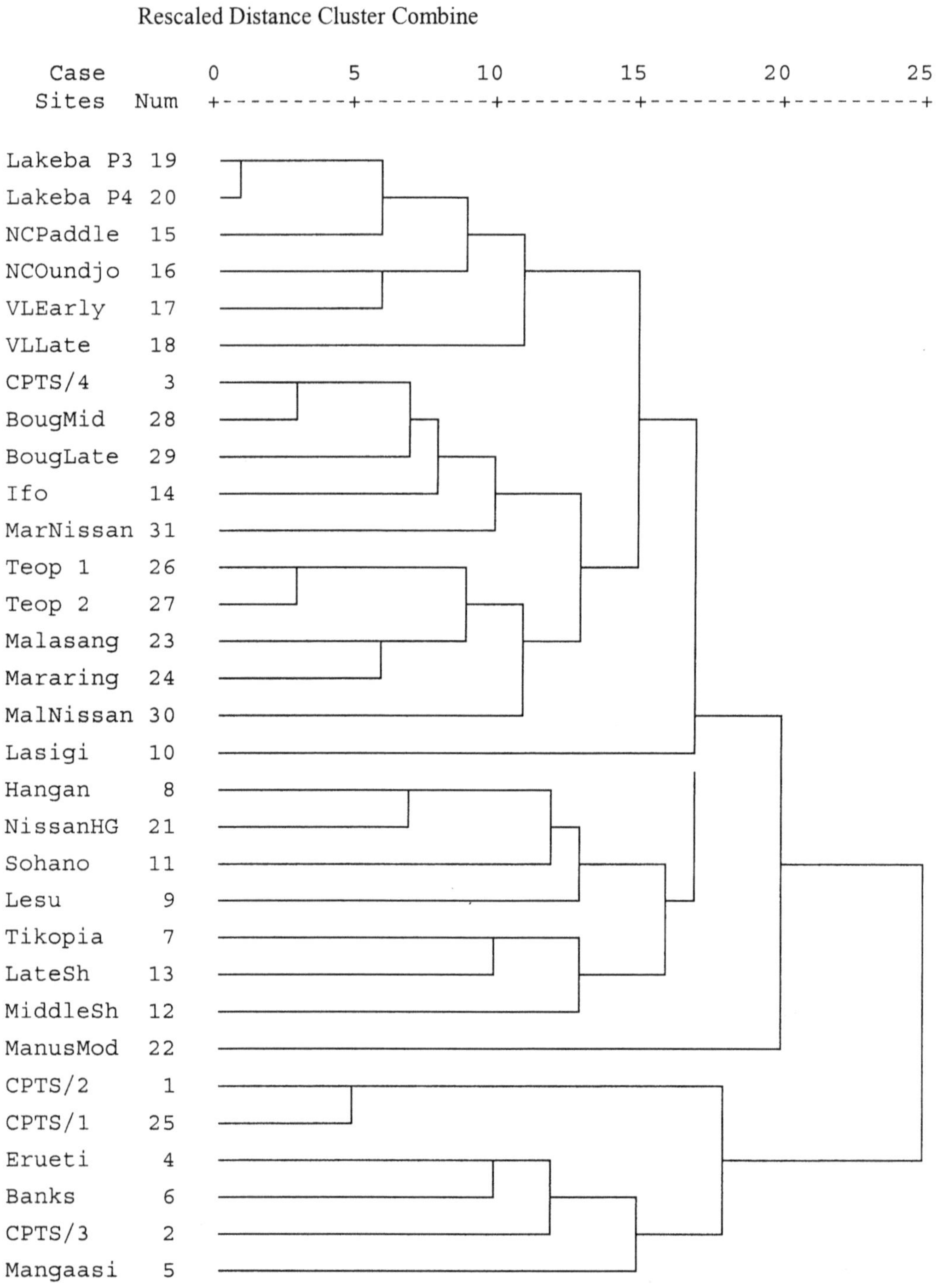

Figure 8.17 Simple matching-Average linkage (Between Groups): all post-Lapita sites as utilised by Wahome (1999).

Rescaled Distance Cluster Combine

```
         C A S E      0         5        10        15        20        25
       Sites    Num   +---------+---------+---------+---------+---------+

       CPTS/2     1
       CPTS/1    25
       CPTS/3     2
       Mangaasi   5
       Erueti     4
       Banks      6
       ManusMod  22
       MiddleSh  12
       LateSh    13
       Tikopia    7
       Hangan     8
       NissanHG  21
       Lesu       9
       Sohano    11
       CPTS/4     3
       BougMid   28
       BougLate  29
       MarNissan 31
       Ifo       14
       MalNissan 30
       Lasigi    10
       Teop 1    26
       Teop 2    27
       Lakeba P3 19
       Lakeba P4 20
       NCOundjo  16
       VLEarly   17
       VLLate    18
       NCPaddle  15
       Malasang  23
       Mararing  24
```

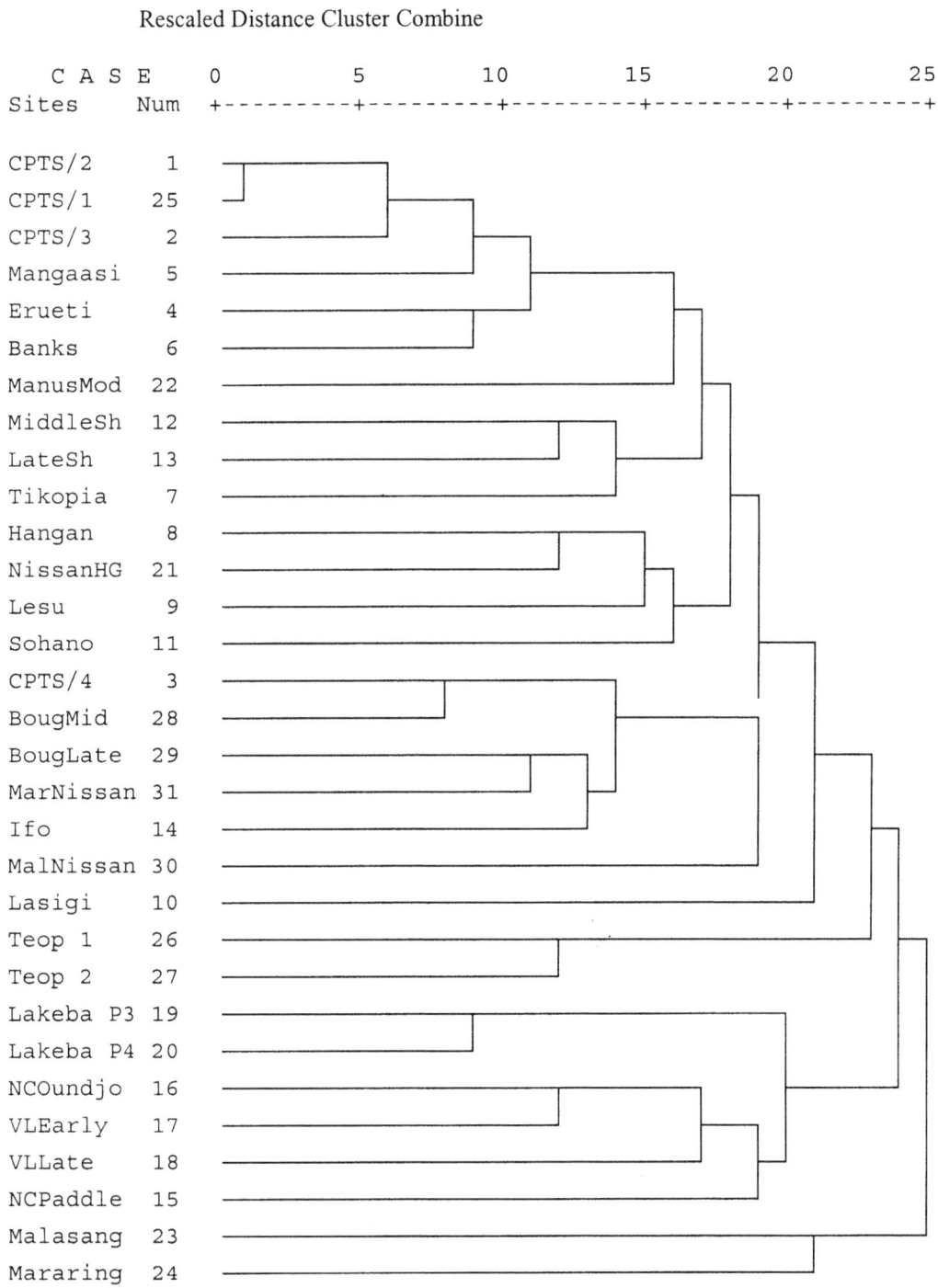

Figure 8.18 Jaccard Average Linkage (Between Groups): all post-Lapita sites as utilised by Wahome (1999).

Table 8.1 Simple Matching Coefficient: all post-Lapita sites utilised by Wahome 1999

Sites	MarNiss	MalNiss	Bouglate	BougMid	Teop 2	Teop 1	CPTS/1	Mararing	Malasang	ManusMod	NissanHG	Lakeba P4	Lakeba P3	VLLate	VLEarly	NCOundjo	NCPaddle	Ifo	LateSh	MiddleSh	Sohano	Lasigi	Lesu	Hangan	Tikopia	Banks	Mangaasi	Erueti	CPTS/4	CPTS/3	CPTS/2
CPTS/2	.524	.381	.452	.452	.333	.333	.881	.310	.310	.595	.429	.262	.310	.452	.429	.381	.405	.476	.548	.619	.571	.381	.452	.500	.429	.595	.643	.619	.405	.738	
CPTS/3	.595	.452	.524	.571	.452	.452	.762	.429	.381	.524	.452	.381	.381	.571	.500	.500	.476	.500	.571	.595	.595	.500	.524	.571	.500	.762	.714	.738	.524		.738
CPTS/4	.738	.738	.810	.905	.786	.786	.333	.714	.762	.667	.738	.714	.714	.619	.643	.643	.667	.833	.619	.643	.643	.738	.667	.762	.595	.667	.524	.690		.524	.405
Erueti	.667	.667	.643	.738	.619	.619	.548	.500	.548	.548	.619	.500	.690	.452	.476	.548	.500	.619	.548	.667	.714	.571	.643	.690	.524	.786	.595		.690	.738	.619
Mangaasi	.548	.667	.476	.571	.548	.548	.714	.500	.381	.548	.548	.381	.381	.571	.429	.429	.429	.548	.524	.643	.548	.452	.667	.619	.643	.595		.595	.524	.714	.643
Banks	.548	.548	.714	.571	.548	.762	.571	.690	.857	.619	.667	.762	.571	.595	.571	.571	.571	.667	.595	.643	.643	.643	.690	.714	.643		.595	.786	.667	.762	.595
Tikopia	.619	.714	.690	.762	.619	.619	.357	.619	.643	.595	.667	.738	.690	.643	.690	.762	.690	.690	.786	.714	.667	.619	.643	.786		.643	.643	.524	.595	.500	.429
Hangan	.619	.714	.762	.714	.738	.786	.429	.857	.619	.667	.833	.810	.714	.714	.714	.738	.738	.738	.714	.738	.738	.762	.762		.786	.714	.619	.690	.762	.571	.500
Lesu	.571	.595	.571	.667	.643	.690	.429	.619	.619	.595	.667	.571	.619	.595	.595	.595	.571	.595	.714	.571	.571	.690		.762	.643	.690	.667	.643	.667	.524	.452
Lasigi	.667	.619	.619	.738	.667	.714	.357	.690	.690	.690	.667	.714	.690	.595	.667	.667	.643	.595	.643	.619	.571		.690	.762	.619	.643	.452	.571	.738	.500	.381
Sohano	.667	.667	.595	.738	.667	.714	.500	.643	.643	.595	.762	.643	.643	.643	.643	.643	.643	.667	.595	.619		.571	.571	.738	.667	.643	.548	.714	.643	.595	.571
MiddleSh	.643	.667	.595	.643	.738	.738	.548	.690	.500	.643	.762	.548	.643	.548	.619	.619	.548	.667	.738		.619	.619	.571	.738	.714	.643	.643	.667	.643	.595	.619
LateSh	.643	.714	.571	.595	.786	.786	.476	.714	.619	.595	.667	.619	.619	.571	.595	.595	.643	.595		.738	.595	.643	.714	.714	.786	.595	.524	.548	.619	.571	.548
Ifo	.762	.690	.833	.690	.667	.714	.405	.595	.643	.690	.667	.643	.690	.548	.667	.667	.643		.595	.667	.667	.595	.595	.738	.690	.667	.548	.619	.833	.500	.476
NCPaddle	.667	.667	.714	.667	.738	.738	.333	.762	.762	.619	.786	.857	.857	.786	.833	.738		.643	.643	.548	.643	.643	.571	.714	.690	.571	.429	.500	.667	.476	.405
NCOundjo	.667	.667	.595	.833	.905	.714	.310	.690	.738	.595	.762	.833	.762	.738	.857		.738	.667	.595	.619	.643	.667	.595	.738	.762	.571	.429	.548	.643	.500	.381
VLEarly	.571	.762	.595	.690	.690	.667	.357	.690	.690	.595	.667	.786	.857	.786		.857	.833	.667	.595	.619	.643	.667	.595	.714	.690	.571	.429	.476	.619	.500	.429
VLLate	.548	.595	.571	.714	.643	.595	.381	.667	.619	.595	.571	.762	.762		.786	.738	.786	.548	.571	.548	.643	.595	.595	.714	.643	.595	.571	.452	.619	.571	.452
Lakeba P3	.690	.810	.738	.690	.786	.786	.238	.762	.762	.571	.762	.952		.762	.857	.762	.857	.690	.619	.643	.643	.690	.619	.714	.690	.571	.381	.690	.714	.381	.310
Lakeba P4	.690	.762	.714	.714	.786	.786	.190	.857	.810	.643	.738		.952	.762	.786	.833	.857	.643	.619	.548	.643	.714	.571	.810	.738	.762	.381	.500	.714	.381	.262
NissanHG	.714	.762	.738	.738	.714	.714	.357	.643	.738	.643		.738	.762	.571	.667	.762	.786	.667	.667	.762	.762	.667	.667	.833	.667	.667	.548	.619	.738	.452	.429
ManusMod	.762	.762	.667	.667	.595	.595	.571	.667	.619		.643	.643	.571	.595	.595	.595	.619	.690	.595	.643	.595	.690	.595	.667	.595	.619	.548	.548	.667	.524	.595
Malasang	.738	.738	.667	.762	.786	.786	.238	.857		.619	.738	.810	.762	.619	.690	.738	.762	.643	.619	.500	.643	.690	.619	.619	.643	.857	.381	.548	.762	.381	.310
Mararing	.690	.714	.714	.714	.881	.786	.238		.857	.619	.643	.857	.762	.667	.690	.690	.762	.595	.714	.690	.643	.690	.619	.857	.619	.690	.429	.500	.714	.429	.310
CPTS/1	.452	.310	.381	.381	.262	.262		.238	.238	.571	.357	.190	.238	.381	.357	.310	.333	.405	.476	.548	.500	.357	.429	.429	.357	.571	.429	.548	.333	.762	.881
Teop 1	.762	.810	.738	.833	.905		.262	.786	.786	.595	.714	.786	.786	.595	.667	.714	.738	.714	.786	.738	.714	.714	.690	.786	.619	.762	.548	.619	.786	.452	.333
Teop 2	.762	.762	.690	.786		.905	.262	.881	.786	.595	.714	.786	.786	.643	.690	.905	.738	.667	.786	.738	.667	.667	.643	.738	.619	.548	.548	.619	.786	.452	.333
BougMid	.714	.762	.857		.786	.833	.262	.714	.762	.667	.738	.714	.714	.714	.690	.833	.667	.690	.595	.643	.738	.738	.667	.714	.762	.571	.571	.738	.905	.571	.452
Bouglate	.833	.786		.857	.690	.738	.262	.714	.667	.667	.738	.714	.738	.571	.595	.595	.714	.833	.714	.595	.595	.619	.571	.762	.690	.714	.476	.643	.810	.524	.452
MalNiss	.762		.786	.833	.762	.810	.310	.714	.738	.667	.762	.762	.810	.595	.762	.667	.667	.690	.714	.595	.667	.619	.595	.714	.714	.548	.667	.667	.738	.452	.381
MarNiss		.762	.833	.833	.714	.762	.452	.690	.738	.595	.714	.690	.690	.548	.571	.667	.667	.762	.643	.667	.667	.667	.571	.619	.619	.595	.548	.667	.738	.595	.524

Key to sites: CPTS/2 (various Manus sites dating from 1900-350 BP PL); CPTS/3 (various Manus sites dating from 2100-1900 BP PL); CPTS/4 (Mouk, Lapita and PL); Erueti (PL); Mangaasi (PL); Banks (Pakea assemblage PL); Tikopia (Sinapupu ware PL); Hangan (Buka PL); Lesu (New Ireland PL); Lasigi (New Ireland PL); Sohano (Buka PL); MiddleSh (Shortlands Middle LP); LateSh (Shortlands Late LP); Ifo (Erromango PL); NCPaddle (New Caledonia Paddle impressed PL); NCOundjo (New Caledonia Ounjdo PL); VLEarly (Viti Levu Early PL); VLLate (Viti Levu Late PL); Lakeba P3 (Lakeba Period 3 PL); Lakeba P4 (Lakeba Period 4 LP); NissanHG (Nissan Hangan PL); ManusMod (Manus Modern LP); Malasang (Buka LP); Mararing (Buka LP); CPTS/1 (various Manus sites post-dating 800 BP LP); Teop 1 (Phase 1 LP); Teop 2 (Phase 2 LP); BougMid (Bougainville Middle LP); Bouglate (Bougainville Late LP); MalNiss (Nissan Malasang LP); MarNiss (Nissan Mararing LP)

PL = Post-Lapita

LP = Late Prehistoric

Table 8.2 Jaccard Similarity Coefficient: all post-Lapita sites utilised by Wahome 1999

Sites	CPTS/2	CPTS/3	CPTS/4	Erueti	Mangaasi	Banks	Tikopia	Hangan	Lesu	Lasigi	Sohano	MiddleSh	LateSh	Ifo	NCPaddle	NCOundjo	VLEarly	VLLate	Lakeba P3	Lakeba P4	NissanHG	ManusMod	Malasang	Mararing	CPTS/1	Teop 1	Teop 2	BougMid	BougLate	MalNiss	MarNiss
CPTS/2		.686	.242	.529	.571	.485	.273	.344	.343	.257	.438	.529	.424	.333	.219	.212	.273	.324	.121	.061	.273	.485	.121	.121	.861	.152	.152	.281	.281	.212	.394
CPTS/3	.686		.286	.621	.613	.630	.276	.357	.355	.300	.414	.469	.400	.300	.214	.250	.276	.379	.103	.071	.233	.375	.103	.143	.722	.179	.179	.333	.286	.207	.414
CPTS/4	.242	.286		.381	.231	.300	.056	.333	.263	.313	.211	.318	.200	.500	.067	.063	.118	.158	.077	.000	.267	.300	.167	.077	.222	.250	.154	.636	.385	.214	.353
Erueti	.529	.621	.381		.433	.609	.200	.409	.400	.280	.478	.481	.296	.333	.125	.217	.154	.179	.087	.045	.304	.321	.136	.087	.486	.238	.182	.450	.318	.333	.417
Mangaasi	.571	.613	.231	.433		.600	.375	.360	.357	.207	.321	.483	.310	.296	.200	.240	.269	.333	.037	.000	.269	.429	.077	.077	.657	.160	.208	.280	.185	.240	.321
Banks	.485	.630	.300	.609	.600		.286	.400	.280	.318	.348	.423	.280	.318	.200	.316	.227	.304	.100	.053	.227	.308	.100	.158	.486	.211	.278	.444	.368	.316	.476
Tikopia	.273	.276	.056	.200	.375	.286		.400	.316	.158	.263	.429	.471	.222	.133	.286	.250	.211	.071	.083	.176	.227	.000	.154	.250	.000	.143	.118	.188	.200	.200
Hangan	.344	.357	.333	.409	.360	.400	.400		.444	.278	.389	.476	.368	.353	.200	.267	.313	.143	.143	.077	.500	.333	.067	.143	.314	.214	.308	.333	.333	.267	.389
Lesu	.343	.355	.263	.400	.357	.280	.316	.444		.350	.381	.458	.429	.227	.100	.211	.250	.217	.111	.059	.389	.280	.053	.111	.316	.235	.313	.263	.143	.150	.208
Lasigi	.257	.300	.313	.280	.207	.318	.158	.278	.350		.182	.280	.227	.200	.118	.176	.222	.190	.133	.071	.294	.160	.063	.133	.270	.059	.125	.235	.313	.176	.182
Sohano	.438	.414	.211	.478	.321	.348	.263	.389	.381	.182		.360	.318	.182	.167	.222	.143	.125	.118	.063	.412	.292	.118	.118	.400	.250	.176	.353	.278	.294	.333
MiddleSh	.529	.469	.318	.481	.483	.423	.429	.476	.458	.280	.360		.522	.391	.227	.273	.304	.269	.136	.095	.304	.423	.087	.087	.486	.130	.238	.318	.261	.273	.417
LateSh	.424	.400	.200	.296	.310	.280	.471	.368	.429	.227	.318	.522		.227	.158	.150	.190	.217	.111	.125	.190	.280	.111	.250	.389	.050	.167	.263	.333	.150	.318
Ifo	.333	.300	.500	.333	.296	.318	.222	.353	.227	.200	.182	.391	.227		.118	.176	.222	.136	.133	.000	.222	.381	.063	.000	.306	.200	.125	.400	.500	.176	.444
NCPaddle	.219	.214	.067	.125	.200	.200	.133	.200	.100	.118	.167	.227	.158	.118		.364	.214	.333	.333	.250	.308	.200	.091	.091	.200	.083	.083	.067	.143	.071	.313
NCOundjo	.212	.250	.063	.217	.240	.316	.286	.267	.211	.176	.222	.273	.150	.176	.364		.500	.313	.300	.222	.286	.190	.083	.000	.194	.077	.077	.063	.133	.067	.222
VLEarly	.273	.276	.118	.154	.269	.227	.250	.313	.250	.222	.143	.304	.190	.222	.214	.500		.438	.364	.182	.176	.227	.071	.071	.250	.067	.143	.056	.118	.059	.143
VLLate	.324	.379	.158	.179	.333	.304	.211	.143	.217	.190	.125	.269	.217	.136	.333	.313	.438		.286	.231	.150	.250	.059	.125	.297	.056	.118	.100	.100	.105	.174
Lakeba P3	.121	.103	.077	.087	.037	.100	.071	.143	.111	.133	.118	.136	.111	.133	.333	.300	.364	.286		.600	.286	.100	.000	.111	.111	.100	.100	.077	.167	.083	.188
Lakeba P4	.061	.071	.000	.045	.000	.053	.083	.077	.059	.071	.063	.095	.125	.000	.250	.222	.182	.231	.600		.083	.053	.000	.143	.056	.000	.100	.000	.091	.100	.133
NissanHG	.273	.233	.267	.304	.269	.227	.176	.500	.389	.294	.412	.304	.190	.222	.308	.286	.176	.150	.286	.083		.286	.154	.000	.250	.000	.067	.267	.267	.286	.333
ManusMod	.485	.375	.300	.321	.429	.308	.227	.333	.280	.160	.292	.423	.280	.381	.200	.190	.227	.250	.100	.053	.286		.158	.048	.486	.143	.150	.300	.300	.190	.476
Malasang	.121	.103	.167	.136	.077	.100	.000	.067	.053	.063	.118	.087	.111	.063	.091	.083	.071	.059	.000	.000	.154	.158		.250	.111	.150	.100	.167	.167	.083	.267
Mararing	.121	.143	.077	.087	.077	.158	.154	.143	.111	.133	.118	.087	.250	.000	.091	.000	.071	.125	.111	.143	.000	.048	.250		.111	.100	.100	.077	.077	.083	.188
CPTS/1	.861	.722	.222	.486	.657	.486	.250	.314	.316	.270	.400	.486	.389	.306	.200	.194	.250	.297	.111	.056	.250	.486	.111	.111		.139	.139	.257	.257	.194	.361
Teop 1	.152	.179	.250	.238	.160	.211	.000	.214	.235	.059	.250	.130	.050	.200	.083	.077	.067	.056	.100	.000	.000	.143	.150	.100	.139		.500	.364	.154	.273	.333
Teop 2	.152	.179	.154	.182	.208	.278	.143	.308	.313	.125	.176	.238	.167	.125	.083	.077	.143	.118	.100	.100	.067	.150	.100	.100	.139	.500		.250	.071	.167	.250
BougMid	.281	.333	.636	.450	.280	.444	.118	.333	.263	.235	.353	.318	.263	.400	.067	.063	.056	.100	.077	.000	.267	.300	.167	.077	.257	.364	.250		.500	.417	.533
BougLate	.281	.286	.385	.318	.185	.368	.188	.333	.143	.313	.278	.261	.333	.500	.143	.133	.118	.100	.167	.091	.267	.300	.167	.077	.257	.154	.071	.500		.308	.533
MalNiss	.212	.207	.214	.333	.240	.316	.200	.267	.150	.176	.294	.273	.150	.176	.071	.067	.059	.105	.083	.100	.286	.190	.083	.083	.194	.273	.167	.417	.308		.375
MarNiss	.394	.414	.353	.417	.321	.476	.200	.389	.208	.182	.333	.417	.318	.444	.313	.222	.143	.174	.188	.133	.333	.476	.267	.188	.361	.333	.250	.533	.533	.375	

A few of the more diverse assemblages that showed some sort of correlation are as follows; Ifo (now known to date no later than 2000 BP) with Bougainville Middle and Late (dating from 900 BP-Modern (Wahome 1999:157)), Tikopia (1200 BP–750 BP) with Late Shortlands (c. 200 BP); CPTS/4 (the Mouk site which included dentate-stamped ceramics) with Bougainville Middle and Late. What the Simple Matching and Jaccard's Indices demonstrate in fact, when all sites from all periods are included in the analysis, is that there appears to be very little separation of any of the assemblages, further highlighting the all-encompassing nature of the attributes and the unreliability of the conclusions. The one cluster that continues to show some variance with the other assemblages (although it now also includes post 800 BP assemblages) is that represented by assemblages from Manus and Vanuatu.

Having argued that the basic data sets were not sufficiently detailed to test the 'reality' of a Melanesia-wide Incised and Applied Relief tradition, a point noted by Wahome himself (see above), further detailed dissection of the separate clusters would seem unnecessary here. Only the Manus-Vanuatu connections will be discussed in finer detail below.

New Caledonia

Postulated homologous ceramic traits and implications of frequent interaction between New Caledonia and Vanuatu date from at least the first archaeological excavations in New Caledonia (Gifford and Shutler 1956:93) and some time before (Avias 1950:122). It continued to be a widely published view (Frimigacci 1975; Frimigacci and Maitre 1981; Golson 1972:568–573; Spriggs 1984:215), which has only rarely been questioned (Green and Mitchell 1983:63), and one which is still generally accepted today (Galipaud 1997:104; Sand 1995:123, 125; 1999:156). In the light of the clear divergence early in the ceramic sequences between Erromango and Efate the claimed connections between the sequences of the nearest neighbours, Vanuatu and New Caledonia, require some closer scrutiny.

It has now been clearly established (Bedford *et al.* 1998; Sand 1997) that the initial settlement of both Vanuatu and New Caledonia was associated with the appearance of Lapita ceramics. Single sherds recovered from Malo and Erromango have even been sourced to New Caledonia (Dickinson and Shutler 1979; Spriggs and Wickler 1989). This is the somewhat more easily explained and unambiguous part of the ceramic sequence. Two distinctive incised motifs associated with immediately post-Lapita ceramics from both New Caledonia and Vanuatu also show clear parallels (e.g. E-motif 28 [Fig. 8.3m], Ef-motif 1 [Fig. 8.9c] and Ef-motif 31 [Fig. 8.8a] from Erromango and Efate respectively with motifs 55 and 56 from the Îles des Pins (Frimigacci 1974:63) and a sherd from Ongoué (Sand and Ouétcho 1993:Fig. 10)).

More significant it seems however, is the overwhelming evidence for dissimilarities in the ceramic record from New Caledonia and Vanuatu which appeared even during the Lapita phase of settlement. Paddle impressed or Podtanéan Ware, so dominant in the recovered ceramics from the Koné period (first 1000 years) in New Caledonia, is completely absent from the archaeological record in Vanuatu, in terms of either the variety of vessel forms or decoration. Appearing soon after Lapita (c. 2600 BP), at least in the south of New Caledonia, is the largely incised Puen tradition (Sand 1995:85–91). Design motifs consist principally of incised chevrons and triangles located under the rim but a number of other incised and appliqué motifs are also known (Sand 1999:148). Although Sand argues that some of the decorative chevrons and triangle patterns found on the Puen Ware are similar to those found on the Mangaasi-style pottery, he is not totally convinced. He also pointed out that there were clear differences and that the Mangaasi motifs were more complex than those associated with the Puen Ware (Sand 1999:149). With the revised ceramic sequence from Efate now placing the Puen Ware in a similar chronological time frame to that of Erueti Ware, any previously argued-for similarities with Mangaasi in vessel form and or decoration appear much less likely.

What then of the Plum tradition, which appeared some time around 1800 BP in the south and west of New Caledonia and has been argued to have greater affinities with Mangaasi-style ceramics (Sand 1999:149)? The vessels tend to be globular with incurving rims along with occasional handles. Again the vessels are largely decorated with incised motifs, comprising chevrons making up simple or complex leaf-like motifs. Incised concentric triangles are also a distinctive motif form. There is general agreement amongst researchers in New Caledonia that these motif forms seem likely to have developed from the earlier Puen or Podtanéan traditions (Galipaud 1997; Sand 1995:123). However to argue for striking similarities between certain decorations found on the Plum and Mangaasi Wares would seem to be overstating the connections (Sand 1995:123), particularly if complete design motifs are compared. Incised concentric triangles as seen on the Puen and Plum Wares are not found amongst the Vanuatu ceramics. Incised chevrons are found but are almost always only a component of more complex motifs. The handles associated with the Plum Ware vessels appear to be quite different morphologically to those originally recovered by Garanger from Mangaasi, although this material is suggested as having some possible influence on the Plum Ware (Sand 1995:125). Galipaud describes the Plum pottery as being 'decorated rarely and only under the rim and around the handles' (1997:103) which is a complete contrast to much of the heavily decorated Mangaasi-style pottery.

By 1000 BP in New Caledonia there is a divergence in the ceramic sequences between the north and south of the Grand Terre. The Oundjo tradition developed in the north while the Nera tradition was restricted to the south. Both these ceramic traditions continued until around the time of European contact (Sand 1999:155). The very distinct and somewhat unique Oundjo vessel forms and sequence as defined by recent research have not inspired any suggestions of affinity with Vanuatu ceramics, whereas the Nera vessel forms have been described as being very similar to certain Mangaasi forms (Sand 1995:151). This is certainly the case but globular incurving rim vessels are a somewhat ubiquitous form throughout much of the New Caledonian and Vanuatu sequences. More to the point is the distinctive decoration found on the Nera vessels which is completely unlike any of the motifs originally illustrated by Garanger (1972) or those that have been subsequently identified. Quite different decorative techniques are also employed with the Nera Ware, most distinctive of these being 'pustules' which are unknown in Vanuatu. The suggested similarities between certain decorations found on the Nera pottery with those on some of the *Naamboi* of Malekula (Sand 1995:152) might be seen by some as a speculative possibility despite the contrast in vessel morphologies and chronology. However if it proved to be the case it would at the same time certainly confirm the unrelated nature of the Mangaasi and Nera traditions. The *Naamboi* are both morphologically and chronologically unrelated to anything found on Efate.

Revised dating of the ceramic sequence on Efate further confirms the unlikely nature of any connections between the Nera and Mangaasi traditions. The appearance of Nera barely coincided with the last phase of the Mangaasi tradition where vessels tended to have outcurving rims and were decorated with a multitude of motifs consisting of applied, incised and punctate techniques. Around the time of the appearance of the Nera tradition in New Caledonia ceramic manufacture on Efate had ceased. Following the clarifications of the ceramic sequences from both Vanuatu and New Caledonia that have occurred over the last ten years, it is now becoming increasingly clear that, starting immediately post-Lapita, they have predominately followed independent trajectories.

Tikopia and Vanikoro

Kirch and Yen (1982) have provided a detailed analysis of the prehistory of Tikopia through extensive archaeological investigations and detailed publication. The prehistory was divided into three broad phases, Kiki (c. 2900 BP–c. 2100 BP), Sinapupu (c. 2100 BP–c. 750 BP) and Tuakamali (c. 750 BP–200 BP). The rigidity of these divisions was tempered somewhat by the authors who pointed out that in defining the cultural sequence and various phases, a great deal of variability may have been easily overlooked (Kirch and Yen 1982:323).

The first two phases were associated with ceramic remains. Kiki Ware ceramics were primarily characterised by plainwares with a limited decorated component comprising dentate stamping, incision and applied bands (Kirch and Yen 1982:197). This largely calcareous-tempered locally produced material was argued to have existed for up to nine hundred years. An abrupt change in the ceramic sequence emerged in the archaeological record from 2000 BP with the appearance of the Sinapupu Ware. These ceramics were argued to be, on a number of grounds, imported into Tikopia, possibly from the Banks Islands in Vanuatu and were tentatively associated with a Mangaasi Ceramic Series (Kirch and Yen 1982:205). The scenario of ceramics being imported into Tikopia from the Banks Islands or other areas of Vanuatu is one that has become accepted in the general literature (Spriggs 1997:176), but there have been some dissenting voices (Ward 1989). The more recent results from Vanuatu tend to support the arguments put forward by Ward questioning the likelihood of importation of ceramics from the Banks Islands.

Several arguments led Kirch and Yen (1982) to suggest that the Sinapupu Ware was imported into Tikopia from Vanuatu. Distinctive temper and the low frequency of sherds in the Sinapupu Phase deposits were put forward as supportive evidence that the pottery was imported (Kirch and Yen 1982:331). The initial temper analysis carried out by Dickinson on sherds from the Sinapupu Phase suggested that on petrographic data alone it seemed likely that they originated from a Tikopian source (Kirch and Yen 1982:371). However later analysis of similar sherds from nearby Vanikoro began to point to a possible Vanuatu (Santo) origin (Kirch and Yen 1982:372). At the time Dickinson had access to only a limited range of ceramic samples from the other islands. In a more recent report, Dickinson (1997) has been able to reject Santo as a possible source but also noted that individual examples of lithic-rich tempers in sherds from Pakea (Banks Islands), Vanikoro and Tikopia could not be sorted petrographically without labels on the slides as a guide. Dickinson noted that overall geologic relations still did not inherently preclude Tikopia as being the source of temper for the Sinapupu Ware (Kirch and Yen 1982:5). Certainly on petrographic grounds alone a Vanuatu source for the Sinapupu Wares now appears less definitive. As regards variable ceramic density between the different phases at the Tikopia sites, there are a whole host of possible explanations other than a low density of sherds implying importation (Kirch and Yen 1982:192).

The importation of basaltic glass from the Banks Islands into Tikopia was another argument for linking the Sinapupu Wares with the Banks Islands. However the vast majority (90%) of the basaltic glass recovered from Tikopia tentatively ascribed to a northern Vanuatu source was recovered from the Tuakamali Phase (568) whereas only four flakes were recovered from the Sinapupu Phase (Kirch and Yen 1982:256). This scenario tends to suggest that the four flakes from the Sinapupu Phase might well have been displaced by gardening or similar activities from the later phase.

Although few of the illustrated Sinapupu Ware sherds appear large enough to determine motif form as opposed to decorative technique, Kirch and Yen (1982:205) claim nearly exact parallels in design motif with Mangaasi ceramics. Arguments over the validity of subjective homologous traits can often be an unproductive and contentious exercise but comment on the nature of the Sinapupu assemblage is required here to further assess the validity of the claimed Mangaasi connections. The proposed c. 1200 year Sinapupu sequence is based on a total of only 152 sherds, of which 9 were rim sherds and 25 (18 of which are illustrated) were decorated body sherds (Kirch and Yen 1982:200–201). The majority of the illustrated decorated sherds are no more than a few square centimetres in area which severely limits motif identification and comparison. The assorted decorative techniques could equally be favourably paralleled with a number of assemblages further west on New Ireland or Buka or to those further east to Fiji. A single paddle impressed sherd from the Sinapupu assemblage was in fact assigned a Fijian origin (Kirch and Yen 1982:202).

The more recent excavations in Vanuatu have demonstrated the complexity of post-Lapita ceramic assemblages particularly in terms of motif forms which were continually changing through time. Rarely is decorative technique alone sufficient to define post-Lapita assemblages. Rather, it requires large assemblages with a healthy percentage of larger decorated sherds for the finer definition of the ceramic sequences.

The origins of the ceramics from the nearby island of Vanikoro were explained in similar terms to those from the Sinapupu Phase on Tikopia (Kirch 1983:108). The conclusions, as stressed by Kirch, were tentative as the archaeological excavation and survey that was carried out was only exploratory in nature (Kirch 1983:71). A total of only 37 sherds were recovered, nine being decorated body sherds, which were assigned to the Mangaasi Ceramic Series. These sherds were also, due to their rarity and petrographic profile, assigned to a foreign source, namely Santo. The recent recovery of Lapita dentate-stamped sherds from Vanikoro which are argued to be indigenous to the island sheds further light on the petrographic debate. Dickinson notes that the Lapita sherds contain similar temper sands to the later Vanikoro sherds suggesting that they too were made on the island (Dickinson 2000:23). He further points out that in this region of the Pacific (Banks-Vanikolo-Backarc) temper analysis is not able to readily resolve the question of sherd origins (Dickinson 2000:23). The same difficulties as seen with the Sinapupu assemblage can be pointed out with the sherds recovered from Vanikoro. Both sample and sherd size preclude anything more than highly speculative comparisons. In addition, what appears to be parallel rows of plain applied bands on several of the sherds from Vanikoro (Kirch 1983:99) is a decorative feature not thus far identified from the Central Vanuatu sequence with which Kirch was making comparisons.

The claimed connections of the Sinapupu Wares from Tikopia and the ceramics recovered from Vanikoro with the Mangaasi ceramics from Vanuatu must now be seen as less than secure. There is also some doubt as to whether those same ceramics can be sourced to the Banks Islands or other islands of Vanuatu. Certainly the revised chronology of the central Efate ceramic sequence in combination with the evidence thus far gleaned from research in the Banks Islands further reduces the likelihood of ceramic imports to Tikopia from at least those two sources in Vanuatu.

Fiji

Claimed linkages or resemblances between the ceramics of Vanuatu and Fiji again have a lengthy history (Garanger 1966:76; Gifford 1951:236; MacLachlan 1939:54; Surridge 1944:21–22), albeit with occasional dissent (Hunt 1987; Palmer 1971:83; Rechtman 1992) and one which has continued to be a persuasive theme (Best 1984:493, 2002:30–32; Frost 1979:79). The re-dating of the Vanuatu sequence and its further refinement in association with recent detailed research in Fiji (Clark 1999) now casts serious doubt on these assertions. The earliest claimed parallels (Gifford 1951; MacLachlan 1939; Schurig 1930; Surridge 1944) relied solely, particularly in the case of Vanuatu, on comparisons of small numbers of poorly provenanced surface collected sherds during a period prior to the establishment of ceramic sequences from either archipelago. Alleged similarities identified both then and more recently (Best 1984; Frost 1979) have been restricted largely to ceramics that can now be associated with the later part (post 1700 BP) of the Fiji sequence. All of the homologous ceramic traits claimed between Fiji and Vanuatu have been restricted to broad parallels in decorative technique rather than any detailed comparisons of motifs or vessel forms. On closer examination these parallels can be seen as less than secure.

Frost associated ceramics from the Vuda and Ra Phases dating from 850 BP with Mangaasi material, describing its appearance in Fiji as a Melanesian ceramic intrusion that could reasonably be attributed to earlier Central Vanuatu sources (Frost 1979:79). Decorative techniques associated with the Vuda Phase (c. 900–450 BP (Best 1984:293–295)) are characterised by a number of different paddle impressed motifs and end-tool impression. However paddle impression is to date unknown in Vanuatu and the illustrated end tool impressed motifs of the Vuda Phase (Frost 1979:69) show little resemblance to any of the Central Vanuatu motif forms. Apart from this clear variance in patterns of surface modification both the Vuda and even more so the Ra Phases (450–200 BP) can now be shown to post-date ceramic manufacture on Efate and the Central Islands of Vanuatu.

Best has suggested that influence from Vanuatu was somewhat earlier at around 1700 BP during the Navatu Phase and that it potentially continued for some hundreds of years as indicated by the recovery in Fiji of basaltic glass from a Banks Islands source. This he argued seemed to coincide with the appearance of three new ceramic decorative techniques. These new techniques were described as 'asymmetric and fingernail incising and finger pinching, cord wrapped paddle impressing and rim notching' (Best 1984:493). Again however, under closer scrutiny these decorative techniques can be seen to have been either utilised very differently or were even non-existent in the Central Vanuatu sequence of the same period.

None of the Fijian motifs that utilised the fingernail as a decorative technique (Birks 1973:131–137; Clark 1999:125) have parallels with the Central Vanuatu sequence where fingernail decoration was extremely rare and largely restricted to pinched bands. Added to the disjunct nature of the sequences is rim notching which by the time it had appeared in Fiji had long disappeared from the decorative repertoire in Central Vanuatu. Further to the dissimilarities of decorative techniques that were utilised during this period is paddle impressing. This was a dominant decorative feature of the Navatu Phase in Fiji which, as noted above, is unknown in Vanuatu.

Best himself warned against placing too much importance on the asymmetric incision or rim-notching techniques arguing that they were hardly diagnostic enough to be of any significance particularly over such long distances. Parallels with the cord marked Mele sherds (Garanger 1971) were seen as potentially more meaningful (Best 1984:493). It appears that even these potential connections can now also be dismissed. The sherds recovered from Mele have now been identified as not being indigenous to Vanuatu and to be associated with the much earlier Japanese Jomon tradition (Dickinson *et al.* 1999). The Fijian cord-marked ceramics are of local manufacture (Best 1984:333–334). In a more recent publication Best acknowledges that following the recent research results in Vanuatu his earlier assertions, at least regarding ceramic connections to central Vanuatu, are now looking less secure. He does however persist with the idea of a Vanuatu-Fiji ceramic connection, albeit as yet undiscovered (Best 2002:30–32), suggesting (as does Spriggs 2004:142) that such evidence may well be found in northern Vanuatu when more detailed research is undertaken there.

The most recent detailed research which has focused specifically on ceramic change post-dating the Lapita phase in Fiji had this to say about external influences or connections, 'a comparison with ceramic assemblages from beyond Fiji failed to identify any close analogues to the Fijian assemblage established during the major shift of 2300–1900 BP' (Clark 1999:219). The absence of ceramic analogues between Fiji and Vanuatu which date to this period can certainly now be further confirmed. There is no evidence to indicate parallels between the Fijian material with either the Central Vanuatu Erueti Phase or the Ifo Phases from Erromango (Bedford and Clark 2001).

It is argued that earlier claims of widespread homologous ceramic traits indicating some sort of cultural interaction and or diffusion between Vanuatu and its nearest neighbours, at least up to 1000 BP, are now looking much less certain based on current knowledge. What then of the more distant claimed connections. The focus is towards sites further west, as the ceramic sequences from islands east of Fiji, post-dating Lapita, comprise primarily of plainware assemblages which have for a long time been argued to have developed largely independently of any influences further west than Fiji (Golson 1961; Green 1963).

Panaivilli

The ceramic assemblage recovered from the inter-tidal reef flat in front of Panaivilli village, Roviana Lagoon, in the Western Solomons is included in this discussion primarily because Reeve (1989) has been one of the few researchers who has questioned the Incised and Applied Relief paradigm. The assemblage was recovered by Reeve and Spriggs from an inter-tidal reef flat and so provided few chronological clues but a number of comparisons were made with other ceramic assemblages. The ceramics displayed a number of decorative techniques including incision, applied relief, stick impression and perforation. Rims were often notched or scalloped (Reeve 1989:49). Much of the assemblage was thought to be related to or at least derived from Lapita, although no dentate-stamped sherds were recovered. Parallels were noted with the sites of Kreslo, Ambitle and Watom. Reeve also noted that many of the post-Lapita ceramics recovered from across Island Melanesia (Admiralties, New Ireland, Buka, Tikopia and Vanuatu) were decorated with similar techniques as those ceramics from Panaivilli. But more importantly he emphasised that in few of these areas had recovered collections of ceramics been large enough to be analysed in terms of their motif systems (Reeve 1989:55). Only two assemblages were cited as providing sufficient detail for such analysis, namely the Mangaasi (Garanger 1972) and Buka sequences (Specht 1969). He further commented that although these two sequences had been used to support a pan-Melanesian ceramic tradition, Panaivilli which was located geographically midway between these supposedly related traditions and displayed the same decorative techniques, appeared to have a quite unrelated design system (Reeve 1989:55). This observation is further confirmed when the Panaivilli sherds are compared directly to the more recently established sequences of central and southern Vanuatu.

Buka

Specht, along with Garanger, was one of first researchers to construct lengthy post-Lapita ceramic sequences in the Southwest Pacific (Specht 1969). The Buka ceramic sequence was divided into six phases which spanned almost 3000 years. The initial appearance of ceramics was associated with the Buka Phase (c. 2500–2200 BP) which was followed by the Sohano (c. 2200–1450 BP), Hangan (c. 1450–1200 BP), Malasang (c. 1200–850 BP), Mararing (c. 850–350 BP) and Recent Phases (c. 350 BP-present).

Specht's Buka sequence has largely stood the test of time and subsequent research by others with the addition of an early Lapita phase and some minor embellishments and shifts in the dating of the various phases which are now as follows; Early Lapita (3200–2500 BP), Late Lapita or Buka (2500–2200 BP), Sohano (2200–1400 BP), Hangan (1400–800 BP), Malasang (800–500 BP) and finally Mararing/Recent (500–0 BP) (Wickler 1995:Table 1.1). In attempting to place the sequence in a regional perspective Specht had few other archaeologically recovered assemblages with which to compare the Buka material (Specht 1969:222). The ceramics that he had excavated from Watom (Specht 1968), were all that was available. Specht noted that in terms of form and decoration the Watom material displayed many similarities with the Buka Phase (i.e. immediately post-Lapita) material but that there was a striking lack of similarity with the Sohano and later styles (Specht 1969:228). At the time much more concentrated research was being carried out in Remote Oceania (New Caledonia, Vanuatu and Fiji) and it was to recovered ceramic collections from these islands that Specht looked for comparative data (Specht 1969:236–259). Cited parallels between Vanuatu, New Caledonian and Fiji materials have already been outlined. Discussion here is restricted to the claimed parallels of the Buka sequence to that from Vanuatu more than 2000 kilometres distant.

It must be pointed out that Specht was at somewhat of a distinct disadvantage in that his doctoral thesis was completed well before the detailed publication of Garanger's material. Rather, comparisons with the Vanuatu ceramics had to be made from an earlier article (Garanger 1966) and a copy of a paper read by Garanger at the Wenner-Gren Conference in Fiji in 1969 which was later published (Garanger 1971). Apart from the ceramics excavated by Garanger, Specht noted that a single sherd illustrated in MacLachlan (1939) also demonstrated some affinities with the Hangan Incised and Relief Substyle (Specht 1969:236). However the provenance of this sherd was originally uncertain and now seems even less likely to have come from Vanuatu as the illustrated parallel wavy applied band motif is not known from any recovered ceramics from the Central or Southern islands of Vanuatu.

Apart from the Buka Style ceramics which Specht tentatively associated with the ceramics from Erueti and Malo, the only other Buka sequence ceramics that he strongly argued were related to the Mangaasi material was the Sohano Incised and Relief Substyle (Specht 1969:240). Other styles were seen as either dissimilar or possibly at best having only vague affinities (Specht 1969:258). Specht suggested that the discontinuous applied relief decoration found in the Sohano Incised and Applied Relief Substyle (Plate XI–11) was almost identical to sherds associated with Garanger's early Mangaasi (Garanger 1971:Figs 4h, i) but he did also note an absence of incision on the Buka sherds which are a feature of this style of Mangaasi material (Specht 1969:240). Also not found associated with any of the paralleled Mangaasi material were perforations or cut-outs (a minor feature at this stage of the Buka sequence), or notched lips which were a standard feature of this substyle.

Combining Garanger's later more detailed publication (1972) with the results from the recent excavations on Efate another less ambiguous parallel with this Buka Substyle can be pinpointed (also noted by Ward 1979:7–28). This is the incised pendant triangle or alternating oblique parallel incision which creates a large zigzag motif (Specht 1969:Plate X1–12). This is one motif which does demonstrate intriguing similarities with motifs from Vanuatu (Garanger 1972:Figs 129, 267; Fig. 8.11, k–n). But do these relatively minor similarities in terms of the styles as a whole warrant 'contact with the industries of the Central New Hebrides' (Specht 1969:257)? It has been argued that this particular motif form or variations of it are found throughout much of the Central Vanuatu sequence, perhaps even having generic connections with Lapita motifs (see for example Anson motifs 157–159, 297, 299). The same motif or variations of it also appear with some frequency from the Sohano at least into the Hangan Style on Buka. In fact the universality of this particular motif is demonstrated across much of the Pacific and it is seen even in the ethnographic period in places such as New Guinea (Arifin 1991:Fig. 5; May and Tuckson 2000:Figs 9.117, 9.214, 9.75), Vanuatu (McLachlan 1939:39) and Fiji (Rossitto 1990:Fig. 8; Surridge 1944:30).

The small size of the Buka-style sherds hamper the possible identification of this motif form in the earlier part of that sequence but it may be that in the cases of both Vanuatu and Buka, continuities from earlier Lapita design motifs are being identified in the post-Lapita material of those islands rather than them being the result necessarily of any direct connections between the industries. Again it might be noted that lip notching, which was absent in the Vanuatu materials of this date, was a modal attribute associated with this motif form both in the Sohano and Hangan traditions. Ward, commenting specifically on perceived similarities between the Mangaasi and Sohano Incised and Applied Relief Substyle, warned that 'superficial similarities and differences can be identified among many different assemblages and it is probably more productive to concentrate attention upon ceramics which are less ambiguously related in time and space' (Ward 1979:7–29).

In summing up any Vanuatu connections Specht did make the point that there appeared to be an absence from the Buka area of many of the common motifs associated with the Mangaasi material (Specht 1969:241). He was justifiably cautious when proposing tentative external long distance links with the Buka sequence, an approach which later researchers have not always been so cognisant.

It was the later research of Wickler (1995) which refined the ceramic sequence on Buka as outlined above. Importantly he was able to include the previously largely missing Lapita component of the sequence and to convincingly demonstrate continuity throughout the sequence from Lapita through to the later traditions. As well as a selection of Lapita motifs Wickler (1995:383) added a total of 35 new motifs to Specht's original tally. Wickler was not so concerned with identifying inter-regional parallels or potential influences for the Buka sequence, which had been an aspect of Specht's research. Rather he took an approach that assumed 'that changes in pottery attributes are derived from local processes unless clear external links are demonstrated in the form of exotic clay, temper or other evidence' (Wickler 1995:443). While not dismissing Specht's claimed links with other post-Lapita traditions in Melanesia Wickler stressed that without the establishment of much finer detailed local ceramic sequences from throughout the region the assessment of widespread stylistic change was premature (Wickler 1995:443).

New Ireland

Moving further west to New Ireland is the site of Lossu (formerly Lesu) where a number of mound features were excavated by White in 1969 (White and Downie 1980). Many thousands of sherds were recovered but most were described as no larger than a thumbnail which severely limited the potential for the reconstruction of vessel form and or decorative motifs (White and Downie 1980:203). A total of 332 sherds were decorated of which nearly half (152) were rims. Plain rims (454) were predominant. The sherds were decorated by a combination of incision and applied relief and scalloping of the rim (White and Downie 1980:205). A single dentate-stamped sherd was found on the surface near the excavations. The stratigraphy of the excavated areas was mixed but the ceramics were initially thought likely to be associated with a date of 2460 BP. However reassessment by Golson (1991:257) has led White (White and Murray-Wallace 1996:42) to also entertain a later date of 1660 BP for one of the excavated mound features.

White and Downie (1980:209) argued that the decorative elements found at Lossu had their closest parallels with ceramics recovered by Garanger (1972) from sites in Vanuatu rather the more nearby sites of Watom or Buka. The Vanuatu connection however also appeared to be somewhat less than definitive because, as pointed out by White and Downie (1980:212), any Vanuatu parallels seemed to be matched by at least an equal number of mismatches. In summing up White and Downie (1980:214) placed the Lossu pottery firmly within the popular contemporary paradigm of the time, the Incised and Applied Relief tradition that could apparently be found throughout Island Melanesia.

Lasigi is another area of New Ireland from which post-Lapita ceramics have been reported (Golson 1991). The two sites of Dori and the Mission returned what was described as a collection of extremely fragmented and abraded sherds (Golson 1991:251). Clear parallels were noted between the 344 diagnostic sherds from the two sites with those from Lossu, both in terms of decoration and rim form, although the range in both was narrower (Golson 1991:251). Golson was circumspect when comparing the Lasigi ceramics with those from other sites from nearby islands, describing the New Ireland coastal sites as a whole (Lasigi and Lossu) as being difficult to fit into the wider picture. Noting that others had made connections with the New Ireland ceramic assemblages Golson emphasised the need for detailed comparison of these post-Lapita assemblages with each other and with Lapita before any conclusions could be drawn (Golson 1991:256).

It may be that there is some sort of connection (albeit often based on mixed, fragmented ceramic assemblages) between the immediately post-Lapita ceramics recovered from some sites in the Bismarck Archipelago, Buka and as far east as Nissan as a number of authors have argued (Gosden *et al.* 1989:571, 582; Green and Anson 1991:179; Kirch *et al.* 1991:152, 160; Spriggs 1991b:241; White and Murray-Wallace 1996:43; Wahome 1997:122) but the extrapolation of somewhat vague homologous ceramic traits, to explain change in the ceramic sequences of islands over 2000 kilometres further east, seems much less certain.

Admiralty Islands

It was Kennedy (1981, 1982) who first attempted to make some sense of the Admiralty Islands ceramic sequence from a somewhat mixed collection of both excavated and surface collected sherds. Lapita was identified as signifying the initial appearance of ceramics on the island, subsequently superseded by ceramics displaying a variety of incised, punctate and/or applied decorations. The non-Lapita ceramics were likened to Specht's (1969) Sohano and Hangan Phases and then by implication to the early Mangaasi ceramics as defined by Garanger (1972). A tentative Melanesia-wide network of communicative links was proposed although mitigated somewhat by the requirement of further definition and confirmation through more intensive archaeological investigation and particularly the location of securely dated sequences (Kennedy 1982:26). This still remains a problem today with post-Lapita sequences in the Admiralties (Ambrose 1991:111) and rest of the Bismarck archipelago where somewhat mixed assemblages, often poorly preserved where the sherds are both small in number and size, have been all that has been available in attempting to establish what is potentially up to a 1500 year sequence post-dating Lapita.

One well dated and preserved ceramic site located near the island of Manus is the Sasi site on Lou Island. The Sasi site, buried beneath several metres of tephra, was excavated by Ambrose (1988, 1991) and has been securely dated to c. 2100 BP. The recovered ceramics, which included often decorated flat horizontal rims, were likened to the Erueti-style ceramics of Vanuatu which had been initially dated to c. 2300 BP (Garanger 1971:61). The re-dating of Erueti material at the Mangaasi site now places the appearance of this vessel form at some 600–700 years earlier than its appearance at Sasi. By 2100 BP it had disappeared from the ceramic sequence on Efate. Apart from the now very disparate dates between the two wares the ceramics themselves overall can also be shown to be not as similar as earlier suggested. Closer examination of the rim and lip forms (i.e. flat horizontal rim/lip) which have been argued to have some close parallels tend to indicate that even this fundamental link is flawed. The Erueti rim forms tend to be abruptly divergent with the exterior wall of the rim outcurving to meet at the lip exterior (Garanger 1972:Figs 21, 22) which in many cases is not horizontal. This is in contrast to the horizontal Sasi rims which generally have an incurving parallel neck with an abruptly outcurving horizontal rim (Wahome 1999:Figs 3.4 u–v, 3.5 j). The Sasi Ware that has been published or illustrated (Ambrose 1991; Wahome 1999) shows decoration centred on the flat horizontal rims as opposed to the Erueti Ware where the limited decoration tends to be on other less common and quite different vessel forms. The heavily decorated flat horizontal rims from Sasi also include applied relief as a decorative technique. This is not found associated with the Erueti material. The Sasi Ware assemblage also includes undecorated narrow-necked flasks, a vessel form unknown in Vanuatu.

Further to the disparate chronologies is the argument that there appears to be a gap of up to 700 years between the Sasi Ware and Lapita (Ambrose 1991:109), whereas the Erueti Ware has now been shown to derive from and immediately post-date Lapita in Vanuatu (Bedford 2000a, 2000b; Bedford and Spriggs 2000; Spriggs and Bedford 2001). Despite Wahome's claim that his seriation of the Manus ceramics provided evidence of continuity from Lapita through to the later Sasi Wares (i.e. that Sasi Wares dated to earlier than 2100 BP) it must be emphasised that this conclusion was reached on less than reliable data. At the undated Mouk site Wahome argued this continuity could be demonstrated but at the same time he pointed out that the sherds were particularly fragmentary at the site (Wahome 1997:121) and some of the conclusions reached were based on only single sherds. The recent dating of the site by Ambrose using obsidian hydration techniques (Ambrose and McEldowney 2000) tends to confirm that there is currently a gap in the ceramic sequence between Lapita and Sasi Wares as he had originally argued. The earliest dates tend to cluster around 3000 BP. Soon after there appears to be a gap in the occupation of the area with another cluster of dates appearing after 2000 BP. The results from Vanuatu have demonstrated that the likelihood of ceramics demonstrating little change over a period of 700–800 years is highly improbable.

How reliable then are Wahome's claimed connections of the Manus ceramic material with that from the sites of Mangaasi, Erueti and Pakea in the Banks Islands (Wahome 1999:176–177)? The same difficulties in the data set, as were outlined earlier with regard to his Melanesia-wide study, can be seen with the Manus and Vanuatu ceramics. The ceramic chronologies were poorly defined, the assemblages were often mixed and in a number of cases small in number, the attribute categories were broadly defined and even then there was a large number of missing values (Wahome 1999:177–178 and Fig. 7.1). The claimed connections were seen in two assemblages from Manus namely CPTS/3 (comprising 'Early post-Lapita' ceramics from Mouk and Lou which were bound together by the unique horizontal rims i.e. Sasi Ware dating from 2100–1900 BP) and CPTS/2 (comprising an eclectic mix of 'Late post-Lapita' ceramics from sites that dated from between 1900–350 BP) with those recovered from Mangaasi, Erueti and the Banks Islands (Pakea). All of these periods and sites demonstrated a high correspondence with the degree of association between members of the group ranging between 59 and 80% (Wahome 1999:Fig. 7.2). By including CPTS/1 from Manus (ceramics that post-date 800 BP) in the statistical analysis it can be shown that they also show some level of correspondence (54–71%) with the Vanuatu sites (see Fig. 8.17). Interestingly CPTS/4 the earliest post-Lapita (and Lapita) ceramics from Manus tended not to cluster with the early Vanuatu material. With the advantage of now having a well defined sequence from central and southern Vanuatu one has to seriously question the relevance of such correspondence. The decorated flat horizontal rims would seem to account for much of the perceived similarities of at least the Sasi, Erueti and Pakea assemblages (75%) but as outlined above this attribute can no longer be seen as reliably comparable and even then belongs to quite different chronological periods. Even the Mangaasi assemblage achieves a 70% correspondence with the Sasi Wares. At the same time Erueti ceramics also manage a high correspondence with the 'Late post-Lapita' material from Manus (63%) as do the Mangaasi ceramics (63%). Even the Mangaasi and Erueti assemblages manage to reach a correspondence level of 60%, a further demonstration of how mixed assemblages can skew the results. Several attributes (simple knobs, incisions and bands) included by Wahome (Wahome 1999:Fig. 7.1) that were originally identified at both the Erueti and Mangaasi sites (Garanger 1972) have now been shown to be unassociated with Erueti Wares. A simple visual comparison of the Early Erueti Wares with the Late Mangaasi Wares clearly demonstrates that they are completely unlike, yet Wahome's results would suggest otherwise. It would seem that the factor of missing values may also have had some influence on the statistical results. It might not be totally coincidental that the assemblages from Manus and Vanuatu were some of larger and the more detailed available, thereby enabling a comprehensive assessment of attributes, which due to their broad nature encouraged parallels. In many of the other assemblages this same level of information was not able to be assessed and attributes were therefore simply recorded as not being present.

What Wahome was comparing was a widely disparate set of often mixed ceramic assemblages which display a multitude of decorative techniques and vessel forms, covering a time period of almost 1800 years in the case of Vanuatu, which tends to facilitate high correspondence levels. Had Wahome had access to the detailed ceramic sequences that have now been identified from Central Vanuatu, rather than poorly dated mixed assemblages, quite different conclusions would have emerged.

Some comment must also be made regarding the later Puian Ware from Manus which dates to around 1600 BP and was a component of Wahome's CPST 2 which showed high correspondence levels with all of the Vanuatu material (Wahome 1999:176–177). Arriving at parallels between the Puian Ware and any of the Vanuatu material would seem to further weaken the validity of Wahome's results. The Puian Ware is characterised by globular vessels with outcurving rolled rims and 'elaborate shell impressed designs on the rim and inner and outer neck area' (Ambrose 1991:107). Linear incision and applied strips were also a regular decorative feature, while fingernail crescents and rim notching were present but rare. A number of clear anomalies highlight the dissimilarities inherent between the Puian Ware and any of the Vanuatu assemblages. Neither shell impressed decoration nor decoration located on the inner neck area of a vessel have been reported from anywhere in Vanuatu. The illustrated Puian Ware applied relief (Ambrose 1991:108) also shows little resemblance to the form or motifs utilised amongst the Vanuatu assemblages.

Summary

The ceramic sequences from Vanuatu have in the past had far-reaching influence on theories relating to the colonisation and settlement of the Pacific. But they have also been seen as increasingly anomalous in some respects as ceramic sequences from other areas in the Pacific have been further refined. This recent research in Vanuatu has now substantially modified both the chronology and definition of the ceramic sequences from the archipelago. The results have important implications for the further understanding of the complex processes which were involved in the initial settlement of the Pacific and the subsequent cultural, linguistic and biological transformation and differentiation that has occurred.

Three issues or theories are especially pertinent here. They are that Vanuatu was settled pre-Lapita or at least contemporaneously by distinct cultural groups (Galipaud 1996a; Gorecki 1992); that Lapita to post-Lapita ceramic sequences can be seen as representing a basic cultural continuity (Kennedy 1982; Spriggs 1984, 1997) and finally that post-Lapita ceramics at least in the Southwest Pacific area demonstrated synchronous change suggesting that some sort of integrated post-Lapita network was in operation (Spriggs 1997:161–162; Wahome 1997, 1999). This research can fairly confidently claim to have clarified the first. There is no evidence thus far on any of the islands of Vanuatu to indicate Pleistocene or early Holocene occupation. It is Lapita colonisation alone that represents initial human settlement across the archipelago. The argument of continuity from Lapita through to the later ceramic traditions in Melanesia is looking increasingly secure and has certainly been indicated by the detailed sequences recovered from Efate and Erromango.

The third argument however, that there is evidence for synchronous change in the ceramic sequences of the Southwest Pacific which would support the concept of some sort of post-Lapita sphere of interconnection, needs further examination and serious revision. A review of the ceramic assemblages from sites outside Vanuatu which have been claimed as having some affiliation with the Mangaasi ceramic tradition as proposed by Garanger (1972) has highlighted the less than secure nature of many of those originally proposed connections in the light of more recent research in Vanuatu.

First it would seem that terminology has often been inter-mixed and wrongly applied. Mangaasi, a term which Garanger specifically coined to define the ceramic assemblages of Central Vanuatu has often been used indiscriminately to simply differentiate ceramics which were not dentate stamped or paddle impressed, or in an attempt to support the concept of a Melanesia-wide interconnected ceramic tradition. It has often been the case that through decorative techniques, without regard for vessel or motif form or firm chronological control, claimed homologous ceramic decorative traits have led to diffusionist implications. But the act of grouping pottery using nebulous and common stylistic criteria does not mean that a close historical relationship need have existed between the pottery makers who might have included incised and or applied relief amongst their decorative repertoire. There has been a tendency to homologise collections rather than highlight differences. This outcome can be partly explained by the poorly defined post-Lapita assemblages which have been excavated to date, the subconscious influence of the Lapita phenomenon and the continuing need to explain the ever vexatious question of human diversity (biological, cultural, linguistic) in the region.

Certainly in the earliest stages of many ceramic sequences from the Southwest Pacific similar changes can be identified. There is a change over time from the founding dentate-stamped decorative wares to an increasing percentage of incised wares and or plainwares, concomitant with a decrease in the variety of vessel form. A few immediately post-Lapita motif forms, decorative techniques and vessel forms can also shown to have some similarities across the sequences but these might equally be explained as continuities from the founding ceramic tradition rather than requiring any need to invoke continued high levels of interaction. The homogenous nature of Lapita ceramics can be seen at least in part as being the result of the frequent interaction, between a small and widely dispersed population (Graves *et al.* 1990:228; Summerhayes 2000:235) or some demonstration of recent common ancestry. But if ceramic assemblages began to show some variation soon after Lapita settlement there are clearly other processes which are influencing the form and composition of those sequences. Graves *et al.* (1990:228) suggest that rapidly changing population dynamics in association with changing environmental conditions might partly explain a diversification in ceramic styles which were reflecting 'conscious efforts to produce and maintain geographically based social distinctions'. The development of distinctive regionalised post-Lapita ceramic traditions can be partially explained in similar terms to those which have been proposed for the development of the emblematic Lapita design system in Near Oceania, that is that it provided 'a symbolic differentiation for contemporary and competing populations' (Irwin 1992:40). It could even be feasible to argue that on occasions the arrival of new groups into areas where there were already established pottery producing communities might have been an inspiration for further diversification rather than encouraging homogeneity. There is now mounting evidence, at least in the ceramic sequences of Remote Oceania, that they began to follow increasingly independent trajectories soon after initial Lapita settlement and up until 1000 BP.

In the five archipelagos (Vanuatu, New Caledonia, Fiji, Tonga and Samoa) reviewed by Clark he argued that there was no convincing evidence for anything other than *in situ* ceramic development in the post-Lapita period (Clark 1999:247). A similar review of sites adjacent to and further west of Vanuatu has arrived at a similar conclusion. Of course, as has been long pointed out Lapita itself was never a universal culture and as populations spread further east, increasing diversification and localisation developed (Green 1978). The fact that the distinctive dentate-stamped decorative technique was dropped so quickly from the ceramic sequences in some regions might also suggest that we can expect increasing variability in the record from that point. The Vanuatu evidence would seem to support this and is particularly highlighted in the case of the Efate and Erromangan sequences where the record from these occasionally inter-visible islands shows differing trajectories almost as soon as the affiliated Lapita ceramics disappeared. The picture is somewhat different from Efate north to the Shepherds where it can be shown that clear parallels in the ceramic record are traceable for almost 1600 years until its disappearance from those islands. Quite different ceramics were present during the last 1000 years but were restricted to Northern Vanuatu. In fact the Northern Vanuatu sequences post-dating 2500 BP remain largely unknown and as yet it is not possible to establish how similar or not they may have been to the ceramics of Central Vanuatu. This missing data also leaves open the prospects that future research may well reveal some level of inter-archipelago ceramic homogeneity, particularly dating to the last 1000 years. Regardless of this gap in our knowledge, on present evidence there seems to be much less empirical verification than previously supposed to support the claim for the existence of an inter-related Melanesia-wide Incised and Applied Relief ceramic tradition.

The detailed sequence on Efate has also highlighted the ever-changing and complex nature of the ceramics found within a single sequence on one island. Rarely are they static or unchanging with both form, decorative techniques and motifs often changing quite dramatically over short periods of time. Proposed ceramic sequences from the Pacific that have in the past been argued to have survived largely unchanged for 1000 years and more (Garanger 1972; Gorecki 1992; Poulsen

1967) have been shown to be unsustainable. In fact a ceramic sequence which showed little change even over 500 years would be a remarkable exception and without the support of detailed *in situ* ceramic assemblages must be regarded with some scepticism.

Wholesale diffusionist or migrationist models to account for cultural change and differentiation in the Southwest Pacific have become less popular over the years, being replaced by attenuated chains of connection (Spriggs 1984:217–8, 1993:196) which it has been argued would have enabled continued communication and the potential for ceramic sequences to change roughly synchronously (1993:196). While this explanation is probably the most parsimonious at this stage of the research to at least account for some aspects of the Melanesian-Polynesian dichotomy, can we realistically expect this level of interaction to have much of an influence on the production of ceramics or even be able to be recognised in the archaeological record ? Its non-recognition does not of course preclude inter-island and archipelago contacts, as emphasised by Hunt (1987:304) in the case of Fiji, but rather the challenge is to actually identify empirical evidence in the archaeological record for such contacts.

For this whole issue to be more adequately addressed there is a crucial need for the excavation and detailed study of well dated ceramic sequences post-dating Lapita. The nature and definition of post-Lapita sequences remain in the Southwest Pacific 'a major research concern with significant implications for the origins of ethnic diversity in the region' (Kirch and Yen 1982:202) and it would seem even more so now with the increasing evidence that Lapita-style ceramics generally comprise at the most only the first few hundred years of any sequence in the region (Anderson and Clark 1999; Burley *et al.* 1999; Sand 1997; Specht and Gosden 1997).

9

Non-Ceramic material culture

'In my experience those who claim Lapita to be nothing more than a certain style of decorated pottery have not done their homework on the range of portable artefacts found in association with this pottery' (Green 1991b:299–300).

A wide assortment of non-ceramic artefacts (Figs 9.1–9.15) were recovered from the excavations on Erromango (Ponamla and Ifo) and Efate (Mangaasi) with smaller numbers being retrieved from Malekula. The most frequent artefact types were *Tridacna* sp. adzes along with armbands and rings made from a variety of shellfish species. The vast majority of the excavated artefacts came from the earliest cultural horizons and can generally be recognised as somewhat ubiquitous artefact forms associated with Lapita and immediately post-Lapita sites across the Pacific. There are a number of less common artefact forms and others which appear to be exclusive to the Vanuatu cultural sequence. In attempting to establish an archipelago-wide chronological framework a comparison is made with the limited number of excavated artefacts from Malo (Hedrick nd), the Banks Islands (Ward 1979), Southern Vanuatu (Shutler and Shutler 1965; Shutler 1969; Shutler *et al.* 2002) and the much more extensive assemblages recovered by Garanger (1972). Garanger's work particularly provides a rich inventory of artefacts relating specifically to the last 600 years of the sequence and by combining it with the ethnographic records of Speiser (1996 [1923]) a detailed picture of the material culture dating to the later period can be established. As with the ceramic remains from the Arapus site, the non-ceramic artefacts are not presented here in any detail, rather key elements only have been noted. A chronological synthesis of non-ceramic material culture is presented at the end of the chapter which summarises the current state of knowledge, gleaned from the numerous sources, regarding the various artefact forms (see also Bedford and Spriggs 2002).

Adzes

Tridacna sp. shell adzes completely dominated the collection of recovered adzes which included 19 from Ponamla (Table 9.1), 20 from Ifo (Table 9.2), six from Mangaasi (Table 9.3) and one from Malekula (Table 9.4). A full inventory of the recovered adzes, whole, part or fragmentary is presented in Tables 9.1–9.4 and almost all are illustrated (Figs 9.1–9.8). Classification and description of the shell adzes was guided by the methodology outlined by Kirch and Yen (1982) although the small number of whole adzes limited any detailed comparative analysis.

Many of the adzes from Ponamla and Ifo (c. 2800-2000 BP) can be assigned to Kirch and Yen's Type 1 or micro-adzes (1982:221). In fact many of the adzes from these two sites are even smaller than those from Tikopia (Kirch and Yen 1982:224). A greater number of the smaller adzes have survived intact. The total length of these small adzes (6) from Ponamla was only 5cm or less (Table 9.1) and at Ifo 5.8cm or less (5). The widths of the same adzes from Ponamla and Ifo were no more than 3.2cm. As far as could be determined the majority of the adzes from both Ponamla and Ifo were made from the dorsal region of the *Tridacna* sp. shell. Only two hinge region adzes were identified from Ponamla (Fig. 9.2 c, d [Type 8]) and three (all Type 7) from Ifo (Figs 9.4 a, b; 9.5 a). Of the three most intact *Tridacna* sp. adzes recovered from Mangaasi one appears to be made from the hinge region of the shell (Fig. 9.5 f [Type 7]) and the other two from the dorsal region (Figs 9.5 d, e). Two larger fragments also appear to be related to the hinge area of the shell (Table 9.3). The adzes from all the above sites were predominantly fully ground.

Table 9.1 Ponamla adzes: Descriptive data

Provenance Area A	Fig.	material	x-sect#	lgth (cm)	width (cm)	thick. (cm)	Degree of grinding*	wght (gm)	comments
Layer 1									
TP1 (342)	9.1d	*Tridacna*	1	-	1.9	.80	1	4.9	butt-frag.
TP1 (333)	9.1e	*Tridacna*	1	3.0	1.5	.80	1	6.1	complete
TP2.2(997)	9.1i	*Tridacna*	1	4.2	1.8	.50	2	8.3	complete
TP1.3 (905)	9.1i	*Tridacna*	1	-	2.8	.80	1	11.6	bevel end
TP2.3 (599)	9.7g	basalt	2	-	2.5	.90	1	9.7	bevel end
Layer 2									
TP2 (146)	9.1a	*Tridacna*	1	-	2.5	1.0	1	7.7	mid-section
TP1.4 (1057)	9.1c	*Tridacna*	1	4.6	2.4	.60	2	8.3	
TP1.4 (1097)	9.1f	*Tridacna*	1	-	2.0		1	8.9	butt-frag.
TP1.2 (968)	9.1j	*Tridacna*	1	-	1.7	.70	1	7.5	chisel
TP1.3 (1095)	9.1k	*Tridacna*	3	-	-	.50	1	4.6	bevel frag.
TP2 (145)	9.1b	*Tridacna*	3	5.0	3.2	1.0	1	27.3	
TP2 (1039)	9.2d	*Tridacna*	2	-	3.2	1.5	1	41.7	butt end
Layer 3a									
TP1.7 (1140)	9.1g	*Tridacna*	1	3.7	2.0	.50	1	7.5	complete
TP1 (53)	9.1h	*Tridacna*	1	3.3	2.3	.80	1	53	
TP1.7 (1139)	9.1n	*Tridacna*	2	-	1.6	.80	1	3.7	bevel end
TP3 (244)	9.2a	*Tridacna*	3	-	5.8	1.1	1	52.3	bevel end
Layer 3b									
TP1 (453)	9.1m	*Tridacna*	1	-	2.0	.60	1	9.4	
TP3 (308)	9.2c	*Tridacna*	2?	-	-	1.1	1	21.1	mid-section
TP1.3 (1027)	9.2e	*Tridacna*	3	-	3.6	.70	2	18.5	
TP7.0									
L.3 (769)	9.1b	*Tridacna*	2	-	1.2	1.3	1	10.5	butt end

1= plano-convex; 2= elliptical/oval; 3= convex; * 1=fully ground; 2=partly ground

The vast majority of the shell adzes discussed above were recovered from cultural horizons associated with the first 1000 years of the cultural sequence. The only exceptions were a *Tridacna* sp. example from Chachara, Malekula (Fig. 9.5g) and a *Lambis* sp. adze from Mangaasi (Fig. 9.6d). The shell adze recovered from Chachara was only partly ground and made from the dorsal area of a *Tridacna* sp. shell (Kirch and Yen Type 2). This combination of attributes has at other sites been suggested as being associated with more recent style adzes (Kirch and Yen 1982:212) which is certainly the case with this adze dated to post-600 BP. The single *Lambis* sp. adze was recovered from the post-600 BP levels at Mangaasi. This adze form appears to be restricted to this later period of the cultural sequence in Vanuatu. Large numbers were recovered by Garanger from surface collections (Garanger 1972:Figs 82, 147, 283, 292) and the upper layers of his excavations throughout Efate and the Shepherd Group. A single *Terebra* sp. shell adze or gouge was found on the surface of the Mangaasi site in 1999. This adze form is again associated with later cultural horizons (post-600 BP) across Vanuatu and along with morphologically similar *Mitra* sp. shell adzes were recovered by Garanger from Efate and the Shepherds (1972:Figs 40, 108, 293). All of these later (post-600 BP) shell adze forms were recorded by Speiser (1996 [1923]:Plate 32) from a number of different islands.

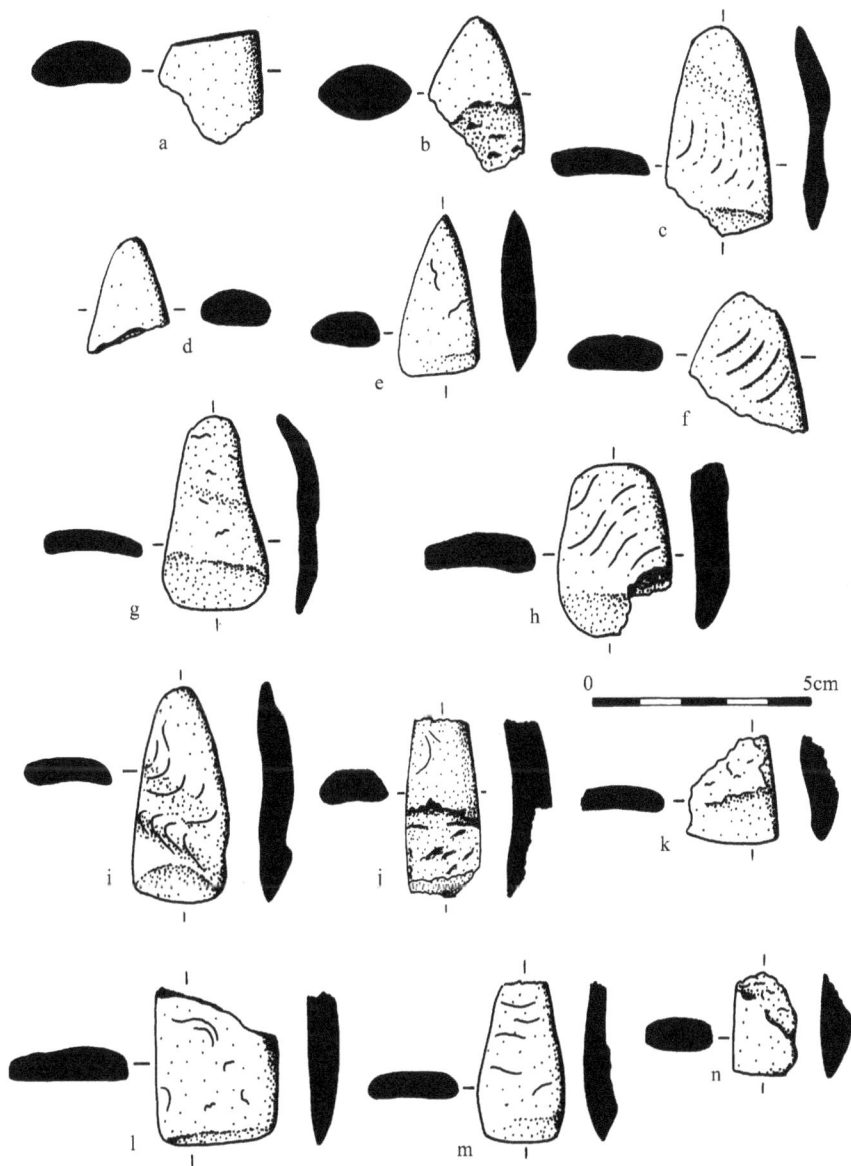

Figure 9.1 Ponamla shell adzes. a. TP2, L.2; b. TP7; c. TP1.4, L.2; d–e. TP1, L.1; f. TP1.4, L.2; g. TP1.7, L.3a; h. TP1, L.3a; i. TP2.2, L.1; j. TP1.2, L.2; k. TP1.3, L.2; l. TP1.3, L.1; m. TP1, L.3b; TP1.7, L.3a.

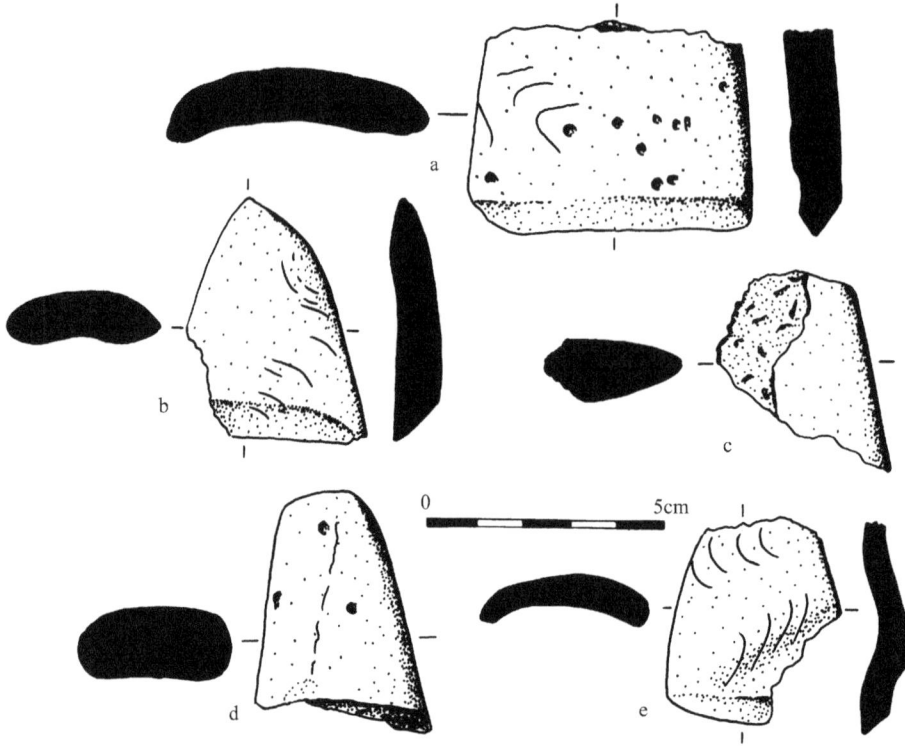

Figure 9.2 Ponamla shell adzes. a. TP3, L.3a; b. TP2, L.2; c. TP3, L3b; d. TP1.4, L.2; e. TP1.3, L.3b.

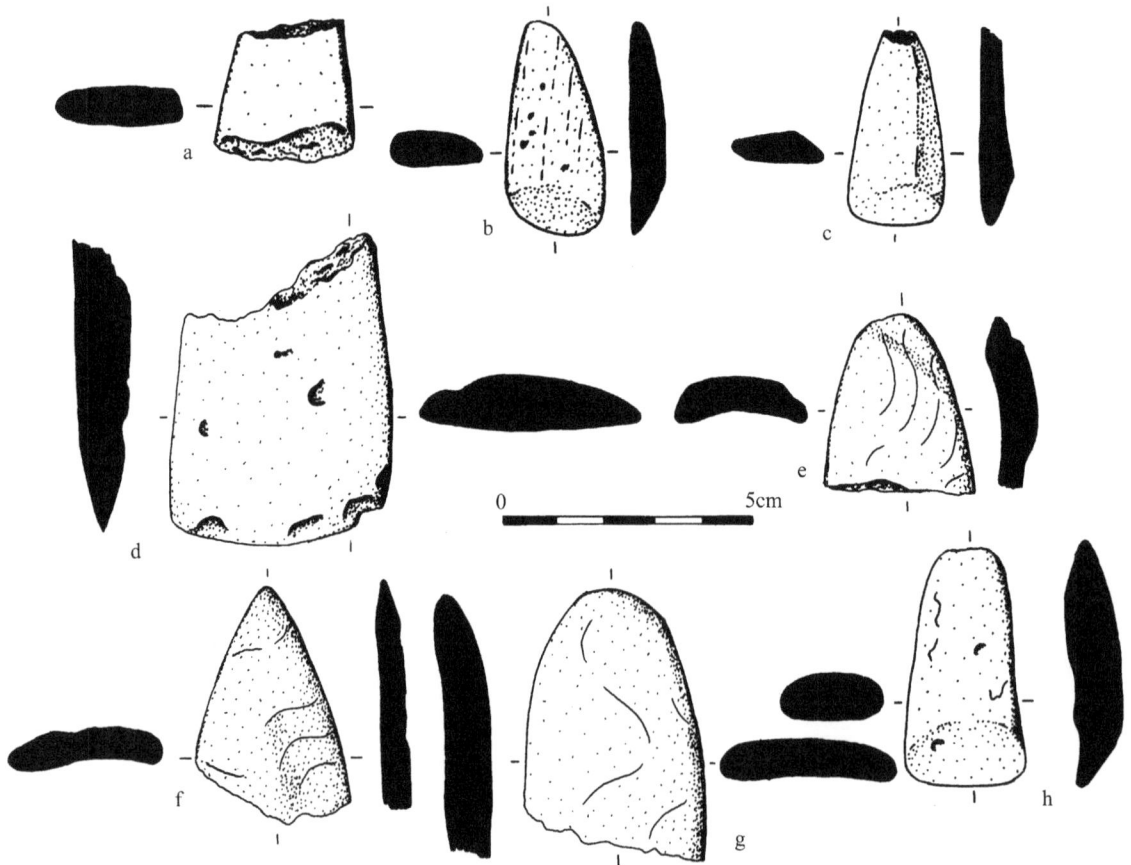

Figure 9.3 Ifo shell adzes. a. TrD, L.2; b. TP3, L.3; c. TrB, l.1; d. TrC, L.4; e. TrD, l.1; f. TP5, l.2; g. TrD, l.1; h. Area A, L.1.

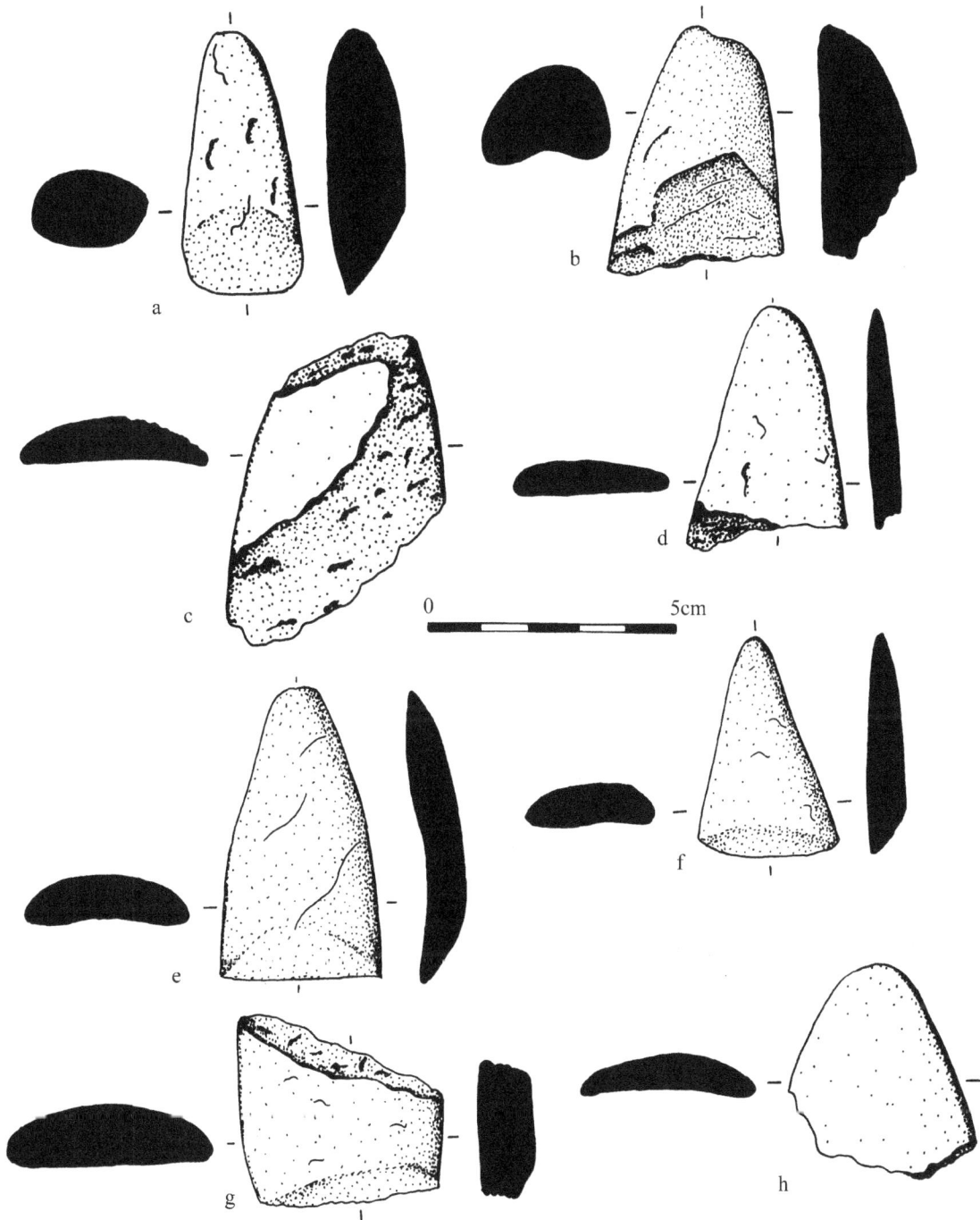

Figure 9.4 Ifo shell adzes. TP7, L.2; b. Area A, L.1; c. TP13, L.2; d. TP3, L.3; e. TrC, L.2; f. TP8, L.2; g. TrB, L.3; h. TP13, L.2.

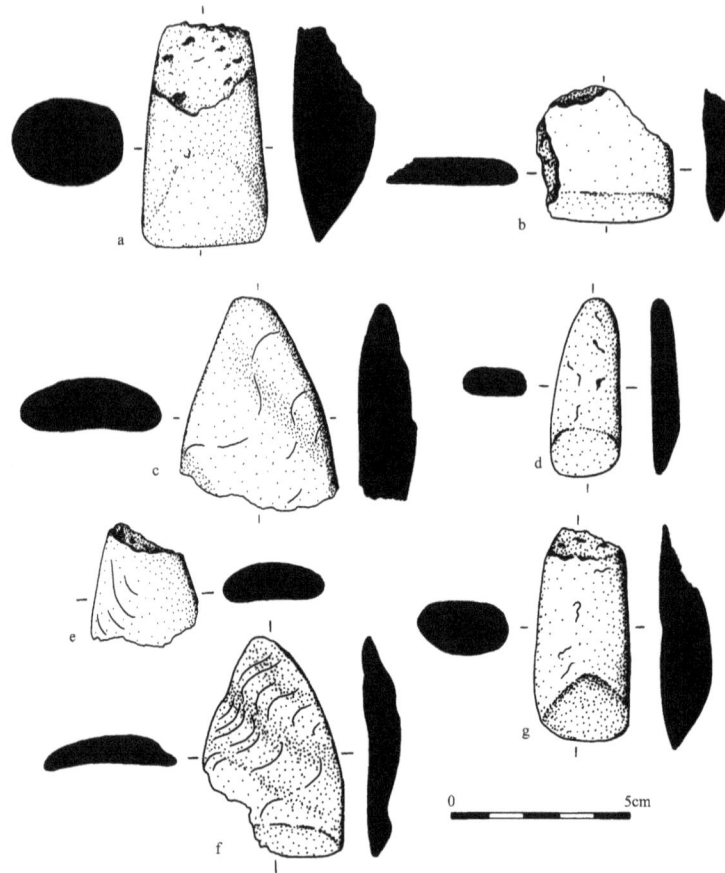

Figure 9.5 Miscellaneous shell adzes. a. Ifo, TrD, L.2; b-c. Ifo, TrD, L.1; d. Mangaasi, TP12, L.9e; e. Mangaasi, TP3, L.1; f. Chachara North, g. Mangaasi, TP9, L.9d.

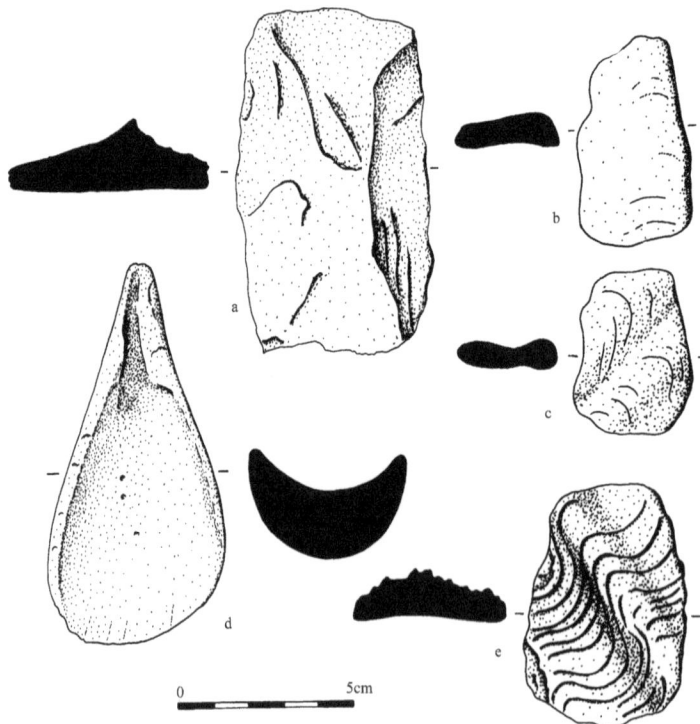

Figure 9.6 Miscellaneous shell artefacts. a. *Tridacna* adze blank (Ifo, TP3, L.3); b–c. *Tri.* adze blanks (Ponamla, TP3, L.3b, TP1.2, L.1); d. *Lambis* adze (Mangaasi, TP4, L.2i); e. *Tri.* adze blank (Ponamla, TP1.8, L.3).

Table 9.2 Ifo adzes: Descriptive data

Provenance Trench BCD	Fig.	material	x-sect#	lgth (cm)	width (cm)	thick. (cm)	Degree of grinding*	wght (gm)	comments
Layer 1									
B.5 (153)	9.3c	*Tridacna*	1	-	1.9	.50	1	7.0	
D.2 (259)	9.5c	*Tridacna*	1	-	4.2	1.5	1	51.3	
D.2 (259)(2)	9.3e	*Tridacna*	3	-	2.8	.60	2	10.7	butt end
D.4 (287)	9.3g	*Tridacna*	1	-	3.6	.80	1	29	
D.6 (401)	9.5b	*Tridacna*	1	-	-	.80	1	17.2	bevel frag.
Layer 2									
C.5 (253)	9.4e	*Tridacna*	3	5.8	3.2	.80	1	31.4	complete
D.2 (328)	9.5a	*Tridacna*	2	-	3.5	3.5	1	79.3	
D.6 (424)	9.3a	*Tridacna*	1	-	2.8	.80	1	11.2	mid-section
Layer 3									
B.6 (354)	9.4g	*Tridacna*	1	-	4.0	1.2	1	28.6	
Layer 4									
C.4 (293)	9.3d	*Tridacna*	1	-	4.5	1.1	1	42.1	
Area A									
L.1 (144)	9.3h	*Tridacna*	2	4.2	2.2	1.0	1	20.7	complete
L.1 (194)	9.4b	*Tridacna*	3	-	3.3	1.7	1	41.9	butt end
L.2 (89)	9.4a	*Tridacna*	2	5.2	2.4	1.5	1	30	complete
TP 3									
L.1 (17)	-	*Tridacna*	-	-	-	-	1	23.3	mid section
L.3 (25)	9.3b	*Tridacna*	1	4.2	2.0	.60	1	9.9	complete
L.3 (25) (2)	9.4d	*Tridacna*	1	-	3.2	.60	1	15.4	
TP 5									
L.2 (59)	9.3f	*Tridacna*	1	-	3.1	.50	1	15.1	
TP 8									
L.2 (112)	9.4f	*Tridacna*	1	4.4	2.7	.80	1	14.1	complete
TP 13									
L.2 (480)	9.4c	*Tridacna*	1	-	4.0	.80	1	27.2	mid section
L.2 (480)(2)	9.4h	*Tridacna*	1	-	17.8	.70	1	17.8	butt end

* See Table 9.1

There were strikingly few stone adzes throughout all the sites. Three only were recovered, one from Malua Bay (Fig. 9.7 f), one from Ponamla (Fig. 9.7 g) and one from the Erueti cultural horizons at the Arapus site (Fig. 9.8 f). The two adzes from Malua Bay and Ponamla were relatively small, fully ground basalt adzes with elliptical cross sections (see Tables 9.1, 9.4). They were both associated with immediately post-Lapita deposits (c. 2700 BP). The adze from the Arapus site was also relatively small with a rectilinear cross-section (Fig. 9.8 f). A very similar stone adze was recovered by Garanger from the Erueti site (Garanger 1972:Fig. 25:4). The adze from Arapus was recovered from immediately post-Lapita cultural horizons (c. 2800 BP) and it seems likely the example recovered by Garanger dates from a similar period. One other stone adze that deserves mention is a lenticular cross-section adze (Fig. 9.8e) that was surface collected from Ponousia, a sheltered bay located south of Ifo on Erromango. These adze forms are somewhat loosely described as a 'Melanesian form' (Gifford and Shutler 1956:95) or 'Melanesian type' (Green 1991b:300) but this categorisation remains somewhat ill-defined. Although these adze forms have been frequently recovered across the Southwest Pacific they generally lack provenance detail (see Crosby 1973 for full discussion). The Shutlers (M.E. and R. Shutler 1965; Shutler *et al.* 2002) illustrated a number of these same adze forms, that had either been surface collected or gifted, from the southern islands of Erromango, including Tanna, Futuna and Aneityum, but also Pentecost in the north (Shutler and Shutler 1975:72). Spriggs *et al.* (1986) have also illustrated and described a number of similar adzes from Aneityum. The same adze forms, again all surface collected or gifted, were recorded by Speiser from throughout Vanuatu (Speiser 1996 [1923]:Plate 32). They have not been found amongst any of the earlier cultural deposits and certainly their regular recovery from surface contexts suggests that they belong to the later part of the cultural sequence.

A total of nine adze roughouts were recovered from the excavations (Fig. 9.6a–c, e) (Tables 9.6–9.9). Most (7) were from the dorsal region of *Tridacna* sp. shells (Ifo-5; Mangaasi-2) while two possible stone examples were recovered from Ifo.

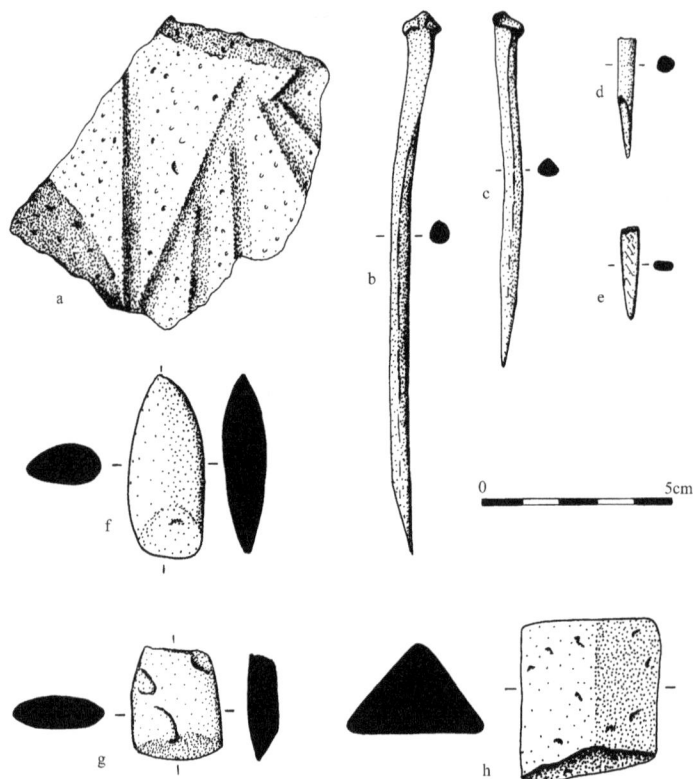

Figure 9.7 Miscellaneous artefacts. a. scoria abrader (Ponamla, TP3, L.3a); b–d. *Pteropus* sp. bone needles (Ponamla, TP1, L.3b, TP1.5, L.1, TP1.5, L.2); e. bone needle (Mangaasi, TP3, L.1); f. basalt adze (Malua Bay, TP3, L.3); g. basalt adze (Ponamla, TP2.3, L.1); h. stone abrader? (Malua Bay TP9, L.3).

Table 9.3 Mangaasi adzes: Descriptive data

Provenance	Fig.	material	x-sect#	lgth (cm)	width (cm)	thick. (cm)	Degree of grinding*	wght (gm)	comments
TP 9									
L.9d (1671)	9.5g	*Tridacna*	2	-	2.7	1.5	1	34.8	
TP 12									
L.9c (1803)	9.5d	*Tridacna*	2	5.0	1.9	.80	1	14.2	complete
TP 1/15									
L.4d (2132)	-	*Tridacna*	-	-	-	-	1	62.9	mid section
TP2									
L3b (314)	-	*Tridacna*	-	-	-	-	1	4.8	mid section
TP 3									
L.1 (161)	9.5e	*Tridacna*	1	-	3.2	1.0	1	16.3	mid section
L.2ii (198)	-	*Tridacna*	-	-	-	-	1	50.8	mid section
TP 4									
L.2i (506)	9.5d	Lambis	3	10.5	5.0	1.8	-	89.2	complete

* See Table 9.1

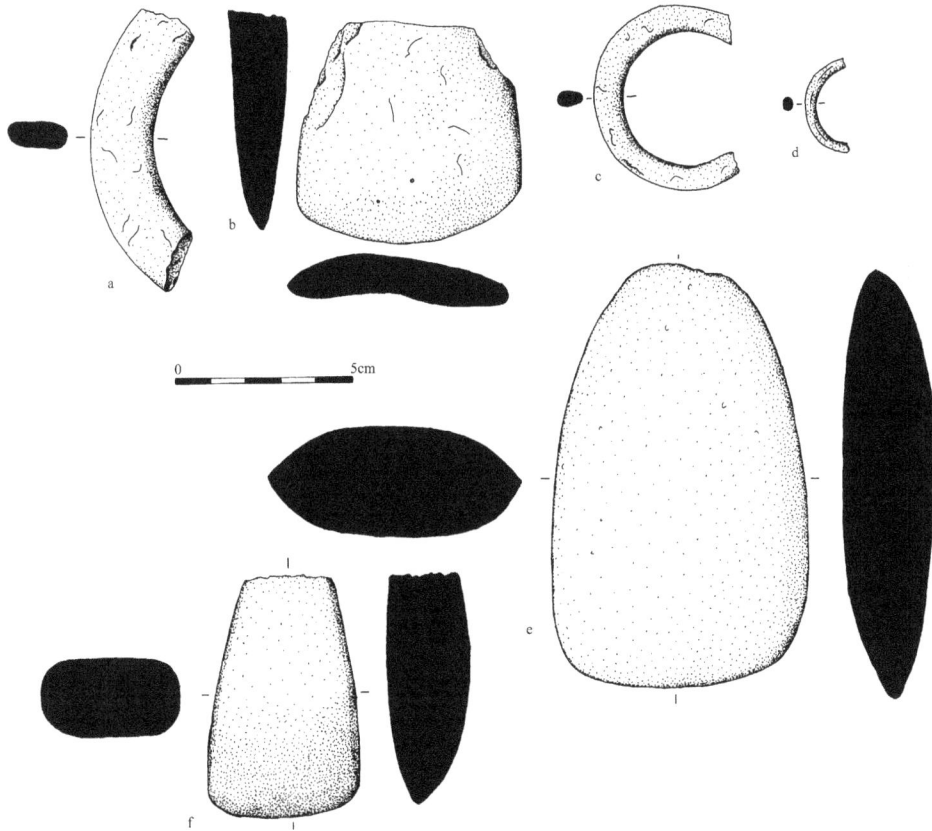

Figure 9.8 Miscellaneous artefacts. a. *Tridacna* ring (TP4, Arapus Phase, Arapus); b. *Tridacna* adze (TP13, Arapus Phase, Arapus); c. *Tridacna* ring (TP4, Arapus Phase, Arapus); d. *Conus* ring (TP13, Arapus Phase, Arapus); e. stone adze (surface, Ponousia, Erromango); f. stone adze (TP13, Early Erueti Phase, Arapus).

Table 9.4 Malekula adzes: Descriptive data

Provenance	Fig.	material	x-sect#	lgth (cm)	width (cm)	thick. (cm)	Degree of grinding*	wght (gm)	comments
Chachara									
L.1 (236)	9.5f	*Tridacna*	1	6.0	3.9	.80	2	31.3	
Malua Bay									
TP 9 L.3 (451)	9.5f	basalt	2	4.8	2.1	1.1	1	22.7	complete

* See Table 9.1

Table 9.5 Ponamla Area A, worked and flaked shell (weight in grams)

	L.1	L.2	L.3A	L.3B	L.4	TOTAL/WEIGHT
Tridacna sp.	8 (107.9)	1 (28)	3 (45.7)	3 (71.8)	3 (12)	18 (265.4gm)
Trochus sp.	-	-	-	-	-	-
Conus sp. 1*	-	-	-	-	-	-
Conus sp. 2*	-	-	-	-	-	-
Conus sp. 3*	1 (22.3)	-	-	-	1 (19.4)	2 (41.7gm)
Conus sp. 4*	-	-	-	-	-	-
Cypraea sp.	-	1 (15.7)	-	-	-	1 (15.7gm)

* Conus sp.1 = axial end of shell ground on both ends (Fig. 9.13 b,d,g). Conus sp. 2 = axial end of shell but ground on outside end only. Conus sp. 3 = axial end of shell ground on the inside edge only (Fig. 9.13 a,c). Conus sp. 4 = rectangular curved bands cut from the body of the shell with often ground edges (Fig. 9.13 i)

Table 9.6 Ifo, worked and flaked shell (weight in grams)

	L.1	L.2	L.3	L.4	TOTAL/WEIGHT
Trenches BCD					
Tridacna sp.	1(22.5)	-	-	-	1(22.5)
Trochus sp.	-	-	-	1 (28.1)	1 (28.1)
Conus sp. 2*	-	1(27.3)	-	-	1(27.3)
Lambis sp.	-	1(20.6)	-	-	1(20.6)
Tri. Adze blank	2(117.4)	-	-	-	2(117.4)
TP 3					
Trochus sp.	-	-	1 (13)	-	1(13)
TP 4					
Tri. Adze blank	2(98.4)	-	-	-	2(98.4)
TP5 *Trid.* sp.	-	1(8.2)	-	-	1(8.2)
TP6 *Trid.* sp.	1(12.1)	-	-	-	1(12.1)
Area A					
Tri. Adze blank	1(41.1)	-	-	-	1(41.1)
TP9 *Trid.* sp.	-	1 (7.1)	-	-	1(7.1)
TP 13					
Tridacna sp.	-	-	1(69.8)	-	1(69.8)
Trochus sp.	-	1(15)	-	-	1(15)

*see Table 9.5

Table 9.7 Mangaasi, worked and flaked shell, TP 9 and 12 (weight in grams)

	L.1/2I	L.9A	L.9B	L.9C	L.9D	TOTAL/WGHT
TP 9						
Tridacna sp.	-	4(93.3)	5(246.6)	20(1677.7)	34(1474.2)	63 (3491.8)
Trochus sp.	-	1(19.2)	-	2(53.4)	1(14.8)	4(87.4)
Conus sp. 1*	-	-	-	3(23.4)	-	3(23.4)
Conus sp. 3*	-	-	-	-	2(25.4)	2(25.4)
Lambis sp.	-	2(114.2)	-	1(57.5)	-	3(171.7)
TP 12						
Tridacna sp.	1(22.4)	1(41.6)	4(515.8)	-	-	6(579.2)
Trochus sp.	-	-	1(14.3)	1(15.3)	-	2(29.6)
Conus sp. 1*	-	1(9.3)	-	1(31.1)	-	2(40.4)
Conus sp. 2*	-	2(18.7)	-	-	-	2(18.7)
Conus sp. 4*	-	1(8.7)	2(14.2)	-	-	3(22.9)
Lambis sp.	-	-	3(287.4)	-	-	3(287.4)

*see Table 9.5

Table 9.8 Mangaasi, worked and flaked shell, TPs 10, 17, 1/15, 2 and 3 (weight in gms)

	L.1/2I/2II	L.3A	L.3B	L.3D	L.4	L.4B	TOTAL/WEIGHT
TP 10							
Tridacna sp.	-	3(458.9)	-	-	-	-	3(458.9)
TP 17							
Tridacna sp.	-	-	-	1(52.2)	-	-	1(52.2)
Trochus sp.	-	-	1(23.2)	-	-	-	1(23.2)
Conus sp. 4*	-	-	1(4.9)	1(8.7)	-	-	1(13.6)
TP 1/15							
Tridacna sp.	1(14.4)	-	11(565.3)	-	-	15 (398.1)	27(977.8)
Trochus sp.	-	-	1(24.8)	-	-	-	1(24.8)
Conus sp. 4*	1(2.8)	-	-	-	-	-	1(2.8)
Lambis sp.	-	-	2(29.7)	-	-	-	2(29.7)
TP 2							
Tridacna sp.	2(70.2)	5(364.5)	1(20.6)	-	1(16)	-	4(471)
Trochus sp.	1(42.4)	1(24.9)	-	-	-	-	2(67.3
Conus sp. 2*	-	1(14.7)	-	-	1(22)	-	2(36.7)
Conus sp. 4*	-	2(15.4)	-	-	-	-	2(15.4)
*Tri.*adze blank	-	2(237)	-	-	-	-	2(237)
TP 3							
Tridacna sp.	3(154.3)	4(56)	-	-	-	-	7(210.3)
Conus sp. 3*	1(15.8)	-	-	-	-	-	1(15.8)

*see Table 9.5

Table 9.9 Mangaasi, worked and flaked shell, TPs 4, 8, 11, 13, 14, 16 (weight in grams)

	L.1/2I/2II	L.3	L.5I	L.5II	L.7	L.9	TOTAL/WEIGHT
TP 4							
Tridacna sp.	-	-	-	-	-	19(527)	19(527)
Trochus sp.	-	-	3(65.7)	-	-	-	3(65.7)
Conus sp. 3*	-	-	1(17.2)	-	-	-	1(17.2)
Conus sp. 4*	-	-	1(7.8)	-	-	-	1(7.8)
TP 8							
Tridacna sp.	1(61.1)	-	-	-	-	-	1(16.1)
TP 11							
Tridacna sp.	2(82.3)	-	-	-	-	-	2(82.3)
Trochus sp.	1(26.8)	-	-	-	-	-	1(26.8)
TP 13							
Tridacna sp.	3(114.9)	-	2(30.2)	1(62.2)	-	-	6(207.3)
Trochus sp.	-	-	1(17.7)	-	-	-	1(17.7)
TP 14							
Tridacna sp.	2(64.4)	-	-	-	3(65.6)	-	5(130)
Trochus sp.	1(12.9)	-	-	-	1(31)	-	2(43.9)
TP 16							
Tridacna sp.	2(147.2)	3(101.9)	-	-	-	-	5(249.1)

* see Table 9.5. No worked or flaked shell was recovered from TPs 5, 6, 7 or 18.

Ornaments

Recovered ornaments were dominated by a varied collection of shell armbands, rings and beads (Figs. 9.9-9.12), made principally from *Tridacna* sp., *Conus* sp. and much less frequently *Trochus* sp.

shells. A number of these artefacts were able to be assigned to Kirch's (1988c) ornament classes. Miscellaneous shell manufacturing debris (Fig. 9.13) associated with at least some of the production of these types of ornaments was recovered from throughout the various excavations (see Tables 9.5–9.9).

Fully ground *Tridacna* sp. shell armbands (Class C2) were recovered from Mangaasi (21) (Fig. 9.9, 9.10), Ifo (7) (Fig. 9.11) and Ponamla (4) (Fig. 9.12). They comprise an eclectic assortment of diameters and widths but were principally recovered from stratigraphic contexts dating to the first 1000 years of settlement. Two fragments (Fig. 9.10 c, g) from Early Mangaasi cultural horizons (c. 2200-1600 BP) hint at a possible shift to wider forms of *Tridacna* sp. armbands over time. A number of shell armring fragments illustrated by Garanger, which were recovered from Mangaasi and can now be more securely dated to post-2200 BP (Garanger 1972:Fig. 105 16,19, 21), also appear to be relatively wide. This ornament form had certainly disappeared from the cultural repertoire at least by 600 BP being superseded by full-circle pig tusks (formed when the upper incisor is removed which then enables the lower incisor to grow unheeded) which have been found associated with numerous burials in Central Vanuatu (Garanger 1972:Figs 192–196) and also recorded in the ethnographic record (Speiser 1996 [1923]:167).

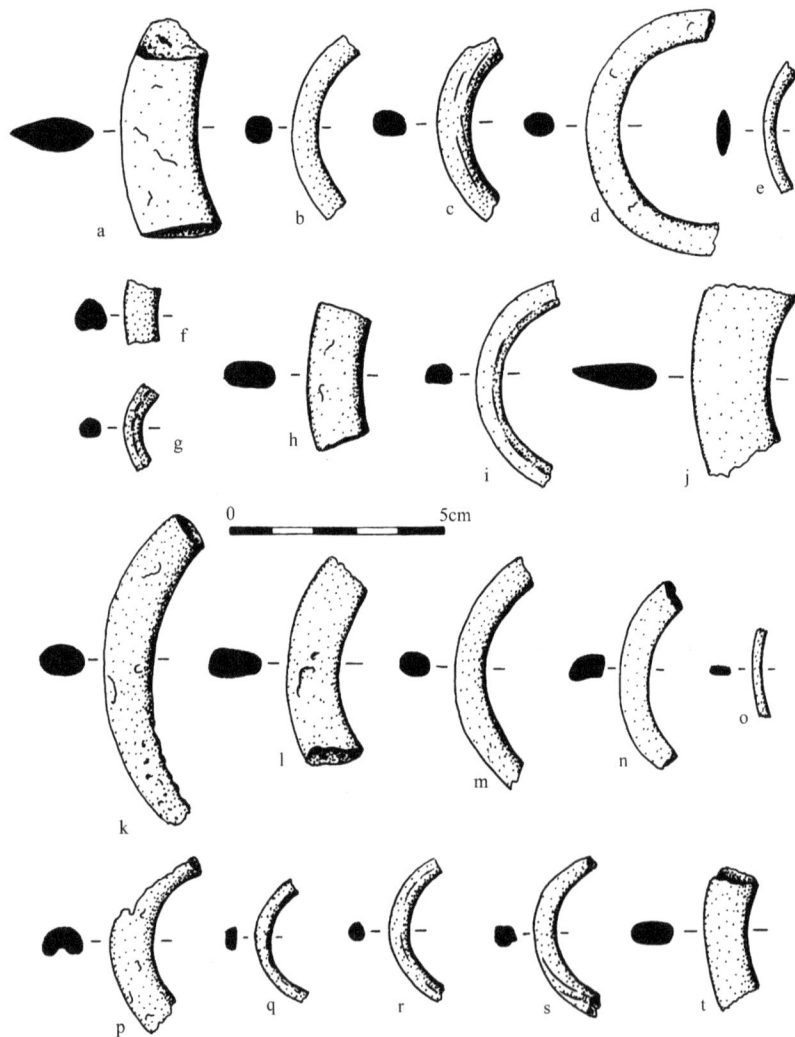

Figure 9.9 Shell armbands and rings, Mangaasi. a-c. TP5, L.9(*Tridacna*); d. TP16, L.8 (Tri.); g. TP12, L.9b(*Conus*); h. TP12, L.9b (Tri.); i. TP15, L.3a (Tri.); j. TP9, 9b (Tri.); k. TP7, L.5ii (Tri.); l. TP12, L.2i (Tri.); m. TP9, L.9c (Tri.); n. TP9, L.9a (Tri.); o. TP12, L9b (Conus); p. TP12, L.9b (Tri.); q. TP17, L3d (Conus); r-s. TP9, L.9ed (Conus); t. TP9, L.9c (Tri.).

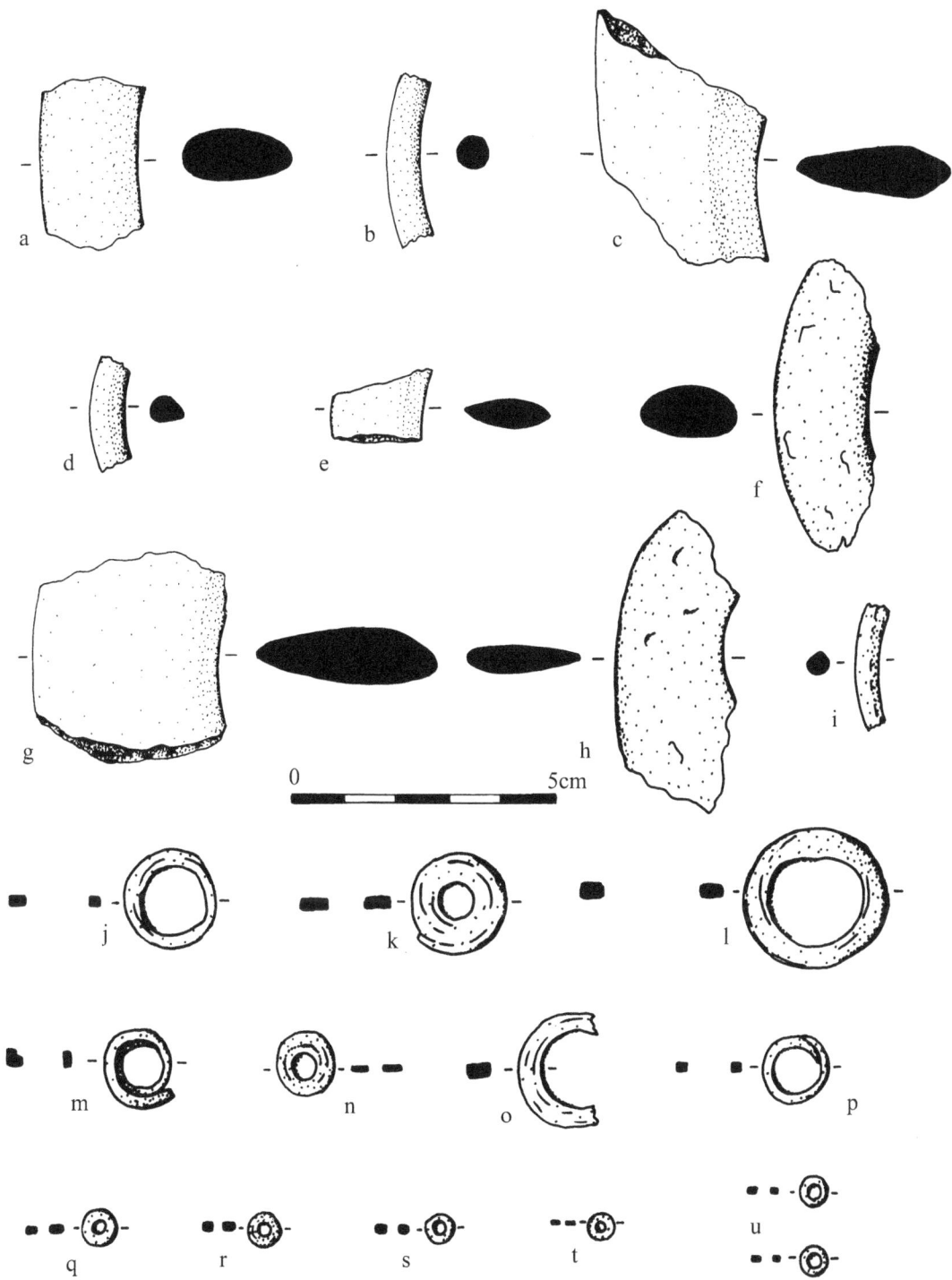

Figure 9.10 Shell armbands, rings and beads, Mangaasi. a-h. *Tridacna* rings (a. TP1, L.3b; b. TP4, L.5i; c. TP1, L.3b; d. TP2, L.3a; e. TP2, L.2ii; f. TP14, L.2ii; g. TP1, L3b); h. TP16, L.8); i-p. *Conus* rings (i. TP15, L.3a; j-k. TP9, L.9a; l. TP9, L.9d; m-n. TP9, L.9c; o. TP17, L.3a; p. TP16, L.8; q-u. *Conus* beads (q. TP14, L.2ii; r. TP18, L.1; s. TP14, L.2i; t. TP10, L.3a; u. TP10, L.3a).

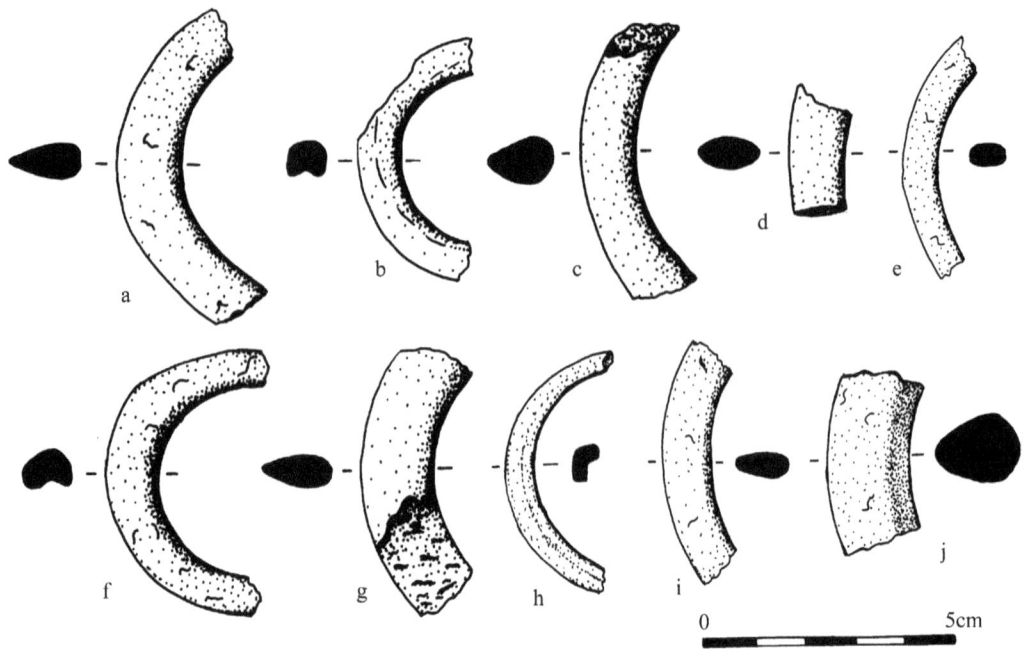

Figure 9.11 Shell rings, Ifo. a. *Tridacna* (TP3, l.3); b. *Conus* (Area A, L.3); c-e. *Tridacna* (c. Area A, L.3; d. TP5, L.2; e. TrD, L.1); f. *Conus* (TrB, L.3); g. *Tridacna* (TrB, L.1); h. *Trochus* (Area A, L.2); i-j. *Tridacna* (TP12, L.3; j. TrD5, L.2).

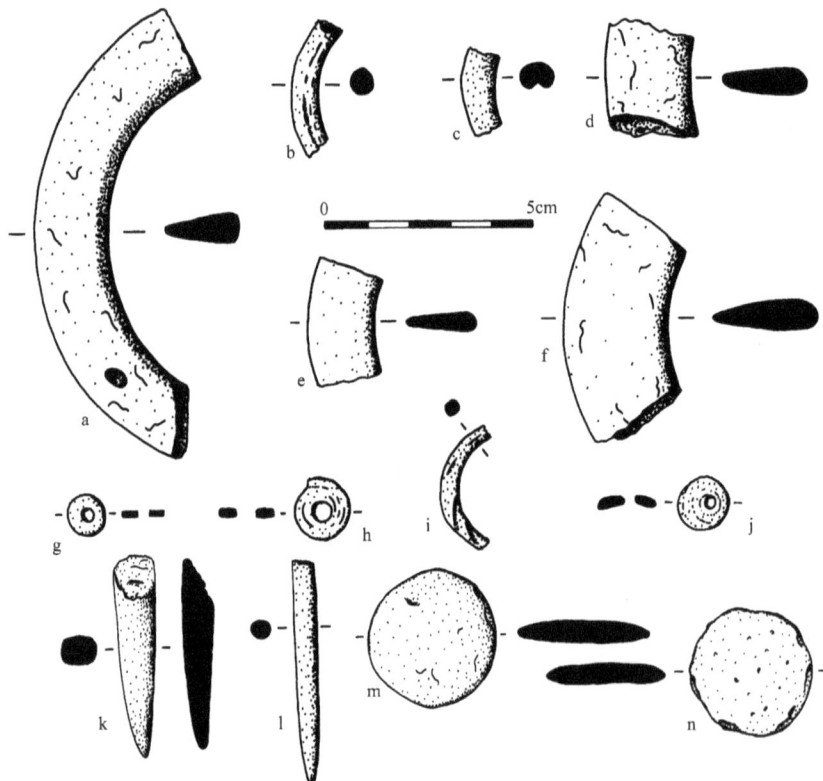

Figure 9.12 Miscellaneous artefacts. a. *Tridacna* ring (Ponamla, TP3, L.3b); b-c. *Conus* ring (b. Ponamla, TP 6, L.6; c. TP1, L.3b); d. *Tridacna* ring (TP1, L.3a); e-f. Tridacna rings (e. Ponamla, TP1.4, L.2; f. TP2, L.2); g-h. Conus bead (g. Ponamla, TP6, L.6; h. Ifo, TrD, L.1); i. *Conus* ring (Malua Bay, TP9, L.3); j. *Conus* bead (Malua Bay, TP9, L.4); k-l. Nasal? Ornament (k. Ifo, TrB, L.1; l. TrC, L.1); m. *Tridacna* disc (Ifo, TrB, L.3); n. stone disc (Ifo, TrD, L.2).

Rings (Class C1) and beads (Class E1) made from *Conus* sp. shell were recovered from all three islands. *Conus* sp. shell rings and beads are also frequently recorded throughout the later prehistoric (post-600 BP) (Garanger 1972:Figs 181, 184) and into the ethnographic period (Speiser 1996 [1923]:167). From Mangaasi a total of sixteen rings (Fig. 9.9e, f, g, o-s; Fig. 9.10 j–p) and six beads (Figs. 9.10 q-u) were retrieved from both the Erueti and Mangaasi cultural horizons (c. 2800–1200 BP). Two *Conus* sp. rings were recovered from the lower levels (c. 2800–2500 BP) of Ifo (Fig. 9.11b. f) and another two examples were recovered from similar dated levels at Ponamla (Fig. 9.12b, c). Single *Conus* sp. beads were recovered from Ponamla (Fig. 9.12 g), Ifo (Fig. 9.12h) and Malua Bay (c. 2700–2500 BP) (Fig. 9.12j). A *Conus* sp. ring dating to the same period was also found at Malua Bay (Fig. 9.12i).

A single example of a *Trochus* sp. shell armring (Class C3) was recovered from Ifo (Fig. 9.11h) from deposits dated to c. 2500-2000 BP. It was more highly ground than the *Trochus* sp. shell arm bands which are found with some frequency in the later prehistoric (post-600 BP) (Garanger 1972:Fig 191) and ethnographic record of Vanuatu (Speiser 1996 [1923]:167).

Shell discs included two examples made from different shellfish. One was a flaked *Tridacna* sp. disc recovered from Ifo (Fig. 9.12 m) which was identical in size and form to a flaked basalt example also from Ifo (Fig. 9.12 n). Garanger recovered these artefact forms from Erueti where they had been fashioned from both *Tridacna* sp. and coral (1972:Figs 27, 28). Their function is not known but they do appear to be fashioned discs rather than manufacturing debris. Garanger linked them to similar artefact forms in Polynesia where they are said to have been used in games (Garanger 1972:30). Those from Vanuatu may also have been gaming pieces but they have been recovered from deposits of a much greater age (c. 3000–2000 BP). Pottery discs of a similar size and associated with a similar chronological period have been noted from, Fiji, Tonga and New Caledonia (e.g. Birks 1973:41). These were often described as gaming pieces but in a number of cases had the added feature of a central hollow or depression. The only other shell disc from the Vanuatu excavations was made from axial end of a *Conus* sp. shell recovered from early deposits (c. 2700–2500 BP) at Malua Bay (Fig. 9.13 k). A similar disc has been noted at Nuiatoputapu recovered from Lapita deposits while perforated versions are known from the late (post-800 BP) prehistory of Tikopia (Kirch 1988b:208; Kirch and Yen 1982:248).

An assortment of shell manufacturing debris was recovered from throughout the excavations (Tables 9.5–9.9). *Tridacna* sp. dominated but little more could be said other than it appeared to have been flaked. *Conus* sp. shell debris could be more easily identified as having been worked. Four categories of worked *Conus* sp. were established;

1)   the axial end of the shell cut and ground on both ends (Fig. 9.13 b, d, g);
2)   the axial end of the shell cut but ground on the outside end surface only;
3)   the axial end of the shell ground on the inside edge only (Fig. 9.13 a, c);
4)   rectangular curved bands cut from the body of the shell often with ground edges (Fig. 9.13 i).
Other recorded flaked shell species included *Cypraea* sp., *Lambis* sp. and *Trochus* sp.. The listed flaked *Trochus* sp. comprised only the base segment of the shell which has at some sites been associated with the first stages in the manufacture of shell fishhooks (Kirch 1997:200, Plate 7.2; Smith 1991) although this activity could not be identified with certainty from any of the recovered Vanuatu material.

Pendants

A variety of pendants (9) made from shell were recovered from all islands, but principally from the Erueti horizons at the Mangaasi site. Pendants which displayed a single drilled hole at one end dominated (Fig. 9.14 b–d, g–i). They were recovered from both Mangaasi on Efate and Waal and

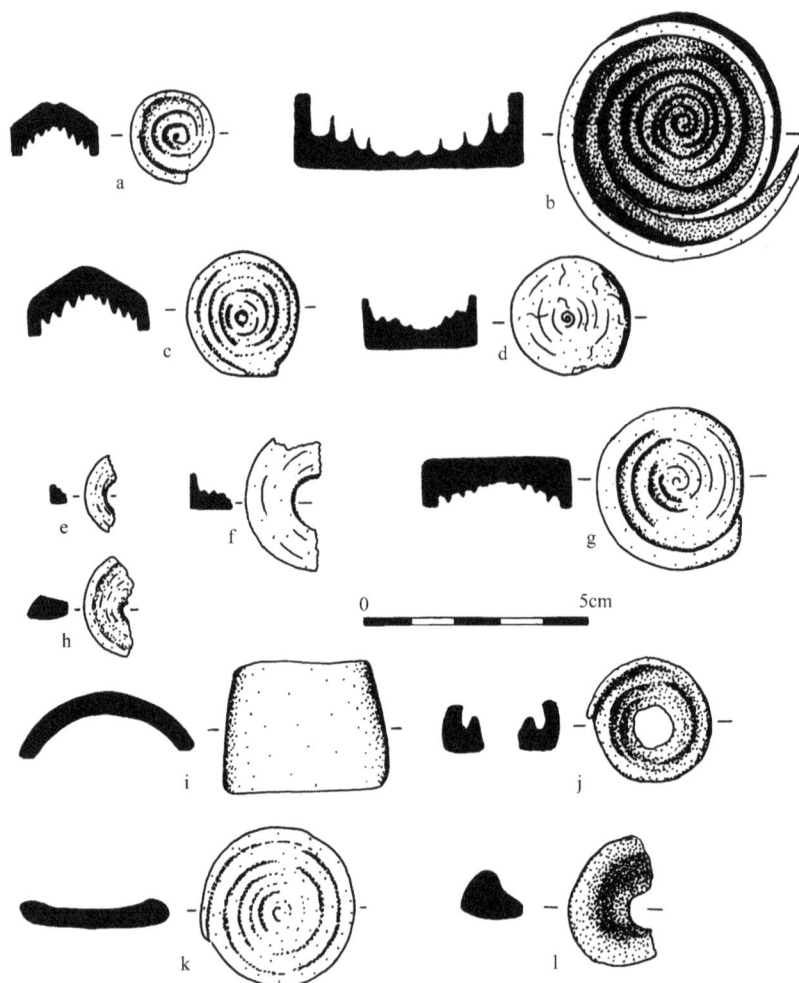

Figure 9.13 Miscellaneous worked shell and coral. a-j. worked *Conus* (a. Type 3, Ponamla, TP2, L3b; b. Type 1, Ifo, Area A, L.2; c. Type 3, Ponamla, TP1.4, L.1; d. Type 1, Mangaasi, TP12, L.9a; e-f. Mangaasi, TP9, L.9b; g. Type 1, Mangaasi, TP7, L.5ii; h. Navaprah, TP3, L.3; i. Type 4, Malua Bay, TP9, L.3; j. Type 1, TP9, L.3); k. *Conus* disc, Malua Bay, TP4, L.3; l. worked coral, possible net sinker (Malua Bay, TP9, L.3).

Malua Bay on Malekula from the early cultural horizons (c. 2800–2200 BP). Identification of the shell used for the manufacture of this form of pendant was difficult due to the highly worked nature of the artefacts, but one (Fig. 9.14 g) was clearly made from the distinctive *Pectinidea* sp. shell. Also made from a fully ground rectangular piece of *Pectinidea* sp. shell was a single bracelet (?) or necklace (?) segment with perforations at both ends (Fig. 9.14 a). A quite unique pendant form was recovered from the Late Erueti cultural horizon (c. 2500–2200 BP) of the Mangaasi site. Possibly cut from a pearl shell it has been fashioned into the shape of a cross with a central perforation (Fig. 9.14 f). This delicate ornament form appears to be thus far unique in archaeological sites across the Pacific. A single drilled shark tooth was recovered from Ponamla (Fig. 9.14 j). Finally a fully ground piece of *Tridacna* sp. recovered from Ifo could also have been a possible pendant (Fig. 9.14 k), where any evidence of perforation has been lost due to breakage.

Also grouped with the pendants is a single grooved sea urchin spine (Fig. 9.14 n) which was recovered from the Arapus cultural horizons (c. 2900 BP) of the Arapus site. These artefact forms are relatively rare but have been recovered from a number of Lapita or immediately post-Lapita sites (c. 2900–2300 BP) including single examples from Ifo (Spriggs 1984:217), Samoa (Janetski 1976:72-3), two from different sites in Fiji (Best 1981:15, 1984:461) and a grand total of three from Tikopia (Kirch and Yen 1982:271).

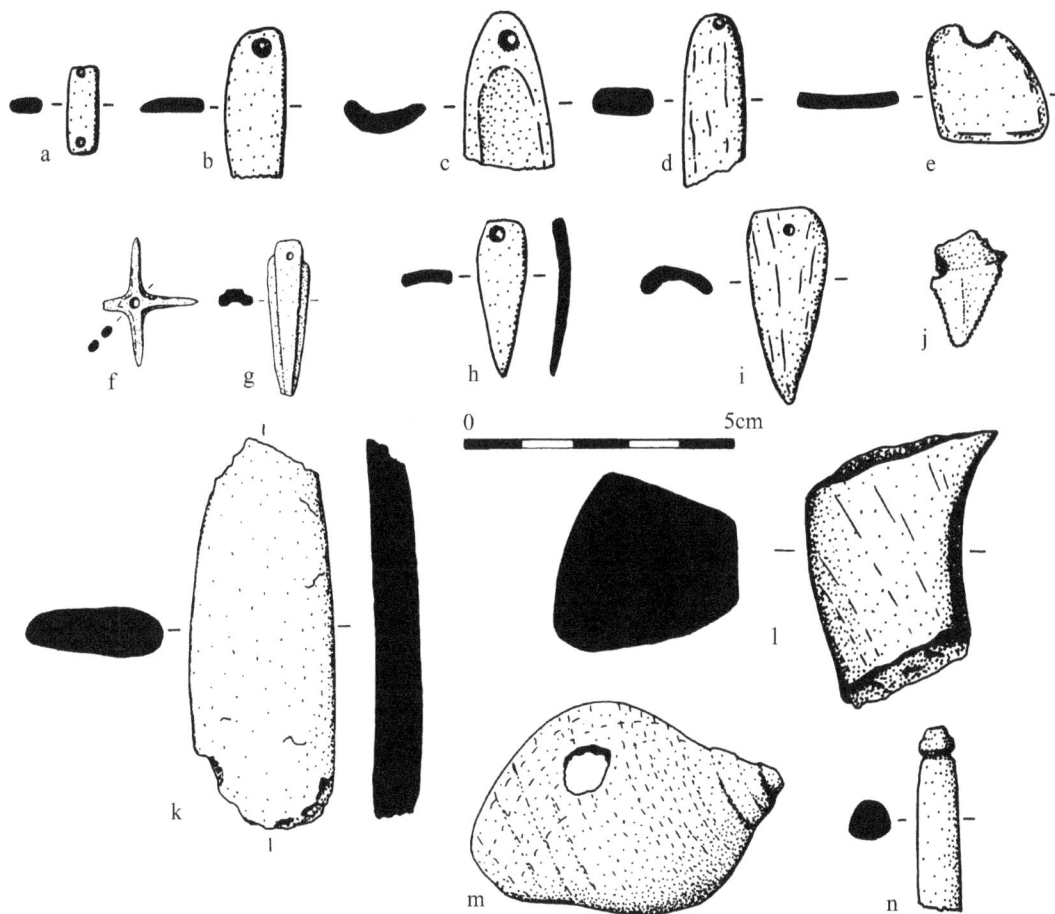

Figure 9.14 Pendants and other miscellaneous artefacts. a. *Pectinidea* shell unit (Mangaasi, TP12, L.9c); b–d. shell pendant (all Mangaasi, b. TP12, L.9b; c. TP10, L.3b; d. TP12, L.9b); e. shell fish hook tab? (Mangaasi, TP5, L.9); f. cross-shaped shell pendant (TP12, L.9b); g. *Pectinidea* shell pendant (TP17, L.4c); h. shell pendant (Waal, L.2); i. shell pendant (Malua Bay, TP9, L.3); j. shark tooth pendant (Ponamla, TP1.9, L.1); k. shell pendant (Ifo, Area A, L.1); k. possible shell pendant (Ifo, Area A, L.1.1); l. navela (Ifo, TP10, L.1); m. perforated gastropod (Ifo, Area A, L.1); n. grooved sea urchin spine (Arapus, TP13, 220–240cm bd).

Stone artefacts

A single flaked basalt stone disc was recovered from Ifo (Fig. 9.12 n). This was identical in size and form to the flaked *Tridacna* sp. shell version mentioned above from Ifo. It also dates to a similar period, c. 2500–2000 BP (Fig. 9.12 m).

Abraders either of scoria (11), pumice (11) and sandstone (3) were found from the excavations on Efate and Erromango (Tables 9.10–9.12). Scoria abraders were only found from the Erromango sites (Ponamla 5, Ifo 6), perhaps a reflection of material availability on that island. Two forms were identified, one which displayed a series of random grooves (Fig. 9.7 a), and the other which had flat ground surfaces. The grooves tended to be 4.5mm or less in width. These two forms of scoria abrader were present both in the Lapita and the post-Lapita deposits at Ifo and Ponamla (i.e. c. 3000–2000 BP). Pumice abraders displaying flat ground surfaces were predominant at Ifo (10), again probably simply a reflection of the availability of the stone at the site. They were recovered from all layers at the Ifo site (c. 3000–2000 BP). A single pumice abrader was recovered from the Mangaasi cultural horizon (c. 1600–1200 BP) at the Mangaasi site. Sandstone abraders displaying only flat ground surfaces were recovered from Ponamla (1) and Mangaasi (2). All of the various abraders seem most likely to have been used in the manufacture of the various shell artefacts. Similar abraders have been recovered from other sites in Vanuatu (Garanger 1972:Fig 207) and throughout the Pacific. They are an ubiquitous artefact form through time and space where any variation appears to be largely restricted to the choice or availability of raw materials.

A single triangular sectioned stone (Fig. 9.7 h) formed by grinding was also seen as a possible abrader but the symmetric form of the artefact casts some doubt. It was recovered from the lower layers of the Malua Bay site dating to c. 2700 BP. A circular piece of coral with a perforated centre (Fig. 9.13 l) was recovered from a similar stratigraphic context at Malua Bay. Its function also is somewhat mysterious but it was possibly a net sinker.

A total of thirteen pieces of ochre, which displayed either ground or striated surfaces, were found from Ponamla (7), Ifo (5) and Mangaasi (1) (see Tables 9.10–9.12). Speiser noted the widespread nature of the use of pigments throughout Vanuatu at the time of contact including for the painting of the body (1996 [1923]:169–170). Ochre was one of the sources of pigment and it seems quite likely that the excavated ochre served a similar purpose. The recovered specimens were from deposits that dated from c. 2800–2000 BP.

Obsidian which could have been sourced from the Banks Islands, was conspicuous by its absence. To date, obsidian whether sourced to the Banks Islands or further afield has only been recovered in Vanuatu from excavations as far south as Malo Island (Hedrick nd.; Galipaud 1998a), although Talasea obsidian is known from Lapita contexts in New Caledonia (Sand 2000:29). Banks Islands obsidian has also been recorded in some quantity in both early (Lapita) and late contexts (post 800 BP) on Tikopia (Kirch and Yen 1982:260–261) and later contexts (probably from around 1000 BP) in Fiji (Best 1984:494). A low grade glassy basalt is commonly found in the river beds of Efate and a total of nine pieces (see Table 9.12) were recovered from the excavations at Mangaasi but none appeared to be flaked or could be argued as being artefactual. In fact the worn nature of the surfaces of most of recovered pieces of basaltic glass suggested that their presence was the result of natural deposition. The material does not appear to be sufficiently vitreous for it to be particularly suitable for flaking. The results of an SEM EDAX analysis carried out by Wallace Ambrose on several samples indicated a likely Efate origin (Ambrose pers. comm.). The only other stone artefacts recovered from the excavations were the stone adzes from Ponamla, Malua Bay, Arapus and Ponousia respectively which have already been discussed (see 9.1 Adzes).

A number of basalt (254), chert (69) and chalcedony (6) flakes were recorded throughout the excavations but few of these were positively identified as tools or necessarily manufacturing debitage, although all were counted and weighed (Tables 9.10–9.12). There are abundant quantities of all these stones to be found from local sources. The vast majority of the stone flakes came from Ponamla (315 [96%]; basalt 240 [94%], chert 69 [100%], chalcedony 6 [100%]) and most of those can be best described as micro-flakes or 'shatter' (Kirch and Yen 1982:262).

Table 9.10 Ponamla Area A, worked, flaked and other stone (weight in grams)

	L.1	L.2	L.3A	L.3B	L.4	TOTAL/WGT
basalt	57(286)	30(164.9)	15(113.6)	108(267)	30(119.8)	240(951)
chert	19(150.4)	17(40.8)	7(4.5)	18(10.4)	8(3.9)	69(210)
chalcedony	2(36.7)	1(9)	1(4.8)	2(4)	-	6(54.5)
ochre#	3(56)	1(20.4)	2(112.7)	1(40.8)	-	7(229.9)
scoria abrader 1*	3(85.8)	-	1(14)	-	-	4(99.8)
scoria abrader 2*	-	1(60.1)	-	-	-	1(60.1)
sandstone abrader	-	-	1(14.5)	-	-	1(14.5)

= ochre displaying ground or striated surfaces; 1*= scoria displaying flat ground surfaces, 2*= scoria displaying ground grooves

Table 9.11 Ifo worked, flaked and other stone (weight in grams)

	L.1	L.2	L.3	L.4	TOTAL/WEIGHT
Trenches BCD					
basalt	–	2(35.7)	–	–	2(35.7)
basalt adze blank?	–	1(214.6)	–	–	1(214.6)
scoria abrader 1*	1(20.5)	1(32.4)	2(43.7)	–	4(96.6)
scoria abrader 2*	–	–	–	1(21.4)	1(21.4)
pumice abrader	2(108.9)	3(32.9)	1(55.7)	4(46.6)	10(244.1)
ochre#	4(61.6)	1(15.1)	–	–	5 (76.7)
TP 2					
scoria abrader 2*	–	–	1(48.8)	–	1(48.8)
TP 3					
basalt	–	–	1(3.5)	–	1(3.5)
TP 6					
basalt	1(7.9)	–	–	–	1(7.9)
Area A					
basalt	4(19.7)	1(4.3)	1(75.7)	–	6(99.7)
TP 9					
basalt adze blank?	–	1(135.6)	–	–	1(135.6)
TP 13					
ochre#	1(11.8)	–	–	–	1(11.8)

* see Table 9.6; # see Table 9.6

Table 9.12 Mangaasi worked, flaked and other stone (weight in grams)

	L.1/2I/2II	L.3A	L.3B	L.5I	L.7	L.9A	L.9B	L.9C	TOTAL/WEIGHT
TP 9									
basalt	–	–	–	–	–		–	1(4)	1(4)
ochre#	–	–	–	–	–		–	1(2.8)	1(2.8)
basaltic glass	–	–	–	–	–	–	–	1(23.4)	1(23.4)
TP 12									
basalt	–	–	–	–	–		1(4.6)	–	1(4.6)
basaltic glass	–	–	–	–	–	1(17.5)	–	–	1(17.5)
TP 1/15									
basalt	–	–	1(1.4)	–	–		–	–	1(1.4)
sandstone abrader	–	1(9.7)	–	–	–		–	–	1(9.7)
pumice abrader	–	1(18.2)	–	–	–		–	–	1(18.2)
TP 2									
basalt	1(.8)	–	–	–	–		–	–	1(.8)
sandstone abrader	–	1(65.3)	–	–	–		–	–	1(65.3)
basaltic glass	1(34.5)	–	–	–	–	–	–	–	1(34.5)
TP 4									
basaltic glass	–	–	–	–	–	1(16.8)	–	–	1(16.8)
TP 7									
basaltic glass	–	–	–	1(36.7)	–	–	–	–	1(36.7)
TP 13									
basaltic glass	–	–	–	1(14.7)	1(18.6)	–	–	–	2(33.3)
TP 14									
basaltic glass	1(22.5)	–	–	–	1(15.2)	–	–	–	2(37.7)
TP 17									
basaltic glass	–	1(17.8)	–	–	–	–	–	–	1(17.8)
TP 18									
basaltic glass	–	–	–	1(22.1)	–	–	–	–	1(22.1)

see Table 9.6

These flakes seem likely to have been the result of the movement and placement of stone during the construction of terracing at the site, an activity that could be expected to inadvertently produce chips or flakes. This explanation was further supported by the almost complete lack of micro-flakes from the other sites. Larger flakes only (all basalt) were recovered from Ifo (10), and Mangaasi (4) and these seemed more likely to be related to adze or other manufacturing activities. They are however strikingly few in number which although correlating well with the rarity of stone adzes in the record, may also reflect the existence of discrete activity areas which were not located during the excavations.

Miscellaneous artefacts

Two fragments of a very distinctive shell artefact, specific to Erromango, were recovered from the Ifo site. These were namely two fragments of the two forms of *navela* or shell money comprising fossil *Tridacna*. *Navela* functioned in the recent past as prestige items on Erromango and were exchanged between chiefs on important occasions and often during marriages (Aubert de la Rue 1945:192a; Humphreys 1926:171–172). One fragment associated with the circular type (Fig. 9.14c) was excavated near the surface of TP 10 and seems likely to date to the recent past. A large portion of the other straight form (Fig. 9.15) was recovered from Layer 2 of Trench D which dates to before 2000 BP, suggesting that this form of artefact has some antiquity on Erromango.

A single perforated *Turbo* sp. shell (Fig. 9.14m) was recovered from Layer 1 of Area A at Ifo, which dates to c. 2000 BP. These perforated shells have been identified at other sites in the Pacific as net sinkers (Kirch 1988b:205) but the frequency of perforated shells associated with burials in later sites in Vanuatu makes that identification in this case somewhat uncertain.

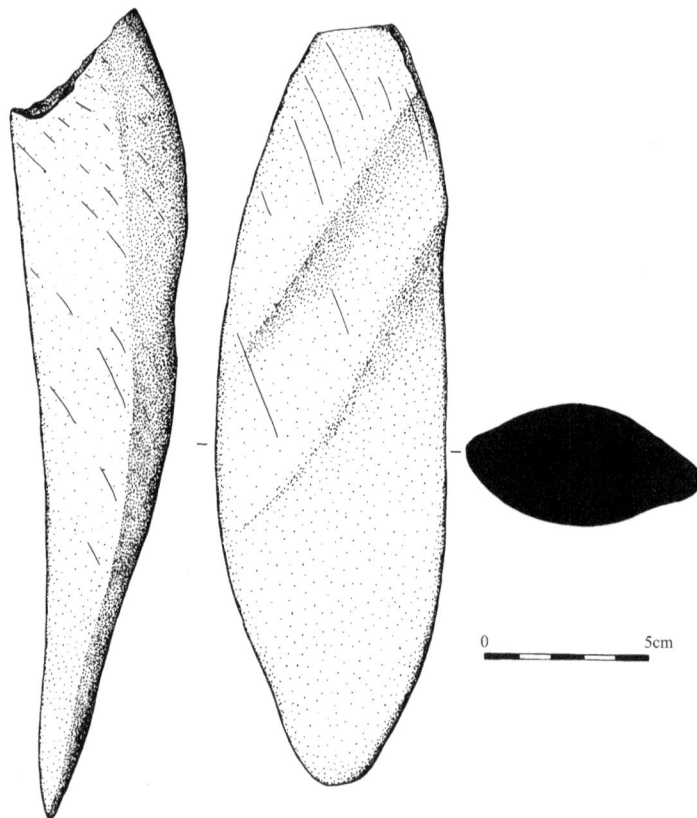

Figure 9.15 *Navela* (*Tridacna* shell money) Ifo (TrD, L.2).

Two spike-like artefacts (Fig. 9.12 k,l) made from fully ground *Tridacna* sp. shell were recovered from Ifo, from deposits dating to between c. 2500 and 2000 BP. These are the only examples thus far to have been recovered from excavated contexts in Vanuatu and may have perceivably been utilised as nasal ornaments. The practice of piercing the septum and inserting an ornament of some form was widespread throughout Vanuatu at the time of European contact (Speiser 1996 [1923]:158–160). Three nasal ornaments made of *Tridacna* (2) and wood (1) illustrated by Speiser (1996 [1923]: Plate 38, 28–30), collected from the islands of Santo and Malekula, are almost identical in form to the excavated examples.

Several bone needles were recovered from Ponamla from deposits dating to c. 2700–2500 BP and a single example was recovered from the post-600 BP levels of the Mangaasi site. The Ponamla examples were all similar in form and made from *Pteropus* sp. ulna bones (Fig. 9.7 b–d). Identical forms were recovered by Best at Lakeba (Best 1984:465). The bone needle recovered from Mangaasi (Fig. 9.7 e) was a slightly different form fashioned from an unidentified bone.

A number of historic artefacts were recovered from the surface of the Area A excavations at Ponamla. They included a metal boot heel, fragments of a tin box with wire handle, a ceramic bottle stopper and a glass bottle top. Another historic artefact that has been frequently observed on Erromango are glass beads which are found in association with early historic burials.

Synthesis and Summary

The non-ceramic items of material culture recovered from the excavations in Vanuatu have provided a glimpse of the rich variety of artefact forms that existed, notwithstanding the potentially much greater assortment of perishable items which do not survive in the archaeological record (Kirch 1997:239). Many of the artefact types and forms are paralleled in numerous archaeological sites across the Pacific and this is particularly the case with those artefacts associated with Lapita and immediately post-Lapita cultural deposits to c. 2500 BP. An initial broad homogeneity of ceramics and, for a longer period, other artefact forms are a feature of sites dating to this period, although there are always a number of anomalies (Sand 2000). In the case of Vanuatu the lack of obsidian in the archaeological record further south than Malo is one such anomaly. Another is the complete absence of fishhooks amongst the recovered materials which do occur, although not in abundance, in other Lapita sites (Kirch 1997:200; Sand 2000:28). However, fishing with hooks may never have been particularly prevalent and certainly by the ethnographic period there are indications that fishhooks were rarely used and spearing and netting were much more common techniques (Speiser 1996 [1923]:141–142).

A chronological synthesis of non-ceramic artefact forms is presented below along with a summary table (Table 9.13). It must be emphasised that this synthesis is only preliminary in nature and is skewed by a number of factors but primarily by a lack of knowledge relating to certain parts of the sequence and the unbalanced sampling of different site types. Lapita period sites in Vanuatu are poorly known, as are any sites dating from c. 1200–600 BP. The sites which have returned a wealth of non-ceramic items of material culture dating to the post-600 BP period are largely burial sites rather than habitation sites (Garanger 1972). Immediately post-Lapita remains to 2000 BP are well represented from both Efate and Erromango but they are specifically habitation sites that have been excavated. The excavations on Malekula were predominately located at cave sites which generally tend to return fewer artefactual remains than open sites. Changing settlement patterns also seem likely to have had some influence on what is recovered archaeologically. Lapita and immediately post-Lapita settlements appear to have generally been in the form of nucleated villages which would have generated a correspondingly greater concentration of midden dumping. Later settlement was often more dispersed which would have generated more diffuse midden remains. It is these factors, along with the often small quantities of specific artefact types, which must be taken into consideration when any conclusions are made regarding chronological definition or variation in the non-ceramic artefact assemblages. Despite these limitations a number of significant patterns can be highlighted.

3000–2800 BP

Non-ceramic artefacts associated with the Lapita period of settlement in Vanuatu are not well represented primarily due to the mixed nature of the deposits or the limited identification and excavation of these site types. The most securely provenanced non-ceramic artefacts (Fig. 9.14n, 9.8a–d) are from the recently excavated Arapus site (Bedford and Spriggs 2000; Spriggs and Bedford 2001). They include *Tridacna* sp. shell arm and smaller rings, a fully ground *Tridacna* sp. adze, a *Conus* sp. shell ring and a grooved sea urchin spine. Added to these artefact forms are abraders of both scoria and pumice. Although the Lapita sites on Malo were often somewhat mixed, Hedrick (nd.) reported recovering flaked tools of both chert and obsidian (sourced to either the Banks Islands, Talasea in New Britain or Lou in the Admiralty Islands), stone adzes (of plano-lateral and plano-convex cross-section) and *Tridacna* sp. adzes cut from the hinge region of the shell. Other artefacts included, shell scrapers, coral files, peelers, anvils and stone burnishing tools, shell rings and discs, and a variety of shell beads. Galipaud (1998a:7) reported recovering both *Tridacna* and *Trochus* sp. rings and or small armbands from his recent excavations on Malo. It could be pointed out that a whole host of artefacts found at other Lapita sites throughout the Pacific have not so far been identified in Vanuatu (e.g. fishhooks, 'long units', decorated bracelets etc. [Kirch 1997; Sand 2000) but the identification of any disjunct distributions or assessment of rarity would be premature considering the lack of research carried out on these site types in Vanuatu to date. This parallels the earlier situation in New Caledonia up to the 1980s at which time Green and Mitchell reported that 'remarkably few non-pottery portable artefacts have been recovered archaeologically' (Green and Mitchell 1983:64). Since that time, further excavation has revealed abundant and varied shell artefact assemblages (Sand 2000:27–29).

2800–2500 BP

Artefacts from this period of the archaeological record are probably some of the best represented and they hint at the wider range of artefacts that might well be expected to be recovered from Lapita sites in Vanuatu in the future. Continuity with the Lapita deposits is clearly demonstrated by the presence of *Tridacna* sp. shell arm and smaller rings, and the *Conus* sp. armrings, smaller rings and beads. Fully ground *Tridacna* sp. shell adzes were also frequently recovered. They generally fell into two categories, very small 'microblades' or larger hinge region adzes. Stone adzes are rare and thus far only two forms have been identified associated with this period. One is a fully ground rectilinear cross-section form (Fig. 9.8f) and the other a stone version of the shell 'microblade' (Figs 9.7 f, g). A wide range of pendants were also present during this period, made from either shell or in one case a shark tooth (Fig. 9.14j). Somewhat less frequent artefacts associated with this time period were the fine bone needles made from the *Pteropus* sp. ulna, recorded on Erromango, a non-perforated *Conus* sp. disc and a possible net sinker, both recovered from Malua Bay on Malekula. Abraders continued through this period and into the next. Exotic stone (e.g. Lapita sites on Malo) has not been found in these later contexts.

2500–2000 BP

A number of artefact forms found in the earlier stratigraphic layers continued through into this later phase of the cultural sequence. They included *Tridacna* sp. armrings, *Conus* sp. armrings and smaller rings and beads and shell pendants. A *Trochus* sp. shell armring was recorded from layers dating to this period but they would appear to have been at this stage of the sequence a rare ornament. Shell adzes, again all largely fully ground and more often the 'micro-adze' form, were also still present. Stone abraders also continued to be recorded. Items recorded which had not been previously recovered from the earlier layers included stone and shell discs (Fig. 9.12 m, n), possible nasal ornaments (Fig. 9.12 k, l), *navela* (Fig. 9.15) and a perforated shell (Fig. 9.14 m).

2000–1600 BP

The only excavations to date that relate to this period of the cultural sequence are those carried out at the Mangaasi site (Garanger 1972; Bedford *et al.* 1998) which greatly reduces the sample in comparison to that recovered from the first 1000 years. In general fewer non-ceramic artefacts were recovered from cultural horizons dating to this period at the Mangaasi site, but that was the case for all classes of midden and other artefactual remains and may relate to changing settlement patterns which in turn affected the composition and pattern of midden dumping, rather than definitive evidence for a decrease in the frequency or variety of non-ceramic artefact forms. The non-ceramic remains were largely restricted to *Tridacna* sp. armrings (which were possibly wider than before) and *Conus* sp. armrings, smaller rings and beads. A single perforated pendant (Fig. 9.14 c) also related to this period. Fragments of *Tridacna* sp. shell adzes signalled their continued presence.

1600–1200 BP

This phase of the cultural sequence again relies solely on the excavations at the Mangaasi site and the sparse non-ceramic remains largely parallel those outlined for the period 2000–1600 BP i.e. *Tridacna* sp. armrings and adzes and *Conus* sp. armrings, smaller rings and beads. A great deal of further archaeological excavation concentrating on this time period is required to further define the cultural sequence, both non-ceramic and ceramic. The sparse non-ceramic remains that were recovered do at least tend to demonstrate continued links with earlier artefact forms rather than those which appear in the sequence over the last 1000 years.

1200–600 BP

This phase of Vanuatu prehistory in general remains largely unknown apart from the sparse ceramic remains recovered from Malekula. This is due both to a general lack of archaeological research in Vanuatu and the difficulty of identifying sites which date to this period, particularly in areas where ceramics were no longer being produced (from the Shepherd Group in the north to Aneityum in the south). A number of artefacts described by Shutler (1969) in his tentative cultural chronology for central and southern Vanuatu might be placed into this time period but the mixed nature of the deposits and the often confusing dates lessen the reliability of their provenance.

600 BP to ethnographic period

It is principally the excavations of Garanger (1972) which provide us with the rich inventory of excavated non-ceramic artefacts that are found in the record of the last 600 years. The excavations of Ward (1979) at Pakea also provide a limited number of artefacts from the far north of Vanuatu. He noted that the greatest similarities to the Pakean artefacts were found in the burials from Central Vanuatu (Ward 1979:10–32) suggesting that the majority of the material from Pakea related to this latter period of prehistory. The recovered non-ceramic artefacts relating to this period demonstrate a striking contrast with much of the material culture that has been outlined earlier. The inventory of ethnographic material culture recorded by Speiser (1996 [1923]:403–404) for Vanuatu is very detailed. It provides further detail for this later period and particularly highlights the wide range of perishable items that are largely missing from the archaeological record. The perishable items are not discussed here in any detail here. Rather, the focus is identifying the range of artefacts which have been recovered from archaeological contexts and continued in use through to European contact.

Post-600 BP shell adzes were still made from *Tridacna* sp. shell but invariably made from the dorsal region of the shell and were not heavily ground. More conspicuous in the archaeological record is the widespread use of a number of other shell species from this period. *Lambis* sp., *Terebra* sp., and *Mitra* sp. (and very occasional *Conus* sp. [Shutler and Shutler 1975:74; Shutler *et al.* 2002]) shell adzes and gouges have been frequently reported from the Shepherds in the north to at least as far south as Efate (Garanger 1972:Figs 292, 293). Sparse evidence of these adze forms was also recorded by Ward at Pakea in the Banks (Ward 1979:9–6) and they were noted by Speiser in Santo and Malekula (1996 [1923]:Plate 32; 26, 27). *Terebra* sp. shell adzes are certainly a widespread artefact form found across much of the Southwest Pacific during this later period of prehistory (Kirch 2000:129).

As noted, the stone adzes associated with this period tend to be lenticular in cross-section but are unfortunately poorly provenanced, primarily surface collected only. This would at least suggest they belong to the later period of the cultural sequence and they have certainly not been found in any of the earlier excavated contexts. They have been recovered from throughout Vanuatu during both archaeological (Shutler and Shutler 1965; Spriggs *et al.* 1986) and ethnographic (Speiser 1996[1923]) surveys. They have in the past been broadly labelled as a 'Melanesian' form of adze but finer definition of their distribution, chronological and cultural associations awaits future research. But certainly this artefact form again hints at intra and inter-archipelago contacts during the latter phase of the region's prehistory.

The array of ornaments is particularly detailed for this period as much of the excavated evidence has come from burials, from either Tongoa, Retoka, Efate (Garanger 1972), Futuna (M.E. and R. Shutler 1965) or Aneityum (Spriggs 1997:218). Necklaces appeared in the sequence which were unknown from earlier contexts. They were often made of strings of cut *Conus* sp. beads or rings with an assortment of different pendants including perforated whale and crocodile teeth (e.g. Garanger 1972:Figs 185, 186, 186,10). Necklaces of pig tooth beads also seem to have been widespread, at least from the centre to the south of Vanuatu (M.E. and R. Shutler 1965:Plate 5e; Spriggs 1997:218). According to Spriggs this artefact type along with the whale-tooth beads is suggestive of Polynesian influence (Spriggs 1997:218). Another bead form associated with this later period are circular *Tridacna* shell beads recovered from a burial on Aneityum (Spriggs 1997:218). Necklaces and anklets of perforated shells (frequently *Cypraea* sp. and more occasionally single *Spondylus* sp.) have also been recorded in association with burials (Garanger 1972:Figs 176, 197) as have conch shell trumpets (Garanger 1972:Fig. 151), and both artefact forms were recorded by Speiser (1996 [1923]:Plates 39, 41 and 106). Ward also noted perforated *Cypraea* sp. shell at Pakea (1979:10–11).

Conus sp. beads were prevalent throughout Vanuatu at this later time and into the ethnographic period. In the Banks Islands they were a form of shell money that were strung together and known as *som* (Ward 1979:10–15; Speiser 1996:242). In other areas they were principally worn as decoration, in the form of necklaces or bracelets, and even incorporated into woven armbands or skirts (e.g. Garanger 1972:Fig. 181). All of these artefact forms continued in use until European contact (Speiser 1996 [1923]:Plates 40). Other necklace forms recorded both archaeologically and in the ethnographic record include combination shell and fishbone vertebrae bead or simply fishbone bead necklaces (Garanger 1972:Fig. 190, 4; Speiser 1996 [1923]:Plate 78).

Restricted thus far to the island of Tanna are a number of unprovenanced serpentine pendants which date from at least the time of Cook's (Beaglehole 1969:505) first visit (1774) to that island and through to the ethnographic period; (Dubois 1996:79–82; Speiser 1996 [1923]:166, Plate 40). The source of the stone is thought to have been New Caledonia (Aubert de la Rue 1938; Dubois 1996). Confirming this connection is a similar artefact that was recovered by Sand (1996:164) from the Loyalty Islands. Thus far restricted to the north are the remains of *Tridacna* sp. shell breast plates reported by Ward at Pakea from the later levels of the site which he likened to similar ornament forms known ethnographically from Santa Cruz (Ward 1979:10–12).

Two other somewhat ubiquitous artefact forms which also appear in this latter phase of the archipelago's prehistory are pig tusk bracelets and *Trochus* sp. bracelets (Garanger 1972:Figs 191, 192). Both bracelet forms have been recorded archaeologically and ethnographically (Speiser 1996 [1923]:144, 165–168). *Trochus* sp. shell rings are rare during the early part of the sequence, (noted on Malo by Galipaud 1998a:7 and one other from Ifo dating to c. 2500 BP), although Ward recorded them as being present throughout his Pakea sequence (Ward 1979:10-8, 10–15). They are certainly more frequent throughout Vanuatu in the latter part of the sequence. The pig tusk bracelets have only generally been recorded archaeologically within this latter phase of the cultural sequence, although Ward (1979:11–26) suggests that circumstantial evidence dates these artefact forms to as early as 1050 BP. They do at least date from 1606 when they were noted by the first European (the Spanish led by Quiros) visitors as being present on both Gaua in the Banks Islands and on Santo (Kelly 1966:I:199, 299).

Summary

It would appear that non-ceramic artefacts were generally not as susceptible to change as were ceramics, as they continued to show continuity in form over much longer time periods. Many of the artefact forms identified in the earliest cultural contexts, such as the shell armrings and other smaller rings along with *Tridacna* sp. adzes, show little change in form for at least up to 1000 years after initial human settlement. Even the limited numbers of artefacts recovered from cultural contexts which dated up to 1500 years after initial settlement showed some affinity with earlier artefact forms. The most perceptible change amongst the non-ceramic artefacts during the first 1500 years is a tendency over time to a restriction of variety and quantity, a feature also noted by Sand (2000:30) in the case of New Caledonia. However it has been argued that this may in part be due to more dispersed settlement patterns which would have affected both midden depositional practices and accumulation processes. Certainly the rich nature of the non-ceramic record from both the earliest and latest periods of the cultural sequence support this above scenario. More detailed archaeological research concentrating on sites dating to this later time period (2000–600 BP) is required to further clarify this issue.

Change in non-ceramic artefact forms may have been slow and gradual but there is certainly ample evidence to indicate that it was substantially changed over the last 600 years. *Tridacna* sp. adzes continued through to European contact but were by that stage only partly ground and invariably fashioned from the dorsal area of the shell. A number of other shells also began to be utilised in the manufacture of adzes during the last 600 years, including *Lambis* sp. and *Terebra* sp. (Garanger 1972:Figs 292, 293). The earliest evidence of the appearance of island specific artefact forms was on Erromango with the *navela* appearing by c. 2000 BP. After 600 BP an assortment of ornaments appeared that were not previously known. These included perforated shellfish necklaces (Garanger 1972: Fig. 197), pendants of stone and other non-shell forms, beads made from whale teeth and pig tusks (Garanger 1972: Figs 184–192; Spriggs 1997:218). Pig tusk and *Trochus* sp. bracelets were also a common artefact form dating to the later period of Vanuatu's history. Less common but equally identified with only the late phase of the sequence are conch shell trumpets and serpentine pendants.

A number of these later artefact forms have been noted as having some widespread distribution across both Near and Remote Oceania and the Melanesian and Polynesian divide at this time (Kirch 2000:129; Spriggs 1997:203) which certainly hints at some form of interaction across this vast area. More specifically, renewed links between Southern Vanuatu and New Caledonia are strongly indicated. There is certainly a lot more archaeological 'homework' to be done on the latter period of Vanuatu's history when quite marked cultural change appeared to have occurred and it will be the non-ceramic artefacts which will play a vital role in its further elucidation, as it is during this period (c. last 1000 years) that ceramics disappeared from much of the archipelago.

Table 9.13 Summary of non-ceramic artefact forms

DATES / ARTEFACT	3000-2800BP	2800-2500	2500-2000	2000-1600	1600-1200	1200-600	600-ethno.	ethno.
Tridacna armring	+	+	+	†	†	?	-	-
Conus armring	+	+	+	†	†	?	-	-
Conus ring	+	+	+	+	+	?	+	+
Conus bead	†	†	†	+	+	+?	+	+
Trochus armring	†	†?	†	†?	†?	†?	+	+
Trid. adze Type 1	+	+	+	-?	-?	-?	-	-
Trid. adze Type 2	-	-	-	-	-?	?	+	+
Trid. adze Type 4	-?	+	+	+	+?	-?	-	-
Trid. adze T 7-8	+	+	+	†?	-	-?	-	-
Lambis adze	-	-	-	-	-	?	+	+
Terebra adze/gouge	-	-	-	-	-	?	+	+
Mitra adze	-	-	-	-	-	?	+	+
stone adze (rectilinear)	†	†	†?	?	?	-?	-	-
stone adze (lenticular)	-	-	-	-	-	-?	+	+
worked shell pendants[2]	+?	+	+	†?	†?	?	†	†
serpentine pendant	-	-	-	-	-	-?	‡	‡
stone abrader	+	+	+	+	+	+?	+	+
grooved sea urchin spine	†	†	-?	-	-	-	-	-
navela	-	-	-	‡	‡?	‡?	‡	‡
pig tusk bead	-	-	-	-	-	?	+	+
pig tusk bracelet	-	-	-	-	-	†?	+	+
whale tooth bead	-	-	-	-	-	?	+	+
perforated shell[1]	-	-	-	-	-	?	+	+
Trid/stone disc	-?	†	†	†	-?	-?	-	-
nasal ornaments	-	-	†‡?	†?‡?	†?‡?	†?‡?	+	+
fishbone vertebrae necklace	-	-	-	-	-	-?	+	+
bone needles[3]	†?	†	†?	†?	†?	†?	†	†
shell trumpet	-	-	-	-	-	?	+	+

+archipelago wide; † rare; ‡ island specific; - absent; ? lacking sufficient data; -? lacking sufficient data but seems likely to be absent. 1. miscellaneous perforated shells used as necklaces or pendants; 2. finer chronological delineation is difficult due to sample size 3. finer differentiation of needles is again seen as premature due to the small sample size

10

Faunal remains

'The principal occupation of the people is gardening, for their diet is predominantly a vegetarian one, yams being the staple food-stuff. In the coastal villages, however, fish are caught and shell-fish and crabs are collected, while everywhere wild pig is hunted; but the products of these activities are regarded as tasty extras to the usual vegetable dish, never as a basis of a meal' (Deacon (1934:16) on the subject of the Malekulan mode of life).

Information to date relating to the prehistoric faunal record in Vanuatu can be described as somewhat sparse or certainly lacking in detail. From his extensive excavations Garanger noted only the presence of pig bones with pottery and several fish vertebrae from a number of sites in the Shepherds (Garanger 1972:82, 86, 87) while at the Mangaasi site fishbone was described as plentiful and pig as having been present from the earliest levels (Garanger 1972:57). In a preliminary report relating to their archaeological research on Efate and the southern islands of Vanuatu the Shutlers noted large quantities of shellfish at many of the excavated sites along with the presence of pig bone on Tanna. The presence of pig on a number of other islands (Futuna and Fila) could also be deduced from the recovery of a number of artefacts which were made either from pig bone or tusks (Shutler and Shutler 1968[1966]). On Malo, at the site of Batuni'urunga, Hedrick (nd) was able to demonstrate the presence of introduced species such as pig and rat in association with Lapita settlement of the island, along with the exploitation of local species such as turtle, bird, fruit bat and fish. Shellfish were described as the predominant component in the middens. Hedrick (1971:10) also listed the shellfish species recovered from the Avunitare Lapita site. The limited faunal remains recovered from the Ifo site on Erromango in 1983 were also briefly outlined by Spriggs and Wickler (1989:80). The most detailed analysis and reporting of faunal remains from Vanuatu have been those from Ward's excavations on Pakea in the Banks Islands (Ward 1979:11-2 -31) which will be referred to below. The excavations that have been carried out over from 1995 to 1999 across the archipelago have returned a large quantity and variety of faunal material which has greatly enriched the archaeological record. All materials were analysed (see Bedford 2000b:213–237 and Appendix 7 for the complete data set) except for the faunal remains from the 1999 excavations at the Arapus site which are not presented here in any detail, although a number of aspects of the recovered assemblage are commented on. The data presented (Tables 10.1–10.43) and discussed here is a refined version of earlier research (Bedford 2000b) where a number of sites which contributed little in terms of the wider faunal debate have not been included.

Analysis of the material has been carried out by a number of researchers from different institutions. Initially all excavated bone was sent to David Steadman (University of Florida) who sorted the bone to taxa level and provided a bone count or NISP (number of identified specimens) for the assemblages. All bone apart from the bird bone was returned to Canberra where confirmation of Steadman's identifications of mammals and turtle were carried out by utilising the Australian National University faunal comparative collection. The bird bone was analysed in detail by Steadman who has provided all the data presented here. Further analysis of the rat bone was undertaken by Peter White and Tim Flannery of the University of Sydney and formerly of the Australian Museum respectively who were able to positively identify *Rattus praetor*. The fishbone was sent to the archaeozoology lab in Wellington, New Zealand, for analysis. Professor Jim Mead of the University of Northern Arizona has identified the *Mekosuchid* bone from the Arapus site. Shellfish analysis was undertaken by Joe Gyngell as part of his BA Hons dissertation at the Australian National University which focused on the shellfish from Erromango (Gyngell 1997). At a later date he analysed samples of the recovered shellfish from both Efate and Malekula. Lyn Schmidt (research assistant at the Department of Archaeology and Natural History, ANU) completed the analysis of the shellfish from Malekula and has summarised the data from all three islands (Schmidt 2000).

Mammals

Recovered mammal remains (Tables 10.1–10.9) included introduced species such as pig (*Sus scrofa*) and rat (*Rattus exulans* and *Rattus praetor*), along with indigenous species of fruit bat (*Pteropus* sp.). Sparse human remains such as teeth and other often single bones were recovered from all islands. None of these appeared to be related to burials except for those from the Mangaasi site where very disturbed burials were recorded. Dog was not positively identified amongst the excavated materials.

Table 10.1 Ponamla Area A bone remains (non fish)

Taxon	Layer 1	Layer 2	Layer 3a	Layer 3b	Layer 4	Total
Mammals						
Rattus exulans	45	54	61	67	69	296
Pteropus sp.	41	25	21	45	27	159
Sus scrofa	7	5	4	5	2	23
Homo sapiens	2	-	2	1	1	6
Small bat	-	-	-	1	2	3
Medium mammal	6	2	1	1	1	11
Birds						
Gallus gallus	7	-	7	6	6	26
Gallirallus philippensis	3	3	2	5	2	15
Porphyrio porphyrio	16	4	8	3	12	43
Ducula sp.	4	9	9	1	5	28
Columba vitiensis	6	3	2	3	-	14
Columbidae sp.	5	-	4	-	2	11
Tyto alba	2	2	1	-	-	5
Aplonis sp.*	-	-	-	-	2	2
Puffinus pacificus	1	-	-	-	-	1
cf. *Gallinula* sp.	-	-	1	-	-	1
Macropygia mackinlayi	-	-	-	-	1	1
Turdus poliocephalus	1	-	-	-	-	1
Trichoglossus haemotodus	1	-	-	-	-	1
Egretta sacra	-	-	-	1	-	1
Halcyon farquhari＊	-	-	-	2	-	2
Bird sp.	21	9	31	14	34	109
Reptile						
Turtle	6	3	3	-	1	13
Lizard	2	-	-	-	-	2
Unidentified	-	1	-	2	-	3
Total NISP	174	120	157	155	169	775

* extirpated species

Table 10.2 Ifo, Trenches B, C, D bone remains (non fish)

	Layer 1	Layer 2	Layer 3	Layer 4	Total
Mammals					
Rattus exulans	3	3	3	5	14
Pteropus sp.	2	13	34	33	82
Sus scrofa	6	-	-	1	7
Homo sapiens	3	-	-	-	3
Med. mammal	4	2	1	1	8
Birds					
Gallus gallus	3	3	6	2	11
Porphyrio porphyrio	-	-	3	-	3
Ptilinopus greyii	-	-	-	1	1
Ducula pacifica	-	-	-	1	1
Ducula sp. (large)+	-	-	1	1	2
Ducula sp.	4	-	5	-	9
Columba vitiensis	1	-	-	-	1
Tyto alba	1	1	-	-	2
Reptiles					
Turtle	-	1	3	2	6
Total NISP	27	23	56	47	153

+ extinct species

Table 10.3 Mangaasi TPs 1/15, 17, 10, 12, 9 bone remains (non fish)

Taxon	Mammals Rattus exulans	Rattus praetor	Pteropus sp.	Sus scrofa	Homo sapien	Medium mammal	Birds Gallus gallus	Gallirallus philippensis	Porzana tabuensis	Porphyrio porphyrio	Columba vitiensis	Tyto alba	Trichoglossus haematodus	Halcyon sp.	Columbidae sp	Ptilinopus greyii	Passeriformes sp.	Bird sp.	Reptile Turtle	Lizard	Unidentified	Total
TP1/15 L.1	4	-	-	5	-	-	-	-	-	-	-	-	-	-	-	-	-	1	-	-	-	9
L.2ii	1	-	1	23	1	-	-	-	-	-	-	-	-	-	-	-	-	-	-	-	-	26
L.3a	2	-	-	10	1	-	-	-	-	-	-	-	-	-	-	-	-	-	-	-	-	13
L.4a	-	-	-	2	1	1	-	-	-	-	-	-	-	-	-	-	-	-	-	-	-	4
L.4b	-	-	-	-	-	1	-	-	-	-	-	-	-	-	-	-	-	-	-	-	-	1
L.3b	4	-	-	2	1	-	-	-	-	-	-	-	-	-	-	-	-	-	-	-	-	7
L.11	-	-	-	-	1	-	-	-	-	-	-	-	-	-	-	-	-	-	-	-	1	2
TP 17 L.1	5	-	-	2	-	-	-	-	-	-	-	-	-	-	-	-	-	-	-	-	-	7
L.3a	3	-	-	7	-	-	-	-	-	-	-	-	-	-	-	-	-	-	-	-	-	10
L.3b	49	-	2	17	-	2	-	1	-	1	-	-	-	-	-	-	-	-	2	1	-	75
L.3c	9	-	-	2	-	-	-	-	-	1	-	-	-	-	-	-	-	1	-	-	-	13
L.4a	3	-	-	-	-	-	-	-	-	-	-	-	-	-	-	-	-	-	-	-	-	3
L.4b	1	-	-	-	-	-	-	1	-	-	-	-	-	-	-	-	-	-	-	-	-	2
L.5iii	8	-	-	-	-	-	-	-	-	-	-	-	-	-	-	-	-	-	-	-	-	8
L.11	1	-	-	-	-	-	-	-	-	-	-	-	-	-	-	-	-	-	-	-	-	1
TP10 L.1	8	-	-	-	-	2	-	-	-	-	-	-	-	-	-	1	-	-	-	-	-	11
L.2i	4	-	-	1	-	1	-	-	-	-	-	-	-	-	-	-	-	-	-	-	-	6
L.3a	7	-	-	-	-	3	1	1	-	-	-	-	-	-	-	-	-	-	-	-	-	12
L.3b	14	-	-	4	-	1	-	-	-	-	-	-	-	-	-	-	-	-	-	-	-	19
L.9a	2	-	1	-	1	-	-	-	-	1	-	-	-	-	-	-	-	-	-	-	-	5
L.8	2	-	-	-	-	-	-	-	-	1	-	-	-	-	-	-	-	-	-	-	-	3
L.9b	3	-	-	-	-	-	-	-	-	1	-	-	-	-	-	-	-	-	-	-	-	4
L.11	1	-	-	-	1	-	-	-	-	-	-	-	-	-	-	-	-	-	-	-	-	2
TP 12 L.1	6	-	-	15	-	-	-	-	-	-	-	-	-	-	-	-	-	2	-	-	-	23
L.2ii	3	-	1	13	-	-	-	-	-	-	-	-	-	-	-	-	-	-	-	-	-	17
L.9a	22	-	5	7	2	1	-	1	-	-	-	-	-	-	-	-	-	6	5	-	-	49
L.9b	65	-	10	-	-	-	2	3	-	3	-	-	-	-	-	-	-	1	-	1	-	85
L.8	1	-	-	-	-	-	-	-	-	-	-	-	-	-	-	-	-	-	-	-	-	1
L.9c	14	-	1	-	-	-	-	-	-	1	-	-	-	-	-	-	-	-	1	-	-	18
L.11	4	-	-	-	-	-	-	-	-	-	-	-	-	-	-	-	-	-	-	-	-	4
TP 9 L.1	-	-	-	2	-	5	-	-	-	-	-	-	-	-	-	-	-	-	-	-	-	7
L.2ii	3	-	-	1	-	7	-	-	-	-	-	-	-	-	-	-	-	-	-	-	-	11
L.9a	140	-	4	3	-	3	3	2	-	2	-	-	2	-	-	-	-	-	-	1	-	160
L.9b	44	-	1	1	-	2	1	1	-	5	-	1	-	-	-	-	-	-	-	-	-	56
L.8a	12	1	1	-	-	-	1	-	-	2	-	-	-	-	-	-	-	-	-	-	-	17
L.9c	94	10	13	5	-	-	1	-	1	28	1	-	1	1	1	-	-	-	-	1	-	163
L.8b	1	-	-	-	-	-	1	1	-	2	-	-	-	-	-	-	-	-	-	-	-	5
L.9d	19	-	3	-	-	-	2	-	1	4	-	-	-	-	-	-	2	-	-	-	-	31
L.11	-	-	-	-	-	-	-	-	-	-	-	-	-	-	-	-	-	-	-	-	-	-
Total NISP	559	11	43	122	9	29	12	11	2	52	1	1	3	1	1	1	2	10	9	3	1	883

Table 10.4 Mangaasi TPs 2, 3, 4, 5, 6, 7, 8, 11, 13, 14, 16, 18 bone remains (non fish)

Taxon	Mammals Rattus exulans	Pteropus sp.	Sus scrofa	Homo sapien	Medium mammal	Birds Gallus gallus	Gallirallus philippensis	Porzana tabuensis	Porphyrio porphyrio	cf. Procellariidae sp.	Megapodius Layardi	Tyto alba	Cuculidae sp.	Anas superciliosa	Halcyon farquhari*	Bird sp.	Reptile Turtle	Lizard	Total
TP 2 L.1	–	–	4	–	–	–	–	–	–	–	–	–	–	–	–	–	–	–	4
L.2i	–	1	3	–	–	–	–	–	–	–	–	–	–	–	–	–	–	–	4
L.2ii	3	–	4	–	–	–	–	–	–	–	–	–	–	–	–	–	–	–	7
L.3a	33	3	2	–	4	1	–	–	5	–	–	–	–	–	1	3	–	–	52
L.4	1	–	–	–	–	–	–	–	–	–	–	–	–	–	–	–	–	–	1
L.11	–	–	1	–	–	–	–	–	–	–	–	–	–	–	–	–	–	–	1
TP 3 L.1	11	1	4	–	–	–	–	–	–	–	–	–	–	–	–	–	–	–	16
L.2i	–	1	7	–	–	–	–	–	–	–	–	–	–	–	–	–	–	–	8
L.2ii	4	1	5	–	–	–	–	–	–	–	–	–	–	–	–	–	1	–	11
L.3	3	–	–	–	–	–	–	–	–	–	–	–	–	–	–	–	3	–	6
L.11	–	–	–	–	–	–	–	–	–	–	–	–	–	–	–	–	3	–	3
TP4 L.1	24	1	3	–	1	–	1	–	–	–	–	–	–	–	–	–	–	–	30
L.2i	8	–	19	–	–	2	–	–	–	–	1	–	–	–	–	1	–	–	31
L.5i	40	2	3	–	–	–	–	–	1	–	–	–	–	–	–	4	–	–	50
L.5ii	34	4	–	–	1	–	–	1	–	–	–	–	–	–	–	3	–	–	43
L.9	5	3	–	–	1	–	–	–	14	–	–	1	–	–	–	–	–	–	24
TP5 L.1	2	2	3	–	–	–	–	–	–	–	–	–	–	–	–	–	–	–	7
L.2ii	2	3	7	–	1	–	–	–	–	–	–	–	–	–	–	–	–	–	13
L.2iii	–	–	–	–	1	–	–	–	–	–	–	–	–	–	–	–	–	–	1
L.5i	11	–	4	–	3	–	–	–	–	–	–	–	–	–	–	–	–	–	18
L.9	10	1	–	–	2	–	–	–	2	–	–	–	–	1	–	–	–	–	16
TP6 L.1	6	–	1	–	1	–	–	–	–	–	–	–	–	–	–	–	–	–	8
L.2i	1	1	–	–	1	–	–	–	–	–	–	–	–	–	–	–	–	–	3
L.5i	1	1	1	–	2	–	–	–	–	–	–	–	–	–	–	–	–	–	5
TP7 L.2iii	–	1	–	–	–	–	–	–	–	–	–	–	–	–	–	–	–	–	1
L.5i	4	–	1	1	3	–	–	–	–	–	–	–	–	–	–	–	–	–	9
TP8 L.2i	–	–	2	–	–	–	–	–	–	–	–	–	–	–	–	–	–	–	2
L.2ii	–	–	–	1	–	1	–	–	–	–	–	–	–	–	–	–	–	–	2
L.5i	4	–	1	–	3	1	–	–	–	–	–	–	–	–	–	–	–	–	9
TP11 L.2i	5	2	5	–	–	–	–	–	–	–	–	–	1	–	–	–	–	–	14
L.11	–	–	–	–	2	–	–	–	–	–	–	–	–	–	–	–	–	–	2
TP13 L.1	–	–	6	–	–	–	–	–	–	–	–	–	–	–	–	–	–	–	6
L.2i	–	–	10	–	–	–	–	–	–	–	–	–	–	–	–	–	–	–	10
L.5ii	1	–	2	–	–	–	–	–	–	–	–	–	–	–	–	–	–	–	3
L.7	–	–	1	–	–	–	–	–	–	–	–	–	–	–	–	–	–	–	1
TP14 L.1	1	–	4	–	–	–	–	–	–	–	–	–	–	–	–	–	–	–	5
L.2i	–	–	3	–	–	–	–	–	–	–	–	–	–	–	–	1	–	–	4
L.2ii	–	1	–	–	–	–	–	–	–	–	–	–	–	–	–	–	–	–	1
L.7	–	–	1	–	–	–	–	–	–	–	–	–	–	–	–	–	–	–	1
TP16 L.3	4	1	22	–	–	–	1	–	–	1	–	–	–	–	–	–	–	–	29
L.4	10	–	1	1	–	–	–	–	–	–	–	–	–	–	–	3	–	1	16
L.8	8	–	1	–	–	–	–	–	–	–	–	–	–	–	–	2	–	–	11
L.5iii	2	1	–	–	–	–	–	–	–	–	–	–	–	–	–	–	–	–	3
TP18 L.1	4	–	14	–	–	–	–	–	–	–	–	–	–	–	–	–	–	–	18
L.2i	2	–	12	–	–	–	–	–	–	–	–	–	–	–	–	–	–	–	14
L.5i	1	–	4	–	–	–	–	–	–	–	–	–	–	–	–	–	–	1	6
L.11	–	–	1	–	–	–	–	–	–	–	–	–	–	–	–	–	–	–	1
Total NISP	245	31	162	3	26	5	2	1	22	1	1	1	1	1	1	17	7	2	529

* extirpated species

Malekula faunal remains (non fish)

Table 10.5 Woplamplam bone remains (non fish)

Taxon	Layer 1	Layer 2	Total
Mammals			
Rattus exulans	347	367	714
Rattus praetor	27	54	81
Sus scrofa	4	1	5
Homo sapien	–	2	2
Birds			
Puffinus cf. *gavia*	–	1	1
Megapodius cf. *layardi*	–	4	4
Gallirallus philippensis	–	2	2
Porzana tabuensis*?	1	2	3
Ptilinopus greyii	–	2	2
Ptilinopus tannensis	–	2	2
Tyto alba	1	1	2
Turdus poliocephalus	–	1	1
Pachycephala pectoralis	–	1	1
Myiagra/Neolalage	1	–	1
Zosterops lateralis	–	3	3
Passeriformes sp.	1	2	3
Bird sp.	1	1	2
Total NISP	383	446	829

*extirpated species

Table 10.6 Navaprah TPs 1,2,3 bone remains (non fish)

TPs / Taxon	TP1 L1	L2	L3	L4	Total	TP2 L1	L3	L4	Total	TP3 L1	L2	L4	TOTAL
Mammals													
Rattus exulans	5	1	14	38	58	3	4	183	190	123	612	904	1639
Rattus praetor	1	1	2	5	9	2	–	14	16	16	25	51	92
Sus scrofa	–	–	–	2	2	–	–	6	6	1	–	2	3
Homo sapien	–	–	–	1	1	–	–	–	–	–	–	–	–
small bat	–	–	1	1	2	–	–	4	4	3	–	1	4
Birds													
Gallus gallus	–	–	1	1	2	–	–	–	–	–	–	–	–
Porphyrio porphyrio	1	–	–	–	1	–	–	–	–	–	–	–	–
Gallirallus philippensis	–	–	–	–	–	–	–	–	–	–	1	–	1
Porzana tabuensis*?	–	–	–	–	–	–	–	1	1	–	2	9	11
cf. large flightless *Porzana*+	–	–	–	–	–	–	–	–	–	–	–	1	1
Ptilinopus greyii	4	–	–	1	5	–	–	–	–	–	1	–	1
Gerygone flavolateralis	1	–	–	–	1	–	–	–					
Ptilinopus tannensis	–	–	–	–	–	–	–	1	1	–	2	–	2
Macropygia mackinlayi	–	–	–	–	–	–	–	–	–	–	1	–	1
Chalcophaps indica	–	–	–	–	–	–	–	–	–	–	–	1	1
Trichoglossus haematodus	–	–	–	–	–	–	–	–	–	–	1	–	1
Eudynamis taitensis	–	–	–	–	–	–	–	–	–	–	1	–	1
Tyto alba	–	–	–	–	–	–	–	1	1	–	4	32	36
Coracina caledonica	–	–	–	–	–	–	–	–	–	–	1	–	1
Lalage sp.	–	–	–	–	–	–	–	–	–	–	2	–	2
Turdus poliocephalus	–	–	–	–	–	–	–	–	–	–	2	3	5
cf. *Phylidonyris notabilis*	–	–	–	–	–	–	–	–	–	–	1	–	1
Myzomela cardinalis	–	–	–	–	–	–	–	–	–	–	1	–	1
Zosterops cf. *lateralis*	–	–	–	–	–	–	–	–	–	–	1	–	1
Aplonis sp.	–	–	–	–	–	–	–	1	1	–	11	1	12
Collocalia esculenta	–	–	–	–	–	–	–	2	2	–	–	–	–
Lichmera incana	–	–	–	–	–	–	–	2	2	–	–	–	–
Passeriformes sp.	–	–	–	–	–	–	–	–	–	–	4	2	6
Passeriformes sp. (ca. size of *Lalage*)	–	–	–	–	–	–	–	–	–	–	1	–	1
Bird sp.	–	–	–	1	1	–	–	2	2	–	2	8	10
Total NISP	12	2	18	50	82	5	4	217	226	143	776	1016	1935

* extirpated species + extinct species

Table 10.7 Chachara and Navepule C remains (non fish)

Taxon	Mammal				Birds			Totals
Sites and layers	*Rattus exulans*	*Pteropus* sp.	*Sus scrofa*	*Homo sapien*	*Porphyrio porphyrio*	*Gallirallus philippensis*	*Chalcophaps indica*	
Chachara Nth								
TP 1 All Layer 2	–	–	2	–	–	–	–	2
TP 2	–	–	1	1	–	–	–	2
TP.3	–	–	1	5	–	–	–	6
TP 4	2	–	10	2	–	–	–	14
TP 5	–	–	5	–	–	–	–	5
TP 6	–	–	3	–	–	–	–	3
Total NISP	2	–	22	8	–	–	–	32
Navepule C								
L.1	–	–	2	–	1	–	–	3
L.2	–	–	3	–	–	–	–	3
L.3	–	–	2	–	–	–	–	2
Total NISP	–	–	7	–	1	–	–	8

Table 10.8 Yalo Sth bone remains (non fish)

Taxon	Layer 1	Layer 2	Layer 3	Layer 4	Layer 5	Total
Mammals						
Rattus exulans	102	650	142	53	4	951
Rattus Praetor	4	44	16	3	–	67
Pteropus sp.	1	–	–	–	–	1
small bat	2	18	2	28	12	62
Sus scrofa	2	12	1	2	–	16
Homo sapien	–	6	–	2	–	8
Med. mammal	–	–	–	1	–	1
Birds						
Gallirallus philippensis	–	7	2	1	–	10
*Porzana tabuensis**	1	1	–	1	–	3
cf. large flightless *Porzana*+	–	1	1	–	–	2
Porphyrio porphyrio	–	–	2	2	–	4
Ptilinopus greyii	–	–	1	–	–	1
Ducula sp.	–	–	2	–	–	2
Columba vitiensis	–	1	–	–	–	1
Chalcophaps indica	–	3	1	–	–	4
Columbidae sp.	–	–	1	–	–	1
Chrysococcyx lucidus	–	1	–	–	–	1
Tyto alba	–	6	–	–	–	6
Collocalia esculenta	–	–	–	1	–	1
Halcyon farquhari	1	–	4	–	–	5
Turdus poliocephalus	–	3	7	–	–	10
Pachycephala pectoralis	–	–	–	1	–	1
Rhipidura sp.	–	–	1	–	–	1
Gerygone flavolateralis	–	–	–	1	–	1
Phylidonyris notabilis	–	–	1	–	–	1
Zosterops flavifrons	–	1	1	–	–	2
Erythrura sp.	–	1	–	–	–	1
Aplonis sp.	–	1	–	–	–	1
Passeriformes sp.	–	3	–	1	–	4
Bird sp.	–	10	4	2	–	16
Total NISP	113	769	189	99	16	1186

* extirpated species + extinct species

Table 10.9 Malua Bay School Site bone remains (non fish)

Taxon	Layer 1	Layer 2	Layer 3	Layer 4	Total
Mammals					
Rattus exulans	–	1	7	1	9
Pteropus sp.	–	–	3	1	4
Sus scrofa	–	3	14	8	25
Homo sapiens	–	–	1	2	3
Med. mammal	–	–	3	1	4
Birds					
Gallus gallus	–	–	1	1	2
Gallirallus philippensis	–	–	1		1
Columbidae sp.	–	–		1	1
Eclectus (new) sp.+	–	–		1	1
Bird sp.	–	–	3	7	10
Reptiles					
turtle	–	–	3	3	6
Total NISP	–	4	36	26	66

+ extinct species

Pig

Pigs play an integral role in contemporary Vanuatu society to the point that they have been described as 'the standard of value and all other values are related to the pig' (Speiser 1996:246, quoted in Rodman 1996:162). Many of the social and political structures of the north and north central islands of Vanuatu are tied to the accumulation and dispatching of boars with modified tusks. Fifteen years ago it was calculated that there were more than 70,000 pigs in Vanuatu an average of three per household (Rodman 1996:161).

The archaeological record from Vanuatu indicates that pigs arrived with the earliest colonisers and it seems likely they spread relatively quickly throughout the archipelago. As noted above, Hedrick (nd) recorded the occurrence of pig bone with dentate stamped pottery on Malo and it has also now been identified at the Ifo site on Erromango in association with a similar Lapita assemblage (Table 10.2). After this initial phase of colonisation pig bone seems to appear relatively consistently, although rarely in large quantity, throughout the archaeological record. On Erromango it is seen in the archaeological record from initial arrival at Ifo, through to c. 2400 BP at the Ponamla site (Table 10.1) and then to at least 2000 BP in the later layers of the Ifo site (Table 10.2) and pig was certainly present at the time of first European contact on that island (Spriggs and Wickler 1989). Pig bone can again be seen from c. 2800 BP on Efate (Table 10.3) and consistently through the record (Tables 10.3, 10.4). The association of full-circle pig tusks in burials on Efate and Tongoa (Garanger 1972:Fig 192, 256) by at least the seventeenth century and possibly as early as the fifteenth century, attest to the increasingly ceremonial role of pigs in Vanuatu society. Malua Bay, the earliest site thus far located on Malekula, shows further evidence of pigs being present from at least 2700 BP on that island (Table 10.9). It was also present in the earliest layers of the coastal cave sites of Woplamplam, Navaprah and Yalo South (Tables 10.5, 10.6, 10.8). Not surprisingly, given its status on Malekula today, pig bone was found in the upper layers of these same cave sites and in somewhat greater density at the inland site of Chachara (Table 10.7). The frequency of pig bone in the archaeological record was also identified by Ward at the Pakea site (Ward 1979:11–21 -30; Horton and Ward 1981) particularly in the early part of the sequence. It has been argued that pigs were an integral part of the diet at this site during the earlier period of settlement but that over time became less important (Horton and Ward 1981:10–11). Full-circle pig tusks appeared as early as 1050 BP at Pakea (Ward 1979:11–26) and were further noted by the Spanish in the Banks on their visit in 1606 (Kelly 1966:I:199,299). These full-circle pig tusks are formed by the removal of the upper canine allowing the lower canines to grow unimpeded in a full circle. The animals are permanently tethered and consequently need to be hand fed, a further indicator of their highly domesticated nature in Vanuatu society.

Rat

Two species of rat have been identified from the excavated faunal assemblages, namely *Rattus exulans* and *Rattus praetor*. *R. exulans* was recovered from all sites and all periods which is no great surprise given its spread throughout the Pacific dating certainly from the initial settlement in Remote Oceania (Kirch 1997:218–219). The remains were found both in association with midden remains (see Tables 10.1–10.9) dating from Lapita settlement and throughout the sequence in Vanuatu and also in some quantity in cave deposits that are related to owl roosting (eg. Woplamplam [Table 10.5], Navaprah [Table 10.6] and Yalo South [Table 10.8]). More of a surprise was the identification of *Rattus praetor* in Vanuatu. It is a native of mainland New Guinea and was previously thought to have been reached (through its association with humans) only as far east as Tikopia, but has now also been identified in Fiji (White *et al.* 2000). Several hundred *R. praetor* bones (Tables 10.5, 10.6, 10.8) were recovered from cave sites on Malekula, by far the largest archaeological assemblage to date. Previously the identification of *R. exulans* and particularly *R. praetor* had focused heavily on cranial measurements but this large assemblage offered the opportunity to attempt to establish the osteometric parameters of *R. exulans* and *R. praetor* using a selection of skeletal elements (mandible, humerus, femur and tibia). Variance statistics were able to demonstrate coherent separation between bones from the different rats. More detailed discussion of this research and presentation of data can be found in White *et al.* 2000. Although there was no direct evidence to suggest that *R. praetor* was a food source it was also recovered from midden deposits at Mangaasi, which confirmed at least its commensal nature.

The earliest occurrence of *R. praetor* can be seen in the lowest cultural layers of the Mangaasi site (Table 10.3) and the cave sites on Malekula. Tracking its presence through the archaeological record is difficult but it can be seen to have been present at least on Malekula until very recently (e.g. upper layers of Woplamplam, Navaprah and Yalo South which date to the last several hundred years). To date it has not been identified on Erromango despite extensive excavation of the two open sites and a number of cave sites and it is no longer extant anywhere in either Vanuatu or Fiji.

Human

Human remains were few and far between largely restricted to single teeth and other often single miscellaneous bones. Only at the Mangaasi site were these clearly associated with disturbed burials that had been cut from near the surface, i.e. dating to the last several hundred years. Several recent less disturbed burials were excavated by Garanger at the site (Garanger 1972:103). The only anomaly amongst the human bone was the small assemblage (as opposed to single occurrences) from the mound site at Chachara (Table 10.7) which did not appear to be a burial but rather midden debris and seems likely to have been associated with ceremonial activities at the adjacent *nasara*. Certainly ritualised cannibalism was observed during the Early European contact period and into the twentieth century on Malekula (Deacon 1934:229–230).

Fruit bat

Pteropus sp. appears to have been heavily exploited from first arrival in the archipelago. This was particularly evident at the Ponamla and Ifo sites on Erromango (Tables 10.1–10.2). Fruit bat bone was also recorded by Hedrick (nd) as being present in the Lapita sites on Malo. Its exploitation clearly continues throughout the archaeological record (Tables 10.1–10.9) and it is a species that could be described as a great survivor at least in Vanuatu as it continues today to be regularly consumed throughout the islands and even in the restaurants of Port Vila. The larger quantity of *Pteropus* sp. bone present in the lower layers of the sites would seem to reflect its initial ease of accessibility which through continued intensive exploitation would have changed over time. Prior to human arrival *Pteropus* sp. would certainly not have been restricted to hanging from only the tallest trees as they are today. The bone was not identified to species level but there exists a total of four species in Vanuatu today (Bregulla 1992:32).

Bones from a small bat were recovered only occasionally from the open site excavations (e.g. Ponamla, Erromango) and more frequently from the caves sites on Malekula (e.g. Yalo South). These were again not identified to species but can be assigned to one of the eight insectivorous varieties found throughout the archipelago (Bregulla 1992:32). These small bat remains, apart from those from the Ponamla site, may well be the result of natural accumulation in the cave sites, due to either simple attrition or predation by owls.

Reptiles

By far the most important component of the reptile remains, and at this stage of the research the only one that has been positively identified as having been consumed by humans, was turtle. Species was not able to be positively identified but probably both the green sea turtle (*Chelonia mydas*) and the hawksbill turtle (*Eretmochelys imbricata*) are represented as they continue to be encountered in Vanuatu today.

There is a consistent pattern in the archaeological record of Lapita settlement throughout the Southwest Pacific of the intensive exploitation of turtle on initial arrival (Kirch 1997:202). It is present at the Lapita levels of the Ifo site (Table 10.2) and also, in somewhat greater quantities, amongst the Lapita sites on Malo (Galipaud 1998a; Hedrick nd). It was also present in the earliest layers of the Malua Bay site on Malekula (Table 10.9). There was no evidence of turtle in any of the cave sites on Malekula or the later layers of the open sites. Turtle remains are recorded throughout the archaeological record at sites on Erromango and Efate although only in small quantities which perhaps reflect a decrease in the population. These remains are likely to simply represent occasional opportunistic kills. Ward recorded small numbers of turtle bone throughout the layers of the Pakea site and concluded that it was an insignificant component of the diet (Ward 1979:11-36). He argued that this was hardly surprising given the limited area of suitable coastal habitat at Pakea and the susceptibility of turtles to over-exploitation from predation on the eggs and both immature and mature individuals (Ward 1979:11–38). These are factors which clearly influence its presence or absence in archaeological sites across the archipelago. Certainly today the once abundant sea turtle has reached the point of being an endangered species in many regions of the Pacific and is now often protected by law, as it is in Vanuatu.

Small lizard bones (as yet not identified to species) were the only other reptile remains recovered from the excavations at Ponamla on Erromango (Tables 10.1) and Mangaasi (Table 10.3, 10.4). Their archaeological provenance (generally not associated with concentrated midden) would tend to suggest that they were not being consumed by humans. Rather, their presence at the sites can be seen as the result of natural attrition.

One other reptile that can be added, is the now-extinct land based crocodile (*Mekosuchid* genus) which was recovered (maxilla and tibia fragment) from the earliest layers of the Arapus site (Mead *et al.* 2002). This crocodile genus has previously been reported from both Fiji (Worthy *et al.* 1999) and New Caledonia (Sand 1995) although direct association with cultural remains at that time remained elusive. The remains from the Arapus site clearly demonstrate a direct association. These are the first and only land crocodile remains thus far recovered from Vanuatu which points to its rapid disappearance on human arrival.

Birds

The quantity and variety of excavated bird bone has provided some detail to a previously substantial gap in Vanuatu's prehistoric faunal past, that of the avian record. The bird bone remains were recovered from a variety of site types and time periods including the concentrated middens associated with the open sites of Ponamla and Ifo on Erromango, Mangaasi on Efate and Malua Bay on Malekula, and cave deposits on Malekula where much of the recovered assemblages appear to be associated with owl roosting. This summary utilises a combination of Steadman's identifications and the modern inventory found in the *Birds of Vanuatu* by Bregulla (1992) to provide a more detailed picture of the prehistoric avian fauna of Vanuatu and a gauge of how human arrival, settlement and the exploitation of various species may have influenced the composition of the extant populations that are seen today (Tables 10.1–10.9).

Evidence of extinct bird bone recovered from archaeological sites in the Pacific is now generally regarded as a sign of first arrival of humans in a pristine landscape. We can 'expect extinction after people arrive on an island. Survival is the exception' (Steadman 1995:1130). It is a consistent pattern found across the Pacific and a scenario that has been likened to a 'blitzkrieg'. Avian fauna were not the only species affected, but their remains thus far have provided the most visible and dramatic evidence to support such an episode. The most recent detailed and quite spectacular evidence for this sort of event comes from research in Polynesia (Steadman 1995, 1997). The instantaneous nature of these extinction events on smaller islands ensures that the evidence will only be found in well-preserved sites dating to initial human arrival. Sites that date to even only a few hundred years after this initial settlement phase rarely contain evidence of extinct bird species. This situation has been highlighted in the recovered faunal remains from Vanuatu.

Few bird bones associated with Lapita settlement in Vanuatu have thus far been recovered. Hedrick (nd) mentioned occasional bird bone in the sites on Malo but they were not identified further. The shallow and disturbed nature of the Malo sites thus far reported would not have been particularly conducive for the preservation of bird bone. Although there have now been a number of Lapita sites identified in Vanuatu (Bedford *et al.* 1998; Bedford 2003) the total excavated area of those same sites amounts to only relatively small areas of excavation. However, despite this moderate sampling they do provide some limited evidence of avian extinctions. On Erromango, from the lowest cultural levels (Lapita) of the Ifo site the bones of a large extinct pigeon (*Ducula* sp. [large]) have been recorded (Tables 10.2). Extirpated species found in the lowest immediately post-Lapita levels of the Ponamla site include the remains of a kingfisher (*Halcyon farquhari*) and a starling (*Aplonis sp.*) (Table 10.1). At the Malua Bay School site on Malekula a new species of now extinct parrot (*Eclectus* sp.) (Table 10.9) was identified in the earliest layers of that site and in the lower cultural levels of the cave sites of Yalo (Table 10.8) and Navaprah (Table 10.6) a large flightless rail (cf. *Porzana* sp.) also appears to have been directly affected by human arrival. *Porzana tabuensis* is thought to be no longer extant on Malekula (although Bregulla notes that its distribution is uncertain [1992:141]) but it appears throughout the archaeological record and if it is no longer present on the island its disappearance may be a relatively recent event. At the Arapus site on Efate the disappearance of two species can be associated with initial human settlement on that part of the island. They are a hawk (*Accipiter* sp.) and a Megapode (cf. *Megapodius alimentum*). A later extirpation on Efate was *Halcyon farquhari*.

While the small number of extinct species associated with initial human arrival in Vanuatu hardly warrants the 'blitzkrieg' label as yet, this can be largely explained by the nature of the sites and the total excavated sample. That is they either all post-date that very initial human arrival phase on particular islands (e.g. Arapus, Ponamla) or that the faunal remains are poorly preserved (e.g. Malo). Future research is likely to greatly increase the extinction inventory. Most of the other bird bone remains from sites which post-date the initial Lapita phase of settlement are associated with species that have survived through to the present.

The recovered bird bone remains from sites post-dating Lapita settlement point to the exploitation of a wide variety of species throughout the archaeological sequence (Tables 10.1–10.9). Rails were one of the most frequent bird species (in terms of NISP) recovered from the excavations on all islands, with *Porphyrio porphyrio* predominant. Other identified, but less frequent, rail species included *Gallirallus philippensis* and *Porzana tabuensis*. Other regularly recorded species were a number of pigeons and doves including *Ducula pacifica*, *Columba vitiensis* and *Ptilinopus greyi*. Less frequently recorded examples restricted to the cave sites of Malekula were *Ptilinopus tannensis*, *Chalcophaps indica* and *Macropygia mackinlayi* (the latter was also recorded at Ponamla).

Five other species which were occasionally recorded at both the open sites and the caves sites were a barn-owl (*Tyto alba*), a scrubfowl (*Megapodius layardi*) a kingfisher (*Halcyon farquhari*), a parrot (*Trichoglossus haematodus*), and a starling (*Aplonis sp.*). A single duck bone (*Anas superciliosa*) was recovered from the Mangaasi site. Sea birds were not common at the sites but did include a heron (*Egretta sacra*) from Ponamla and shearwaters (*Puffinus pacificus* and cf. *Puffinus gavia*) from Ponamla and Woplamplam respectively.

The high number and variety of small bird bones recovered from the caves sites on Malekula further confirm that these remains were principally derived from owl predation. The identified smaller birds included cuckoos (*Chrysococcyx lucidus, Eudynamis taitensis*), cuckoo-shrike/triller (*Coracina caledonica, Lalage* sp.), fantails (*Rhipidura* sp.), honeyeaters (*Phylidonyris notabilis, Lichmera incana* and *Myzomela cardinalis*), monarchs (*Myiagra/Neolalage*), parrotfinch (*Erythrura* sp.), swifts (*Collocalia esculenta*), thrushes (*Turdus poliocephalus* [also recorded from Layer 1 Ponamla]), robins/whistlers (*Pachycephalaides pectoralis*), warblers (*Gerygone flavolateralis*) and white-eyes (*Zosterops flavifrons* and *lateralis*).

Chicken (*Gallus gallus*), the only introduced bird species, was a consistent component in the middens from the earliest layers of the sites (e.g. Lapita levels at Ifo, Malua Bay and Arapus) and throughout the cultural sequences, suggesting that it was present in some quantity from initial arrival, that it was kept as a domesticate and continued, as it does today, to comprise a reliable although modest component of the diet.

Clear patterns of avian exploitation have emerged from this research which are paralleled in other parts of the Pacific (Steadman 1995). There is a general trend which indicates that any bird species that were unable to adapt to the arrival of humans (i.e. large and flightless) were very quickly driven to extinction. In a few cases other species were, over a longer period, extirpated from particular islands. After this initial colonisation phase the extant bird populations appear to be relatively stable up until the present when they have again begun to come under increasing pressure from forest clearance and modern weaponry (Bregulla 1992:194). There is evidence that a wide variety of bird species have been exploited throughout the archaeological sequence, including the introduced domesticate *Gallus gallus*. Apart from the initial colonisation phase it appears that they would have provided an occasional protein source rather than acting as a mainstay in the diet.

Fish

The fishbone was analysed by Foss Leach, Janet Davidson, Karen Fraser and Aglaja Budec at the Archaeozoology Laboratory, Museum of New Zealand Te Papa Tongarewa. The methodology and analysis is detailed in an unpublished technical report (Leach *et al.* 1998) and the bone is now held on permanent loan at Te Papa Tongarewa. Fishbone was recovered from ten sites (Tables 10.10–10.17), the largest samples by far coming from Ponamla and Mangaasi. Some 465 fishbones were able to be identified from nine of the sites producing a minimum number of 336 individuals (Table 10.17). As is often typical with fishbone remains, much of the bone was not identifiable even to family but is included in the presentation of the remains. The most common fish caught were parrot fish (Scaridae) which inhabit the reef flats around coral outcrops and are usually caught by netting. The next most common fish amongst the midden remains was the porcupine fish (Diodontidae), again a reef dwelling fish caught using a variety of methods including netting, spearing and other foraging techniques. Surgeon (Acanthuridae) and trigger fish (Balistidae) inhabit similar environments and are likely to have been caught by similar methods. Together these fish families made up nearly 60% of the overall identified fishbone (Leach *et al.* 1998:11;Table 10.17).

A further 20% of the assemblage was made up of wrasses (Coridae/Labridae families) and rock cod or coral trout (Epinephelidae). These benthic species normally occupy demersal waters and are usually caught with baited hooks (Leach *et al.* 1998). Other species included in this category, although recovered in much lesser quantities are the squirrel fish (Holocentridae) and snapper (Lutjanidae). Pelagic or open ocean species were conspicuous by their absence. The predominance of inshore fish is a feature found in most archaeological sites and sequences across Vanuatu and the Southwest Pacific (Green 1986; Kirch 1988b; Kirch and Yen 1982; Ward 1979:12–9) and one that dates from initial Lapita settlement (Kirch 1997). The complete absence of pelagic species from the Vanuatu sites (apart from a single shark tooth pendant recovered from Ponamla) confirms the overwhelming dominance of inshore and to a lesser extent benthic fish species.

The potential for inter-site comparisons was limited by the small samples (Leach *et al.* 1998) but the results from both Mangaasi and Ponamla demonstrate a similar reliance on inshore species (Table 10.16). No fishbone was recorded from the inland cave sites on Malekula. Potential for observations regarding temporal variation within sites was also limited by sample size. Tentative conclusions can be drawn which suggest fish were a consistent component of the diet throughout the sequence and that methods of procurement and the targeted species changed little through time. The fishbone recovered from the Arapus site is not presented in detail here but it does show the same predominance of reef and inshore species, although on initial arrival fish may have contributed less to the diet. Further elaboration on this possibility is discussed below.

Table 10.10 Mangaasi TPs 1/15, 17, 10, 12, 9 fishbone remains

Taxon	Scaridae	Diodontidae	Coridae	Epinephelidae	Acanthuridae	Balistidae	Holocentridae	Elasmobranchii	Lethrinidae	Teleostomi	Scorpaenidae	Lutjanidae	Muraenidae	Mullidae	Nemipteridae	Monotaxis Gran.	Caranx sp.	Leptocephalidae	Unidentified	Total
TP1/15 L.1	-	-	2	-	-	-	-	-	-	-	-	-	-	-	-	-	-	-	-	2
L.2i	-	-	-	-	-	-	-	-	-	-	-	-	-	-	-	-	-	-	6	6
L.3a	2	-	-	-	-	-	-	-	-	-	-	-	-	-	-	-	-	-	2	4
L.4	-	-	-	-	-	-	-	-	-	-	-	-	-	-	-	-	-	-	1	1
L.3b	1	2	-	-	-	-	-	-	-	-	-	-	-	-	-	1	-	-	191	195
L.11	1	2	-	-	-	-	-	-	-	-	-	-	-	-	-	-	-	-	4	7
TP 17 L.1	-	-	-	-	-	-	-	-	-	-	-	-	-	-	-	-	-	-	2	2
L.2i	1	-	-	-	-	-	-	-	-	-	-	-	-	-	-	-	-	-	3	4
L.3a	2	-	1	-	-	-	-	1	-	-	-	-	-	-	-	-	-	-	19	23
L.3b	12	-	-	2	-	-	-	-	1	-	-	-	-	-	-	-	1	-	71	87
L.3c	-	-	1	-	1	-	-	-	-	1	-	-	-	-	-	-	-	-	17	20
L.4a	-	-	-	-	-	-	-	-	-	-	-	-	-	-	-	-	-	-	6	6
L.3d	-	-	-	1	-	-	-	-	-	-	-	-	-	-	-	-	-	-	8	9
L.11	-	-	-	-	-	-	-	-	-	-	-	-	-	-	-	-	-	-	3	3
TP10 L.1	-	-	-	-	-	-	-	-	-	-	-	-	-	-	-	-	-	-	1	1
L.3a	2	-	-	-	-	-	-	-	-	-	-	-	-	-	-	-	-	-	25	27
L.3b	3	1	-	1	-	-	1	-	-	1	-	-	-	-	-	-	-	-	77	84
L.9a	2	1	-	-	-	-	-	-	-	-	-	-	-	-	-	-	-	-	10	13
L.8	-	-	-	-	-	-	-	-	-	-	-	-	-	-	-	-	-	-	4	4
L.9b	-	-	-	-	-	-	-	-	-	-	-	-	-	-	-	-	-	-	5	5
TP 12 L.1	-	-	1	-	-	-	-	-	-	-	-	-	-	-	-	-	-	-	7	8
L.2i	1	-	-	-	-	-	-	1	-	-	-	-	-	-	-	-	-	-	8	10
L.9a	6	2	1	1	-	-	-	-	1	-	-	-	-	-	-	-	-	1	110	122
L.9b	4	1	1	2	-	-	2	-	-	-	-	-	2	1	-	-	-	-	91	104
L.9c	1	-	-	-	-	1	-	-	-	-	-	-	-	1	-	-	-	-	52	55
L.11	-	-	-	-	-	-	-	-	-	-	-	-	-	-	-	-	-	-	10	10
TP 9 L.9a	3	-	1	-	2	-	-	1	-	-	-	-	-	-	1	-	-	-	165	173
L.9b	8	2	1	2	2	-	-	-	1	-	-	-	-	-	-	-	-	-	330	346
L.8a	-	2	-	-	-	-	-	-	-	-	-	-	-	-	-	-	-	-	24	26
L.9c	31	5	6	8	5	1	3	1	2	-	2	1	-	-	2	-	-	-	728	791
L.8b	-	-	-	-	-	-	-	-	-	-	-	-	-	-	-	-	-	-	31	31
L.9d	6	1	-	2	2	2	-	-	-	-	-	-	-	-	-	-	-	-	282	295
Total NISP	86	19	15	19	11	4	6	4	5	2	2	1	2	2	3	1	1	1	2293	2474

Table 10.11 Mangaasi TPs 2, 3, 4, 5, 6, 7, 8, 11, 14, 16, 18 fishbone remains

Taxon	Scaridae	Diodontidae	Coridae	Epinephelidae	Acanthuridae	Balistidae	Holocentridae	Elasmobranchii	Lethrinidae	Caranx sp.	Muraenidae	Nemipteridae	Unidentified	Total
TP 2 L.1	–	–	–	–	–	–	–	1	–	–	–	–	–	1
L.2ii	–	–	–	–	–	–	–	–	–	–	–	–	7	7
L.3a	5	–	–	1	–	–	–	1	–	–	–	–	101	108
L.4	–	–	–	–	–	–	–	–	–	–	–	–	4	4
TP 3 L.1	–	–	–	–	–	–	–	–	–	–	–	–	3	3
L.2i	–	–	–	–	–	–	–	–	–	–	–	–	6	6
L.2ii	2	–	–	–	–	–	–	–	–	–	–	–	3	5
L.3	–	–	–	–	–	–	–	–	–	–	–	–	4	4
L.11	–	–	–	–	–	–	–	–	–	–	–	–	1	1
TP4 L.1	–	–	–	–	–	1	1	2	–	–	1	–	24	29
L.2i	–	–	4	–	–	–	–	–	–	–	–	–	109	113
L.5i	1	–	–	1	–	–	–	–	–	–	–	–	118	120
L.5ii	7	1	–	3	–	1	–	–	–	–	–	–	95	107
L.9	7	1	2	1	–	–	–	2	1	–	–	2	130	146
TP5 L.1	2	1	–	–	–	–	–	–	–	–	–	–	9	12
L.2ii	–	–	–	–	–	–	–	–	–	–	–	–	13	13
L.2iii	–	–	–	–	–	–	–	–	–	–	–	–	1	1
L.5i	2	–	1	–	1	–	–	–	–	–	–	–	42	46
L.9	3	3	–	1	–	1	–	–	–	–	1	–	66	75
TP6 L.1	–	–	–	–	–	–	–	–	–	–	–	–	5	5
L.2i	1	–	–	–	–	–	–	–	–	–	–	–	1	2
L.5i	–	–	–	–	–	–	–	–	–	–	–	–	1	1
TP7 L.5i	–	–	1	1	1	–	–	–	–	–	1	–	9	13
L.5ii	–	–	–	–	–	–	–	–	–	–	–	–	5	5
TP8 L.5i	–	–	–	–	1	–	–	–	–	–	–	–	2	3
TP11 L.2i	–	–	–	–	–	–	–	–	–	–	–	–	14	14
TP14 L.1	–	–	–	1	–	–	–	–	–	–	–	–	2	3
TP16 L.3	–	2	–	–	–	–	–	–	–	–	–	–	19	21
L.4	–	–	–	–	–	–	–	–	–	–	–	–	16	16
L.8	2	–	–	–	–	–	–	–	–	–	–	–	20	22
TP18 L.1	–	–	–	–	–	–	–	–	–	1	–	–	–	1
L.2i	–	–	–	–	–	–	–	–	–	–	–	–	8	8
Total NISP	32	8	8	6	3	3	1	6	1	1	3	2	838	915

Table 10.12 Ponamla Area A fishbone

Taxon	Layer 1	Layer 2	Layer 3a	Layer 3b	Layer 4	Total
Scaridae	11	6	3	5	2	27
Diodontidae	13	10	8	5	-	36
Coridae	4	3	2	2	-	13
Epinephelidae	4	3	1	1	1	10
Acanthuridae	-	2	-	3	2	7
Balistidae	1	-	-	-	-	1
Holocentridae	4	-	2	1	1	8
Elasmobranchii	-	-	-	-	-	-
Lethrinidae	1	-	-	3	1	5
Teleostomi	1	1	-	1	-	3
Carangidae	-	1	-	-	-	1
Lutjanidae	-	-	1	-	-	1
Muraenidae	-	-	-	-	-	-
Ostraciidae	1	2	-	-	1	4
Belonidae	1	-	-	-	1	2
Myliobatiformes	1	-	-	-	-	1
Unidentified	450	436	328	396	208	1818
Totals NISP	491	464	344	417	217	1937

Table 10.13 Ifo Trenches A, B, C, D fishbone

Taxon	Layer 1	Layer 2	Layer 3	Layer 4	Total
Scaridae	2	-	-	-	2
Diodontidae	5	1	11	3	20
Acanthuridae	1	-	-	-	1
Balistidae	-	-	3	-	3
Coridae	-	-	1	-	1
Unidentified	18	21	39	24	101
Totals NISP	26	22	53	27	128

Table 10.14 Malekula fishbone

Site / Taxon	Scaridae	Diodontidae	Coridae	Acanthuridae	Balistidae	Elasmobranchii	Teleostomi	Scorpaenidae	Lutjanidae	Muraenidae	Nemipteridae	Unidentified	Total
Malua Bay School													
Layer 1	-	-	-	-	-	-	-	-	-	-	-	-	-
Layer 2	-	-	-	-	-	-	-	-	-	-	-	-	-
Layer 3	-	-	1	-	-	-	-	-	-	-	-	22	23
Layer 4	1	1	-	-	1	-	1	-	-	-	1	27	32
Total	1	1	1	-	1	-	1	-	-	-	1	49	55
Navaprah													
Layer 1	-	-	-	-	-	-	-	-	-	-	-	10	10
Layer 2	-	-	-	-	-	-	-	-	-	-	-	13	13
Layer 3	-	-	-	-	-	-	-	-	-	-	-		
Layer 4	2	-	1	1	-	7	-	-	-	-	-	22	33
Total	2	-	1	1	-	7	-	-	-	-	-	55	66
Woplamplam													
Layer 1a	1	-	-	2	-	-	-	-	-	-	-	31	34
Layer 1b	-	-	-	-	-	-	-	-	-	-	-	5	5
Layer 1c	-	-	-	-	-	-	-	-	-	-	-	1	1
Layer 2	-	8	-	-	-	-	-	-	1	-	-	32	41
Total	1	8	-	2	-	-	-	-	1	-	-	69	81
Yalo													
Layer 1	-	-	-	-	-	-	-	-	-	-	-	-	-
Layer 2	1	-	-	-	-	-	-	-	-	1	-	15	17
Layer 3	-	-	-	-	-	-	-	-	-	-	-	2	2
Layer 4	-	-	-	-	-	-	-	-	-	-	-	4	4
Total	1	-	-	-	-	-	-	-	-	1	-	21	23
Wambraf													
Layer 1	-	-	-	-	1	-	-	-	-	-	-	10	11
Total	-	-	-	-	1	-	-	-	-	-	-	10	11
Ndavru													
Layer 1	-	-	-	1	-	-	-	-	-	-	-	1	2
Layer 2a	-	-	-	-	-	-	-	-	-	-	-	1	1
Total	-	-	-	1	-	-	-	-	-	-	-	2	3
Total NISP	5	9	2	4	2	7	1	1	1	1	1	199	239

Table 10.15 Number of fishbones identified by family for each site

Family	Mangaasi	Ifo	Ponamla	Ndavru	Malua Bay	Woplam	Wambraf	Yalo	Navaprah
Scaridae	118	2	28	–	1	2	–	1	2
Diodontidae	28	24	41	–	1	8	–	–	–
Coridae	23	1	15	–	1	–	–	–	1
Epinephelidae	27	–	12	–	–	–	–	–	–
Acanthuridae	15	1	7	1	–	2	–	–	1
Balistidae	7	3	5	–	1	–	1	–	–
Holocentridae	7	–	8	–	–	–	–	–	–
Elasmobranchii	10	–	2	–	–	–	–	–	7
Lethrinidae	6	–	5	–	–	–	–	–	–
Teleostomi	2	–	5	–	1	–	–	–	–
Carangidae	2	–	4	–	–	–	–	–	–
Lutjanidae	1	–	1	–	–	1	–	–	–
Muraenidae	1	–	1	–	–	–	–	1	–
Ostraciidae	–	–	4	–	–	–	–	–	–
Belonidae	–	–	2	–	–	–	–	–	–
Myliobatiformes	–	–	1	–	–	–	–	–	–
Scorpaenidae	2	–	–	–	–	–	–	–	–
Nemipteridae	5	–	–	–	1	–	–	–	–
Mullidae	4	–	–	–	–	–	–	–	–
Total	**259**	**31**	**141**	**1**	**6**	**13**	**1**	**2**	**11**

Table 10.16 Fish MNI by site

Family	Mangaasi	Ifo	Ponamla	Ndavru	Malua	Woplam.	Wambraf	Yalo	Navaprah
Scaridae	71	2	23	–	1	2	–	1	2
Diodontidae	18	13	24	–	1	2	–	–	–
Coridae	18	1	12	–	1	–	–	–	1
Epinephelidae	23	–	11	–	–	–	–	–	–
Acanthuridae	11	1	7	1	–	1	–	–	1
Balistidae	6	2	4	–	1	–	1	–	–
Holocentridae	7	–	7	–	–	–	–	–	–
Elasmobranchii	9	–	2	–	–	–	–	–	2
Lethrinidae	6	–	4	–	–	–	–	–	–
Teleostomi	2	–	5	–	1	–	–	–	–
Carangidae	2	–	3	–	–	–	–	–	–
Lutjanidae	1	–	1	–	–	1	–	–	–
Muraenidae	1	–	1	–	–	–	–	1	–
Ostraciidae	–	–	3	–	–	–	–	–	–
Belonidae	–	–	2	–	–	–	–	–	–
Myliobatiformes	–	–	1	–	–	–	–	–	–
Scorpaenidae	2	–	–	–	–	–	–	–	–
Nemipteridae	4	–	–	–	1	–	–	–	–
Mullidae	4	–	–	–	–	–	–	–	–
Total	**186**	**19**	**110**	**1**	**6**	**6**	**1**	**2**	**6**

Table 10.17 Total fish MNI by family for Vanuatu sites

Family	MNI	%
Scaridae	102	30.35
Diodontidae	58	17.26
Coridae	33	9.8
Epinephelidae	34	10.11
Acanthuridae	22	6.5
Balistidae	14	4.1
Holocentridae	14	4.1
Elasmobranchii	13	3.8
Lethrinidae	10	3.3
Teleostomi	8	2.3
Nemipteridae	5	1.4
Carangidae	5	1.4
Lutjanidae	3	0.9
Muraenidae	3	0.9
Ostraciidae	3	0.9
Belonidae	2	0.6
Mullidae	4	1.1
Scorpaenidae	2	0.6
Myliobatiformes	1	0.3
Total	**336**	**100**

Shellfish

The analysis of the recovered shellfish remains (Tables 10.18–10.43) focused on defining the range of species that had been exploited at the different sites over time and also attempted to establish the nature of the collection strategies that had been employed. Some comparison is made with the only other detailed shellfish analysis reported from Vanuatu, that from the Banks Islands (Ward 1979). Large samples of shellfish from most of the excavated sites were transported to Canberra where they were further subdivided into what were regarded as representative samples for analysis. Although sample size varied substantially from site to site (primarily due to variance in site type and site depositional history) the samples spanned the full chronological history of any one site. Sample sizes from the open sites comprised total samples from excavated areas that generally measured at least $1m^2$ in area, while the cave sites on Malekula included the entire excavated sample (although a number of the cave sites in Malekula returned no shellfish or so little that they were not included in the analysis). The samples on which this analysis and conclusions were based included the following; Ponamla, TP 1, Area A [$1m^2$] (Gyngell analysed a total of $5m^2$); Ifo, B5 [$1m^2$] (Gyngell analysed a total of $11m^2$); Mangaasi, TPs 9 [$2m^2$] and 10 [total $3m^2$]; Malua Bay School, TPs 3 and 9.1 [$2m^2$]; Fiowl, entire sample; Nuas, entire sample; Chachara, entire sample; Navaprah, TP3; Woplamplam, entire sample; Wambraf, entire sample; Navepule A, entire sample and Navepule C, entire sample. *Trochus* sp. shells from the Arapus cultural horizon levels of the Arapus site were also included in the analysis as a size comparison with those from the earliest levels of the Mangaasi site.

The recovered material from the different sites was sorted and identified according to Cernohorsky (1971, 1972) and Coleman (1981) to species level where possible and to family level when this was not. The material was then weighed and minimum numbers of individual (MNI) counts were carried out for the various species groupings enabling them to be ranked in order of dominance. This order provided information on what species were being collected and what environmental zones were being exploited (Reef intertidal, Sand intertidal or Rock intertidal).

Inferences were then able to be made regarding collection strategies, and the testing of an optimal foraging model (Stephens and Krebs 1986). Comparisons of the rates and nature of exploitation zones within sites and between sites and different islands was carried out to determine whether differences or consistencies were present. Once these patterns were identified some assessment as to the various influences which produced them could be assessed. These included variables such as environmental factors, cultural influence and effects of human exploitation on the shellfish populations. Related tables (Tables 10.18–10.43) are to be found throughout the text and more detailed discussion of analytical procedure and theoretical perspectives are found in Gyngell (1997) and Schmidt (2000).

Shellfish collection was assumed to fit an optimal foraging model within differing patch zones. Optimal foraging theory (OFT) assumes that choices will be made in regards to collection strategies that maximize foraging efficiency in terms of energy capture per unit of energy expended by the collecting group (Stephens and Krebs 1986). If the collection strategy did not fit the OFT model then it could be argued that it may represent the influence of cultural choice that had somehow swayed the collection strategy. There are several foraging theory models that are used to predict collecting behaviour in general, but not all are applicable to the study of shellfish which, as a stationary resource, do not generally involve substantial difference in energy expenditure. The two models that are most applicable to the situation in the Vanuatu sites are patch choice and central place foraging models (Schmidt 2000:2).

The patch choice model deals with differing environmental zones as distinct patches of resources. Patch types are ranked on the basis of the resources that are located from within a patch. The number of patches that are exploited increases with the depletion of the higher ranked patches until the point is reached where the addition of a new patch adds nothing to the foraging result. As the community moves from one patch to the next they leave behind them a string of patches that have been depleted to a marginal return rate for the time spent collecting (Charnov 1976; Nagaoka 2000).

Central place foraging on the other hand assumes that the majority of any collected catch is returned to the base camp of the community for consumption (Meehan 1982). In this model resources that are first depleted are those close to the central base (Hamilton and Watt 1970). This depleted zone then expands to cover patches further from the base with the radius of the depleted zone dependent on the size of the population at the central base. For testing and differentiating these models we need to be able to identify the following trends in the archaeological material;

If a patch choice model was practiced then a main patch would have been focused on with an expanding range of zones being exploited as returns dropped. Distance to the resources from the site should be only a minor factor and the range of zones exploited should reflect the range of resources available.

If the central place model was practiced then patches that are the same distance from the main base should be exploited on the basis of which ones were the most productive. On depletion of these zones, areas further afield will be added to the collection strategy until distance made them unprofitable. In this case the resources collected would reflect the range of resources available close to the central base.

Differentiating between these two models from archaeological assemblages is often difficult. The models differ only in the detail. Both predict that as the return rate from any one patch decreases then another patch will be added unless it is deemed not to be profitable. The decision not to collect a new resource could be due to distance from the base camp, difficulty of collecting in that zone, or difficulty in processing resources from that zone. Identifying these factors archaeologically is difficult and only when clear patterns emerge can the above models be differentiated. Both models are, however, optimal foraging models.

Cultural choice may also play a part in collection strategies with some groups expending more energy to collect a seemingly unprofitable resource simply due to its taste or cultural significance. If species appear in the record that are not profitable in terms of the optimal foraging strategy then cultural choice might be used to explain the discrepancy. However examples of this type of inefficient collecting are rare for shellfish where patches are available for efficient exploitation (Schmidt 2000:4).

Discussion of Shellfish remains by site

Erromango

Ponamla

The main species collected at Ponamla were *Trochus* sp. from the sandy intertidal zone, *Vasum*, *Cypraea and Turbo* sp. from the reef and *Nerita* and *Patella* sp. from the rocky zone (Tables 10.18, 10.19). The MNI information demonstrates a change in the species being collected over time, with the main emphasis being *Nerita* sp. in the earliest levels of the site, to *Vasum* and *Trochus* sp. in the later levels (Table 10.18).

Shell density increased from Layer 5 to Layer 3b, which contained the highest density of shellfish, and then decreased again towards Layer 1. Table 10.19 shows the different habitation zones of exploited shellfish. Over the period of the site's occupation the major zone of exploitation according to the MNI was the rock intertidal zone while by weight the reef intertidal zone is predominant. In Layers 1 to 3a the reef zone has the highest levels in all the categories but from the 3b level to Layer 5 while it predominates on weight, on MNI counts the rocky intertidal zone has greatly increased in importance. The small size of the rocky intertidal species requires that a higher number must be collected to account for the same amount of weight and thus its importance tends to be under represented in an analysis based solely on weight. This is especially so for Layer 3b when the MNI for *Nerita* and *Patella* sp. dramatically increased and then returned in later layers to the number that had been collected earlier. The greater reliance on the rocky zone in the earliest levels indicates a more diverse collection strategy, initially with the rocky zone supplemental to the reef. This is further supported by the greater quantity of species collected from the sandy intertidal zone in Layer 4. One of the main species collected in all levels was *Trochus* sp. which was collected from the sandy zone. The change in collection strategy from a more diverse reliance on the sandy and rocky intertidal zones in the earliest levels may indicate a reduced yield from these zones, leading to a greater reliance on the reef intertidal platform.

Over the period of the site's occupation there is a general decrease in the quantity of the shell discarded. This is most noticeable between Layers 3a and 2 when the amount decreases by 989gm. This drop in shell quantity appears to correspond to a period of less intensive occupation of the site prior to abandonment. The pattern of collection can be summarised as a mixed strategy utilising all three zones at all times with some variation in intensity which demonstrates support for the optimal foraging model. The sandy zone peaked in Layer 4 while the rocky zone peaked in Layer 3b, but overall in weight the reef is dominant in all levels reflecting the predominance of this environmental zone at Ponamla.

Ifo

The main species that were being collected at Ifo were *Turbo* sp. and *Astrea stellare* from the reef intertidal zone and *Trochus* and *Tectus* sp. from the sandy intertidal zone. The most numerous species in each level changed from *Tectus* sp. in the earliest levels of the site to *Trochus* sp. in the later levels of the site (Tables 10.20, 10.21). Over the period of the site's occupation the following changes occurred in the shellfish collection strategies. The pattern of zone exploitation reflects a mixed strategy with the reef and sandy zones dominant rather than one of concentration on a single zone (Table 10.21). Layer 1 demonstrated a concentration of reef intertidal zone species

Table 10.18 Ponamla (TP 1.1) shellfish species, weight, MNI and habitat

Species	Wgt	MNI	Habitat
Layer 1			
Cypraea sp.	52.8	5	Reef intertidal
Drupa sp.	12.5	3	Reef intertidal
Tridacna maxima	74.9	1	Reef intertidal
Turbo sp.	104.8	5	Reef intertidal
Vasum ceramicum	48.6	4	Reef intertidal
Vasum turbinellus	122.7	31	Reef intertidal
Total	416.3	49	
Chitonida sp.	0.2	1	Rock intertidal
Nerita sp.	8.1	6	Rock intertidal
Patelloida saccharina	1.9	4	Rock intertidal
Total	10.2	11	
Conus sp.	46.7	8	Sand intertidal
Tectus pyramis	66.3	7	Sand intertidal
Tellina staurella	1.2	1	Sand intertidal
Trochus maculatus	89.4	24	Sand intertidal
Total	203.6	40	
Fragments	187.4		
Layer total	817.5	100	
Layer 2			
Barbatia sp.	2.1	1	Reef intertidal
Cypraea sp.	42.8	11	Reef intertidal
Drupa sp.	15.6	2	Reef intertidal
Tridacna maxima	21.2	2	Reef intertidal
Turbo sp.	121	9	Reef intertidal
Vasum ceramicum	36.3	3	Reef intertidal
Vasum turbinellus	92.2	24	Reef intertidal
Total	331.2	52	
Chitonida sp.	0.6	1	Rock intertidal
Haliotis varia	2.7	1	Rock intertidal
Latrius paetelianus	3.2	1	Rock intertidal
Nerita sp.	8.7	9	Rock intertidal
Patelloida saccharina	5.9	11	Rock intertidal
Total	21.1	23	
Conus sp.	14.7	12	Sand intertidal
Tectus pyramis	45.9	3	Sand intertidal
Trochus maculatus	141.2	30	Sand intertidal
Total	201.8	45	
Fragments	140.9		
Layer total	695	120	
Layer 3A			
Astrea stellare	11.4	1	Reef intertidal
Barbatia sp.	3.2	2	Reef intertidal
Cypraea arabica	9.8	2	Reef intertidal
Cypraea sp.	145.5	30	Reef intertidal
Drupa sp.	17.9	2	Reef intertidal
Tridacna maxima	62.8	4	Reef intertidal
Turbo sp.	451.8	16	Reef intertidal
Vasum ceramicum	49	9	Reef intertidal
Vasum turbinellus	241.7	81	Reef intertidal
Total	993.1	147	
Chitonida sp.	0.4	1	Rock intertidal
Haliotis varia	9.2	2	Rock intertidal
Nerita sp.	0.9	82	Rock intertidal
Patella flexuosa	13	12	Rock intertidal
Patelloida saccharina	28.1	35	Rock intertidal
Turbo cinereus	7.5	1	Rock intertidal
Total	59.1	133	
Conus sp.	92.1	27	Sand intertidal
Fragum sp.	11.2	2	Sand intertidal
Tectus pyramis	81.7	9	Sand intertidal

continued over

Table 10.18 continued

Species	Wgt	MNI	Habitat
Tellina virgata	20.7	3	Sand intertidal
Trochus maculatus	236.8	40	Sand intertidal
Total	442.5	81	
Fragments	474.5		
Layer total	2018.4	361	
Layer 3B			
Barbatia sp.	4.6	3	Reef intertidal
Cypraea arabica	8.9	2	Reef intertidal
Cypraea sp.	202.4	32	Reef intertidal
Tridacna maxima	28.9	1	Reef intertidal
Turbo argyrostomus	5.2	1	Reef intertidal
Turbo marmoratus	3.2	1	Reef intertidal
Turbo sp.	509.6	22	Reef intertidal
Vasum ceramicum	27.6	1	Reef intertidal
Vasum turbinellus	206.1	60	Reef intertidal
Total	996.5	123	
Chitonida sp.	3.2	1	Rock intertidal
Latrius paetelianus	4.1	1	Rock intertidal
Nerita sp.	3.2	188	Rock intertidal
Patella flexuosa	8.4	9	Rock intertidal
Patelloida saccharina	86.1	97	Rock intertidal
Turbo cinereus	6.4	1	Rock intertidal
Total	111.4	297	
Asaphis violascens	4.8	1	Sand intertidal
Conus sp.	86.3	21	Sand intertidal
Fragum sp.	3.3	1	Sand intertidal
Harpa harpa	15.2	2	Sand intertidal
Tectus pyramis	23.1	3	Sand intertidal
Tellina virgata	26.5	2	Sand intertidal
Trochus maculatus	341.5	58	Sand intertidal
Total	500.7	88	
Fragments	347.4		
Layer total	2075.9	508	
Layer 4			
Cypraea arabica	9.5	1	Reef intertidal
Cypraea sp.	69	13	Reef intertidal
Drupa sp.	9	1	Reef intertidal
Tridacna maxima	23.4	2	Reef intertidal
Turbo marmoratus	5.1	1	Reef intertidal
Turbo sp.	669.3	18	Reef intertidal
Vasum ceramicum	15.3	1	Reef intertidal
Vasum turbinellus	152.8	31	Reef intertidal
Total	953.4	68	
Chitonida sp.	0.2	1	Rock intertidal
Nerita sp.	13.8	73	Rock intertidal
Patelloida saccharina	32.4	35	Rock intertidal
Tectarius tectumpersicum	4.9	3	Rock intertidal
Total	51.3	112	
Conus sp.	100	12	Sand intertidal
Tectus pyramis	266	25	Sand intertidal
Trochus maculatus	493.1	51	Sand intertidal
Total	859.1	88	
Fragments	190.7		
Layer total	2103.5	268	
Layer 5			
Turbo sp.	146.5	4	Reef intertidal
Haliotis varia	1.7	1	Rock intertidal
Latrius paetelianus	3.1	1	Rock intertidal
Nerita polita	6.7	6	Rock intertidal
Tectarius tectumpersicum	6	1	Rock intertidal
Total	17.5	9	
Conus sp.	16.2	1	Sand intertidal
Layer total	180.2	14	

followed closely by the sandy intertidal zone. This changed in Layer 2 to a predominance of shellfish from the sandy intertidal zone. In Layer 3 the reef intertidal zone was predominant while at the earliest level of the site, Layer 4, the sandy intertidal zone species were once again predominant. This pattern reflects a combined collection strategy concentrating on the sandy and reef intertidal zones.

Table 10.19 Ponamla (TP 1.1) shellfish habitat zone usage, MNI and habitat

Habitat	No of sp.	% of sp.	Wgt	% of wgt	MNI	% of MNI
Layer 1						
Reef intertidal	6	46.2%	416.3	66.1%	49	49.0%
Sand intertidal	4	30.8%	203.6	32.3%	40	40.0%
Rock intertidal	3	23.1%	10.2	1.6%	11	11.0%
Totals	13		630.1		100	
Layer 2						
Reef intertidal	7	46.7%	331.2	59.8%	52	43.3%
Sand intertidal	3	20.0%	201.8	36.4%	45	37.5%
Rock intertidal	5	33.3%	21.1	3.8%	23	19.2%
Totals	15		554.1		120	
Layer 3A						
Reef intertidal	9	45%	993.1	66.4%	147	40.7
Sand intertidal	5	25%	442.5	29.6%	81	22.4
Rock intertidal	6	30%	59.1	4%	133	36.6
Totals	20		1494.7		361	
Layer 3B						
Reef intertidal	9	40.9%	996.5	61.9%	123	24.2%
Sand intertidal	7	31.8%	500.7	31.1%	88	17.3%
Rock intertidal	6	27.3%	111.4	6.9%	297	58.5%
Total	22		1608.6		508	
Layer 4						
Reef intertidal	8	53.3%	953.4	51.2%	68	25.4%
Sand intertidal	3	20%	859.1	46.1%	88	32.8%
Rock intertidal	4	26.7%	51.3	5.4%	112	41.8%
Total	15		1912.6		268	
Layer 5						
Reef	1	16.7%	146.5	81.3%	4	28.6%
Sand	1	16.7%	16.2	9.0%	1	7.1%
Rock	4	66.7%	17.5	9.7%	9	64.3%
Total	6		180.2		14	

Rock intertidal species did not figure strongly in the assemblages at Ifo but this would seem to reflect the comparatively smaller areas of this zone type in the immediate surrounding environment. Mangrove zone species were also present in the Ifo assemblages but the small quantity suggests that they were probably harvested opportunistically while other resources, such as crabs, were being collected. Bivalve species appeared to be somewhat more sensitive to heavy exploitation than other species at the Ifo site, especially those that were restricted to the sandy beach zone of which there is only a very limited area at Ifo. Local informants explained that *Periglypta reticulata*, which were recovered from the lower levels of the site, were no longer available in the area and they would certainly appear to be one case of shellfish extirpation from this region of the island.

Table 10.20 Ifo (B5) shellfish species, weight, MNI and habitat

Species	Wgt	MNI	Habitat
Layer 1			
Astrea stellare	198.5	11	Reef intertidal
Barbatia sp.	4.1	5	Reef intertidal
Cypraea helvola	4.6	1	Reef intertidal
Cypraea sp.	40.3	11	Reef intertidal
Drupa morum	8.3	1	Reef intertidal
Hippopus hippopus	373.4	1	Reef intertidal
Tridacna maxima	375.2	8	Reef intertidal
Turbo argyrostomus	100.2	1	Reef intertidal
Turbo laminiferus	31.1	1	Reef intertidal
Turbo sp.	2465.2	40	Reef intertidal
Vasum ceramicum	47.6	2	Reef intertidal
Vasum turbinellus	37.8	5	Reef intertidal
Layer total	3686.3	87	
Cellana sp.	7.4	6	Rock intertidal
Osteridae sp.	144.6	7	Rock intertidal
Tectarius tectumpersicum	6.2	4	Rock intertidal
Turbo cinereus	12.9	5	Rock intertidal
Total	171.1	22	
Asaphis violascens	8.5	1	Sand intertidal
Conus sp.	307.4	16	Sand intertidal
Rhinoclavis vertagus	6.3	1	Sand intertidal
Tectus pyramis	393.1	16	Sand intertidal
Tellina perna	22.3	1	Sand intertidal
Tellina virgata	68.2	1	Sand intertidal
Trochus maculatus	790.4	48	Sand intertidal
Trochus niloticus	208.2	4	Sand intertidal
Total	1804.4	88	
Geloina erosa	67.5	3	Mangrove
Fragments	1031.9		
Layer Total	6761.2	200	
Layer 2			
Astrea stellare	97.7	6	Reef intertidal
Barbatia sp	3.6	2	Reef intertidal
Drupa sp.	6.6	1	Reef intertidal
Tridacna maxima	16.5	1	Reef intertidal
Turbo laminiferus	13.2	1	Reef intertidal
Turbo sp.	743.7	14	Reef intertidal
Total	881.3	25	
Cellana sp.	4.9	3	Rock intertidal
Nerita albicilla	2.2	1	Rock intertidal
Osteridae sp.	96.5	4	Rock intertidal
Tectarius tectumpersicum	3.1	1	Rock intertidal
Total	106.7	9	
Periglypta reticulata	7.1	1	Sand intertidal
Tectus pyramis	505.4	18	Sand intertidal
Tellina perna	17.6	1	Sand intertidal
Trochus maculatus	571.2	23	Sand intertidal
Trochus niloticus	384.1	1	Sand intertidal
Total	1485.4	44	
Geloina erosa	45.7	1	Mangrove
Fragments	471.4		
Layer total	2990.5	79	
Layer 3			
Astrea stellare	825.7	22	Reef intertidal
Barbatia sp.	0.6	1	Reef intertidal
Cypraea helvola	127.3	1	Reef intertidal
Cypraea sp.	48.9	3	Reef intertidal
Operculum	190.7	31	Reef intertidal
Spondylus ducalis	29.8	3	Reef intertidal
Tridacna maxima	349.8	2	Reef intertidal

continued over

Table 10.20 continued

Species	Wgt	MNI	Habitat
Turbo argyrostomus	90.4	1	Reef intertidal
Turbo laminiferus	28.2	2	Reef intertidal
Turbo sp.	2362.5	34	Reef intertidal
Vasum ceramicum	300.9	7	Reef intertidal
Vasum turbinellus	16.4	2	Reef intertidal
Total	4371.2	109	
Cellana sp.	10	6	Rock intertidal
Chiton sp.	6.7	1	Rock intertidal
Nerita polita	7.2	8	Rock intertidal
Osteridae sp.	205.8	9	Rock intertidal
Tectarius tectumpersicum	26	7	Rock intertidal
Turbo cinereus	9.5	3	Rock Intertidal
Total	265.2	34	
Asaphis violascens	7.5	1	Sand intertidal
Cerithium nodulosus	17.6	7	Sand intertidal
Cerithium sp.	1.9	1	Sand intertidal
Conus sp.	202.1	7	Sand intertidal
Harpa articularis	2.9	1	Sand intertidal
Rhinoclavis fasciatus	2.2	1	Sand intertidal
Rhinoclavis vertagus	6.7	2	Sand intertidal
Tectus pyramis	1710.6	51	Sand intertidal
Tellina virgata	23.4	3	Sand intertidal
Trochus maculatus	987.6	32	Sand intertidal
Trochus niloticus	80.8	1	Sand intertidal
Total	3043.3	107	
Geloina erosa	181.3	5	Mangrove
Fragments	385.9		
Layer Total	8236.3	257.6	
Layer 4			
Astrea stellare	497.6	14	Reef intertidal
Barbatia sp.	69.1	4	Reef intertidal
Cypraea helvola	12.7	1	Reef intertidal
Cypraea sp.	70.4	7	Reef intertidal
Drupa morum	3.1	1	Reef intertidal
Operculum	459.3	27	Reef intertidal
Tridacna maxima	97.2	2	Reef intertidal
Turbo argyrostomus	107.8	2	Reef intertidal
Turbo laminiferus	69.7	4	Reef intertidal
Turbo sp.	1689.3	18	Reef intertidal
Total	3076.2	80	
Chitonida sp.	2.1	1	Rock intertidal
Nerita albicilla	4.4	2	Rock intertidal
Nerita plicata	19.6	6	Rock intertidal
Nerita polita	22.2	11	Rock intertidal
Neritina sp.	1.9	1	Rock intertidal
Osteridae sp.	7.2	1	Rock intertidal
Tectarius tectumpersicum	85.2	15	Rock intertidal
Total	142.6	37	
Cerithium nodulosus	6.2	2	Sand intertidal
Cerithium sp.	12.8	3	Sand intertidal
Conus sp.	109.8	7	Sand intertidal
Harpa harpa	4.1	1	Sand intertidal
Periglypta reticulata	12.8	2	Sand intertidal
Rhinoclavis vertagus	20.5	6	Sand intertidal
Strombus luhuanus	8.6	3	Sand intertidal
Tectus pyramis	1304.9	40	Sand intertidal
Tellina virgata	3.1	1	Sand intertidal
Trochus maculatus	606.5	22	Sand intertidal
Trochus niloticus	579.4	7	Sand intertidal
Vexillum sp.	5.1	3	Sand intertidal
Total	2673.8	97	
Geloina erosa	18.5	1	Mangrove
Fragments	844.9		
Layer Total	6756	215	

Table 10.21 Ifo (B5) shellfish habitat usage zones

Habitat	No of sp.	% of sp.	Wgt	% of wgt	MNI	% of MNI
Layer 1						
Reef intertidal	12	48.0%	3686.3	64.3%	87	43.5%
Sand intertidal	8	32.0%	1804.4	31.5%	88	44.0%
Rock intertidal	4	16.0%	171.1	3.0%	22	11.0%
Mangrove	1	4.0%	67.5	1.2%	3	1.5%
Total	25		5729.3		200	
Layer 2						
Reef intertidal	6	37.5%	881.3	35.0%	25	31.6%
Sand intertidal	5	31.3%	1485.4	59.0%	44	55.7%
Rock intertidal	4	25.0%	106.7	4.2%	9	11.4%
Mangrove	1	6.3%	45.7	1.8%	1	1.3%
Total	16		2519.1	100.0%	79	100.0%
Layer 3						
Reef intertidal	12	40.0%	4371.2	55.6%	109	42.7%
Sand intertidal	11	36.7%	3043.3	38.7%	107	42.0%
Rock intertidal	6	20.0%	265.2	3.4%	34	13.3%
Mangrove	1	3.3%	181.3	2.3%	5	2.0%
Total	30		7861	100.0%	255	100.0%
Layer 4						
Reef intertidal	10	33.3%	3076.2	52.0%	80	37.2%
Sand intertidal	12	40.0%	2673.8	45.2%	97	45.1%
Rock intertidal	7	23.3%	142.6	2.4%	37	17.2%
Mangrove	1	3.3%	18.5	0.3%	1	0.5%
Total	30		5911.1	100.0%	215	100.0%

A comparison of these two open sites on Erromango reveals the following. The two sites are situated in slightly different environmental locations and this is reflected in the composition of the assemblages. A common feature of both sites is that all the zones were utilised at all periods, although there is indications that there was some slight variation through time. At the lower levels of the Ponamla site there are indications of a diverse strategy of zone usage which included the rocky and sandy zones which over time changed to a concentration on the reef intertidal zone in the later levels. At Ifo the same pattern seems to have occurred with a concentration on the sandy and reef intertidal zones, but only a minor role for the rocky zone. This difference is also reflected in the main species that were being collected at each site. One species that was common to both sites was the *Trochus* sp. which is recognised for its mean high meat weight. At Ponamla there is a general decrease in the amount of shellfish collected over time while at Ifo the total weight of shell remains stable. Mangrove species were only present at Ifo, a reflection of environmental variability between the two sites.

Further to the effects of human predation are the data gleaned by Gyngell (1997:42–44) from these two sites, where he measured the length and/or width of four targeted species (*Trochus maculatus*, *Tectus pyramis*, *Astrea stellare* and *Echinoderm* sp. (spine width proximal end)). Quite dramatic decreases in shell size through time were demonstrated for all of the targeted species. Although both sites fit the optimal foraging model the limited sample precluded any differentiation between the patch choice or central place models.

Efate

Mangaasi

Two separate shellfish samples were analysed from the Mangaasi site, namely TP9 (2m²) and TP10 (1m²). These two test pits spanned the first 1500 years of settlement at the site. A full list of the species from the two test pits are presented (Table 10.22, 10.24) as is a summary of zonal utilisation (Table 10.23, 10.25). The samples from these two test pits showed a predominance of shellfish species recovered from the sandy intertidal zone. This zone is the habitat for large fleshy bivalve species that are easily collected, very cost effective in terms of expended energy and are also relatively easy to process (Schmidt 2000:25). This zone predominated in all levels of the site in both weight and MNI categories.

The main species being collected here were *Gafrarium* sp., *Conus* sp. and *Strombus* sp.. The next zone of importance was the reef, with collection concentrating on *Barbatia* sp., *Cypraea* sp. and *Turbo* sp. The rocky intertidal zone species largely constituted the remainder of the assemblage. Mangrove species were also recorded at Mangaasi with collection that focused on the large bivalve *Geloina* sp. In TP 9 the highest density of shell occurred in the earliest layers (Layer 9d) after which the concentration decreased slightly but remained fairly constant. In TP 10 the greatest density was found in Layer 3b. The shellfish from both of the test pits showed a similar pattern both in terms of the most abundant species and the targeted zones apart from the lowest levels of TP 10 which demonstrated greater dominance of reef zone species (Table 10.25). Again the shellfish collection strategy at Mangaasi reflects the efficient gathering of the closest available resources (i.e. optimal foraging), the composition of which is directly related to the predominant local environmental zones.

Arapus

While excavating the earliest levels of the Arapus site all those involved commented on the very large size of the recovered shellfish, particularly the *Trochus niloticus* which were also one of the more frequent species. This part of the site was associated with initial human arrival in this area of the island and therefore some size comparison of the recovered shellfish with that from the Mangaasi site was seen as necessary. The first suture width on the *Trochus* shells was the measurement used to enable a comparison (Swadling 1986). The mean first suture width of the shells from the Arapus site was 66.1mm while those from the earliest levels of the Mangaasi site was 37.6mm, representing an almost 50% reduction in size (Schmidt 2000:23). Dramatically illustrated is the sudden impact of human arrival in a pristine marine environment. A similar method and result, using the same shellfish species, has been demonstrated at the Reefs Santa Cruz sites where there was a dramatic change from 'an established older population plateau, which is characteristic of most unexploited shellfish populations, to a population dominated by younger individuals' (Swadling 1986:146).

Malekula

Many of the shell samples from the excavated sites on Malekula were relatively small, thereby restricting analysis to little more than the identification of species. Any conclusions regarding collection strategies and zone utilisation must thus be seen as only speculative and need to be treated with some caution.

Table 10.22 Mangaasi (TP 9) shellfish species, weight, MNI and habitat

Species	wgt	MNI	Habitat
Layer 9a			
Geloina sp.	80.5	6	Mangrove
Astrea stellare	12.4	2	Reef intertidal
Barbatia sp.	215.1	62	Reef intertidal
Cardita sp.	28.6	10	Reef intertidal
Cypraea sp.	115.7	23	Reef intertidal
Spondylus ducalis	41.3	1	Reef intertidal
Tridacna maxima	213.5	8	Reef intertidal
Turbo sp.	309.7	13	Reef intertidal
Vasum turbinellus	73.1	7	Reef intertidal
Total	1009.4	126	
Haliotis varia	2.7	1	Rock intertidal
Nerita sp.	28.2	11	Rock intertidal
Osteridae sp.	12.3	1	Rock intertidal
Saccostrea sp.	34.6	1	Rock intertidal
Total	77.8	14	
Anadara antiquata	117.7	6	Sand intertidal
Cerithium nodulosum	291.8	9	Sand intertidal
Conus sp.	417.9	52	Sand intertidal
Garafrium sp.	618.6	123	Sand intertidal
Quidnipagus palatum	181.6	16	Sand intertidal
Strombus sp.	21.4	8	Sand intertidal
Tectus pyramis	61.4	2	Sand intertidal
Trochus maculatus	174.2	11	Sand intertidal
Trochus niloticus	44.4	1	Sand intertidal
Total	1929	228	
Layer total	3096.7	374	
Layer 9b			
Geloina sp.	49.6	2	Mangrove
Barbatia sp.	118.5	22	Reef intertidal
Cardita sp.	3.6	3	Reef intertidal
Cypraea sp.	73.7	16	Reef intertidal
Tridacna maxima	124.3	4	Reef intertidal
Turbo sp.	60.1	4	Reef intertidal
Vasum turbinellus	48.5	3	Reef intertidal
Total	428.7	52	
Haliotis varia	3.4	1	Rock intertidal
Patella flexuosa	1.1	1	Rock intertidal
Saccostrea sp.	60.2	1	Rock intertidal
Total	64.7	3	
Anadara antiquita	75.7	2	Sand intertidal
Conus sp.	92.8	17	Sand intertidal
Garafrium sp.	66.1	23	Sand intertidal
Quidnipagus palatum	14.8	5	Sand intertidal
Strombus sp.	15.1	5	Sand intertidal
Tectus pyramis	11.6	3	Sand intertidal
Terebra sp.	26.9	1	Sand intertidal
Trochus maculatus	233.7	8	Sand intertidal
Trochus niloticus	7.9	1	Sand intertidal
Total	544.6	65	
Layer total	1087.6	122	
Layer 9c			
Geloina sp.	71.6	4	Mangrove
Astrea stellare	48.4	4	Reef intertidal
Barbatia sp.	206.2	48	Reef intertidal
Cardita sp.	30.9	10	Reef intertidal
Cypraea sp.	115.3	36	Reef intertidal
Hippopus hippopus	27.4	1	Reef intertidal
Tridacna maxima	477.7	8	Reef intertidal

continued over

Table 10.22 continued over

Species	wgt	MNI	Habitat
Turbo sp.	341	18	Reef intertidal
Vasum turbinellus	117.6	7	Reef intertidal
Total	1364.5	132	
Chiton sp.	1.4	1	Rock intertidal
Haliotis varia	3.7	1	Rock intertidal
Nerita sp.	28.5	11	Rock intertidal
Saccostrea sp.	22.1	1	Rock intertidal
Total	55.7	14	
Anadara antiquata	36.7	3	Sand intertidal
Cardium sp.	87.3	2	Sand intertidal
Cerithium sp.	46.7	1	Sand intertidal
Conus sp.	476.8	43	Sand intertidal
Garafrium sp.	278.1	60	Sand intertidal
Nassarius sp.	43.3	4	Sand intertidal
Periglypta sp.	26.7	1	Sand intertidal
Quidnipagus palatum	215.7	14	Sand intertidal
Strombus sp.	59.7	26	Sand intertidal
Tectus pyramis	193	8	Sand intertidal
Trochus maculatus	469.3	13	Sand intertidal
Trochus niloticus	137.8	1	Sand intertidal
Total	2071.1	176	
Layer total	3562.9	326	
Layer 9d			
Geloina sp.	93.9	7	Mangrove
Astrea stellare	123.4	10	Reef intertidal
Barbatia sp.	306.5	98	Reef intertidal
Cardita sp.	36.9	16	Reef intertidal
Cypraea helvola	72.9	37	Reef intertidal
Cypraea sp.	116.6	21	Reef intertidal
Lambis lambis	87.4	2	Reef intertidal
Latrius paetinellus	21.7	2	Reef intertidal
Tridacna maxima	344.1	12	Reef intertidal
Turbo sp.	592.7	21	Reef intertidal
Total	1702.2	219	
Cellana sp.	0.4	1	Rock intertidal
Chiton sp.	0.8	1	Rock intertidal
Nerita sp.	41.9	14	Rock intertidal
Ostreadae sp.	31.9	4	Rock intertidal
Saccostrea sp.	66.2	3	Rock intertidal
Total	141.2	23	
Anadara antiquata	22.7	3	Sand intertidal
Cerithium nodulosum	53.8	3	Sand intertidal
Conus sp.	344.7	75	Sand intertidal
Fragum unedo	27.9	1	Sand intertidal
Garafrium sp.	455.8	102	Sand intertidal
Nassarius sp.	8.7	7	Sand intertidal
Quidnapagus palatum	109.4	11	Sand intertidal
Strombus sp.	165.4	69	Sand intertidal
Tectus pyramis	72.1	8	Sand intertidal
Trochus maculatus	595.5	33	Sand intertidal
Trochus niloticus	39	2	Sand intertidal
Vasum turbinellus	234.9	19	Sand intertidal
Total	2129.9	333	
Layer total	4067.2	582	

Table 10.23 Mangaasi TP 9 shellfish habitat zone usage

Habitat	No of sp.	% of sp.	Wgt	% of wgt	MNI	% of MNI
Layer 9a						
Reef intertidal	8	36.4%	1009.4	32.6%	126	33.7%
Sand intertidal	9	40.9%	1929	62.3%	228	61.0%
Rock intertidal	4	18.2%	77.8	2.5%	14	3.7%
Mangrove	1	4.5%	80.5	2.6%	6	1.6%
Total	22		3096.7		374	
Layer 9b						
Reef intertidal	6	31.6%	428.7	39.4%	52	42.6%
Sand intertidal	9	47.4%	544.6	50.1%	65	53.3%
Rock intertidal	3	15.8%	64.7	5.9%	3	2.5%
Mangrove	1	5.3%	49.6	4.6%	2	1.6%
Total	19		1087.6		122	
Layer 9c						
Reef intertidal	8	32.0%	1364.5	38.3%	132	40.5%
Sand intertidal	12	48.0%	2071.1	58.1%	176	54.0%
Rock intertidal	4	16.0%	55.7	1.6%	14	4.3%
Mangrove	1	4.0%	71.6	2.0%	4	1.2%
Total	25		3562.9		326	
Layer 9d						
Reef intertidal	9	33.3%	1702.2	40.6%	219	37.6%
Sand intertidal	12	44.4%	2129.9	52.4%	333	57.2%
Rock intertidal	5	18.5%	141.2	3.5%	23	4.0%
Mangrove	1	3.7%	93.9	2.3%	7	1.2%
Total	27		4067.2		582	

Open sites

Malua Bay School Site

The shell sample analysed from this site comprised total samples from two test pits namely TPs 3 and 9.1 (2m^2). The MNI counts show some preference for rocky zone species (skewed by the quantity of *Nerita* sp.) in the lowest layer (Table 10.26), although weights tend to indicate a major reliance on reef species. The assemblages showed a reliance on shellfish species predominantly from the reef and rock intertidal zones (Table 10.27). The species present in all occupation layers of the site are generally the same as are their relative proportions, *Nerita* sp. being the obvious exception. The sudden decrease in the numbers of *Nerita* sp. might reflect resource depletion on the rock platform. The next most common species was *Turbo* sp., followed by *Cypraea* sp. which are both derived from the reef zone. A mangrove zone species, *Geloina* sp., was also present at the site, as were the sand/mud zone species *Quidnipagus palatam*, but their small number would suggest that species from these zones were not being specifically targeted.

Table 10.24 Mangaasi (TP10) shellfish species, weight, MNI and habitat

Species	Wgt	MNI	Habitat
Layer 1			
Cardita sp.	1.2	1	Reef intertidal
Cypraea sp.	4.7	1	Reef intertidal
Turbo sp.	10.7	1	Reef intertidal
Total	16.6	3	
Chiton sp.	4.8	1	Rock intertidal
Nerita sp.	11.3	12	Rock intertidal
Total	16.1	13	
Conus sp.	34.2	15	Sand intertidal
Garafrium sp.	2.8	2	Sand intertidal
Tridacna maxima	9.5	1	Sand intertidal
Trochus maculatus	7.8	1	Sand intertidal
Total	54.3	19	
Layer total	87	35	
Layer 3a			
Geloina sp.	92	8	Mangrove
Astrea stellare	29.3	1	Reef intertidal
Barbatia sp.	53.7	14	Reef intertidal
Cardita sp.	11.1	5	Reef intertidal
Cypraea sp.	71	19	Reef intertidal
Tridacna maxima	174.9	4	Reef intertidal
Turbo sp.	70.4	7	Reef intertidal
Vasum turbinellus	10.4	2	Reef intertidal
Total	420.8	52	
Chiton sp.	2.8	2	Rock platform
Nerita sp.	32	17	Rock platform
Saccostrea sp.	11.6	1	Rock platform
Total	46.4	20	
Anadara antiquata	135.1	9	Sand intertidal
Cerithium sp.	12	1	Sand intertidal
Conus sp.	207.4	29	Sand intertidal
Garafrium sp.	146.1	38	Sand intertidal
Nassarius sp.	31.9	35	Sand intertidal
Quidnapagus palatum	115.2	10	Sand intertidal
Strombus sp.	76.8	34	Sand intertidal
Terebra sp.	13.1	8	Sand intertidal
Trochus maculatus	77.3	6	Sand intertidal
Total	814.9	170	
Layer total	1374.1	250	
Layer 3b			
Geloina sp.	72	5	Mangrove
Astrea stellarre	20	2	Reef intertidal
Barbatia sp.	105.4	36	Reef intertidal
Cardita sp.	11.5	7	Reef intertidal
Cypraea sp.	95.2	19	Reef intertidal
Hippopus hippopus	18.8	1	Reef intertidal
Tridacna gigas	98.8	1	Reef intertidal
Tridacna maxima	804.1	10	Reef intertidal
Turbo sp.	232.4	13	Reef intertidal
Vasum turbinellus	103.8	6	Reef intertidal
Total	1490	95	
Latrius sp.	4.3	2	Rock platform
Nerita sp.	40	17	Rock platform
Saccostrea sp.	96.2	5	Rock platform
Total	140.5	24	

continued over

Table 10.24 continued

Species	Wgt	MNI	Habitat
Anadara antiquata	204.1	18	Sand intertidal
Cerithium sp.	54.2	3	Sand intertidal
Conus sp.	161.6	27	Sand intertidal
Garafrium sp.	316.3	81	Sand intertidal
Nassarius sp.	3.6	1	Sand intertidal
Quidnapagus palatum	279	24	Sand intertidal
Strombus sp.	37.4	16	Sand intertidal
Tectus pyramis	15.1	2	Sand intertidal
Terebra sp.	1.4	1	Sand intertidal
Trochus maculatus	167.3	12	Sand intertidal
Trochus niloticus	24.8	1	Sand intertidal
Total	1264.8	186	
Layer total	2967.3	310	
Layer 9			
Geloina sp.	59.5	4	Mangrove
Astrea stellare	32.4	2	Reef intertidal
Barbatia sp.	78.6	37	Reef intertidal
Cardita sp.	16.9	12	Reef intertidal
Cellana sp.	1.3	1	Reef intertidal
Cypraea sp.	122.5	27	Reef intertidal
Tridacna maxima	346.3	5	Reef intertidal
Turbo sp.	135.8	13	Reef intertidal
Vasum turbinellus	32.4	4	Reef intertidal
Total	766.2	101	
Chiton sp.	3.4	2	Rock platform
Nerita sp.	34.2	14	Rock platform
Osteridae sp.	18.9	1	Rock platform
Saccostrea sp.	6.5	1	Rock platform
Total	63	18	
Anadara antiquata	70.4	5	Sand intertidal
Cerithium sp.	49.8	2	Sand intertidal
Conus sp.	116	27	Sand intertidal
Garafrium sp.	78.2	26	Sand intertidal
Quidnapagus palatum	67.6	10	Sand intertidal
Strombus sp.	26.5	15	Sand intertidal
Trochus maculatus	149.7	14	Sand intertidal
Total	558.2	99	
Layer total	1446.9	222	
Layer 11			
Geloina sp.	9	2	Mangrove
Astrea stellare	37.2	3	Reef intertidal
Barbatia sp.	47.8	24	Reef intertidal
Cardita sp.	17.1	10	Reef intertidal
Cypraea sp.	42.2	11	Reef intertidal
Lambis lambis	46.7	1	Reef intertidal
Tridacna maxima	50.8	2	Reef intertidal
Turbo sp.	41.8	2	Reef intertidal
Vasum turbinellus	32.2	2	Reef intertidal
Total	315.8	55	
Chiton	1.1	1	Rock platform
Saccostrea sp.	9.2	1	Rock platform
Total	10.3	2	
Conus sp.	46.1	15	Sand intertidal
Garafrium sp.	51.6	15	Sand intertidal
Quidnapagus palatum	2.1	1	Sand intertidal
Strombus sp.	8.5	4	Sand intertidal
Tectus pyramis	10.9	1	Sand intertidal
Trochus maculatus	24.5	1	Sand intertidal
Total	143.7	37	
Layer total	478.8	96	

Table 10.25 Mangaasi TP10 shellfish habitat zone usage

Habitat	No of sp.	% of sp.	Wgt	% of wgt	MNI	% of MNI
Layer 1						
Reef intertidal	3	33.3%	16.6	19.1%	3	8.6%
Sand intertidal	4	44.4%	54.3	62.4%	19	54.3%
Rock intertidal	2	22.2%	16.1	18.5%	13	37.1%
Mangrove	0		0		0	
Total	9		87		35	
Layer 3a						
Reef intertidal	7	35.0%	420.8	30.6%	52	20.8%
Sand intertidal	9	45.0%	814.9	59.3%	170	68.0%
Rock intertidal	3	15.0%	46.4	3.4%	20	8.0%
Mangrove	1	5.0%	92		8	3.2%
Total	20		1374.1		250	
Layer 3b						
Reef intertidal	9	37.5%	1490	50.2%	95	30.6%
Sand intertidal	11	45.8%	1264.8	42.6%	186	60.0%
Rock intertidal	3	12.5%	140.5	4.7%	24	7.7%
Mangrove	1	4.2%	72		5	1.6%
Total	24		2967.3		310	
Layer 9						
Reef intertidal	8	40.0%	766.2	53.0%	101	45.5%
Sand intertidal	7	35.0%	558.2	38.6%	99	44.6%
Rock intertidal	4	20.0%	63	4.4%	18	8.1%
Mangrove	1	5.0%	59.5	4.1%	4	1.8%
Total	20		1446.9		222	
Layer 11						
Reef intertidal	8	47.1%	315.8	66.0%	55	57.3%
Sand intertidal	6	35.3%	143.7	30.0%	37	38.5%
Rock intertidal	2	11.8%	10.3	2.2%	2	2.1%
Mangrove	1	5.9%	9	1.9%	2	2.1%
	17		478.8		96	

Table 10.26 Malua Bay School (TPs 3 and 9.1) shellfish species, MNI, weight and habitat

Species	Wgt	MNI	Habitat
Layer 3			
Geloina sp.	3	1	Mangrove
Astrea stellare	66	3	Reef Intertidal
Barbatia sp.	27.1	6	Reef intertidal
Cypraea sp.	73.8	10	Reef intertidal
Tridacna maxima	169.7	3	Reef intertidal
Turbo sp.	368.7	16	Reef intertidal
Vasum ceramicum	4.4	1	Reef intertidal
Vasum turbinellus	94.8	9	Reef intertidal
Total	804.5	48	
Chiton plates	0.7	1	Rock platform
Latrius sp.	3.4	1	Rock platform
Nerita sp.	46.9	32	Rock platform
Total	51	34	
Cerithium sp.	4.7	1	Sand intertidal
Conus sp.	74.6	9	Sand intertidal
Garafrium sp.	24.8	4	Sand intertidal
Harpa harpa	24.4	1	Sand intertidal
Quidnapagus palatum	41	4	Sand intertidal
Rhinoclavis sp.	22.6	1	Sand intertidal
Trochus maculatus	13.8	2	Sand intertidal
Total	205.9	22	

contined over

Table 10.26 continued

Species	Wgt	MNI	Habitat
Layer total	1064.4	105	
Layer 4			
Geloina sp.	10.9	2	Mangrove
Astrea stellare	27.1	1	Reef intertidal
Barbatia sp.	23.5	8	Reef intertidal
Bursidae sp.	10.3	1	Reef intertidal
Cypraea sp.	103.4	12	Reef intertidal
Tridacna maxima	217.3	5	Reef intertidal
Turbo sp.	368.9	14	Reef intertidal
Vasum turbinellus	25.1	3	Reef intertidal
Total	775.6	44	
Chiton plates	5	3	Rock platform
Latrius sp.	19.2	5	Rock platform
Nerita sp.	129.6	99	Rock platform
Total	153.8	107	
Conus sp.	129.5	13	Sand intertidal
Garafrium sp.	34	3	Sand intertidal
Periglypta puerpera	44.4	3	Sand intertidal
Quidnapagus palatum	27.6	4	Sand intertidal
Tectus pyramis	20.9	2	Sand intertidal
Trochus maculatus	62.5	5	Sand intertidal
Total	318.9	30	
Fragments	53.8		
Layer total	1313	183	

Table 10.27 Malua Bay School (TPs 3 and 9.1) shellfish habitat zone usage

Habitat	No of sp.	% of sp.	Wgt	% of wgt	MNI	% of MNI
Layer 3						
Reef intertidal	7	38.9%	804.5	75.6%	48	45.7%
Sand intertidal	7	38.9%	205.9	19.3%	22	21.0%
Rock intertidal	3	16.7%	51	4.8%	34	32.4%
Mangrove	1	5.6%	3	0.3%	1	1.0%
Total	18		1064.4		105	
Layer 4						
Reef intertidal	7	41.2%	775.6	61.6%	44	24.0%
Sand intertidal	6	35.3%	318.9	25.3%	30	16.4%
Rock intertidal	3	17.6%	153.8	12.2%	107	58.5%
Mangrove	1	5.9%	10.9	0.9%	2	1.1%
Total	17		1259.2		183	

While excavating the site local informants pointed out that *Periglypta puerpera* which was found only in the lower layers of the site was no longer available in the area. This species appears to be particularly sensitive to intensive exploitation and was a favoured species, as it appears in many of the earliest layers of sites on Malekula (e.g. Navaprah, Woplamplam, Fiowl). It also demonstrates an ability to repopulate an area as it was also found in some later sites (e.g. Chachara, Nuas) perhaps indicating that its current status (i.e. extirpated) may be due only to more recent over-exploitation.

Fiowl

All layers at the Fiowl site show a reliance on species predominantly recovered from the reef intertidal zone. There were no significant differences between the various layers with all of them showing the largest number of MNI having been sourced from the reef intertidal zone (Table 10.28). When the species MNI are compared there appears to have been a major decrease in the collection of *Turbo* sp. from the earliest layer of the site (Layer 3) to the later layers (none). Sudden disappearance of a shellfish species in the archaeological record is not a normal indicator of resource depletion. This seems more likely to be a factor of small sample size as it is most unlikely that *Turbo* sp. ceased being collected while viable stocks remained. Also present in the assemblages, albeit in minimal quantity, were *Quidnipagus palatam* and *Melina ephippium*, two species which prefer brackish conditions.

The shellfish remains from both of the open sites located on the coast (Malua Bay School and Fiowl) showed there was a preference for reef intertidal species. This can be seen as directly related to the fact that it is the predominant coastal environmental zone. Shellfish from the Malua Bay site indicate a wider resource collection strategy than that practiced at Fiowl. This can be largely explained by the fact that the Malua Bay site was located in a more varied environmental zone, which included a large sandy bay and a perennial river. *Nerita* sp. were the main species collected at both sites followed by *Turbo* sp.

Table 10.28 Fiowl, shellfish species, MNI, weight and habitat

Species	Wgt	MNI	Habitat
Layer 1			
Chicoreus torrefactus	1	4.3	Reef intertidal
Conus litteratus	1	4.6	Reef intertidal
Cypraea erosa	1	3.3	Reef intertidal
Eucithara sp.	1	1.7	Reef intertidal
Turbo crassus	1	49.5	Reef intertidal
Tridacna sp.	1	65.5	Reef subtidal
Total	6	128.9	
Chiton plates	1	4.5	Rock intertidal
Nerita albicilla	2	1.6	Rock intertidal
Total	3	6.1	
Quidnapagus palatum	1	2.6	Sand intertidal
Trochus niloticus	1	29	Sand intertidal
Total	2	31.6	
Fragments		17.6	
Layer total	11	184.2	
Layer 2			
Astrea stellare	3	4.1	Reef intertidal
Barbatia sp.	6	11.4	Reef intertidal
Cypraea erosa	1	30.3	Reef intertidal
Spondylus sp.	1	1.4	Reef intertidal
Tridacna sp.	1	6.5	Reef intertidal
Turbo crassus	3	27.9	Reef intertidal
Total	15	81.6	
Nerita albicilla	7	9.1	Rock intertidal
Conus litteratus	1	3.5	Sand intertidal
Melina ephippium	3	1.2	Sand intertidal
Nassarius sp.	1	0.8	Sand intertidal

contined over

Table 10.28 continued

Species	Wgt	MNI	Habitat
Quidnapagus palatum	1	7.2	Sand intertidal
Trochus niloticus	1	12.3	Sand intertidal
Total	7	25	
Fragments		22.8	
Layer total	22	138.5	
Layer 3			
Astrea stellare	1	3.4	Reef intertidal
Barbatia sp.	7	16.1	Reef intertidal
Chicoreus torrefactus	1	14	Reef intertidal
Cypraea erosa	1	13.1	Reef intertidal
Tridacna sp.	1	3.7	Reef intertidal
Turbo chrysostomus	6	5.5	Reef intertidal
Turbo crassus	3	32.8	Reef intertidal
Turbo sp.	2	0.5	Reef intertidal
Total	22	89.1	
Capulus sp.	3	3.1	Rock intertidal
Chiton plates	1	6.8	Rock intertidal
Nerita albicilla	1	24	Rock intertidal
Nerita undata	5	5.3	Rock intertidal
Total	10	39.2	
Gafrarium	1	0.7	Sand intertidal
Melina ephippium	2	6.1	Sand intertidal
Nassarius sp.	2	2.5	Sand intertidal
Periglypta purpera	1	11.5	Sand intertidal
Quidnapagus palatum	1	5.7	Sand intertidal
Strombus sp	1	8.3	Sand intertidal
Trochus niloticus	1	95.7	Sand intertidal
Total	9	130.5	
Fragments		128.5	
Layer total	41	387.3	

Table 10.29 Fiowl shellfish habitat zone usage

Habitat	No of sp.	% of sp.	Wgt	% of wgt	MNI	% of MNI
Layer 1						
Reef intertidal	6	60.0%	128.9	77.4%	6	54.5%
Sand intertidal	2	20.0%	31.6	19.0%	2	18.2%
Rock intertidal	2	20.0%	6.1	3.7%	3	27.3%
Total	10		166.6		11	
Layer 2						
Reef intertidal	6	50.0%	81.6	70.5%	15	65.2%
Sand intertidal	5	41.7%	25	21.6%	7	30.4%
Rock intertidal	1	8.3%	9.1	7.9%	1	4.3%
Total	12		115.7		23	
Layer 3						
Reef intertidal	8	42.1%	89.1	34.4%	22	53.7%
Sand intertidal	7	36.8%	130.5	50.4%	9	22.0%
Rock intertidal	4	21.1%	39.2	15.1%	10	24.4%
Total	19		258.8		41	

Chachara and Nuas

The assemblages recovered from these similar site types (total of 54 shells [Tables 10.30, 10.32]) were very small and any summary regarding collection strategies must be treated with caution. The major zone of exploitation was the reef intertidal. This predominance is reflected in both MNI and weight, making it unambiguous. The pattern would reflect a collection strategy of concentration on the reef with some more opportunistic collecting from the sand and rock intertidal zones. However the percentage of collected shellfish deriving from these zones was almost negligible. The shellfish remains from both sites fit a collection strategy of optimal efficiency in that the reef intertidal zone is a rich resource patch and is the dominant coastal environment. The main species being collected was by far and away *Turbo* sp. from the reef zone. A greater number of species overall were recorded at Chachara possibly reflecting a wider ranging pattern of collection which may have been inspired by the distance of the site from the coast (some 2km). Nuas is located some 125m from the coast. At both sites the shellfish component of the diet would seem to be minor, but interestingly, even for those living towards the centre of the island they were at least an occasional addition to the menu.

Table 10.30 Nuas, shellfish species, weight, MNI and habitat

Species	Wgt	MNI	Habitat
Layer 1			
Tridacna maxima	220.2	3	Reef intertidal
Turbo sp.	560.4	19	Reef intertidal
Vasum ceramicum	12.2	1	Reef intertidal
Total	792.8	23	
Tectarrius sp.	9.3	1	Rock intertidal
Conus sp.	32.3	2	Sand intertidal
Periglypta puerpera	16.2	1	Sand intertidal
Total	48.5	3	
Layer total	850.6	27	

Table 10.31 Nuas shellfish habitat zone usage

Habitat	No of sp.	% of sp.	Wgt	% of wgt	MNI	% of MNI
Reef intertidal	3	50%	792.8	93.2%	23	85.2%
Sand intertidal	2	33.3%	48.5	5.7%	3	11.1%
Rock intertidal	1	16.7%	9.3	1.1%	1	3.7%
Total	6		850.6		27	

Table10.32 Chachara shellfish species, weight, MNI and habitat

Species	Wgt	MNI	Habitat
Layer 1			
Tridacna maxima	15.1	2	Reef intertidal
Turbo sp.	153.4	10	Reef intertidal
Vasum ceramicum	168	1	Reef intertidal
Total	336.5	13	
Nerita sp.	1.4	1	Rock intertidal
Nerita sp.	5.4	5	Rock intertidal
Total	6.8	6	
Cerithium sp.	3.1	3	Sand intertidal
Conus sp.	28.7	3	Sand intertidal
Periglypta puerpera	1.3	1	Sand intertidal
Tectus pyramis	4	1	Sand intertidal
Total	37.1	8	
Layer total	380.4	27	

Table 10.33 Chachara shellfish habitat zone usage

Habitat	No of sp.	% of sp.	Wgt	% of wgt	MNI	% of MNI
Reef intertidal	3	33.3%	336.5	88.5%	13	48.1%
Sand intertidal	4	44.4%	37.1	9.8%	8	29.6%
Rock intertidal	2	22.2%	6.8	1.8%	6	22.2%
Total	9		380.4		27	

Malekula cave sites

Navaprah

The shellfish sample (TP 3) analysed from the Navaprah cave site reflected a concentration of species from the rock and reef intertidal zones, with the main species being collected comprising *Nerita* and *Turbo* sp. (Tables 10.34, 10.35). Lesser numbers of other species from other zones suggest more opportunistic collection.

Table 10.34 Navaprah (TP3) shellfish species, weight, MNI and habitat

Species	Wgt	MNI	Habitat
All layers			
Geloina sp.	13.5	2	Mangrove
Astrea stellare	13.4	2	Reef intertidal
Barbatia sp.	39.7	14	Reef intertidal
Cypraea sp.	37.1	8	Reef intertidal
Tridacna maxima	37.9	3	Reef intertidal
Turbo sp.	154.6	91	Reef intertidal
Total	282.7	118	
Chiton sp.	10.6	4	Rock intertidal
Nerita sp.	151.9	115	Rock intertidal
Pateloida flexuosa	1.9	2	Rock intertidal
Total	164.4	121	
Cerithium sp.	7.2	10	Sand intertidal
Conus sp.	23.1	4	Sand intertidal
Drupa sp	3.4	3	Sand intertidal
Periglypta puerpera	21.3	2	Sand intertidal
Quidnapagus palatum	6.8	3	Sand intertidal
Trochus maculatus	10.4	2	Sand intertidal
Total	72.2	24	
Grand total	532.8	265	

Table 10.35 Navaprah shellfish habitat zone usage

Habitat	No of sp.	% of sp.	Wgt	% of wgt	MNI	% of MNI
Reef intertidal	4	26.7%	269.3	50.5%	116	43.8%
Sand intertidal	7	46.7%	85.6	16.1%	26	9.8%
Rock intertidal	3	20.0%	164.4	30.9%	121	45.7%
Mangrove	1	6.7%	13.5	2.5%	2	0.8%
Total	15		532.8		265	

Woplamplam

The sample from Woplamplam shows a very similar pattern to that of Navaprah, i.e. a greater reliance on both the rock and reef intertidal zones, with a lesser concentration on the sand intertidal. *Nerita* and *Turbo* species were again the most numerically dominant (Tables 10.36, 10.37). A single mangrove species was also present in the upper layer of the site suggests that this zone was not being specifically targeted or that it was a zone of limited area.

Table 10.36 Woplamplam shellfish species, weight, MNI and habitat

Layer 1	wgt	MNI	Habitat
Geloina sp.	16.8	1	Mangrove
Astrea stellare	9	2	Reef intertidal
Bursidae sp.	3.4	1	Reef intertidal
Cypraea sp.	18	1	Reef intertidal
Turbo chrysostomus	2	1	Reef intertidal
Turbo sp.	82.7	5	Reef intertidal
Total	**115.1**	**10**	
Chiton sp.	5.6	1	Rock intertidal
Haliotis varia	1.9	1	Rock intertidal
Latrius sp.	5.7	1	Rock intertidal
Nerita sp.	43.1	32	Rock intertidal
Turbo cinereus	2.5	2	Rock intertidal
Total	**58.8**	**37**	
Cerithium sp.	6.8	10	Sand intertidal
Garafrium sp.	1.9	1	Sand intertidal
Periglypta puerpera	3.4	1	Sand intertidal
Quidnapagus palatum	11.1	3	Sand intertidal
Tectus pyramis	14	1	Sand intertidal
Trochus maculatus	10.2	1	Sand intertidal
Total	**47.4**	**17**	
Layer total	**238.1**	**65**	

contined over

Table 10.36 continued

Layer 1	wgt	MNI	Habitat
Layer 2			
Barbatia sp.	3.4	1	Reef intertidal
Cellana sp.	0.4	1	Reef intertidal
Cypraea sp.	0.5	1	Reef intertidal
Tridacna maxima	24.6	1	Reef intertidal
Turbo chrysostomus	2.6	1	Reef intertidal
Turbo sp.	48.4	9	Reef intertidal
Total	**79.9**	**14**	
Chiton sp.	2.8	1	Rock intertidal
Nerita sp.	44.2	21	Rock intertidal
Tectarrius sp.	3.7	1	Rock intertidal
Total	**50.7**	**23**	
Cerithium sp.	2	3	Sand intertidal
Conus sp.	5.5	3	Sand intertidal
Garafrium sp.	24.9	4	Sand intertidal
Periglypta puerpera	7.6	2	Sand intertidal
Quidnapagus palatum	16.8	3	Sand intertidal
Total	**56.8**	**15**	
Layer total	**187.4**	**52**	

Table 10.37 Woplamplam shellfish habitat zone usage

Habitat	No of sp.	% of sp.	Wgt	% of wgt	MNI	% of MNI
Layer 1						
Reef intertidal	5	29.4%	108.6	45.6%	10	15.4%
Sand intertidal	7	41.2%	56.4	23.7%	19	29.2%
Rock intertidal	4	23.5%	56.3	23.6%	35	53.8%
Mangrove	1	5.9%	16.8	7.1%	1	1.5%
Total	**17**		**238.1**		**65**	
Layer 2						
Reef intertidal	6	42.9%	79.9	42.6%	14	26.9%
Sand intertidal	5	35.7%	56.8	30.3%	15	28.8%
Rock intertidal	3	21.4%	50.7	27.1%	23	44.2%
Mangrove	0		0		0	
Total	**14**		**187.4**		**52**	

Wambraf

The changes over time at the site of Wambraf in the proportion of species from each zone can be analysed using the MNI counts (Table 10.38). They indicate that a mixed strategy was practiced with some change in the proportion of each zone over time. Shellfish from the earliest layers of the site were dominated by species from the reef intertidal zone. There was then a shift to a dominance of rock intertidal species (at 70cm) with lesser quantities of reef and sand species, then finally in the uppermost layer of the site the assemblage comprised species derived from the reef and rock intertidal zones (Table 10.39). Explanations to account for the changing emphasis on the various zones must be regarded as only speculative with such a small sample, but such factors such as altered environmental conditions (i.e. uplift) or depletion of particular zone resources are likely to have been involved. There were no mangrove species present at Wambraf. The predominant species collected at Wambraf were *Nerita* sp., followed by *Turbo* sp. Although there was a component of sandy zone species at the site there was unusually no *Trochus* sp. recorded.

Table 10.38 Wambraf, shellfish species MNI, weight and habitat

Species	Wgt	MNI	Habitat
5-20cm			
Cypraea tigris	3	21.5	Reef intertidal
Morula sp.	1	4	Reef intertidal
Turbo chrysostomus	1	1.2	Reef intertidal
Turbo crassus	1	94.3	Reef intertidal
Total	6	121	
Chiton plates	1	3.7	Rock intertidal
Nerita albicilla	8	26.7	Rock intertidal
Nerita plicata	13	18.6	Rock intertidal
Vulsella sp.	1	0.4	Rock intertidal
Total	23	49.4	
Asaphis violascens	4	6.3	Sand intertidal
Cerithium sp.	1	0.5	Sand intertidal
Garfrarium sp.	2	4.8	Sand intertidal
Quidnipagus palatum	1	6.8	Sand intertidal
Rhinoclavis sp.	1	0.6	Sand intertidal
Tectus pyramus	2	4.4	Sand intertidal
Trochus sp.	2	11.3	Sand intertidal
Total	13	34.7	
Layer total	42	205.1	
20-35cm			
Astrea stellare	1	15.5	Reef intertidal
Turbo crassus	1	35.1	Reef intertidal
Turbo sp.	12	1.1	Reef intertidal
Total	14	51.7	
Chiton plates	1	3.6	Rock intertidal
Nerita albicilla	6	24.6	Rock intertidal
Nerita undata	15	14.9	Rock intertidal
Total	22	43.1	
Asaphis violascens	2	6	Sand intertidal
Gafrarium sp.	1	2.1	Sand intertidal
Melina ephippium	1	0.4	Sand intertidal
Nassarius luridus	2	1.3	Sand intertidal
Periglypta puerpera	1	4.1	Sand intertidal
Quidnipagus palatum	5	2.4	Sand intertidal
Strombus sp.	11	6.1	Sand intertidal
Trochus niloticus	1	44.2	Sand intertidal
Total	24	66.6	
Fragments		26.4	
Layer total	60	187.8	
35-50			
Cypraea tigris	4	6.2	Reef intertidal
Turbo chrysostomus	1	0.9	Reef intertidal
Turbo crassus	21	117.8	Reef intertidal
Turbo sp.	2	1.2	Reef intertidal
Total	28	126.1	
Chiton plates	1	1.3	Rock intertidal
Nerita plicata	5	5.4	Rock intertidal
Nerita sp.	2	12.5	Rock intertidal
Total	8	19.2	
Conus sp.	1	3.9	Sand intertidal
Melina ephippium	1	0.9	Sand intertidal
Trochus niloticus	7	9.4	Sand intertidal
Trochus sp.	4	17.4	Sand intertidal
Total	13	31.6	
Fragments		31.6	
Total	49	208.5	
50-70cm			
Barbatia sp.	1	3.8	Reef intertidal
Turbo crassus	1	20.8	Reef intertidal
Turbo sp.	4	0.2	Reef intertidal
Total	6	24.8	

contined over

Table 10.38 continued

Species	Wgt	MNI	Habitat
Nerita albicilla	6	36.4	Rock intertidal
Nerita undata	13	16.1	Rock intertidal
Planaxis sulcata	1	1.3	Rock intertidal
Total	20	53.8	
Conus litteratus	1	3.5	Sand intertidal
Melina ephippium	1	0.4	Sand intertidal
Nassarius sp.	3	1.9	Sand intertidal
Quidnipagus palatum	1	0.6	Sand intertidal
Trochus niloticus	1	6.3	Sand intertidal
Total	7	12.7	
Frags		25.3	
Total	33	116.6	
70-90cm			
Barbatia sp.	2	8.2	Reef intertidal
Turbo crassus	6	6.7	Reef intertidal
Total	8	14.9	
Nerita plicata	9	15.6	Rock intertidal
Nerita sp.	2	14.3	Rock intertidal
Planaxis sulcata	1	1.6	Rock intertidal
Total	12	31.5	
Melina ephippium	1	0.6	Sand intertidal
Fragments		13.4	
Total	21	60.4	
90-100cm			
Barbatia sp.	1	4.9	Reef intertidal
Chicoreus torrefactus	1	1.5	Reef intertidal
Cypraea erosa	1	7.5	Reef intertidal
Morula sp.	1	1.7	Reef intertidal
Turbo chrysostomus	1	7.5	Reef intertidal
Turbo crassus	8	24.3	Reef intertidal
Turbo petholatus	1	12.2	Reef intertidal
Total	14	59.6	
Hipponix conicus	1	0.7	Rock intertidal
Nerita albicilla	1	6.1	Rock intertidal
Nerita undata	4	7.3	Rock intertidal
Planaxis sulcata	1	0.4	Rock intertidal
Total	7	14.5	
Cardita sp.	1	3.2	Sand intertidal
Trochus niloticus	1	34.4	Sand intertidal
Total	2	37.6	
Fragments		21.7	
Total	23	133.4	
1.0-1.2m			
Astrea stellare	1	2.7	Reef intertidal
Barbatia sp.	3	6.8	Reef intertidal
Cypraea sp.	3	4.8	Reef intertidal
Turbo crassus	7	17.5	Reef intertidal
Total	14	31.8	
Nerita albicilla	1	10.7	Rock intertidal
Nerita plicata	4	8.7	Rock intertidal
Planaxis sulcata	1	0.9	Rock intertidal
Total	6	20.3	
Melina ephippium	2	1.6	Sand intertidal
Periglypta puerpera	1	19	Sand intertidal
Quidnipagus palatum	1	1.3	Sand intertidal
Trochus niloticus	2	7.4	Sand intertidal
Total	6	29.3	
Fragments		21.6	
Total	26	103	
120-140cm			
Barbatia sp.	4	4.8	Reef intertidal
Conus litteratus	1	5.8	Reef intertidal
Cypraea erosa	1	15.8	Reef intertidal

contined over

Table 10.38 continued

Species	Wgt	MNI	Habitat
Spondylus sp.	1	7	Reef intertidal
Tridacna sp.	1	2.4	Reef intertidal
Turbo chrysostomus	2	2.4	Reef intertidal
Turbo crassus	10	17	Reef intertidal
Turbo crassus	1	30.2	Reef intertidal
Turbo petholatus	1	4.7	Reef intertidal
Total	22	90.1	
Chiton plates	1	6.4	Rock intertidal
Nerita albicilla	9	11.3	Rock intertidal
Nerita undata	4	6.6	Rock intertidal
Total	14	24.3	

contined over

Table 10.38 continued

Species	Wgt	MNI	Habitat
Gafrarium sp.	1	1.1	Sand intertidal
Nassarius sp.	9	9.5	Sand intertidal
Periglypta puerpera	1	4.8	Sand intertidal
Pythia sp.	5	5.3	Sand intertidal
Quidnipagus palatum	1	1.3	Sand intertidal
Total	17	22	
Fragments		37.2	
Total	53	173.6	

Table 10.39 Wambraf shellfish habitat zone usage

Habitat	No of sp.	% of sp.	Wgt	% of wgt	MNI	% of MNI
5-20cm						
Reef intertidal	4	26.7%	121	59.0%	6	14.3%
Sand intertidal	7	46.7%	34.7	16.9%	13	31.0%
Rock intertidal	4	26.7%	49.4	24.1%	23	54.8%
Total	15		205.1		42	
20-35cm						
Reef intertidal	3	21.4%	51.7	32.0%	14	23.3%
Sand intertidal	8	57.1%	66.6	41.3%	24	40.0%
Rock intertidal	3	21.4%	43.1	26.7%	22	36.7%
Total	14		161.4		60	
35-50cm						
Reef intertidal	4	36.4%	126.1	71.3%	28	57.1%
Sand intertidal	4	36.4%	31.6	17.9%	13	26.5%
Rock intertidal	3	27.3%	19.2	10.9%	8	16.3%
Total	11		176.9		49	
50-70cm						
Reef intertidal	3	27.3%	24.8	27.2%	6	18.2%
Sand intertidal	5	45.5%	12.7	13.9%	7	21.2%
Rock intertidal	3	27.3%	53.8	58.9%	20	60.6%
Total	11		91.3		33	
70-90cm						
Reef intertidal	2	33.3%	14.9	31.7%	8	38.1%
Sand intertidal	1	16.7%	0.6	1.3%	1	4.8%
Rock intertidal	3	50.0%	31.5	67.0%	12	57.1%
Total	6		47		21	
90-100cm						
Reef intertidal	7	53.8%	59.6	53.4%	14	60.9%
Sand intertidal	2	15.4%	37.6	33.7%	2	8.7%
Rock intertidal	4	30.8%	14.5	13.0%	7	30.4%
Total	13		111.7		23	
100-120cm						
Reef intertidal	4	36.4%	31.8	39.1%	14	53.8%
Sand intertidal	4	36.4%	29.3	36.0%	6	23.1%
Rock intertidal	3	27.3%	20.3	24.9%	6	23.1%
Total	11		81.4		26	
120-140cm						
Reef intertidal	9	52.9%	90.1	66.1%	22	41.5%
Sand intertidal	5	29.4%	22	16.1%	17	32.1%
Rock intertidal	3	17.6%	24.3	17.8%	14	26.4%
Total	17		136.4		53	

Navepule

The three inland caves at Navepule returned very small amounts of shell and the assemblages are too limited to provide any meaningful insights into shellfish collection strategies or behavioural patterns. The distance to the resource is reflected in the amount returned to the site. Only three shells in total were recovered from Navepule B. All the shellfish from the three caves were reef intertidal zone species (Tables 10.40–10.43). As far as species composition is concerned there was no difference between the cave sites, the most numerous species being *Nerita* sp.

When all the cave sites are compared a mixed rock/reef collection strategy is clearly predominant. The sandy and mangrove zone species comprise a negligible percentage of the assemblages. The main species that were recorded at the cave sites were *Nerita* sp. followed by *Turbo* sp. There is a difference between the shellfish assemblages recorded for the coastal and inland caves but it is seen in the quantity of shell returned to the individual sites rather than in the particular species being collected. In fact for all excavated site types the shellfish remains displayed little variation apart from quantity. The open coastal sites comprised species that indicated a mixed collection strategy based primarily on reef and rock intertidal zones with the main species being *Nerita* sp and *Turbo* sp. At the open inland sites the main zone was the reef intertidal zone and the main species collected was *Turbo* sp. The shellfish from the cave sites (for which no species difference was found between inland and coastal sites) again indicated a mixed collection strategy focusing on *Nerita* and *Turbo* sp. Some extirpation or intensive exploitation of bivalve species was also hinted at.

Table 10.40 Navepule A shellfish species, MNI, weight and habitat

Species	Wgt	MNI	Habitat
Layer 1			
Geloina sp.	1	8.5	Mangrove
Astrae stellare	1	16.5	Reef intertidal
Tridacna maxima	1	6.3	Reef intertidal
Turbo sp.	15	366.3	Reef intertidal
Total	17	389.1	
Latrius sp.	1	4	Rock intertidal
Nerita sp.	103	221.5	Rock intertidal
Tectarrius sp.	1	4.5	Rock intertidal
Total	105	230	
Cerithium sp.	1	0.9	Sand intertidal
Tectus pyramis	1	7.2	Sand intertidal
Trochus maculatus	3	30.3	Sand intertidal
Total	5	38.4	

contined over

Table 10.40 continued

Species	Wgt	MNI	Habitat
Layer total	128	666	
Layer 2			
Barbatia sp.	1	3.1	Sand intertidal
Cypraea sp.	1	6.9	Reef intertidal
Nerita sp.	2	2.2	Rock intertidal
Total	4	12.2	
Tridacna maxima	1	172.6	Reef intertidal
Turbo chrysostomus	1	3.9	Reef intertidal
Cypraea sp.	1	1	Reef intertidal
Total	3	177.5	

Table 10.41 Navepule A shellfish habitat zone usage

Habitat	No of sp.	% of sp.	Wgt	% of wgt	MNI	% of MNI
Layer 1						
Reef intertidal	2	20.0%	372.6	55.9%	16	12.5%
Sand intertidal	4	40.0%	54.9	8.2%	6	4.7%
Rock intertidal	3	30.0%	230	34.5%	105	84.0%
Mangrove	1	10.0%	8.5	1.3%	1	0.8%
Total			666		128	
Layer 2						
Reef intertidal	1	33.3%	6.9	56.6%	1	25%
Sand intertidal	1	33.3%	3.1	25.4%	1	25%
Rock intertidal	1	33.3%	2.2	18.0%	2	50%
Total			12.2			

Table 10.42 Navepule C shellfish species, MNI, weight and habitat

Species	Wgt	MNI	Habitat
Layer 1			
Barbatia sp.		6.1	Reef intertidal
Cypraea tigris	1	19.7	Reef intertidal
Tridacna maxima	1	7.6	Reef intertidal
Turbo crassus	2	25.1	Reef intertidal
Turbo sp.	1	8.4	Reef intertidal
Total	4	66.9	
Capulus danieli	1	1.1	Rock intertidal
Nerita albicilla	1	2.1	Rock intertidal
Nerita undata	3	5.3	Rock intertidal
Total	5	8.5	
Gafrarium	1	5.6	Sand intertidal
Rhinoclavis sp.	1	1.1	Sand intertidal
Total	2	6.7	
Fragments		4.2	
Layer total	11	86.3	
Layer 2			
Conus litteratus	1	30.7	Sand intertidal
Periglypta puerpera	1	2.1	Sand intertidal
Vasum turbinellus	1	4	Reef intertidal
Layer total	3	36.8	

Table 10.43 Navepule C shellfish habitat zone usage

Habitat	No of sp.	% of sp.	Wgt	% of wgt
Layer 1				
Reef intertidal	5	50.0%	66.9	81.5%
Sand intertidal	2	20.0%	6.7	8.2%
Rock intertidal	3	30.0%	8.5	10.4%
Total		82.1		
Layer 2				
Reef intertidal	1	33.3%	4	10.9%
Sand intertidal	2	66.7%	32.8	89.1%
Total			36.8	

Shellfish summary

The analysis of the shellfish collections from the sites discussed above was carried out to ascertain the collection strategies that were being followed at the various sites and to what degree they conformed to optimal foraging models. The analysis of the shellfish remains indicates that the employed collection strategies fit well with the theory of patch optimality where species that are most abundant and numerous within each patch being targeted. The inland sites that are further from the resource contain less shell density, shell weight and number of species, which also conforms to optimality theory.

Collection strategies did not fit a model of differing strategies for beach, inland or cave sites. Rather it was found that collection strategies employed at the various sites were largely influenced by their topographic setting. Sites that were situated in similar environmental locations were found to contain broadly similar shellfish assemblages, although some fine tuning of the collection strategies (i.e. focus on the most abundant patches) could be identified. Overall the shellfish collection strategy employed was one of efficient gathering of the closest available resources which would of course vary depending on the environment surroundings. This was a pattern also identified by Ward at the Pakea site (Ward 1979:12-9) with some variation. Ward argued that the shellfish from Pakea showed greater variety in the earlier levels of the site and that over time shellfish consumption became an increasingly important component of the diet (Ward 1979:11–38). However this apparent increasing reliance on shellfish seems likely to be directly related to the fact that Pakea is a very small island where the richest resources by far are marine. Ward also noted a restriction in the range of species being collected over time which he suggested was due to either targeting and/or depletion.

Heavy predation on initial arrival in an area and gradual reduction in species size over time has been confirmed at the Arapus site and the sites of Ifo and Ponamla on Erromango (Gyngell 1997). This is a pattern that has parallels with other sites in the Pacific (Kirch and Yen 1982:296; Swadling 1986).

Initial verses later settlement

The faunal remains from the Arapus site have not yet been analysed in any detail, but certain striking features of the assemblage relating to the earliest levels of the site (c. 2900 BP) require comment. The remains from the Arapus site dramatically demonstrate the subsistence strategy of a population on first arrival and how that very quickly that strategy changed following the depletion of pristine resources and a certain level of equilibrium was reached. As of 1999 only c. 6m^2 of the earliest cultural deposits had been excavated at the Arapus site but the pattern seen in this initial excavation has been further replicated over much larger excavated areas of the site during field seasons in 2001-2003. The first sign of human arrival in the area is evidenced by the buildup of midden deposits directly on top of the former foreshore. The lowest levels of the cultural stratigraphy in TPs 8, 14, 17, 20 and 23 are characterised by concentrated deposits of shellfish, with a preference for *Trochus niloticus*. Associated with the shellfish were concentrations of bone, principally turtle, fruit bat, bird and to a lesser extent fish (Bedford and Spriggs 2000). Pottery was relatively sparse. This evidence gives a glimpse of a short time period, before the establishment of horticultural systems, when the population on first arrival was initially heavily reliant on readily procurable local marine and terrestrial resources (e.g. Kirch 1988b:238). As mentioned a number of extinct species have been identified from the earliest levels of the site (birds and land crocodile) which highlights the almost instantaneous nature of the disappearance of certain species on human arrival. There is also very quickly a significant reduction in the size of shellfish (i.e. comparison of *Trochus niloticus* from the earliest deposits of Arapus and Mangaasi).

The faunal remains from the Arapus site would seem to support an argument that on initial arrival in a pristine faunal environment colonisers took an optimising approach, exploiting the high-return protein-rich species (i.e. flightless birds, turtles and land crocodiles) before moving onto those that were less profitable (Anderson 1997; Martin and Steadman 1999). Once the protein-rich easily-caught resources were depleted the procurement strategy began to favour a broader selection criteria. Further evidence for this scenario is hinted at in the earliest layers of the Arapus site (eg. TP14, TP 17, TP 20) where although bone remains were frequent, the percentage of fishbone was relatively small (total for the three test pits only 21 bones). All cultural horizons at the Mangaasi site contained significantly greater quantities of fishbone (Tables 10.10, 10.11). This scenario, along with other observations, has only been identified by contrasting a site associated with initial arrival with later sites. It might well not be identifiable or demonstrated in a site that dated to even 50 years after initial settlement.

Summary

The archaeological faunal remains from Vanuatu have provided a general outline of the non-horticultural prehistoric subsistence patterns. Those species which have always been a regular component of the diet include fish, shellfish, fruit bat and to a lesser extent wild birds, along with the introduced domestics, pig and chicken. Two species of rat completed the list of introduced species. Dog was not identified in the faunal assemblages. As outlined above there are a number of changing subsistence strategies that can be identified in the record. The 'blitzkrieg' scenario in Vanuatu has been hinted at but remains to be further confirmed. As yet the sub-fossil record of Vanuatu is non-existent compared to its nearest neighbours, New Caledonia (Balouet 1991) and Fiji (Worthy *et al.* 1999) but it will provide essential data to assess further the significance of faunal extinction on human arrival. Evidence from the Arapus site demonstrates the devastating effect that the arrival of humans had on naïve faunal populations. It was these abundant natural resources which would have initially sustained a founding population prior to the establishment of stable horticultural systems (Kirch 1988b:238).

But this 'first contact' phase (Martin and Steadman 1999) was short-lived and soon afterwards some stabilisation of subsistence strategies, which undoubtedly comprised a substantial horticultural component, was reached. Those stabilised subsistence strategies, involving 'the interaction of a number of practices targeted at a wide range of different exploitation zones' (Walter 1998:104), appear to have largely remained unchanged until the latter part of the archipelago's prehistory when there is clear evidence of increased intensification of horticultural activities (Spriggs 1981).

Turtle was keenly exploited on initial arrival and appears to have experienced a rapid reduction in the breeding population due to human predation as evidenced by its much rarer occurrence in sites post-dating Lapita settlement (e.g. Kirch 1988b:235; Kirch and Yen 1982). Pigs were brought by the first colonisers and have provided a reliable source of sustenance from that period to the present. Evidence of their increasingly ceremonial role appears in the archaeological record from about 1000 BP.

Wild birds were exploited consistently through the sequence but would appear to have provided an occasional addition to the protein intake rather than a substantial component. A similar scenario is seen with the introduced domestic species *Gallus gallus*. Fruit bats appear to have been heavily exploited on first arrival and more moderately throughout the rest of the sequence. Inshore and reef species of fish have been exploited from the earliest settlement period and onwards with little evidence for changing preference or procurement strategies.

Shellfish remains indicate a general pattern of optimal gathering of the closest resources. Over-exploitation to the point of extirpation was evidenced in a number of areas where particularly sand-dwelling bivalve species recovered from early cultural layers of sites were no longer available in the area (e.g. Ifo and Malua Bay). The effect of initial human arrival on pristine shellfish populations has also been dramatically demonstrated at the Arapus and Mangaasi sites. Continued exploitation and progressive size reduction are also highlighted at the sites of Ponamla and Ifo on Erromango (Gyngell 1997:42–44). Once people moved beyond the initial arrival phase and horticultural produce was readily available, reliability on shellfish would have lessened. This is reflected at all of the excavated sites where shellfish appeared to provide a regular, albeit minor, protein-rich component of a varied diet.

The inclusion of the faunal remains from the Arapus site, which is still under analysis, was instructive in showing that the postulated 'blitzkrieg' scenario can only be seen at sites which date to the very beginning of human occupation of any island. The remains post-dating that phase demonstrate the utilisation of a broad spectrum of exploitation zones where both opportunistic and selective hunting and gathering of indigenous resources was practiced. Animal husbandry was also in evidence. All of these activities were carried out against a background of horticultural and aboricultural production and utilisation. This broad pattern of subsistence activity remained largely unchanged (as did the extant biota), albeit with some regional variation, until European contact.

11

Discussion and Conclusion

'From time to time and from many places, new movements of people have brought changes to the gene pools, to the languages and the customs, and to the subsistence patterns of the people of the Melanesian islands. New ways of exploiting island environments continue to give impetus to movements within the island groups themselves and to further modify the life of the people. Shifting patterns of contact and isolation have brought and continue to bring other changes. Each new excavation brings something of this nature to light and, we suspect, will continue to do so' (Shutler and Shutler 1975:77).

Discussion and Conclusion

Forty years ago when Golson (1959) summarised the archaeological prospects for the Pacific it could be accommodated within 49 pages. Since that time an extraordinary flood of information of increasing complexity and sophistication contained within a myriad of articles and books has radically changed our perceptions of the region. This more recent research into the archaeology of Vanuatu has also gone some way in both modifying earlier assertions and providing further data with which to assess a number of more recent hypotheses. The five research objectives outlined in the introduction to this thesis were as follows: 1) testing for evidence of pre-Lapita settlement in Vanuatu, 2) clarification of pioneering research and the establishment of basic cultural sequences, 3) establish basic outlines of prehistoric subsistence patterns and 4) provisional conclusions regarding settlement pattern and finally 5) an assessment of the evidence for a Melanesia-wide incised and applied relief ceramic tradition. The results of the research and to what level these objectives were achieved are summarised below.

1) Testing for evidence of pre-Lapita settlement

The idea of pre-Lapita settlement in Vanuatu can now be laid to rest. There was no evidence in any of the excavated cave sites on Malekula in the north or Erromango in the south that indicated human settlement prior to 3000 BP. All the earliest dates (3000–2900 BP) thus far for Vanuatu are associated with Lapita sites. Further in support of this assertion is the emerging evidence of faunal extinctions related to the initial settlement of at least Erromango and Efate, a feature that is widely recognised as being associated with first human arrival in Remote Oceania (Steadman 1995). In the case of Erromango the extinct bird bone was associated with dentate-stamped pottery. On Efate the extinct fauna was identified at the Arapus site, which although it may post-date initial settlement of Efate by up to 100 years (Bedford *et al.* 2004) it seems likely that it is associated with the earliest settlement of west Efate. Vanuatu's initial settlement lies firmly within the expansion into Remote Oceania by members of the Lapita Cultural Complex. Dentate-stamped ceramics were associated with the initial settlement phase in Vanuatu but appeared to survive for a relatively short time period of only a few hundred years being replaced by plain and/or a variety of decorated wares. This concurs with a pattern which has now been confirmed for most other regions of Remote Oceania, from New Caledonia to Tonga (Anderson and Clark 1999; Burley *et al.* 1999; Sand 1997) and one which suggests that whatever connections this homogeneity of Lapita ceramics may have represented they were somewhat short-lived.

2) Clarification of pioneering research and establishment of basic cultural sequences.

The Mangaasi ceramic sequence of Central Vanuatu, as originally outlined by Garanger (1972), has been substantially revised. No longer can it be seen as a lengthy conservative tradition separate from Lapita. Nor is there any evidence to support putative connections with ceramic traditions from New Guinea (Galipaud 1996a; Garanger 1972; Gorecki 1992). Rather the ceramic sequence from Central Vanuatu began with the appearance of Lapita and can be shown to have demonstrated an evolutionary transition to the Early (c. 2800–2500 BP) and Late Erueti (c. 2500–2200 BP) Phases and on to the Early (c. 2200–1600 BP) and Late Mangaasi (c. 1600–1200 BP) Phases and finally terminating at around 1200 BP on current evidence.

The ceramic remains from Erromango parallel those from Efate in a number of aspects, namely the primacy of Lapita and evolutionary nature of the sequence, but that is where the similarities largely begin and end. Lapita connections were short-lived. Rather abruptly the post-Lapita ceramic sequence on Erromango demonstrated a quite independent trajectory to the sequences from Efate. Lapita was superseded on Erromango by the Ponamla Plainware Phase (c. 2800-2600 BP) which in turn transformed into the Early (c. 2600–2400 BP) and Late (c. 2400–2000 BP) Ifo Phases, at which point ceramics disappeared from the cultural repertoire.

The archaeology of Northern Vanuatu remains somewhat of an archaeological *terra incognita* although the combined results of Ward in the Banks Islands (1979), Hedrick on Malo (Hedrick nd) and Galipaud on Santo, Malo and the Torres Islands (Galipaud 1996c, 1998a, 1998b), along with this recent research which probed the archaeology of Malekula, have begun to shed several rays of light. Malekula, and almost certainly the other islands of Northern Vanuatu, can now be securely associated with Lapita expansion in the region. More recent research on Malekula has confirmed this as a number of Lapita sites associated with dentate-stamped ceramics have been investigated (Bedford 2003). The earliest settlement found so far on the main island is on the northwest coast of Malekula at Malua Bay which dates to around 2700 BP and is best identified in the archaeological record by calcareous-tempered sherds associated with plain pots of the proposed Malua Phase (c. 2700–2500 BP). The central part of the Malekula ceramic sequence remains largely uncharted. The surface collected ceramics hint at the possibility of a lengthy ceramic sequence, albeit unevenly distributed across the island, which may or may not demonstrate some parallels with the Central Vanuatu sequence.

The last phase of ceramic production on Malekula is characterised by bullet shaped, coil-made pots which appeared in the sequence no earlier and possibly sometime later than 1000 BP. Research to date suggests that these vessel forms have no apparent antecedents in Vanuatu. Thus far this distinctive Chachara Ware, which has been found across most of Malekula, has been tentatively dated to between 600–200 BP. It is found in association with the frequently recorded *nasara* or ceremonial structures which were and still are an integral component of the political and social structure in the north and central north of Vanuatu which is based on a chiefly system that is consolidated through a hierarchy of grades. Clearly these site types and the political activities associated with them have some pedigree in Vanuatu, albeit restricted to the north.

The same Chachara Wares have also been identified as being present on a number of the other northern islands of Vanuatu indicating regular contact and interaction amongst the inhabitants of this region over the last 500 years. Appearing to overlap with these Chachara Wares and clearly developing from them are the *Naamboi* Wares which are unique to Malekula. These vessel forms which were recorded ethnographically represent the final phase and subsequent demise of ceramics on the island. Their appearance also coincided with a breakdown in the above noted pattern of ceramic similarity in the northern islands. This points to a shrinking sphere of interaction in the region and it seems likely that it is from this period that the extant cultural variability, found both on single islands and between islands, has developed. Increased territoriality and cultural differentiation across Malekula may also date to this period.

Glimpses of the latter part of the cultural sequence on Malekula have indicated that quite substantial social and cultural change occurred over the last 500 or more years on the island and in the region known as Vanuatu's 'Oceanic Mediterranean' (Bonnemaison 1996:208). Green (1999) has argued that these changes can at least partly be explained by influence from further west in the form of sustained contact or even migrations of non-Austronesians, a scenario that fits well with some of the archaeological, biological and linguistic evidence. However, finer resolution of the long history associated with this social and cultural transformation, which defined the ethnographic period and comprises the cultural heritage of much of the present population of Northern Vanuatu, awaits future research.

The wide variety and quantity of non-ceramic artefacts recovered during the excavations have also contributed further insights into the archaeology of Vanuatu. The recovered materials indicate that non-ceramic artefacts were generally not as susceptible to change as were the ceramics. They tended to show greater conservatism for longer periods. Artefacts such as *Tridacna* and *Conus* armrings showed little change in form over at least the first 1000 years of settlement. The same could be seen with the shell adze assemblages which were exclusively made of *Tridacna* shell and were almost always fully ground. The most perceptible change amongst the non-ceramic artefact assemblages during the first 1500 years was an apparent reduction in both quantity and variety. However this apparent trend seems more likely to relate to increasingly dispersed settlement patterns and, related to this, less concentrated midden deposits. Certainly the wealth of non-ceramic artefacts associated with both the earliest and latest periods of the cultural sequence support this scenario. The lack of archaeological research focusing on this middle section (2000-600 BP) of the sequence has clearly contributed to this perceived variability.

While change in non-ceramic artefact forms may have been gradual there is overwhelming evidence at least in the central and southern islands that by 600 BP or earlier it was substantially changed. These changes included a variety of different shells being utilised in the production of adzes (*Lambis* and *Terebra* sp.) and a whole host of ornament forms that had not been previously recorded. Pig tusk and *Trochus* sp. bracelets were also a commonly recorded artefact dating to this later period. While change in some of the artefact forms can be ascribed to Polynesian influence, other artefacts indicate some contact with New Caledonia and/or areas to the north west of Vanuatu. Further focus on the non-ceramic artefact forms dating to this latter period has the potential to provide valuable information into the patterns of interaction and cultural transformation which occurred in the region.

3) Prehistoric subsistence patterns.

The largely absent faunal component of Vanuatu's archaeological record has now been substantially augmented. There are hints of a 'blitzkrieg-like' scenario on initial arrival, although further excavation of sites dating to this period is required for confirmation. Extinct birds were identified in only the lowest layers of a number of sites on all islands. The remains of an extinct land crocodile have also been identified at the Arapus site, which dates from initial human arrival on Efate. Further to the impact of human arrival on naïve faunas is the dramatic reduction in the size of shellfish at a number of sites. There are even indications that certain species may have been extirpated from particular areas on different islands.

After this short-lived 'first contact' phase some stabilisation of subsistence strategies was reached and horticultural produce would have begun to play its role as the major supplier of sustenance. The record indicates a wide range of procurement strategies were employed which targeted a broad range of species, both introduced and indigenous, from all accessible zones of exploitation. Those species which contributed regularly to the diet from first arrival and throughout included fish, shellfish, fruit bats and to a lesser extent wild birds, along with the introduced domesticates, pig and chicken. Two species of rat were also identified amongst the faunal remains namely *Rattus exulans* and *praetor*. No dog was identified archaeologically. Turtle appeared more frequently in the earliest layers of sites indicating it was heavily exploited from initial arrival which would have led to a rapid reduction in the breeding population. This is certainly reflected in the archaeological record with turtle bone being much rarer in later layers. Fruit bats also appeared to be heavily exploited during the initial settlement phase but their continued consumption throughout the archaeological record is indicative of their adaptability. Inshore and reef species of fish were targeted from initial arrival and continued to be so throughout the sequence, with no evidence for any change in preference or procurement strategies. The shellfish remains indicated a collection strategy that fitted an optimal foraging model, i.e., efficient gathering of the closest available resources.

All of the faunal data and its relative importance in the strategies of subsistence must of course be assessed against the thus-far largely assumed horticultural component of the diet. If ethnographic evidence is anything to go by, it would have consistently provided the most substantial quota.

4) Settlement pattern

Some consistency in the archaeological record is now beginning to emerge regarding the history of settlement pattern in Vanuatu. The pattern of initial settlement appears to fit a situation where Lapita colonists were moving into an uninhabited landscape and were thus able to select prime locations for habitation. During the initial colonisation phase certain islands and their environs, i.e., Malo, Aore and south Efate, may have acted as some sort of focal point or 'metropolis' (Bedford *et al.* 1999:21, 2004; Green 1978) from which other islands were investigated. Although few excavations in Vanuatu that have focused on Lapita and immediately post-Lapita remains have revealed more than the occasional posthole or hearth feature, the concentrated deposits of midden tend to suggest a pattern of nucleated settlement. The zones targeted for settlement were overwhelmingly coastal, with a preference for areas which included easy canoe access, fringing reefs and/or lagoonal environments and an easily accessible fresh water source (e.g. Ponamla and Ifo on Erromango, Arapus, Mangaasi and Erueti on Efate, Malua Bay on Malekula, the small islands off the northeast coast of Malekula and much of coastal Malo). These factors enabled maximum utilisation of marine and other faunal resources. There is evidence that this phase of settlement was very fluid. Occupation may have been short-lived (e.g. Ponamla on Erromango, Malua Bay and coastal cave sites on Malekula) due to changing environmental factors. Populations still had the option of moving to other pristine areas of the same island or indeed to other islands as resources became depleted. Variability in environmental situations may have been one of the key factors governing the viability of settlements. The environmentally more fragile leeward side of the islands may not have been able to sustain long-term intensive settlement and horticultural practices. On the other hand the tephra-rich soils found on a number of islands would undoubtedly have been an attractive feature that would have facilitated long-term horticultural production.

The archaeological record indicates that it was only much later that populations spread out from these prime location areas. Exploratory visits into the interior of the larger islands may have been a regular feature (e.g. Navepule C, Malekula dated to c. 2000 BP) but more permanent settlement does not seem to have occurred until the last 1000 years.

5) Assessment of the evidence for a Melanesia-wide incised and applied relief tradition

The establishment of fine grained ceramic sequences from Efate and Erromango have enabled a reassessment of earlier results and conclusions relating to the archaeological history of Vanuatu and the broader region. A widely accepted theory used to explain post-Lapita change across the Southwest Pacific is that it was related to a secondary wave and/or continued contact with Non-Austronesian populations further west, which had the effect of 'Melanesianising' the region as far east as Fiji (Bellwood 1979; Golson 1961; Green 1963; Spriggs 1997). It has been argued that evidence for this scenario can be seen in the ceramic record of the region. Some authors maintain that there is evidence for a Pan-Melanesian Incised and Applied Relief tradition which demonstrated synchronous change throughout much of the post-Lapita period (Spriggs 1984, 1993, 1997, 2000), possibly even right up to 800 BP (Wahome 1997, 1999).

A re-appraisal of the ceramic assemblages that have been used to support such an argument suggest that at least in terms of claimed homologous ceramic traits the evidence is less than secure. It is argued that the concept of an Incised and Applied Relief tradition existing across the Southwest Pacific has since its inception been based more on theoretical grounds than empirical data. The term has often simply been used to differentiate ceramics that were neither dentate-stamped nor paddle impressed. Implied ceramic stylistic unity, arranged under the broad Incised and Applied Relief-Mangaasi rubric, has been used to support the concept of a Melanesia-wide inter-connected ceramic tradition. With it has gone a tendency to lump collections together rather than highlight differences. A comparison of common decorative techniques like incision or applique, without regard to vessel form, motif design, or firm chronological control, tends to support scenarios of widespread population migration and diffusion and such is the case with the Incised and Applied Relief tradition. But the act of grouping pottery using common yet ill-defined stylistic criteria does not mean that a close historical relationship need have existed between pottery makers who might have included incised and or applied relief amongst their decorative repertoire (Bedford and Clark 2001).

In the very earliest stages of many ceramic sequences from the Southwest Pacific similar changes can certainly be identified. There is a change from the first dentate stamped decorative wares to an increasing percentage of incised wares and/or plainwares, along with a decrease in the range of vessel forms. A few immediately post-Lapita motifs, decorative methods and vessel forms share a degree of similarity across sequences but these might equally be explained as continuities inherited from the founding ceramic tradition rather than continued inter-archipelago contact.

The relatively homogeneous nature of Lapita ceramics can be seen in part as being the result of frequent interaction over a relatively short time period between a small, widely dispersed and highly mobile colonising population (Graves *et al.* 1990:228; Summerhayes 2000). But Lapita expansion and colonisation itself was not the result of any monolithic state-sponsored settlement program that was totally homogenous across time and space. When ceramic assemblages begin to show island-specific variation soon after Lapita settlement, as seen in Vanuatu, there are clearly other processes that influenced the form and composition of those sequences. Different islands may have been settled by clan groups that were only distantly related and over a short period rapidly changing population dynamics and environmental conditions may have encouraged further delineation. This might partly explain the diversification in ceramic styles which began to reflect efforts to produce and maintain geographically based social distinctions (Graves *et al.* 1990: 228). In other words, the ceramic remains tend to demonstrate the 'evolutionary primacy of local processes' (Hunt 1987:330) rather than inter-archipelago connections. In line with this argument is the mounting evidence that ceramic sequences, at least in Remote Oceania, began to follow increasingly independent trajectories soon after initial Lapita settlement and up until at least 1000 BP. From this date there is evidence, at least in the recovered non-ceramic artefacts (see below), for renewed inter-archipelago contacts.

The concept of a Melanesia-wide Incised and Applied Relief tradition was a useful concept during the infancy of archaeological research in the region. But as new evidence has revealed it is no longer supported by the empirical data. The term and its implications have begun to cloud the archaeological picture and has led to some rather simplistic explanations of cultural change. That is not to say that the study of post-Lapita pottery should be abandoned because inter-regional links might not be evidenced in the ceramic record. Rather, the focus now needs to be on tracking patterns of stylistic variation within each archipelago. Pacific archaeologists who deal with ceramics need to stop packaging the post-Lapita period as a devolved and reduced version of Lapita by making connections between island groups on the basis of individual ceramic traits. Also highlighted from this research is that post-Lapita assemblages that comprise a small number of small sherds must be treated cautiously, particularly those that are claimed to cover long time periods with little change.

These arguments do not preclude the possibility that future research may well reveal low levels of inter-archipelago contact in the post-Lapita period and considerably higher rates during the last 1000 or so years. A greater frequency of long-distance contact in the recent past is an emerging result in Vanuatu where some non-ceramic artefact forms point to increased long-distance interaction. These include the presence of Banks Island basaltic glass in Tikopia and Fiji, the widespread occurrence of artefacts such as *Terebra* shell and lenticular stone adze forms across Vanuatu (and much of the Southwest Pacific), New Caledonian serpentine on Tanna and Polynesian-style ornaments and burial practices in the centre and south of Vanuatu.

Future Directions

Great progress has been made in our understanding of the archaeology of the Southwest Pacific region over the last 50 years of investigation. But the state of knowledge varies greatly from region to region and between specific time periods. While the sustained focus on research into Lapita in the region has been essential and justified one of the more recent emerging revelations from that research is that in Remote Oceania, the highly identifiable dentate-stamped ceramic component of that Cultural Complex comprises no more than 10% of any of this region's prehistory. Archaeologists working in the Pacific often claim that one of their primary objectives is the further elucidation of the origins of the biological, cultural and linguistic variation that is found in the contemporary Pacific. If this is so then we must start focusing on what happened after Lapita and particularly on the archaeology of the last 1000 years. This period remains one of the least well known throughout the entire region but is likely to prove to be the most profitable in terms of detailing more recent episodes of human interaction which have substantially shaped the contemporary human landscape.

Ironically, in Vanuatu it is the post-Lapita period that has been a major of focus of research whilst the archaeology of the Lapita period remains poorly known. This is clearly a significant gap in current knowledge that requires a great deal of archaeological attention. Whilst the post-Lapita period in Vanuatu has been a major focus of research it has been somewhat unevenly concentrated.

Clarification of the archaeology of the central and southern islands is beginning to be achieved, although finer detail particularly of Mangaasi Phase (2200–1200 BP) of the ceramic sequence is still required. Also needed is more research into the period post-dating ceramics, i.e. the last 1200 years. It is as yet a poorly understood period, but as shown by the work of Garanger (1972), it can be seen as a time of substantial change to cultural forms markedly different from what had been seen before. The largest archaeological gap remains Northern Vanuatu and particularly the North Central region, the 'Oceanic Mediterranean' comprising Santo, Maewo, Pentecost and Ambae which has been virtually untouched archaeologically. The last 1000 years in this region has also witnessed significant cultural change. It is during this time that we can see differentiation occurring between the islands of the north and centre. It may also have been during this period that increased inter-regional contacts occurred.

Many of the same research questions and gaps in archaeological knowledge pertaining to Vanuatu are also seen in other areas of the Southwest Pacific. Archaeological research in the region needs to focus on the establishment of much more detailed regional sequences which will ultimately provide a platform for more informed theoretical debate.

What is abundantly clear is that the pieces of the archaeological puzzle that comprise Vanuatu's *long durée* are far greater in number and present a far more difficult task in their assembly than has appeared to be the case hitherto through the limited archaeological research that has been carried out to date. But this research, building on earlier research, can at least claim to have made some contribution to the on-going process of the construction of the myriad of potential historical plots (Kirch 2000:xix).

Bibliography

Adams, R. 1984. *In the Land of Strangers. A century of European contact with Tanna, 1774–1874.* Pacific Research Monograph Number Nine. Canberra: Australian National University.

Adams Y. W. and E. W. Adams 1991. *Archaeological Typology and Practical Reality.* Cambridge: Cambridge University Press.

Allen, J and C. Gosden (eds.) 1991. *Report of the Lapita Homeland Project.* Occasional Papers in Prehistory 20. Canberra, Department of Prehistory, Research School of Pacific Studies, Australian National University.

Allen, J and C. Gosden 1996. Spheres of Interaction and Integration: Modelling the Culture History of the Bismarck Archipelago. In J. Davidson, G. Irwin, F. Leach, A Pawley and D. Brown (eds.) *Oceanic Culture History. Essays in Honour of Roger Green,* pp. 183–97. Dunedin: New Zealand Journal of Archaeology Special Publication.

Ambrose, W. R. 1976. Obsidian and its Distribution in Melanesia. In Barnard, N. (ed.) *Ancient Chinese Bronzes and South East Asian Metal and Other Archaeological Artefacts,* pp. 351–78. Melbourne: National Gallery of Victoria.

Ambrose, W. R. 1988. An early bronze artefact from Papua New Guinea. *Antiquity,* 62:483–91.

Ambrose, W. R. 1991. Lapita or not Lapita: the case of the Manus pots. In J. Allen and C. Gosden (eds.) *Report of the Lapita Homeland Project.* Occasional Papers in Prehistory 20, pp. 103–12. Canberra: Department of Prehistory, Research School of Pacific Studies, Australian National University.

Ambrose, W. R. 1992. Clays and Sands in Melanesian pottery analysis. In J-C. Galipaud (ed.) *Poterie Lapita et Peuplement,* pp. 169–76. Noumea: ORSTOM.

Ambrose, W. R. 2002. Big pots on a small Lou Island. In Bedford, S.,C. Sand and D. Burley (eds). 2002. *Fifty Years in the Field: Essays in Honour and Celebration of Richard Shutler Jrs Archaeological Career,* pp.59–68. Auckland: New Zealand Archaeological Association. NZAA Monograph 25.

Ambrose, W. R. and H. P. McEldowney 2000. An age assessment for Lapita from obsidian at the Mouk Island Site, Manus. In A. Anderson and T. Murray (eds) *Australian Archaeologist. Collected Papers in honour of Jim Allen.* Centre for Archaeological Research and Department of Archaeology and Natural History, pp. 268–78. Canberra; Australian National University.

Anderson, A. 1997. Prehistoric Polynesian Impact on the New Zealand Environment: Te Whenua Hou. In P. Kirch and T. Hunt (eds.) *Historical Ecology in the Pacific Islands. Prehistoric Environmental and Landscape Change,* pp. 271–83. New Haven: Yale University Press.

Anderson, A., S. Bedford, G. Clark, I. Lilley, C. Sand, G. Summerhayes and R. Torrence 2001. An Inventory of Lapita Sites containing dentate-stamped pottery. In Clark, G., A. Anderson and T. Sorovi-Vunidilo (eds.) *The Archaeology of Lapita Dispersal in Oceania*: Papers from the Fourth Lapita Conference, June 2000, Canberra, Australia, pp. 1–14. Terra Australis 17. Canberra: Centre for Archaeological Research and Department of Archaeology and Natural History, RSPAS, ANU.

Anderson, A. and G. Clark 1999. The Age of Lapita Settlement in Fiji. *Archaeology in Oceania* 34:31–39.

Anson, D. 1983. Lapita Pottery of the Bismarck Archipelago and its Affinities. Unpublished Ph.D. thesis. University of Sydney, Australia.

Anson, D. 1990. Aspiring to Paradise. In M.J.T. Spriggs (ed.), *Lapita design, form and composition: proceedings of the Lapita design workshop, Canberra, Australia, December 1988*, pp. 53–58. Occasional Papers in Prehistory 18. Canberra: Department of Prehistory, Research School of Pacific and Asian Studies, Australian National University.

Arifin, K. 1991. Social aspects of pottery manufacture in Boera, Papua New Guinea. In P. S. Bellwood (ed.) *Indo-Pacific Prehistory 1990*, pp. 373–87. Canberra and Jakarta: IPPA and Asosiasi Prehistorisi Indonesia. Bulletin of the Indo-Pacific Prehistory Association 11.

Arnold, D. 1985. *Ceramic Theory and Cultural Process*. Cambridge: Cambridge University Press.

Aubert de la Rüe, E. 1938. Sur la nature et l'origine probable des pierres portées en pendentifs à l'îles Tanna (Nouvelles Hébrides). *L'Anthropologie* 48:249–60.

Aubert de la Rüe, E. 1945. *Les Nouvelles Hébrides. Iles de cendre et de corail*. Montreal: L'Arbre.

Avias, J. 1950. Poteries Canaques et poteries préhistoriques en Nouvelle Calédonie. *Journal de la Société des Océanistes* 6:111–40.

Balouet, J. 1991. The fossil vertebrate record of New Caledonia, pp. 1381–1409. In P. Vickers-Rich, Monaghan, J., Baird, R. and T. Rich (eds.) *Vertebrate palaeontology of Australasia*. Melbourne: Monash University.

Beaglehole, J. C. (ed.) 1969. *The Journals of Captain James Cook on his Voyages of Discovery, Edited from the Original Manuscripts. Volume II: The Voyage of the Resolution and Adventure 1772–1775*. Cambridge: Cambridge University Press for the Hakluyt Society.

Bedford, S. 1999. Lapita and Post–Lapita Ceramic Sequences from Erromango, Southern Vanuatu. In J-C Galipaud and I. Lilley (eds.), *The Western Pacific from 5000 to 2000 BP. Colonisation and transformations*, pp. 127–38. Paris: IRD Éditions

Bedford, S. 2000a. Results from excavations at the Mangaasi type site: a re-assessment of the ceramic sequence and its implications for Melanesian prehistory. *Bulletin of the Indo-Pacific Prehistory Association*, 19(3):159–66. Canberra.

Bedford, S. 2000b. Pieces of the Vanuatu Puzzle: archaeology of the North, South and Centre. Two Volumes. PhD dissertation, Australian National University, Canberra.

Bedford, S. 2001. Ceramics from Malekula, Northern Vanuatu: The two ends of a potential 3000 year sequence. In Clark, G., A. Anderson and T. Sorovi-Vunidilo (eds.) *The Archaeology of Lapita Dispersal in Oceania*: Papers from the Fourth Lapita Conference, June 2000, Canberra, Australia, pp. 105–14. Terra Australis 17. Canberra: Centre for Archaeological Research and Department of Archaeology and Natural History, RSPAS, ANU.

Bedford, S. 2003. The timing and nature of Lapita colonisation in Vanuatu: the haze begins to clear. In C. Sand (ed), *Pacific Archaeology: assessments and prospects*, pp. 147–58. Noumea: Les Cahiers de l'archeologie en Nouvelle-Calédonie 15.

Bedford, S. and G. Clark, 2001. The Rise and Rise of the Incised and Applied Relief Tradition: a review and reassessment. In Clark, G., A. Anderson and T. Sorovi-Vunidilo (eds.) *The Archaeology of Lapita Dispersal in Oceania*, pp. 61–74. Terra Australis 17. Canberra: Centre for Archaeological Research and Department of Archaeology and Natural History, RSPAS, ANU.

Bedford, S., A. Hoffman, M. Kaltal, R. Regenvanu and R. Shing. 2004. Dentate-stamped Lapita reappears on Efate, Central Vanuatu: a four decade-long drought is broken. *Archaeology in New Zealand*, 47(1):39–49.

Bedford, S., M. Spriggs, M. Wilson and R. Regenvanu, 1998. The Australian National University-National Museum of Vanuatu Archaeological Project 1994–7: A Preliminary Report on the Establishment of Cultural Sequences and Rock Art Research. *Asian Perspectives* 37(2):165–93.

Bedford, S., M. Spriggs and R. Regenvanu 1999. The Australian National University-Vanuatu Cultural Centre Archaeology Project 1994–7: Aims and Results. *Oceania* 70(1):16–24.

Bedford, S. and M. Spriggs 2000. Crossing the Pwanmwou: Preliminary Report on recent excavations adjacent to and south west of Mangaasi, Efate Vanuatu. *Archaeology in Oceania* 35:120–26.

Bedford, S and M. Spriggs 2002. Of Shell, Stone and Bone. A review of non-ceramic artefacts recovered from the first 1000 years of Vanuatu's archaeological record. In Bedford, S.,C. Sand and D. Burley (eds). 2002. *Fifty Years in the Field: Essays in Honour and Celebration of Richard Shutler Jrs Archaeological Career*, pp.135–52. Auckland: New Zealand Archaeological Association. NZAA Monograph 25.

Bellwood, P. 1978. *Man's Conquest of the Pacific. The Prehistory of Southeast Asia and Oceania*. Auckland: Collins Publishers.

Bellwood, P. 1979. The Oceanic Context. In J.D. Jennings (ed.) *The Prehistory of Polynesia*, pp. 6–26. Canberra: Australian National University Press.

Bellwood, P. and P. Koon 1989. 'Lapita Colonists leave Boats Unburned!' The Question of Lapita Links with Island Southeast Asia. *Antiquity*, 63:613–32.

Best, S. 1981. Excavations at Site VL 21/5 Naigani Island, Fiji, a preliminary report. Department of Anthropology, University of Auckland.

Best, S. 1984. Lakeba: the prehistory of a Fijian Island. Unpublished PhD thesis. Department of Anthropology, University of Auckland.

Best, S., 2002. *Lapita: A View from the East.* New Zealand Archaeological Association Monograph No. 24.

Birks, L. 1973. *Archaeological Excavations at Sigatoka Dune Site, Fiji.* Bulletin of the Fiji Museum, No.1. Fiji: Suva.

Bonnemaison, J. 1996. Graded societies and societies based on title: forms and rites of traditional political power in Vanuatu. In J. Bonnemaison, K. Huffman, C. Kaufmann, and D. Tryon (eds) *Arts of Vanuatu*, pp. 200–03. Bathurst: Crawford House Press.

Braudel, F. 1980. *On History.* Chicago: University of Chicago Press.

Bregulla, H. 1992. *The Birds of Vanuatu.* Shropshire: Anthony Nelson.

Bulmer, S. 1999. Revisiting Red Slip: the Laloki style pottery of Southern Papua and its possible relationship to Lapita. In J-C Galipaud and I. Lilley (eds) *The Western Pacific from 5000 to 2000 BP. Colonisation and transformations*, pp. 543–77. Paris: IRD Éditions

Burley, D. 1998. Tongan archaeology and the Tongan Past, 2850–150 BP. *Journal of World Prehistory* 12(3):337–92.

Burley, D., D. E. Nelson and R. Shutler Jr. 1999. A radiocarbon chronology for the Eastern Lapita frontier in Tonga. *Archaeology in Oceania* 34:59–70.

Cayrol, F. 1992. La Ceramique en Melanesie du Sud, Fonction ou Statut? Le cas des Naamboi de Malekula, une approche ethno-archaeologique. Thèse de Doctorat, Université de Paris I. 3 vols.

Cernohorsky, W.O. 1971. *Marine Shells of the Pacific.* Volume 1. Sydney: Pacific Publications.

Cernohorsky, W.O. 1972. *Marine Shells of the Pacific.* Volume 2. Sydney: Pacific Publications.

Charnov, E.L. 1976. Optimal foraging, the marginal value theorem. *Theoretical Population Biology* 9:129–36.

Clark, J. and J. Terrell 1978. Archaeology in Oceania. *Annual Review of Anthropology* 7:293–319.

Clark, G. 1999. Post-Lapita Fiji: Cultural transformation in the mid-sequence. Unpublished PhD thesis. Australian National University, Canberra.

Clark, G. 2003. Shards of Meaning. Archaeology and the Melanesia-Polynesia Divide. *The Journal of Pacific History*, 38(2):197–215.

Clark, G., A. Anderson and T. Sorovi-Vunidilo (eds) 2001. *The Archaeology of Lapita Dispersal in Oceania*: Papers from the Fourth Lapita Conference, June 2000, Canberra, Australia. Terra Australis 17. Canberra: Centre for Archaeological Research and Department of Archaeology and Natural History, RSPAS, ANU.

Coleman, N. 1981. *What Shell is That?* Sydney: Lansdowne Press.

Colley, H and R. Ash 1971. *The Geology of Erromango.* Port Vila: British Service.

Corris, P. 1973. *Passage, Port and Plantation: a History of Solomons Islands Labour Migration, 1870-1914.* Melbourne: Melbourne University Press.

Crosby, E. 1973. A Comparative Study of Melanesian Hafted Edge-Tools and Other Percussive Cutting Implements. Unpublished PhD thesis. Australian National University, Canberra.

Deacon, B. 1934. *Malekula: A Vanishing People in the New Hebrides.* London: Routledge and Sons.

Denham, T. and C. Ballard (eds) 2003. Perspectives on prehistoric agriculture in the New Guinea Highlands. A tribute to Jack Golson. *Archaeology in Oceania* 38 (3).

Dickinson, W. R. 1997. Petrographic Report WRD 137. Sand tempers in Sherds from the Banks Islands, Northern Vanuatu. Unpublished report.

Dickinson, W. R. 1998. Petrographic Temper Provinces of Prehistoric Pottery in Oceania. *Records of the Australian Museum* 50:263–76.

Dickinson, W. R. 2000. Petrography and Geologic Provenance of Sand Tempers in Prehistoric Potsherds from Fiji and Vanuatu, South Pacific. Unpublished paper.

Dickinson, W. R. pers. comm. Emeritus Professor, Department of Geosciences, University of Arizona, Tucson, Arizona, USA

Dickinson, W. R. and R. Shutler 1979. Petrography of Sand Tempers in Pacific Island Potsherds. *Geological Society of America Bulletin* 90 (Part II):1644–1701.

Dickinson, W. R., Sinoto, Y., R. Shutler, M. E. Shutler, J. Garanger and T. M. Teska 1999. Japanese Jomon sherds in artefacts collections from Mele Plain on Efate in Vanuatu. *Archaeology in Oceania* 34:15–24.

Donovan, L. J. 1973. A Study of the Decorative System of the Lapita Potters in Reefs and Santa Cruz Islands. Unpublished M.A. Research Essay, University of Auckland.

Douceré, Mgr. V. 1922. *Les Populations Indigènes des Nouvelles-Hébrides.* Paris: Émile Larose.

Dubois, M-J. 1996. Vanuatu seen from Maré. In J. Bonnemaison, K. Huffman, C. Kaufmann, and D. Tryon (eds.) *Arts of Vanuatu*, pp. 79–82. Bathurst: Crawford House Press.

Etheridge, R. 1917. Additions to the ethnological collections, chiefly from the New Hebrides. *Records of the Australian Museum* 11(8):189–203.

Eissen, J-P., M. Monzier and C. Robin 1994. Kuwae, l'Eruption Volcanique Oubliée. *La Recherche* 270:1200-1202.

Felgate, M. 2003. Reading Lapita in Near Oceania: Intertidal and shallow-water pottery scatters, Roviana Lagoon, New Georgia, Solomon Islands. Unpublished PhD thesis. University of Auckland, Auckland.

Frimigacci, D., 1974. Les Deux Niveaux à Poterie du Site de Vatcha (Ile des Pines). *Journal de la Société des Océanistes* 30(42–3):25–70.

Frimigacci, D. 1975. *La préhistoire néo-calédonienne*. Thèse de Doctorat, Université Paris 1.

Frimigacci, D. 1980. Localisation Eco-Geographique et Utilisation de l'Espace de Quelques Sites Lapita de Nouvelle Calédonie. *Journal de la Société des Océanistes* 37(66–7):5–11.

Frimigacci, D. 1981. La poterie imprimee au battoir en Nouvelle-Calédonie. *Journal de la Société des Océanistes* 37(70–71):111–18.

Frimigacci, D. et J. P. Maitre 1981. Archéologie et préhistoire-Planche 16. In Sautter, G. (coordination generale), Atlas de la Nouvelle-Calédonie et Dépendances. Paris: Editions de l'ORSTOM.

Frost, E. L. 1979. Fiji. In J.D. Jennings (ed.) *The Prehistory of Polynesia*, pp. 61–81. Cambridge, Mass.: Harvard University Press.

Galipaud, J-C. 1990. The physico-chemical analysis of ancient pottery from New Caledonia. In M.J.T Spriggs (ed) *Lapita design, form and composition: proceedings of the Lapita design workshop, Canberra, Australia, December 1988.* Canberra: Australian National University.

Galipaud, J-C. 1996a. Le rouge et le noir: la poterie Mangaasi et le peuplement des îles de Mélanésie. M. Julien, M. et C. Orliac et B. Gerard, A. Lavondès, H. Lavondès et C. Robineau (eds.) *Mémoire de Pierre, Mémoire d'Homme, Tradition et archéologie en Océanie*, pp. 115–30. Paris: Publication de la Sorbonne.

Galipaud, J-C. 1996b. Pottery and Potters of Vanuatu. In J. Bonnemaison, K. Huffman, C. Kaufmann, and D. Tryon (eds.) *Arts of Vanuatu*, pp. 94–9. Bathurst: Crawford House Press.

Galipaud, J-C. 1996c. Premiers résultats de la mission archéologique Wusi-Hokua. Field Report No. 4. Port Vila, ORSTOM.

Galipaud, J-C. 1997. A Revision of the Archaeological Sequence of Southern New Caledonia. *New Zealand Journal of Archaeology* 17(1995): 77–109

Galipaud, J-C. 1998a. The Lapita Site of Atanoasao Malo, Vanuatu. Field Report No. 8. Port Vila, ORSTOM.

Galipaud, J-C. 1998b. Recherches archéologiques aux îles Torres. *Journal de la Société des Océanistes* 107 (2):159–68

Galipaud, J-C. pers. comm. Archaeologist, IRD, Noumea, New Caledonia.

Garanger, J. 1966. Recherches Archéologiques aux Nouvelles-Hébrides. *L'Homme* 6:59–81.

Garanger, J. 1971. Incised and applied relief pottery, its chronology and development in southeastern Melanesia, and the extra areal comparisons. In R. Green and M. Kelly (eds.) *Studies in Oceanic Culture History*, Volume 2, pp.53–66. Honolulu: Pacific Anthropological Records Number 12.

Garanger, J. 1972. *Archéologie des Nouvelles-Hébrides: contribution à la connaissance des îles du centre*. Publications de la Société des Océanistes, No.30. Paris: ORSTOM.

Garanger, J. 1982. *Archaeology of the New Hebrides. Contribution to the Knowledge of the Central Islands*. Translated by Rosemary Groube. Oceania Monograph 24.

Garanger, J. 1996. Tongoa, Mangaasi and Retoka: History of a Prehistory, In J. Bonnemaison, K. Huffman, C. Kaufmann, and D. Tryon (eds.) *Arts of Vanuatu*, pp. 66–73. Bathurst: Crawford House Press.

Gifford, E.W. 1951. *Archaeological Excavations in Fiji*. Anthropological Records 13 (3). Berkeley and Los Angeles: University of California Press.

Gifford, E.W. and R. J. Shutler Jr. 1956. *Archaeological Excavations in New Caledonia*. Anthropological Records 18 (1). Berkeley and Los Angeles: University of California Press.

Glaumont, M. 1899. *Voyage d'Exploration aux Nouvelles-Hebrides*. Niort: Lemercier et Alliot.

Golson, J. 1959. L'Archéologie du Pacifique Sud. Resultats et Perspectives. *Journal de la Société des Océanistes* 15:5-54.

Golson, J. 1961. Report on New Zealand, Western Polynesia, New Caledonia and Fiji. *Asian Perspectives* 5(2):166–80.

Golson, J. 1968. Archaeological Prospects for Melanesia. In I. Yawata and Y.H. Sinoto (eds.) *Prehistoric Culture in Oceania*, A Symposium, pp. 3–14. Honolulu: Bishop Museum Press.

Golson, J. 1971. Lapita Ware and Its Transformations. In R. Green and M. Kelly (eds.) *Studies in Oceanic Culture History*, Volume 2, pp.67–76. Pacific Anthropological Records Number 12. Honolulu: Bishop Museum.

Golson, J. 1972. Both Sides of the Wallace Line: New Guinea, Australia, Island Melanesia and Asian Prehistory. In N. Barnard (ed.) *Early Chinese Art and Its Possible Influence in the Pacific Basin*, Vol. 3, pp. 533–95. New York: Intercultural Arts Press.

Golson, J. 1991. Two sites at Lasigi, New Ireland. In J. Allen and C. Gosden (eds.) *Report of the Lapita Homeland Project*, pp. 244–59. Occasional Papers in Prehistory 20. Canberra: Department of Prehistory, Research School of Pacific Studies, Australian National University.

Golson, J. 1992. The ceramic sequence from Lasigi, New Ireland. In J-C Galipaud (ed.) *Poterie Lapita et Peuplement*. Actes du Colloque Lapita, Nouméa, Janvier 1992, pp. 155–67. Nouméa: ORSTOM

Gorecki, P. 1992. A Lapita Smoke Screen? In J-C Galipaud (ed.) *Poterie Lapita et Peuplement*. Actes du Colloque Lapita, Nouméa, Janvier 1992, pp. 27–47. Nouméa: ORSTOM.

Gorecki, P. 1996. The Initial Colonisation of Vanuatu. In J. Bonnemaison, K. Huffman, C. Kaufmann, and D. Tryon (eds.) *Arts of Vanuatu*, pp. 62–5. Bathurst: Crawford House Press.

Gorecki, P., M. Mabin and J. Campbell 1991. Archaeology and geomorphology of the Vanimo coast, Papua New Guinea: preliminary results. *Archaeology in Oceania* 26 (3):119–22.

Gosden, C. 1991. Long term trends in the colonisation of the Pacific: putting Lapita in its place. In P.S. Bellwood (ed.) *Indo-Pacific Prehistory 1990*, Vol. 2, pp. 333–38. Canberra and Jakarta: IPPA and Asosiasi Prehistorisi Indonesia. Bulletin of the Indo-Pacific Prehistory Association 11.

Gosden, C., J. Allen, W. Ambrose, D. Anson, J. Golson, R. Green, P. Kirch, I. Lilley, J. Specht and M. Spriggs 1989. Lapita sites in the Bismarck Archipelago. *Antiquity* 63:561–86.

Graves, M., T. Hunt and D. Moore 1990. Ceramic production in the Mariana Islands: Explaining Change and Diversity in Prehistoric Interaction and Exchange. *Asian Perspectives* 29(2):227–233.

Green, R. C. 1961. Pacific Commentary. The Tenth Pacific Science Congress, Archaeology. *Journal of the Polynesian Society* 70: 477–81.

Green, R. C. 1963. A suggested revision of the Fiji sequence. *Journal of the Polynesian Society* 72:235–53.

Green, R. C. 1973. Southeast Solomons Fieldwork. *Asian Perspectives* (for 1972) 15(2):197–99.

Green, R. 1978. New sites with Lapita pottery and their implications for an understanding of the settlement of the Western Pacific. *Working Papers in Anthropology, Archaeology and Maori Studies, No. 51*. Auckland: Department of Anthropology, University of Auckland.

Green, R. C. 1979. Lapita. In J.D. Jennings (ed.) *The Prehistory of Polynesia*, pp. 27–60. Cambridge, Mass.: Harvard University Press.

Green, R. C. 1982. Models for the Lapita cultural complex: an evaluation of some current proposals. *New Zealand Journal of Archaeology* 4:7–20.

Green, R. C. 1985. Comment: Sprigg's 'The Lapita Cultural Complex'. *The Journal of Pacific History*: 220–24.

Green, R. C. 1986. Lapita fishing: The evidence from Site SE-RF-2 of the Main Reef Islands, Santa Cruz Group, Solomons. In A. Anderson (ed.) *Traditional Fishing in the Pacific*. Pacific Anthropological Records 37:119–35. Honolulu: Department of Anthropology, Bishop Museum.

Green, R. C. 1988. Those Mysterious Mounds Are for the Birds. *Archaeology in New Zealand* 31:153–58.

Green, R. C. 1990. Lapita design analysis: the Mead System and its use; a potted history. In M.J.T. Spriggs (ed.) *Lapita design, form and composition: proceedings of the Lapita design workshop, Canberra, Australia, December 1988*, pp.33–52. Occasional Papers in Prehistory 18. Canberra: Department of Prehistory, Research School of Pacific and Asian Studies, Australian National University.

Green, R. C. 1991a. Near and Remote Oceania- disestablishing 'Melanesia' in culture history. In A. Pawley (ed.) *Man and a Half: Essays in Honour of Ralph Bulmer*, pp. 491–502. Auckland: Polynesian Society.

Green, R. C. 1991b. The Lapita Cultural Complex: Current Evidence and Proposed Models. *Bulletin of the Indo-Pacific Prehistory Association* 11(2):295–305. Canberra.

Green, R. C. 1994. Changes over Time. Recent Advances in dating Human Colonisation of the Pacific Basin Area. In D.G. Sutton (ed.) *The Origins of the First New Zealanders*, pp. 19–51. Auckland: Auckland University Press.

Green, R. C. 1997. Linguistic, Biological and Cultural Origins of the Initial Inhabitants of Remote Oceania. *New Zealand Journal of Archaeology* 17 (1995):5–27.

Green, R. C. 1999. Evidence suggesting inputs of non-Austronesian culture and genes into Vanuatu in the last 1000 years. Paper presented at the New Zealand Archaeological Association Conference, Auckland 1999.

Green, R.C. 2003. The Lapita horizon and traditions – Signature for one set of oceanic migrations. In C. Sand (ed), Pacific Archaeology: assessments and prospects (Proceedings of the Conference for the 50th anniversary of the first Lapita excavation. Kone-Noumea 2002), pp. 95–120. Noumea: Les Cahiers de l'archeologie en Nouvelle-Calédonie 15.

Green, R. C. and D. Anson 1991. The Reber-Rakival Lapita site on Watom. Implications of the 1985 excavations at SAC and SDI localities. In J. Allen and C. Gosden (eds.) *Report of the Lapita Homeland Project*, pp. 170–81. Occasional Papers in Prehistory 20. Canberra: Department of Prehistory, Research School of Pacific Studies, Australian National University.

Green, R. C. and J. Davidson 1974. *Archaeology in Western Samoa, Vol. II*. Bulletin of the Auckland Institute and Museum 7. Auckland: Auckland Institute and Museum.

Green, R. C. and P. V. Kirch 1997. Lapita exchange systems and their Polynesian transformations: seeking explanatory models. In M. Weisler (ed.) *Prehistoric Long-Distance Interaction in Oceania: An Interdisciplinary Approach*, pp. 19–37. Auckland: New Zealand Archaeological Association Monograph 21.

Green, R. C. and J. S. Mitchell 1983. New Caledonian culture history: a review of the archaeological sequence. *New Zealand Journal of Archaeology* 5:19–68.

Groube, L. 1972. Unpublished Erromango and Aneityum fieldnotes, held at Division of Archaeology and Natural History, RSPAS, Australian National University, Canberra.

Groube, L . 1975. Archaeological Research in Aneityum. *South Pacific Bulletin* 25 (3):27–30.

Guiart, J. 1956. Unité culturelle et variations locales dans le centre nord des Nouvelles Hébrides. *Journal de la Société des Océanistes* 12:217–25.

Guiart, J. 1958. *Espiritu Santo, Nouvelles Hébrides*. L'Homme, cahiers d'ethnologie, de geographie et de linguistique. Nouvelle Series No.2. Paris: Plon.

Guiart, J. 1963. *Structure de la chefferie en Mélanesie du sud*. Paris: Institut d'ethnologie.

Guiart, J. 1973. 'Le Dossier Rassemble'. In J. Espirat, Guiart, J., Lagrange, M-S et M. Renaud (eds.) *Systèmes des titres dans les Nouvelles Hébrides Centrales, d'Éfate aux îles Shepherd*, pp. 47–370. Mémoires de l'Institut d'Ethnologie 10. Paris: Museum National d'Histoire Naturelle.

Guiart, J. 1983. *La Terre est le Sang des Morts: la Confrontation entre Blancs et Noirs dans le Pacifique Sud Français*. Paris: Editions Anthropos.

Gyngell, J. 1997. Shellfish Exploitation at the Ifo and Ponamla Sites: Erromango Island, Vanuatu. Unpublished BA Hons. dissertation, Department of Archaeology and Anthropology, Australian National University.

Hagelburg, E. 2001. Genetic affinities of the principal human lineages in the Pacific. In Clark, G., A. Anderson and T. Sorovi-Vunidilo (eds.) *The Archaeology of Lapita Dispersal in Oceania*: Papers from the Fourth Lapita Conference, June 2000, Canberra, Australia, pp. 167–176. Terra Australis 17. Canberra: Centre for Archaeological Research and Department of Archaeology and Natural History, RSPAS, ANU.

Hamilton, W. and K. Watt. 1970. Refuging. *Annual Review of Ecology and Systematics* 1:263–286.

Harrisson, T. 1937. *Savage Civilisation*. London: Victor Gollancz.

Hébert, B. 1965. Nouvelles Hébrides. Contribution à l'Étude Archeologique de l'Île Éfaté et des Îles Avoisantes. *Études Mélanésiennes* 18–20:71–98.

Hedrick, J. 1971. Lapita Style Pottery from Malo Island. *Journal of the Polynesian Society* 80 (1):5–19.

Hedrick, J. nd. Archaeological Investigations of Malo Prehistory. Lapita Settlement Strategy in the Northern New Hebrides. Manuscript draft of PhD dissertation, University of Pennsylvania.

Hedrick, J. and K. G. Hedrick 1975. An Expedition to the New Hebrides. *Expedition* 17 (3):2–15.

Hedrick, J. and M. E. Shutler 1969. Report on 'Lapita Style' Pottery From Malo Island, Northern New Hebrides. *Journal of the Polynesian Society* 78 (2):262–65.

Hoffman, A. 2003. A reanalysis of Mary and Richard Shutler's pioneering archaeological fieldwork in southern and central Vanuatu, 1963/64. BA Hons. Dissertation. Department of Archaeology and Anthropology, Australian National University.

Howe, K. 1984. *Where the Waves Fall*. Honolulu: University of Hawaii Press.

Hope, G. and M. Spriggs 1982. A Preliminary Pollen Sequence from Aneityum Island, Southern Vanuatu. *Bulletin of the Indo-Pacific Prehistory Association* 3:88–94.

Horton, D. and G. Ward 1981. The Role of Pigs in the Economy of Pakea Islet, Banks Islands, Southwest Pacific. *New Zealand Journal of Archaeology* 3:3–13.

Huffman, K. 1996. Trading, cultural exchange and copyright: important aspects of Vanuatu arts. In J. Bonnemaison, K. Huffman, C. Kaufmann, and D. Tryon (eds.) *Arts of Vanuatu*, pp.182–194. Bathurst: Crawford House Press.

Humphreys, C. B. 1926. *The Southern New Hebrides: an ethnological record*. Cambridge: Cambridge University Press.

Hunt, T. 1987. Patterns of Human Interaction and Evolutionary Divergence in the Fiji Islands. *Journal of the Polynesian Society* 96:299–334.

Hunt, T. 1988. Lapita Ceramic Technological and Compositional Studies: A Critical Review. In P. Kirch and T. Hunt (eds.) *Archaeology of the Lapita Cultural Complex: A Critical Review*, pp. 49–60. Thomas Burke Memorial Washington State Museum Research Report No.5. Seattle: The Burke Museum.

Hunt, T. 1993. Ceramic provenance studies in Oceania: Methodological issues. In B. Fankhauser and Bird, J.R. (eds.) *Archaeometry: Current Australasian Research*, pp. 51–56. Canberra: Department of Prehistory, Research School of Pacific and Asian Studies, Australian National University. Occasional Papers in Prehistory 22.

Irwin, G. 1985. *The Emergence of Mailu*. Terra Australis 10. Canberra: Department of Prehistory, Research School of Pacific and Asian Studies, Australian National University.

Irwin, G. 1992. *The Prehistoric Exploration and Colonisation of the Pacific*. Cambridge: Cambridge University Press.

Janetski, C. 1976. Artifacts of shell, bone, coral and sea urchin spines. In J.D Jennings, R.N. Holmer, J.C. Janetski, H.L. Smith (eds.) *Excavations on Upolu, Western Samoa*. Pacific Anthropological Records 25, pp. 72–3. Hawaii: Bishop Museum.

Kaeppler, A.L. 1973. Pottery sherds from Tungua, Ha'apai; and remarks on pottery and social structure in Tonga. *Journal of the Polynesian Society* 82:218–222.

Kelly, C. (ed.) 1966. *La Austrialia del Espiritu Santo. The Journal of Fray Martin de Munilla O.F.M. and other Documents Relating to the Voyage of Pedro Fernandez de Quiros to the South Sea (1605–1606) and the Franciscan Missionary Plan (1617–1627).* 2 volumes. London: Hakluyt Society.

Kennedy, J. 1981. Lapita Colonisation of the Admiralty Islands? *Science* 213:757–759.

Kennedy, J. 1982. Archaeology in the Admiralty Islands: Some Excursions. *Bulletin of the Indo-Pacific Prehistory Association* 3:22–35.

Kirch, P.V. 1975. Excavations at sites A1–3 and A1–4: early settlement and ecology in Halawa Valley. In P. V. Kirch and M. Kelly (eds.) *Prehistory and Ecology in a windward Hawaiian valley: Halawa Valley, Molokai,* pp. 17–70. Pacific Anthropological Records No. 24. Honolulu: Bishop Museum.

Kirch, P.V. 1983. An archaeological exploration of Vanikoro, Santa Cruz Islands, Eastern Melanesia. *New Zealand Journal of Archaeology* 5:69–114.

Kirch, P.V. 1984. *The Evolution of Polynesian Chiefdoms.* Cambridge: Cambridge University Press.

Kirch, P.V. 1988a. A Brief History of Lapita Archaeology. In P.V. Kirch and T. Hunt, (eds.) 1988. *Archaeology of the Lapita Cultural Complex: A Critical Review,* pp.1–8. Thomas Burke Memorial Washington State Museum Research Report No.5. Seattle: The Burke Museum.

Kirch, P.V. 1988b. *Niuatoputapu. The Prehistory of a Polynesian Chiefdom.* Thomas Burke Memorial Washington State Museum Monograph No.5. Seattle: Burke Museum.

Kirch, P.V. 1998c. Long-distance Exchange and Island Colonisation: The Lapita Case. *Norwegian Archaeological Review* 21(2):103–17.

Kirch, P.V. 1997. *The Lapita Peoples. Ancestors of the Oceanic World.* Oxford: Blackwell.

Kirch, P.V. 2000. *On the Road of the Winds. An Archaeological History of the Pacific Islands before European Contact.* Berkeley: University of California Press.

Kirch, P.V. and T. Hunt 1988. The Spatial and Temporal Boundaries of Lapita. In P.V. Kirch and T. Hunt (eds.) *Archaeology of the Lapita Cultural Complex: A Critical Review,* pp. 9–32. Thomas Burke Memorial Washington State Museum Research Report No.5. Seattle: The Burke Museum.

Kirch, P.V., T.L. Hunt, M. Weisler, V. Butler and M.S. Allen 1991. Mussau Islands prehistory: results of the 1985–86 excavations. In J. Allen and C. Gosden (eds.) *Report of the Lapita Homeland Project,* pp. 144–163. Occasional Papers in Prehistory 20. Canberra: Department of Prehistory, Research School of Pacific Studies, Australian National University.

Kirch, P.V. and P. H. Rosendahl 1973. Archaeological investigations of Anuta. In D. E. Yen and J. Gorden (eds.) *Anuta: a Polynesian Outlier in the Solomon Islands,* pp. 25–108. Pacific Anthropological Records 21. Honolulu: Bishop Museum.

Kirch, P.V. and D. E. Yen 1982. *Tikopia: The Prehistory and Ecology of a Polynesian Outlier.* Bernice P. Bishop Museum Bulletin 238. Honolulu: Bishop Museum Press.

Lawrie, J. H. 1892 The New Hebrideans. *Scottish Geographical Magazine* 8:302–11.

Layard, J. W. 1928. Degree-taking rites in South West Bay, Malekula. *Journal of the Royal Anthropological Institute* 58: 139–223

Layard, J. W. 1942. *The Stone Men of Malekula: the small island of Vao.* London: Chatto and Windus.

Leach, F., J. Davidson and K. Fraser 1998. Analysis of fishbone from archaeological sites in Vanuatu. Museum of New Zealand Te Papa Tongarewa, Unpublished Technical Report 32.

Leach, H. 1982. Cooking without pots: aspects of prehistoric and traditional Polynesian cooking. *New Zealand Journal of Archaeology* 4:149–56.

Leaney, C. 1965. A Preliminary Archaeological Survey of Malekula Island. Unpublished typescript.

Lebot, V., M. Merlin and L. Lindstrom 1997. *Kava. The Pacific Elixir.* Rochester: Healing Arts Press.

Le Moine, G. 1987. The loss of pottery in Polynesia. *New Zealand Journal of Archaeology* 9:25–32.

Lenormond, M. 1948. Découverte d'un gisement de poteries indigènes à l'île des Pins. *Études Mélanésiennes* 3:54–58.

Lyman, R., M. O'Brien and R. Dunnell 1997. *The Rise and Fall of Culture History.* New York: Plenum Press.

McArthur, N. 1967. Island Populations of the Pacific. Canberra: Australian National University Press.

McArthur, N. 1974. Population and Prehistory: Aneityum. Unpublished Ph.D. thesis, Australian National University, Canberra.

MaCall, G., R. LeMaitre, A. Malahoff, G. Robinson and P. Stephenson 1970. The Geology and Geophysics of the Ambrym Caldera, New Hebrides. *Bulletin of Volcanology* 34:681–96.

McClancy, J. 1981. *To Kill a Bird with Two Stones: a Short History of Vanuatu.* Port Vila: Vanuatu Cultural Centre, Publication 1.

McCoy, P. and P. Cleghorn 1988. Archaeological Excavations on Santa Cruz (Nendo), Southeast Solomon Islands: Summary Report. *Archaeology in Oceania* 23(3):104–14.

MacLachlan, R. C. 1939. Native Pottery of the New Hebrides. *Journal of the Polynesian Society* 48: 32–55.

Markham, S. C. (ed.) 1904. *The Voyages of Pedro Fernandez de Quiros, 1595–1606.* 2 volumes. London: Hakluyt Society.

Marshall, Y. 1985. Who made the Lapita pots? A case study in gender archaeology. *Journal of the Polynesian Society* 94:205–33.

Martin, P and D. Steadman 1999. Prehistoric Extinctions on Islands and Continents. In R. MacPhee (ed.) *Extinctions in Near Time*, pp. 17–55. New York: Kluwer Academic/Plenum Publishers.

May, P and M. Tuckson 2000. *The Traditional Pottery of Papua New Guinea*. Adelaide: Crawford House.

Mead, J, D. Steadman, S Bedford, C. Bell and M. Spriggs 2002. New Extinct Mekosuchine Crocodile from Vanuatu, South Pacific. *Copeia* 2002(3):632–41.

Mead, S. M., L. Birks, H. Birks, and E Shaw. 1975. *The Lapita Pottery Style of Fiji and its Associations*. Polynesian Society Memoir No. 38. Wellington.

Meehan, B. 1982. *Shell Bed to Shell Midden*. Canberra, Australian Institute of Aboriginal Studies.

Michelsen, O. 1893. *Cannibals Won for Christ: a story of missionary perils and triumphs in Tongoa, New Hebrides*. London: Morgan and Scott.

Munsell Soil Color Charts 1975. Maryland: Baltimore.

Nagaoka, L. 2000. Resource Depression, Extinction, and Subsistence Change in Prehistoric Southern New Zealand. Unpublished Phd dissertation. Dept of Anthropology, University of Washington.

Noury, A. 1998. La poterie Lapita au Vanuatu: étude des décors. Maîtrise de préhistoire océanienne. Université PARIS-1 Panthéon-Sorbonne.

Palmer, B. 1971. Fijian Pottery Technologies: Their Relevance to Certain Problems of Southwest Pacific Prehistory, pp. 77–103. In R. Green and M. Kelly (eds.) *Studies in Oceanic Culture History Volume 2*. Pacific Anthropological Records Number 12. Honolulu: Bishop Museum.

Pineda, R. et J-C. Galipaud 1998. Évidences archéologiques d'une surrection différentielle de l'île de Malo (archipel du Vanuatu) au cours de l'Holocène récent. *C. R. Acad. Sci. Paris, Sciences de la terre et des planètes* 327:777–79.

Poulsen, J. 1967. A Contribution to the Prehistory of the Tongan Islands. Unpublished PhD dissertation. Australian National University, Canberra.

Poulsen, J. 1987. *Early Tongan Prehistory*. 2 Vols. Terra Australis 12. Canberra: Department of Prehistory, Research School of Pacific and Asian Studies, Australian National University.

Rechtman, R. B. 1992. The Evolution of Sociopolitical Complexity in the Fiji Islands. Unpublished PhD dissertation, University of California (Los Angeles).

Reeve, R. 1989. Recent work on the prehistory of the Western Solomons, Melanesia. *Bulletin of the Indo-Pacific Prehistory Association* 9:46–67.

Rice, P. M. 1987. *Pottery Analysis: A Sourcebook*. Chicago: University of Chicago Press.

Robin, C., M. Monzier and J-P. Eissen 1993. Giant Tuff Cone and 12 km–wide Associated Caldera at Ambrym Volcano (Vanuatu, New Hebrides Arc). *Journal of Volcanology and Geothermal Research* 55:225–28.

Robin, C., M. Monzier and J-P. Eissen., 1994. Formation of the mid-fifteenth century Kuwae caldera (Vanuatu) by an initial hydroclastic and subsequent ignimbritic eruption. *Bulletin of Volcanology* 56:170–83.

Rodman, W. 1996. The Boars of Bali Ha'i: Pigs in Paradise. In J. Bonnemaison, K. Huffman, C. Kaufmann, and D. Tryon (eds) *Arts of Vanuatu*, pp.158–67. Bathurst: Crawford House Press.

Rossito, R. 1990. Stylistic change of Fijian pottery. Part II: decoration. *Domodomo* (1–4):48–69.

Sand, C. 1995. *'Le Temps d'Avant' la préhistoire de la Nouvelle-Calédonie*. Paris: L'Harmattan.

Sand, C. 1997. The chronology of Lapita ware in New Caledonia. *Antiquity* 71:539–47.

Sand, C. 1999. Lapita and non-Lapita ware during New Caledonia's first millennium of Austronesian settlement. In J-C Galipaud and I. Lilley (eds) *The Western Pacific from 5000 to 2000 BP. Colonisation and transformations*, pp. 139–59. Paris: IRD Éditions.

Sand, C. 2000. The specificities of the 'Southern Lapita Province': the New Caledonian case. *Archaeology in Oceania* 35(1):20–33.

Sand, C. (ed.) 2003. *Pacific Archaeology: assessments and prospects* (Proceedings of the Conference for the 50th anniversary of the first Lapita excavation. Kone-Noumea 2002). Noumea: Les Cahiers de l'archeologie en Nouvelle-Calédonie 15.

Sand, C. and A. Ouétcho 1993. Three thousand years of settlement in the south of New Caledonia: Some recent results from the region of Païta. *New Zealand Journal of Archaeology* 15:107–30.

Scarr, D. 1990. *The History of the Pacific Islands: Kingdoms of the Reefs*. Melbourne: Melbourne University Press.

Schmidt, L. 2000. Shellfish collection strategies in the Vanuatu Islands. Unpublished Technical Report, Department of Archaeology and Natural History, Research School of Pacific and Asian Studies, Australian National University, Canberra.

Schurig, M. 1930. *Die Südseetöpferei*. Leipzig: Druckerei der Werkgemeinschaft.

Sharp, N. 1988. Style and Substance: A reconsideration of the Lapita decorative system. In P.V. Kirch and T. Hunt (eds) *Archaeology of the Lapita Cultural Complex: A Critical Review*, pp. 61–82. Thomas Burke Memorial Washington State Museum Research Report No.5. Seattle: The Burke Museum.

Sharp, N. 1991. Lapita as text: the meaning of pottery in Melanesian prehistory. *Indo-Pacific Prehistory Association* 11(2):323–32. Canberra.

Shepard, A. 1963. *Ceramics for the Archaeologist*. Carnegie Institution of Washington, Publication 609.

Shineburg, D. 1967. *They Came for Sandalwood: a Study of the Sandalwood Trade in the South West Pacific, 1830–1865.* Melbourne: Melbourne University Press.

Shutler, M.E. 1968. Pottery making at Wusi, New Hebrides. *South Pacific Bulletin* 8(4):15–18.

Shutler, M.E. 1973. Pottery making in Espiritu Santo. *Asian Perspectives* (for 1971) XIV:81–83.

Shutler, M.E. and R. Shutler Jr. 1965. Preliminary Report of Archaeological Explorations in the Southern New Hebrides, 1963–1964. Honolulu, Bishop Museum (Typescript plus plates).

Shutler, M.E. and R. Shutler 1967. Origins of the Melanesians. *Archaeology and Physical Anthropology in Oceania* II (2):91–99.

Shutler, M.E. and R.Shutler Jr. 1968. A preliminary report of archaeological explorations in the southern New Hebrides. *Asian Perspectives* (for 1966) 9:157–66.

Shutler, M.E., R. Shutler Jr. and S. Bedford 2002. Further detail on the Archaeological Explorations in the Southern New Hebrides, 1963–1964. In Bedford, S.,C. Sand and D. Burley (eds) 2002. *Fifty Years in the Field: Essays in Honour and Celebration of Richard Shutler Jrs Archaeological Career*, pp.189–206. Auckland: New Zealand Archaeological Association. NZAA Monograph 25.

Shutler, R. Jr. 1969. A Radiocarbon Chronology for the New Hebrides. *Proceedings of the Eighth International Congress of Anthropological and Ethnological Sciences, September 3–10, 1968*, pp. 135–57. Tokyo: Science Council of Japan.

Shutler, R. Jr. 1970. Regional Report: Melanesia 1966–1967. *Asian Perspectives* (for 1967) 10:123–26.

Shutler, R. Jr. 1973. New Hebrides Radiocarbon Dates, 1968. *Asian Perspectives* (for 1971) 14:84–7.

Shutler, R. Jr. 1998. pers. comm. Department of Archaeology, Simon Fraser University, Burnaby BC.

Shutler, R. Jr. and J. Marck 1975. On the dispersal of the Austronesian Horticulturalists. *Archaeology and Physical Anthropology in Oceania* 13 (2&3):215–28.

Shutler, R. Jr. and M. E. Shutler 1968. Archaeological excavations in Southern Melanesia. In I. Yawata and Y. Sinoto (eds) *Prehistoric Culture in Oceania, A Symposium*, pp.15–17. Papers from the Eleventh Pacific Science Congress, Tokyo, Japan 1966. Honolulu: Bishop Museum Press.

Shutler, R. Jr. and M.E. Shutler 1975. *Oceanic Prehistory*. California: Cummings Publishing.

Siorat, J-P. 1990. A technological analysis of Lapita pottery decoration. In M.J.T. Spriggs (ed.), *Lapita design, form and composition: proceedings of the Lapita design workshop, Canberra, Australia, December 1988*, pp.59–82. Occasional Papers in Prehistory 18. Canberra: Department of Prehistory, Research School of Pacific and Asian Studies, Australian National University.

Smith, A. 1991. Molluscs of the Ancient Mariner: Shell Technology, Typology and Pacific Prehistory. BA (Hons) dissertation, Department of Archaeology, La Trobe University, Bundoora.

Solheim, W. G. 1961. Oceania. *Asian Perspectives* 5(1):71–8.

Specht, J. 1968. Preliminary Report of Excavations on Watom Island. *Journal of the Polynesian Society* 77(2): 117–34.

Specht, J. 1969. Prehistoric and Modern Pottery Industries of Buka Island, T.P.N.G. Unpublished Ph.D dissertation, Australian National University, Canberra.

Specht, J. 1972. The Pottery Industry of Buka Island, T.P.N.G. *Archaeology and Physical Anthropology in Oceania* VII(2):125–144.

Specht, J. 1977. Review: S. M. Mead, L. Birks, H. Birks and E. Shaw. The Lapita Pottery Style of Fiji and its associations. *Mankind* 11(1):75–6.

Specht, J. and C. Gosden 1997. Dating Lapita pottery in the Bismarck Archipelago, Papua New Guinea. *Asian Perspectives* 36:175–89.

Speiser, F. 1923. *Ethnographische Materialien aus den Neuen Hebriden und den Banks Inseln*. Berlin: C.W. Kreidel.

Speiser, F. 1996. *Ethnology of Vanuatu. An early twentieth century study*. Bathurst, Crawford House Press.

Spriggs, M.J.T. 1981. Vegetable Kingdoms: taro irrigation and Pacific prehistory. Ph.D. dissertation, Australian National University, Canberra.

Spriggs, M.J.T. 1984. The Lapita Cultural Complex: origins, distribution, contemporaries and successors. *Journal of Pacific History* 19 4:202–23.

Spriggs, M.J.T. 1986. Landscape, land use, and political transformation in southern Melanesia. In P.V. Kirch (ed.) *Island Societies. Archaeological approaches to evolution and transformation*, pp.6–19. Cambridge: Cambridge University Press.

Spriggs, M.J.T. 1988. Cultural resources of the proposed Erromango Kauri Reserve and adjacent areas. In B. Leaver and M. Spriggs, The Erromango Kauri Reserve. FAO Working Document 1, Program VCP/VAN/6755. Rome: FAO.

Spriggs, M.J.T. 1990a. Dating Lapita: Another View. In M.J.T. Spriggs (ed.), *Lapita design, form and composition: proceedings of the Lapita design workshop, Canberra, Australia, December 1988*, pp.6–27. Occasional Papers in Prehistory 18. Canberra: Department of Prehistory, Research School of Pacific and Asian Studies, Australian National University.

Spriggs, M.J.T. 1990b. The Changing face of Lapita: the transformation of a design. In M.J.T. Spriggs (ed.), *Lapita design, form and composition: proceedings of the Lapita design workshop, Canberra, Australia, December 1988*, pp.83–122. Occasional Papers in Prehistory 18. Canberra: Department of Prehistory, Research School of Pacific and Asian Studies, Australian National University.

Spriggs, M.J.T. 1991a. Lapita Origins, Distribution, Contemporaries and Successors revisited. In P. S. Bellwood (ed.) *Indo-Pacific Prehistory 1990*, pp. 306–12. Canberra and Jakarta: IPPA and Asosiasi Prehistorisi Indonesia. Bulletin of the Indo-Pacific Prehistory Association 11.

Spriggs, M.J.T. 1991b. Nissan, the island in the middle. Summary report on excavations at the north end of the Solomons and the south end of the Bismarcks. In J. Allen and C. Gosden (eds) *Report of the Lapita Homeland Project*, pp. 222–43. Occasional Papers in Prehistory 20. Canberra: Department of Prehistory, Research School of Pacific Studies, Australian National University.

Spriggs, M.J.T. 1993. Island Melanesia: the last 10,000 years. In M. Spriggs, D. Yen, W. Ambrose, R. Jones, A Thorne and A. Andrews (eds) *A Community of Culture. The People and Prehistory of the Pacific*, pp. 187–205. Occasional Papers in Prehistory No.21. Canberra: Department of Prehistory, Research School of Pacific Studies, Australian National University.

Spriggs, M.J.T. 1996a. Chronology and colonisation in Island Southeast Asia and the Pacific: New data and an evaluation. In J. Davidson, G. Irwin, F. Leach, A Pawley and D. Brown (eds.) *Oceanic Culture History. Essays in Honour of Roger Green*, pp. 33–50. New Zealand Journal of Archaeology Special Publication.

Spriggs, M.J.T. 1996b. What is southeast Asian about Lapita? In T. Akazawa and E. Szathmáry (eds.) *Prehistoric Mongoloid Dispersals*, pp. 324–48. Oxford: Oxford University Press.

Spriggs, M.J.T. 1997. *The Island Melanesians*. Oxford: Blackwell.

Spriggs, M.J.T. 1999. The Stratigraphy of the Ponamla Site, North West Erromango: Suggested Evidence for 2700 year old stone structures. In J-C Galipaud and I. Lilley (eds.) *The Western Pacific from 5000 to 2000 BP. Colonisation and transformations*, pp. 323–43. Paris: IRD Éditions.

Spriggs, M.J.T. 2000. The Solomon Islands as Bridge and Barrier in the Settlement of the Pacific. In A. Anderson and T. Murray (eds) *Australian Archaeologist: Collected Papers in Honour of Jim Allen*, pp. 348–64. Canberra: Coombs Academic Publishing.

Spriggs, M.J.T. 2004. Is there life after Lapita, and do you remember the 60s? The post-Lapita sequences of the western Pacific. In V. Attenbrow and R. Fullagar (eds) *A Pacific Odessey: Archaeology and Anthropology in the Western Pacific. Papers in Honour of Jim Specht*, pp. 139–44. Records of the Australian Museum Supplement 29.

Spriggs, M.J.T. and A. Anderson 1993. Late Colonisation of East Polynesia. *Antiquity* 67: 200–17.

Spriggs, M.J.T and S. Bedford, 2001. Arapus: a Lapita site at Mangaasi in Central Vanuatu 2001. In Clark, G., A. Anderson and T. Sorovi-Vunidilo (eds.) *The Archaeology of Lapita Dispersal in Oceania*: Papers from the Fourth Lapita Conference, June 2000, Canberra, Australia, pp. 93–104. Terra Australis 17. Canberra: Centre for Archaeological Research and Department of Archaeology and Natural History, RSPAS, ANU.

Spriggs, M.J.T., J. Carney and S. Williams 1986. Prehistoric Stone Tools of Aneitum. *Naika* 23:18–33.

Spriggs, M.J.T. and D. Roe 1989. Planning for preservation: a general evaluation of the cultural resources of Erromango. Report prepared for Moores and Rowland. Canberra: National Heritage Studies.

Spriggs, M.J.T. and S. Wickler 1989. Archaeological Research on Erromango: Recent Data on Southern Melanesian Prehistory. *Bulletin of the Indo-Pacific Prehistory Association* 9:68–91.

Steadman, D. 1995. Prehistoric Extinctions of Pacific Island Birds: Biodiversity Meets Zooarchaeology. *Science* 267:1123–31.

Steadman, D. 1997. Extinctions of Polynesian birds: reciprocal impacts of birds and people. In P. V. Kirch and T. L. Hunt (eds.) *Historical Ecology in the Pacific Islands: Prehistoric Environmental Change*, pp. 51–79. New Haven: Yale University Press.

Steadman, D. pers. comm. Assistant Curator of Ornithology, Florida Museum of Natural History.

Stephens, D. and J. Krebs. 1986. *Foraging Theory*. Princeton: Princeton University Press.

Stuiver, M., P.J. Reimer, E. Bard, J.W. Beck, G.S. Burr, K.A. Hughen, B. Kromer, F.G. McCormac, J. v.d. Plicht and M. Spurk 1998. INTCAL 98 Radiocarbon Age Calibration, 24000–0 cal BP. *Radiocarbon* 40:1041–83.

Summerhayes, G. 1987. Aspects of Melanesian Ceramics. Unpublished M.A. dissertation. University of Sydney, Sydney.

Summerhayes, G. 1996. Interaction in Pacific Prehistory: An Approach based on the production, distribution and use of pottery. Ph.D dissertation, La Trobe University, Bundoora.

Summerhayes, G. 1997. Losing your temper: the effect of mineral inclusions on pottery analyses. *Archaeology in Oceania* 32:108–17.

Summerhayes, G. 2000. *Lapita Interaction*. Terra Australis 15. Canberra: Department of Archaeology and Natural History and the Centre for Archaeological Research, Australian National University.

Surridge, M. 1944. Decoration of Fiji Water-Jars. *Journal of the Polynesian Society* 53:17–36.

Swadling, P. 1986. Lapita Shellfishing: Evidence from sites in the Reef/Santa Cruz Group, Southeast Solomons. In A. Anderson (ed.) *Traditional Fishing in the Pacific*. Ethnographical and Archaeological Papers from the 15th Pacific Science Congress, pp. 137–48. Pacific Anthropological Records 37. Honolulu: Bishop Museum.

Swadling, P., N. Araho and B. Ivuyo 1991. Settlements associated with the inland Sepik-Ramu sea. In P.S. Bellwood (ed.) *Indo-Pacific Prehistory 1990*, Vol. 2, pp. 92–112. Canberra and Jakarta: IPPA and Asosiasi Prehistorisi Indonesia. Bulletin of the Indo-Pacific Prehistory Association 11.

Swadling, P., Chappell, J., Francis, G., Araho, N. and B. Ivuyo 1989. A late Quaternary inland sea and early pottery in Papua New Guinea. *Archaeology in Oceania* 24 (3): 106–9.

Taylor, F. W., B. L. Isacks, C. Jouannic, A. L. Bloom and J. Dubois 1980. Coseismic and Quaternary Vertical Tectonic Movements, Santo and Malekula, New Hebrides Arc. *Journal of Geophysical Research* 85, B10: 5367–81.

Tryon, D. 1996. Dialect Chaining and the Use of Geographical Space. In J. Bonnemaison, K. Huffman, C. Kaufmann, and D. Tryon (eds.) *Arts of Vanuatu*, pp.170–81. Bathurst: Crawford House Press.

Tryon, D. pers. comm. Linguist, Research School of Asian and Pacific Studies, Australian National University, Canberra.

Wahome, E. 1997. Continuity and Change in Lapita and post-Lapita ceramics: a review of evidence from the Admiralty Islands and New Ireland, Papua New Guinea. *Archaeology in Oceania* 32(1):118–23.

Wahome, E. 1999. Ceramics and Prehistoric Exchange in the Admiralty Islands, Papua New Guinea. Unpublished Ph.D. dissertation. Australian National University, Canberra.

Walter, R. 1998. *Anai'o: The Archaeology of a Fourteenth Century Polynesian Community in the Cook Islands*. Dunedin: New Zealand Archaeological Association Monograph 22.

Ward, G. K. 1979. Prehistoric Settlement and Economy of a Tropical Small Island Environment: The Banks Islands, Insular Melanesia. Unpublished Ph.D. dissertation. Australian National University, Canberra.

Ward, G. K. 1989. The Mangaasi pottery and the Mangaasi site. In D. G. Sutton (ed.) *Saying So Doesn't Make It So. Papers in Honour of B. Foss Leach*, pp. 153–67. Dunedin: New Zealand Archaeological Association Monograph 17.

Ward, G. K. pers. comm. Archaeologist, Australian Institute of Aboriginal Studies, Canberra, Australia.

Weisler, M. 1997. Introduction. In M. Weisler (ed.) *Prehistoric Long-distance Interaction in Oceania: an interdisciplinary approach*, pp. 7–18. New Zealand Archaeological Association Monograph 21.

White, J.P. 1971. New Guinea: The First Phase in Oceanic Settlement. In R.C. Green and M. Kelly (eds), *Studies in Oceanic Culture History*, Vol .2, pp. 45–52. Pacific Anthropological Papers 12. Honolulu: Bishop Museum.

White, J.P., J. Allen and J. Specht 1988. Peopling of the Pacific: the Lapita Homeland Project. *Australian Natural History* 22:410–16.

White, J.P and J. E. Downie 1980. Excavations at Lesu, New Ireland. *Asian Perspectives* 23 (2):193–220.

White, J.P and C.V. Murray-Wallace 1996. Site ENX (Fissoa) and the Incised and Applied Pottery Tradition in New Ireland, Papua New Guinea. *Man and Culture in Oceania* 12:31–46.

White, J.P, G. Clark and S. Bedford 2000. Distribution, Present and Past, of *Rattus praetor* in the Pacific and Its Implications. *Pacific Science* 54 (2):105–17.

Wickler, S. K. 1985. Early Human settlement on the island of Erromanga, Southern Vanuatu. Unpublished MA dissertation, University of Hawaii, Honolulu.

Wickler, S. K. 1995. Twenty-nine Thousand Years on Buka: Long-term Cultural Change in the Northern Solomon Islands. Ph.D dissertation, University of Hawaii, Honolulu.

Wickler, S. K and M. Spriggs 1988. Pleistocene Human Occupation of the Solomons Islands, Melanesia. *Antiquity* 62:703–6.

Wilson, A. L. 1978. Elemental Analysis of Pottery in the Study of its Provenance: A Review. *Journal of Archaeological Science* 5:219–36.

Wilson, M. 2002. Picturing Pacific Prehistory: The Rock-art of Vanuatu in a Western Pacific Context. Unpublished PhD thesis, Australian National University, Canberra.

Wilson, M. and B. David. in prep. Rock-art of the post-Lapita era: temporal patterning in the rock-art of Northwest Malakula (Vanuatu) and Island Melanesia.

Worthy, T., A. Anderson and R. Molnar 1999. Megafaunal expression in a land without mammals – the first fossil faunas from terrestrial deposits in Fiji. *Senckenbergiana biologica* 79(2):237–42.

Appendices

Appendix One: Inventory of radiocarbon dates

Table 1. Radiocarbon dates associated with ANU-Vanuatu National Museum Archaeological Project 1994-1999

Lab Number	Age BP	Material	Calibrated[#] age BP (2s.d)	Site and context
Erromango				
ANU-9710	4290±50	marine shell	4515 (4399) 4253	Velilo 1994 (natural deposit coral reef)
ANU-9709	1220±130	charcoal	1350 (1170,1160,1150, 1100, 1090) 802	Velilo 1994 (110–128cms bd)
ANU-9708	1460±80	charcoal	1526 (1340,1320,1310) 1189	Velilo 1994 (oven rakeout, 90–100 cms bd)
ANU-9707	1030±60	charcoal	1059 (950,940,930) 792	Velilo 1994 (oven feature, 60–70 cms bd)
ANU-9706	200±70	charcoal	427 (280,170,150,0) 0	Velilo 1994 (oven feature, 30–40 cms bd)
ANU-9703	810±80	charcoal	922 (730,720,700,690) 567	Raowalai 1994 (55–65 cms bd)
ANU-9705	910±70	marine shell	626 (510) 421	Raowalai 1994 (burial, 65cms bd)
ANU-9704	350±70	charcoal	520 (430,360,330) 0	Raowalai 1994 (65–75 cms bd)
ANU-9702	200±60	charcoal	419 (280,170,150,0) 0	Raowalai 1994 (35–40 cms bd)
ANU-9712	400±120	charcoal	646 (500,490,470) 0	Ilpin 1994 (oven fill, 54–57 cms bd)
ANU-9711	130±48	charcoal	285 (255,223,137,30,0) 0	Ilpin 1994 (oven fill, 30–40cms bd)
ANU-9714	370±60	charcoal	521 (460,350,340) 294	Velemendi 1994 (oven sect. 53–58 cms bd)
ANU-9716	230±60	charcoal	431 (290) 0	Velemendi 1994 (oven sect. 45–52 cms bd)
ANU-9717	139±50	charcoal	291 (259,221,139,27,0) 0	Velemendi 1994 (oven sect. 87–93 cms bd)
ANU-9718	181±63	charcoal	311 (270,170,150,10,0) 0	Velemendi 1994 (oven sect. 74–78 cms bd)
ANU-9713	177±49	charcoal	302 (273,174,146,11,0) 0	Velemendi 1994 (oven sect. 64–73 cms bd)
ANU-9719	96±53	charcoal	280 (60,40,0) 0	Velemendi 1994 (oven 30–50cms bd)
ANU-9507	2560±140	charcoal	2950 (2738) 2333	Ponamla 1994 (TP 1.1, 80 cms bd)
ANU-9508	2840±70	marine shell	2742 (2650) 2348	Ponamla 1994 (TP 1.1, 80 cms bd)
ANU-9509	2470±90	charcoal	2758 (2706, 2644, 2490) 2336	Ponamla 1994 (TP 1.1, 60–80 cms bd)
ANU-9510	1660±90	charcoal	1816 (1543) 1349	Ponamla 1994 (TP 1.1, 40–60 cms bd)
ANU-10078	2550±70	charcoal	2779 (2735) 2359	Ponamla (TP 1.1, 190–200 cms bd)
ANU-10077	3040±90	marine shell	3006 (2776) 2673	Ponamla (TP 1.1, 150–160 cms bd)
ANU-10079	2550±70	charcoal	2779 (2735) 2359	Ponamla (TP 1.1, 100–110 cms bd)
ANU-10073	2620±70	marine shell	2452 (2305) 2114	Ponamla (TP 1.1, layer 1 [lower])

continued over

Table 1. continued

Lab Number	Age BP	Material	Calibrated[#] age BP (2s.d)	Site and context
ANU-10297	2750±70	marine shell	2698 (2430) 2300	Ponamla (TP 1.2, layer 1)
ANU-10299	2590±80	marine shell	2435 (2290) 2049	Ponamla (TP 1.1, layer 1, 46 cms bd)
ANU-10293	1670±80	charcoal	1806 (1553) 1390	Ponamla (TP 2.1, layer 1, lower)
ANU-10294	2690±60	charcoal	2921 (2777) 2741	Ponamla (TP 3, layer 3, 220–240 cms bd)
ANU-10295	2050±70	charcoal	2298 (1995) 1833	Ponamla (TP 5.4, layer 3)
ANU-10296	2680±70	charcoal	2945 (2775) 2715	Ponamla (TP 6, layer 6, 245–270 cms bd)
ANU-10298*	4060±180	marine shell	4544 (4080) 3602	Ponamla (TP 2.1, layer 1)
ANU-9723	1650±70	marine shell	1312 (1200) 1045	Ifo 1994 (shell from bulldozed road)
ANU-9722	114.4±0.8%M	marine shell	modern	Ifo 1994 (shell from bulldozed road)
ANU-10521	3770±70	marine shell	3877 (3920) 3526	Ifo (sq D2, former coral reef)
ANU-10520	2700±80	charcoal	2868 (2779) 2750	Ifo (sq D2, 130–140 cms bd)
ANU-10523	2630±50	marine shell	2407 (2315) 2147	Ifo (sq D2, layer 3, 100 cms bd)
ANU-10533	2170±70	charcoal	2334 (2149) 1952	Ifo (sq D2, 85–110 cms bd)
ANU-10534	2510±60	charcoal	2753 (2711,2627,2620, 2556, 2551) 2355	Ifo (sq D2, 65–85 cms bd)
ANU-10535	2690±70	charcoal	2948 (2777) 2736	Ifo (sq D4, 50–70 cms bd)
ANU-10536	2650±70	marine shell	2498 (2327) 2149	Ifo (sq D2, L.1, 65 cms bd)
ANU-10537	2780±60	marine shell	2704 (2470) 2327	Ifo (sq D2, 85 cms bd)
ANU-10680	3120±60	marine shell	3047 (2870) 2751	Ifo (sq D.4, L.4, 140 cms bd)
ANU-10681	3100±70	marine shell	3047 (2850) 2731	Ifo (TP 3, 150–180 cms bd)
Malekula				
ANU-10075	2450±80	charcoal	2749 (2485,2481, 2469) 2336	Navaprah (TP A, 155–175 cms bd)
ANU-10538	1240±70	charcoal	1293 (1174) 973	Navaprah (TP A, 115–130 cms bd)
ANU-10539	630±50	charcoal	669 (646,583,565) 536	Navaprah (TP A, 75–95 cms bd)
ANU-10540	510±50	charcoal	629 (529) 497	Navaprah (TP A, 50–75 cms bd)
ANU-10525	980±60	marine shell	650 (550) 481	Chachara TP C1
ANU-10526	145±74	charcoal	311 (267,217,143,22,0) 0	Woplamplam (TP 1, 80–105 cms bd)
ANU-10529	1030±70	charcoal	1063 (946,942,933) 788	Wambraf (TP 1, 100–120cm bd)
ANU-10527	950±70	charcoal	972 (913,807,802) 707	Navepule A (TP 1, 55–65 cms bd)
ANU-10528	2080±70	charcoal	2305 (2040,2023,2010) 1880	Navepule C (TP 1, 80–100 cms bd)
ANU-10530	570±50	marine shell	292 (240) 0	Nuas (TP 4, 0–15 cms bd)
ANU-10522	2860±70	marine shell	2752 (2670) 2360	Malua Bay (TP 5.2, L. 4, 90 cms bd)
ANU-10532	2400±70	charcoal	2733 (2357) 2213	Malua Bay (TP 5.2, L. 4, 80–90 cms bd)
ANU-10524	1900±80	charcoal	2000 (1864,1848,1828) 1627	Malua Bay (TP 5.1, L. 3, 65–80 cms bd)
ANU-10531	1030±70	charcoal	1063 (946,942,933) 788	Malua Bay (TP 5.2, 40–65 cms bd)
ANU-10076	1930±80	charcoal	2055 (1875) 1634	Woapraf (TP 1, 175–200 cms bd)
ANU-10074	980±80	charcoal	1056 (925) 729	Bartnator (TP 1, 110–135 cms bd)
Table 1 cont				
Efate				
ANU-10658	2380±60	charcoal	2711 (2353) 2213	Mangaasi (TP 4, 170–190 cms bd)
ANU-10659	2520±60	charcoal	2754 (2712,2625,2623) 2357	Mangaasi (TP 4, 210–230 cms bd)
ANU-10657	2410±60	marine shell	2161 (2020) 1876	Mangaasi (TP 4, 110–130 cms bd)
ANU-10656	220±60	charcoal	431 (288) 0	Mangaasi (TP 4, 70–85 cms bd)
ANU-10655*	3690±60	marine shell	3723 (3590) 3443	Mangaasi (TP 3, 220 cms bd)
ANU-10649*	98.5±1.5%M	charcoal	modern	Mangaasi (TP 3 90–100 cms bd)
ANU-10650	2220±130	charcoal	2708, 2303,2240,2205, 2200, 2181, 2166,2162) 1897	Mangaasi (TP 3, 130–140 cms bd)
ANU-10651*	6190±80	charcoal	7268 (7156,7124,7090,7063, 7029) 6807	Mangaasi (TP 3, 150-160 cms bd)
ANU-10652	2850±50	marine shell	2727(2663) 2429	Mangaasi (TP 3, 170 cms bd)
ANU-10653*	130.0±2.4%M	charcoal	modern	Mangaasi (TP 3, 170-180 cms bd)
ANU-10654	2290±100	charcoal	2710 (2336) 2044	Mangaasi (TP 3, 210-230 cms bd)
ANU-10648	330±60	charcoal	510 (428,377,323) 160	Mangaasi (TP 3, 30-50 cms bd)

continued over

Table 1. continued

Lab Number	Age BP	Material	Calibrated[#] age BP (2s.d)	Site and context
ANU-10647	490±60	charcoal	629 (520) 463	Mangaasi (TP 3, 25-30 cms bd)
ANU-10646	1660±90	charcoal	1816 (1543) 1349	Mangaasi (TP 2, 120-130 cms bd)
ANU-10644*	103.3±1.0%M	charcoal	modern	Mangaasi (TP 1.2-1.3)
ANU-10641*	106.5±2.4%M	charcoal	modern	Mangaasi (TP 1, 110-130 cms bd)
ANU-10640*	1310±130	charcoal	1509 (1263) 953	Mangaasi (TP 1, 70-90 cms bd)
ANU-10642*	3040±140	charcoal	3566 (3315,3312,3257) 2851	Mangaasi (TP 1, 160-170 cms bd)
ANU-10643	2480±60	marine shell	2295 (2120) 1958	Mangaasi (TP 1, 160-170 cms bd)
ANU-10645*	1050±60	marine shell	698 (620) 517	Mangaasi (TP 1.2-1.3)
ANU-10801	2180±130	charcoal	2467 (2291,2273,2151) 1871	Mangaasi (TP 9, 60-80 cms bd)
ANU-10802	2960±140	charcoal	3468(3158,3149,3140,3126, 3112, 3089,3080) 2776	Mangaasi (TP 9, 90-100)
ANU-10803	2550±110	charcoal	2855 (2735) 2347	Mangaasi (TP 9, 95-105 cms bd)
ANU-10800	2550±130	charcoal	2921 (2735) 2336	Mangaasi (TP 9, 155-165 cms bd)
ANU-10798	2420±80	charcoal	2744 (2428,2420,2394,2360) 2214	Mangaasi (TP 9, 175-185 cms bd)
ANU-10796	3050±80	marine shell	2990 (2790) 2695	Mangaasi (TP 9, 195-205 cms bd)
ANU-10799	2790±110	charcoal	3236 (2915,2913,2870) 2741	Mangaasi (TP 9, 195-205 cms bd)
ANU-10797*	1350±180	charcoal	1685 (1287) 927	Mangaasi (TP 9, 205-215 cms bd)
Wk-6602	2820±50	marine shell	2716 (2603) 2359	Mangaasi (TP 9.4, 80-100 cm bd)
Wk-6603	2790±50	marine shell	2720 (2496) 2344	Mangaasi (TP 9.4, 110-130 cm bd)
Wk-6598	2670±50	marine shell	2473 (2336) 2280	Mangaasi (TP10, 110-120 cms bd)
Wk-6599	2740±50	marine shell	2665 (2428) 2321	Mangaasi (TP10, 140-150 cm bd)
Wk-6600	2910±45	marine shell	2755 (2708) 2601	Mangaasi (TP10, 150-160 cms bd)
Wk-6601	3160±50	marine shell	3077 (2937) 2805	Mangaasi (TP10, 180-190 cms bd)
ANU-11158*	2020±110	charcoal	2310 (1987,1979,1970,1960, 1950)1710	Arapus STP 8 (190-200 cm bd)
ANU-11159	3200±50	marine shell	3152 (2985) 2853	Arapus STP 17 (140 cms bd)
ANU-11160	3010±50	marine shell	2868 (2759) 2706	Arapus STP 14 (160 cms bd)
Retoka				
ANU-10257	720±70	marine shell	485 (330) 245	shell ornament from Roy Mata burials

* Denotes that dates are anomalous and have been rejected as unreliable.

All dates have been calibrated using the Calib. Program REV 4.1.2 of Stuiver et al., 1998 with delta R as 0 for marine samples.

Table 2. AMS dates from the ANU-Vanuatu National Museum Archaeological Project

LAB. NO.	AGE BP	MATERIAL	CALIBRATED AGE BP (2S.D)	SITE AND CONTEXT
Efate				
OZC829	2340±50	charcoal	2467 (2347) 2208	Mangaasi (TP 2, 210-220 cms bd)
OZC830	2450±50	charcoal	2736 (2485,2481,2469) 2348	Mangaasi (TP 3, 270-290 cms bd)
OZC831	2090±50	charcoal	2297 (2043) 1928	Mangaasi (TP 4, 110-130 cms bd)
OZD580	1670±60	charcoal	1710 (1553) 1413	Mangaasi (TP 7, 145-170 cm bd)
OZD581	2540±50	charcoal	2753 (2728) 2363	Mangaasi (TP 7, 270-290 cm bd)
OZD582	2580±50	charcoal	2776 (2743) 2492	Mangaasi (TP 9.4, 180-190 cm bd)
OZD578	2250±60	charcoal	2353 (2312,2218,2211) 2119	Mangaasi (TP 10, 90-100 cm bd)
OZD579	2620±110	charcoal	2950 (2750) 2357	Mangaasi (TP 10, 190-200cm bd)
Retoka				
OZC784	690±80	marine shell	479 (300) 136	shell ornament from Roy Mata burials
OZC785	990±125	marine shell	762 (550) 386	shell ornament from Roy Mata burials
Erromango				
OZC828	140±45	charcoal	291 (265,218,142,24,2) 0	charcoal sample from hand stencil,
Velemendi cave				
OZC827	100.86±0.59%M	human bone	modern	Velemendi, bone from oven.

Appendix Two: Stratigraphic details of Ponamla, Ifo, Mangaasi and Arapus test pits

The datums for the test pits were generally established adjacent to the excavated areas and depths were consistently measured from those points. The relative position of the datum points to the surrounding topography were calculated either in relation to their height above sea level (asl) which in all cases apart from Mangaasi was above the mean high tide mark. In the case of Mangaasi, datum points were calculated above a live coral datum representing mean sea level. The layer depths for the various test pits are given as centimetres below datum (cm bd).

Ponamla test pits

A total of nine test pits were excavated across the site. Initially five test pits (TPs 5.1–5.5) were excavated along a north-south transect A–A', followed by another four test pits in various locations (TPs 6–9) (Fig. 3.1). The test pits were spread across the site and therefore often demonstrated quite distinctive stratigraphies. Although several corresponding layers can be identified between the various test pits, the stratigraphy is labelled separately and discussed as per each test pit (Fig. 3.3).

TP 5.1 (13.59m asl), the most southern test pit was located on the flat area of the river terrace, some 15m from the edge of the riverbank. The stratigraphy was largely sterile and comprised four layers.

Layer 1 (10–50cm bd): Small limestone boulders were removed from the surface. Beneath the boulders, the matrix consisted of a black topsoil (10 YR 2/1) with very occasional limestone cobbles. Flecks of charcoal were noted and very degraded shellfish and pottery (1 sherd) were recovered. Layer 1 graded into Layer 2.

Layer 2 (50–60cm bd): A very dark grayish brown (10YR 2/1) sediment dominated the Layer 2 matrix with frequent lumps of yellow/brown clay also noted. Increasing charcoal with depth and again worn pottery (1 sherd) was recovered. There was a sharp break between this layer and Layer 3.

Layer 3 (60–95cm bd): This layer consisted of a very hard-packed concentrated cobbles and very dark brown (10YR 2/2) sandy silt. Flecks of charcoal were recovered along with worn sherds (6). This layer marked the interface of the sterile and initial human settlement.

Layer 4 (95–130cm bd): Very compacted, sterile dark yellowish brown (10YR 3/6) river silt plus concentrated basalt pebbles characterised this layer. This is the remnant former alluvial terrace which has been formed due to the downcutting of the Ponamla river. This layer was identified across the whole site at the base of all the test pits.

TP 5.1 appears to be outside the area of the early settlement site. A total of 10 worn sherds were found through Layers 1 to 3 and were clearly in secondary deposition. Layers 1 and 2 comprise of later humic accumulation, post-dating the initial human settlement. Some slopewash would also have contributed to the buildup.

TP 5.2 (11.68m asl) was located 20m north of TP 5.1, again on the flatter area of the river terrace but adjacent to one of the mound features that were concentrated on the east side of the terrace. The stratigraphy closely resembled that of TP 5.1.

Layer 1 (10–60cm bd): This layer comprised a black sediment (topsoil) (10YR 2/1) with few cobbles or pebbles. No artefactual material was recovered. It graded into Layer 2.

Layer 2 (60–110cm bd): This layer was a very compacted black (10YR 2/1) sediment mixed with frequent small river cobbles with increasing sand content with depth. Worn pottery (17 sherds) and charcoal were recovered. An abrupt delineation was noted between this layer and that below.

Layer 3 (110–150cm bd): This was a sterile layer consisting of river cobbles, boulders and dark yellowish brown (10YR 3/6) silty sand. The remnant former alluvial terrace signalled the base of the test pit.

There was little evidence uncovered in this test pit which indicated *in situ* remains of the early settlement. Layers 1 and 2 are later accumulations from both slopewash and humic buildup and Layer 3 the sterile former river terrace.

TP 5.3 (11.85m asl) was located a further 20m north of TP 5.2 on the top of the tapering end of a mound feature. Four layers were identified.

Layer 1 (10–30cm bd): This layer comprised a black (10YR 2/1) topsoil plus a moderate concentration of river cobbles and silt content. Layer 1 graded into Layer 2 with a noticeable increase in the concentration of cobbles. Sparse worn midden was recorded which included 8 sherds.

Layer 2 (30–70cm bd): A black (10YR 2/1) sediment amongst very concentrated small to medium basalt cobbles and occasional limestone boulders characterised this layer. Worn pottery (25 sherds) and shell were recovered with increasing frequency with depth. A sharp delineation was clear between this layer and Layer 3.

Layer 3 (70–90cm bd): This very hard-packed layer consisted largely of river sand and silt with a lesser percentage of black (10YR 2/1) sediment and only occasional river cobbles. At 90cm the layer became increasingly sterile and sandy marking the appearance of the former alluvial terrace. Four sherds were retrieved.

Layer 4 (90–120cm bd): The basal sterile dark grayish brown (10YR 3/6) sandy silt appeared at this depth along with frequent limestone cobbles and occasional boulders. Four sherds, clearly originating from a higher level, were recovered.

This test pit appears to mark the western periphery of the occupation activity area, with very sparse midden being recovered throughout. Pottery (48 sherds) bone and charcoal were recorded. Much of the material from Layers 1 and 2 appeared to be in secondary context. The midden content appearing at a depth of between 50–90cm, although far from concentrated, indicated the edge of the *in situ* remains of the early settlement sitting on top of the sterile basal river silt and cobbles.

TP 5.4 (9.63m asl) was located north and slightly down-slope of the areal excavation, Area A. Midden material was recovered throughout the four identified layers, although much of the recovered pottery appears to have been washed down-slope from the *in situ* deposit to the south, although at a depth of between 120–150cm bd remnant *in situ* deposits were identified.

Layer 1 (10–70cm bd): The matrix of Layer 1 comprised frequent basalt cobbles amongst a black topsoil (10YR 2/1). Frequent pottery (388 sherds) was recovered along with a number of historic artefacts (nail and glass) which confirmed the mixed nature of this upper layer. Further evidence of the churned nature of the layer was the retrieval of the only other dentate stamped sherd from the entire excavation. A sharp boundary could be identified between this and the layer below.

Layer 2 (70–130cm bd): Increased silt and sand with much less midden and little very dark grayish brown (10YR 2/1) soil content characterised Layer 2. Small to medium cobbles were also present but again in much lesser quantities than Layer 1. A greater concentration of cobbles along with ash and charcoal was noted at a depth of 120cm bd which indicated the level from which the *in situ* remains could be identified. Pottery (123 sherds), shell, charcoal and bone were noted from this level.

Layer 3 (130–160cm bd): Layer 3 consisted largely of a black (10YR 2/1) silty sediment, some ash and midden. This layer represents the initial human occupation on top of former foreshore.

Large limestone boulders also began to appear. A clear boundary could be identified between the layers. Seventeen sherds were recovered.

Layer 4 (160–200cm bd): Loosely compacted sterile very dark brown (10YR 2/2) silt and gravel with small to large cobbles and boulders signalled the surface of the former alluvial terrace.

This test pit was located close to the northern limit of the site. *In situ* remains were recorded only at a depth of 120–160cm bd. The stratigraphy above this level contained midden material in secondary deposition. A total of 528 sherds were recovered from TP 5.4, by far the largest number of any of the test pits. This reflects both the proximity of the test pit to the midden rich mound to the south and the consequent large number of sherds accumulated through secondary deposition and the actual *in situ* deposit at the lower level. A dispersed charcoal sample which returned a date of 2050±70 BP (ANU-10295) 2298–1833 BP* for Layer 3 would seem not to relate to the *in situ* cultural material. The submitted charcoal may well have been contaminated with later material (slopewash) that had mixed with the earlier deposit.

TP 5.5 (7.7m asl) was located a further 20m north of TP 5.4 and was almost 2m lower. Three layers were identified.

Layer 1 (15–25cm bd): This layer consisted of black (10YR 2/1) silty sediment. Infrequent worn shell and one sherd were recovered along with a glass marble. This layer appears to represent the more recent thin humic accumulation on top of the former foreshore with midden materials associated with recent occupation.

Layer 2 (25–35cm bd): Increased pebble and cobble content amongst a black (10YR 2/1) sandy sediment characterised this layer. Infrequent midden consisted of worn sherds (2) shell and bone.

Layer 3 (35–80cm bd): This basal layer consisted of larger river cobbles and boulders with loose very dark gray (10YR 3/1) silt and gravel representing the sterile former foreshore.

The proximity of this test pit to the beach would have precluded it from being suitable for habitation 3000 years ago. This was certainly reflected in the stratigraphy. There is also the possibility that evidence of the earlier settlement may have been destroyed by cyclonic activity.

TP 6.0 (15.06m asl) was located on the top of the southern most mound feature. The ground surface at TP 6.0, was by several metres, the highest of all the excavated test pits. It was closer to the adjacent slope and was therefore subject to increased rates of slopewash. The sterile former foreshore was reached at a depth of 3 metres below datum. A total of 8 layers were identified.

Layer 1 (10–50cm bd): This layer consisted of very concentrated small to medium sized limestone pebbles amongst a black (10YR 2/1) (topsoil) soil matrix and appears to have been formed through recent slopewash. Two worn sherds only were recovered.

Layer 2 (50–85cm bd): This layer was similar to Layer 1 but the concentrated limestone pebbles become small to medium cobbles and boulders within a black (10YR 2/1) soil matrix. Again relatively sterile midden-wise with only two sherds being recovered.

Layer 3 (85–140cm bd): The frequent limestone cobbles and boulders abruptly disappeared which delineated this layer from Layer 2. A black (10YR 2/1) sediment matrix made up the bulk of this layer which was interspersed with frequent limestone pebbles. Seven sherds recovered.

Layer 4 (140–160cm bd): At this depth the limestone pebbles largely disappeared and were replaced by a sand component within a very dark brown (10YR 2/2) sediment. Increased midden (pottery (9 sherds), shell, charcoal and bone) appear from this level but the worn nature of the ceramics suggests that they were still largely in secondary deposition.

Layer 5 (160–195cm bd): Increasingly black (10YR 2/1) sediment with patches of ash and frequent small to medium pebbles and cobbles characterised this layer. The midden content (pottery (29 sherds), shell, bone and charcoal) also increased and appeared to be *in situ*. The layer was primarily made up of rake-out from cooking fires and ovens and was very similar in nature to the stratigraphy of the lower layers of Area A. A concentration of shell was recorded at the bottom level of the layer.

Layer 6 (195–240cm bd): This layer was largely a continuation of the layer above but with an increased content of cobbles. Midden was again frequent and *in situ*. A further increase in the quantity of sherds (127) can be noted.

Layer 7 (240–260cm bd): Layer 7 consisted largely of a hardpacked dark grayish-brown (10YR 4/2) silty sediment. Few cobbles were in evidence. Midden included shell, bone and charcoal. A dispersed charcoal sample from this layer returned a radiocarbon date of 2680±70 BP (ANU-10296) 2945–2715 BP which corresponds to the earliest layers in Area A and confirmed this as the layer representing initial human habitation.

Layer 8 (260–300cm bd): This basal layer consisted of a dark yellowish brown (10YR 4/4) silt and gravel with frequent medium sized cobbles and large limestone boulders, indicating the level of the sterile former foreshore and/or alluvial terrace.

The excavation of TP 6.0 dramatically demonstrated the slopewash concentrated on the east of the site with the *in situ* midden material being identified at a depth of 160cm. Also that the area of the initial habitation continues for some distance further east. The increasing depth of the overburden however prevented any testing further into the slope. The ceramics were consistent with those from Area A as was the date recovered from Layer 7.

TP 7.0 (12.89m asl) was located just below TP 6.0 in a trough between two mound features to establish the nature of the stratigraphy at that point. Four distinct layers were identified.

Layer 1 (5–30cm bd): This layer consisted of a hard-packed black (10YR 2/1) sediment (topsoil) with frequent basalt cobbles and sparse midden content. It graded into Layer 2.

Layer 2 (30–100cm bd): This was again a very hard-packed matrix, made up of black (10YR 2/1) sediment with increasing silt content. Frequent small to medium cobbles were also noted. Increased midden content (pottery (76 sherds), charcoal, shell and bone) was recorded in the lower 20cm of this layer. This material appeared to be *in situ* midden although in much less dense concentration than found within the mound features. A clear boundary could be seen between this and the layer below.

Layer 3 (100–160cm bd): This layer consisted of very dark gray (10YR 3/1) silt and gravel with large limestone boulders. Sparse midden material including 53 sherds were also recovered. This layer marked the interface of the sterile and initial human habitation layers.

Layer 4 (160–170cm bd): This sterile basal layer was made up of a dark brown (7.5YR 3/2) river sand and silt with frequent pebbles and cobbles.

This test pit, although positioned in the same area as the mound features, highlights the variability found between the mounds and flatter areas. The mounds appear to represent areas where stone terraces were build which were foci for cooking activities and the consequent dumping of midden. In other areas the midden appears to be generally in secondary deposition, the pottery was often well worn and has been moved around over long periods of time through various post-depositional processes.

TP 8.0 (10.91m asl) was located on the western area of the terrace to test a small circular mound feature. Four layers were identified.

Layer 1 (5–40cm bd): This layer comprised a black (10YR 2/1) humic soil (topsoil) with limited silt content and frequent river cobbles. Infrequent worn midden was also recovered (9 sherds).

Layer 2 (40–70cm bd): Layer 2 consisted of a black (10YR 2/1) silty sediment with a medium concentration of basalt cobbles. Well worn and weathered midden was recovered from this layer (20 sherds).

Layer 3 (70–105cm bd): Very dark brown (10YR 2/2) river silt with little soil content and few cobbles and even sparser midden (13 sherds) characterised this transitional layer situated on top of the more sterile layer below.

Layer 4 (105–150cm bd): This basal layer represented the sterile former alluvial terrace and comprised a very dark grayish brown (10YR 3/2) sterile silt and gravel. Large limestone boulders were also noted.

The stratigraphy of TP 8.0 indicated that this area of the terrace was peripheral to the early habitation site and the recovered midden material associated with the earlier occupation was in secondary deposition. There was no *in situ* evidence of the early settlement from this test pit. It would appear to be a later feature associated with the clearance of the adjacent areas of stone along with evidence of succeeding humic accumulation.

TP 9.0 (7.2m asl) was located in the north western area of the site amongst a series of historic house terraces. Three layers only were identified.

Layer 1 (10–50cm bd): This layer comprised a black (10YR 2/1) sandy river silt with minimal humic content. Frequent small to medium basalt cobbles were also noted. Worn shell, charcoal flecks and bone were recovered along with an unidentified metal object.

Layer 2 (50–90cm bd): Layer 2 was a layer of very dark brown (10YR 2/2) river silt and sand with no cultural remains present. It represented the interface of the sterile and the later human occupation. This layer graded into the sterile layer below.

Layer 3 (90–130 cm bd): The basal layer consisted of a very dark grayish brown (10YR 3/2) sandy silt with medium concentrations of pebbles and cobbles. This layer is the former riverbank that was only utilised late in the cultural sequence.

No pottery was recovered from TP 9.0. The sparse midden consisted only of weathered shellfish, charcoal and bone from Layer 1. This part of the alluvial terrace was likely to have been much closer to sea level at the time of initial settlement and less suitable for settlement.

Ifo test pits

The Ifo site is concentrated on a series of linear mound features. A number of the ridges run parallel to the river and appear to be former beach ridges while others run at right angles and are primarily made up of cultural material. These cultural (linear) mounds have been formed through the accumulated dumping of midden and other debris over a long period of time. A total of thirteen, 1 by 1m test pits were excavated across the Ifo site to determine the area of the site and any temporal/spatial variation (Fig. 3.4, 3.5). All but two (TP8 and 14) showed similar stratigraphic accumulation and the layers can be directly correlated. The recovered midden remains are outlined in detail in other relevant chapters.

TP 2 (8.94m asl) was located on the top of a north-south aligned mound feature that had been truncated during the construction of a logging road in 1988. This was the most inland, in relation to the Ifo River, of all the test pits. It marked the northern limit of the Lapita settlement, but was outside the area of later settlement associated with distinctive fingernail decorated ceramics.

Layer 1 (10–60cm bd): This layer comprised a black (10YR 2/1) humic topsoil with frequent basalt and coral cobbles. Its accumulation was due to a combination of intermittent clearance of the area interspersed with humic accumulation when the area was under forest. Recovered midden remains were sparse and comprised only occasional worn shellfish. It graded into the layer below.

Layer 2 (60–105cm bd): The matrix of this layer was very similar to that of the layer above apart from an increasing sand component. The black (10YR 2/1) sediment contained less frequent cobbles but increasing quantities of shellfish. No other midden remains were recovered.

Layer 3/4 (105–120cm bd): A distinctive boundary was reached at 105cm bd with the appearance of a dark yellowish brown (10YR 4/4) coral sand. It appeared to mark the interface of the degraded coral terrace and initial human settlement. Occasional coral boulders and flat blocks were also recorded. Midden remains included calcareous-tempered (11 sherds) pottery (generally associated with dentate-stamped Lapita ceramics at Ifo), shellfish and a scoria abrader.

Layer 5 (120cm bd): This hard packed yellowish brown (10YR 4/5) coral sand represented the sterile basal uplifted coral terrace.

TP 3 (9.09m asl) was located on the central area of an east west aligned mound feature (Fig.3.5) to the east of the areal (Trench B,C and D) excavation to which the layers could be directly related. Frequent midden remains including pottery, shell, bone and a number of non-ceramic artefacts were recovered throughout the stratigraphy. A total of six layers were identified amongst the 180cm of stratigraphy.

Layer 1 (0–50cm bd): This layer consisted of a humic black (10YR 2/1) topsoil with concentrated water worn coral and basalt cobbles. It graded into the layer below with a lessening of cobble content with depth. Midden remains included 13 sherds, 6 of which were decorated with fingernail impression.

Layer 2 (50–70cm bd): This layer was similar to the layer above but as noted there was a dramatically decreased cobble content. The black (10YR 2/1) sediment was also increasingly silty. Eight sherds only were recovered.

Layer 3 (70–150cm bd): A distinct stratigraphic change at 70cm bd made Layer 3 easily identifiable. It consisted primarily of concentrated water worn basalt and coral cobbles within a very dark gray (10YR 3/1) sandy sediment. Flat coral blocks were recorded at the lower level of the layer. Frequent shellfish, bone and fingernail decorated pottery (17 sherds) were recovered from throughout this layer.

Layer 4 (150–170cm bd): Decreased cobble and soil content characterised Layer 4. It consisted largely of dark yellowish brown (10YR 4/4) coral sand. Recovered midden included occasional calcareous-tempered pottery (6 sherds), bone and charcoal. The layer was clearly associated with initial human settlement of the area. A radiocarbon determination on a marine shell returned a date of 3100±70 BP (ANU-10681) 3059–2735 BP which corresponds with dates from the earliest layers of Trench D.

Layer 4a (170–180cm bd): This layer which lay directly on top of the degraded coral terrace consisted of concentrated pumice pebbles. There were no midden remains contained within the layer. It would appear to be a natural water borne deposit related to regional volcanic activity prior to human settlement at the site.

Layer 5 (180cm bd): The basal coral terrace was encountered at a depth of 180cm bd at which point excavation was halted.

TP 4 (8.52m asl) was located to the west of the areal excavation on the lower south facing slope of the same mound feature (Fig. 3.5). The stratigraphy was relatively shallow due to the test pit being positioned on the periphery of the mound. Several layers that had been identified in Trenches B–D were not present in TP 4 and midden remains were sparse.

Layer 1 (0–30cm bd): This layer comprised a black (10YR 2/1) humic topsoil with frequent coral and basalt cobbles. Three plain sherds were recovered. A sudden decrease in the concentration of cobbles signalled the appearance of Layer 2.

Layer 2 (30–60cm bd): This was similar to the above layer but the black (10YR) sediment was increasingly silty with a corresponding dramatic decrease in cobble content. At the interface of this layer and the sterile basal coral terrace a thin tapering remnant of the lowest cultural layers (3 and 4), seen in detail in the areal excavation, was recorded. A further two plain sherds were recovered from this layer.

Layer 5 (70cm bd): This layer was the hard packed sterile basal layer characterised by yellowish brown (10YR 5/4) silt/sand from the degraded coral terrace.

TP 5 (9.2m asl) was located on the eastern slope of a north-south aligned mound feature in the south western area of the site. The test pit would appear to be located outside the area of settlement associated with ceramic use. Evidence of the Lapita phase was not present and the stratigraphy returned only sparse midden remains.

Layer 1 (10–50cm bd): This layer consisted of the ubiquitous black humic topsoil (10YR 2/1) with concentrated coral cobbles. Sparse midden remains included worn pottery (11 sherds) and occasional shellfish.

Layer 2 (50–75cm bd): A sudden drop in the concentration of cobbles characterised Layer 2. It was primarily a black (10YR 2/1) silty sediment but with an increasing coral sand content.

Layer 5 (75cm bd): At this level the degraded basal coral terrace was recorded.

TP 6 (8.35m asl) was located to the east of the areal excavation on the western slope of a north-south aligned mound feature. The location of the test pit on the periphery of the mound limited the depth of the stratigraphy. Midden remains were sparse.

Layer 1 (10–50cm bd): This layer consisted of black (10YR 2/1) humic topsoil with concentrated coral and basalt cobbles. Occasional large coral boulders were also encountered. Two plain sherds were retrieved from this layer.

Layer 2 (50–85cm bd): A decrease in the frequency of cobbles and increasing coral sand content amongst a black (10YR 2/1) silty sediment characterised Layer 2. Eight sherds were recovered from this layer only two of which were decorated (fingernail impressed).

Layer 5 (85cm bd): At 85cm bd the hardpacked degraded coral terrace was encountered.

TP 8 (8.24m asl) was located in the eastern part of the site, closer to the Ifo River, on the tip of an east-west aligned mound feature. The stratigraphy proved to be largely sterile made up principally of remnant beach deposits and certainly outside the area of settlement that was associated with ceramics.

Layer 1 (0–20cm bd): This layer consisted of the ubiquitous black (10YR 2/1) humic topsoil along with concentrated basalt and coral cobbles and represented more recent accumulation. No midden remains were recovered.

Layer 2 (20–90cm bd): At 20cm bd the soil content virtually disappeared to simply leave concentrated water worn coral and basalt cobbles, the remains of a remnant beach ridge.

Layer 3 (90–140cm bd): The concentrated coral and basalt cobbles included an increasing content of coral sand from this depth. Excavation was terminated at 1.40m bd due to the unstable nature of the test pit walls. Frequent small pumice pebbles were also recorded.

TP 9 (8.32m asl) was located to the west and downslope of the areal (Trench B–D) excavation. Its location on the periphery of the mound feature limited the depth of stratigraphy. Midden remains were sparse.

Layer 1 (0–30cm bd): This layer consisted of a black (10YR 2/1) humic topsoil with a concentrated coral and basalt cobble component. Very worn pottery (7 sherds) and occasional shellfish were recovered.

Layer 2 (30–85cm bd): Decreasing concentrations of cobbles and an increasing coral sand content characterised Layer 2. Pumice pebbles were also noted amongst the black (10YR 2/1) silty sediment. Sparse shell and worn pottery (11 sherds) comprised the recovered materials.

Layer 5 (85cm bd): At 85cm bd the sterile degraded coral terrace was encountered.

TP 10 (8.25m asl) was located on the northern side of the logging road on the east side of a low north-south aligned mound feature. A recent house site was said have been located in the vicinity. This was almost immediately confirmed with the recording of a coral pebble floor in the first few centimetres of excavation. Fragments of *navela* (shell money) were also recovered near the surface. This part of the site appeared to be largely outside the area of settlement associated with ceramic use.

Layer 1 (10–35cm bd): This layer was characterised by a black (10YR 2/1) humic topsoil with frequent coral and basalt cobbles. Pumice pebbles were also present. Sparse worn pottery (8 sherds) and shellfish were all that was recovered. It graded into the layer below.

Layer 2 (35–80cm bd): This layer was very similar to Layer 1 but with an increased sand content and disappearance of the concentrated cobbles. Pumice pebbles were still in evidence. At the lower level of this layer large coral blocks were recorded. These large coral blocks would appear to mark the initial clearance of the area for settlement, as is seen in the stratigraphy of the areal excavation. In the case of this test pit however the associated midden remains are lacking and there is nothing to suggest that this part of the site was associated with the initial Lapita settlement.

Layer 5 (80cm bd): At this depth the basal degraded coral terrace was encountered.

TP 11 (7.84m asl) was located on top of a low mound south of the areal excavations (Trenches A–D). The stratigraphy was relatively shallow with sparse midden remains, indicating that it was largely outside the area of settlement associated with ceramic use.

Layer 1 (10–40cm bd): This layer consisted predominantly of concentrated coral and basalt cobbles within a black (10YR 2/1) humic topsoil. Sparse worn pottery (4 plain sherds) and occasional shellfish were recovered. This layer graded into the layer below with decreasing cobble content.

Layer 2 (40–70cm bd): This layer was characterised by a black (10YR 2/1) silty sediment with an increasing coral sand content. No midden was recovered.

Layer 5 (70cm bd): Excavation terminated at this depth once the sterile degraded coral terrace was revealed.

TP 12 (9.14m asl) was located adjacent to and above TP 6, on top of the most eastern, north-south aligned mound feature (Fig. 3.5). The stratigraphy, which was up to 170cm deep, revealed five distinct layers which could be directly related to the stratigraphy identified in the areal excavations to the west. The mound feature where this test pit was located appeared to mark the eastern boundary of the site associated with ceramic use.

Layer 1 (5–55cm bd): This layer consisted of a black (10YR 2/1) humic topsoil with concentrated water worn coral cobbles and pebbles. Frequent worn pottery (50 sherds of which 10 were decorated with fingernail impression) and shellfish made up the recovered materials. The layer graded into the layer below.

Layer 2 (55–100cm bd): This layer was similar to the above layer but with a greatly decreased cobble content and an increasingly coral sand component set within a black (10YR 2/1) sediment. Frequent pumice pebbles were also encountered at this level. Midden materials were less frequent with pottery (28 sherds, 5 of which were decorated with fingernail impression) making up the bulk of the recovered materials.

Layer 3 (100–150cm bd): A sudden increase in the concentration of basalt cobbles and occasional coral blocks set within a very dark gray (10YR 3/1) sandy sediment characterised Layer 3. Recovered midden materials again included pottery (6 sherds) and increasing quantities of shellfish and charcoal. This layer was associated with the initial human settlement of the area and can be correlated with Layer 3 in Trenches B–D.

Layer 4 (150–170cm bd): Layer 4, made up of a dark yellowish brown (10YR 4/4) coral sand with sparse silt content, marked the interface of the sterile uplifted coral terrace and the lowest cultural layer.

Layer 5 (170cm bd): The degraded coral terrace was encountered at this level.

TP 13 (9.24m asl) was located on top of a large square mound feature to the west of the areal excavation. A total of six distinct layers were identified in the 170cm of stratigraphy which could be directly related to layers identified in the areal excavation.

Layer 1 (0–50cm bd): This layer comprised a black (10YR 2/1) humic topsoil with concentrated water worn coral and basalt cobbles. Frequent worn pottery (21 sherds, 2 decorated with fingernail impression) and shellfish were recovered. This layer graded into the layer below with decreasing cobble content.

Layer 2 (50–70cm bd): This was again a black (10YR 2/1) sediment but with much less frequent cobbles and an increasing sand content. Patches of ash and charcoal were also noted. Recovered midden included occasional worn pottery (5 sherds and shellfish.

Layer 2a (70–100cm bd): This 30cm thick layer consisted almost solely of concentrated water worn coral and basalt cobbles and was largely devoid of any silt content or midden remains except for occasional worn pottery (3 sherds). It would appear to represent an intensive clearance event. A similar event and stratigraphy were recorded in Layer 2 of Trench A.

Layer 3 (100–150cm bd): This layer consisted of an increased content of very dark gray (10YR 3/1) sandy sediment with concentrated basalt and coral cobbles. Midden remains including pottery (6 sherds) and shellfish were also recovered in association with charcoal and patches of ash. At the lower level of this layer frequent pumice pebbles were recorded.

Layer 4 (130–170cm bd): Layer 4 represented the interface of the earliest cultural layer and the sterile basal coral terrace. It consisted of dark yellowish brown (10YR 4/4) coral sand, gravel and occasional pebbles. Recovered midden although sparse included occasional calcareous-tempered sherds (3) along with shellfish and charcoal. This debris was associated with the initial human settlement of the area.

Layer 5 (170cm bd): Excavation ceased once the degraded coral terrace was reached at this level.

TP 14 (6.53m asl) was located in the far east of the site at the north-eastern corner of the current *siman-lo*, close to the Ifo River (Fig. 3.4). Its height above sea level was the lowest of any of the test pits by some margin. This lack of elevation proved to be somewhat revealing as soon as excavation began. Beneath a thin humic layer of topsoil, concentrated water worn coral cobbles and boulders were revealed which continued to a depth of 100cm bd, at which point the excavation was terminated. The matrix was completely devoid of midden materials. Apart from the more recent humic accumulation the test pit comprised solely of remnant beach deposit.

TP 15 (8.88m asl) was the most northern of the test pits, located on the top of a low lying north-south aligned mound feature that had been cut by the logging road in 1988 (Fig. 3.4). In 1996 this part of the site was covered by secondary forest. Midden remains were sparse and worn. Three distinct layers were identified (Fig. 3.5). Excavation results from this part of the site indicate that it is largely outside the area of settlement associated with ceramic use.

Layer 1 (10–50cm bd): This layer comprised a black (10YR 2/1) humic topsoil with frequent water worn coral and basalt cobbles. Little midden was recovered (2 plain sherds).

Layer 2 (50–100cm bd): From 50cm bd decreasing cobble content and increasing sand characterised Layer 2. Five plain sherds only were recovered.

Layer 5 (100cm bd): Abruptly at 100cm bd the coral terrace was revealed.

Mangaasi test pits

A total of 18 test pits (23.5m²) were excavated at the Mangaasi site from 1996 through to 1999. Due to the complex nature of the stratigraphy and the chosen excavation strategy it was necessary to assign all layers that were found across the site a numeric label. In the case of the tephra layers (2 and 5) Roman numerals have been added to further distinguish variation. These layers only can be correlated across the whole site. This is in contrast to other layers where differentiation within a layer is shown by the addition of a letter to a layer's numeric designation. These letter designations are not correlated across the site but are specific to individual test pits. The layers are not labelled serially and in some cases numeric inversion was inevitable. All 18 test pits have been illustrated (Figs. 3.10–3.14). The descriptions of the test pit stratigraphy below have been established through a combination of sedimentary and culturally defined layers. A total of 18 distinctive layers and/or primary cultural horizons were identified across the site (see Chapter 3).

TP 1 (9.22m asl), the first test pit excavated in 1996, was located adjacent to Garanger's 1967 excavation. It was designed to locate and further clarify the stratigraphy which had been defined during the earlier excavations (Fig. 3.12). In 1998 TP 15 was excavated adjacent to TP 1 in order to collect a greater sample of the Mangaasi ceramics and conduct a further reassessment of the stratigraphy. The descriptions of the layers below correlate directly with those recorded in TP 15 (Fig. 3.12).

Layer 1 (15–40cm bd): This layer consisted of a black (10YR 2/1) humic topsoil (as with all test pits) with concentrations of coral pebbles from former house floors. It is associated with an aceramic period of occupation post dating the eruption of Kuwae. This layer graded into the layer below with an increasing tephra content with depth.

Layer 2ii (40–70cm bd): From this depth sediment consisted of a very dark gray (10YR 3/1) developing soil with a high tephra content (weathered Kuwae). The concentrated coral pebbles were no longer present. Midden material was sparse and the worn nature of the ceramics indicated that they were in secondary deposition.

Layer 3a (70–100cm bd): This layer consisted of an increasingly sandy very dark gray (10YR 3/1) sediment with an increased midden and coral cobble content. Ash lenses were also recorded towards the lower level of this layer. This was the first indication of an *in situ* occupational horizon associated with ceramics, all of which are Mangaasi-style sherds. A dispersed charcoal sample from this layer (70–90cm bd) returned a radiocarbon determination of 1310±130 BP (ANU–10640) 1509–953 BP. This is one of the few dates related to this later period of ceramic use and quite probably very close to the termination of ceramic use in the area.

Layer 4a (100–110cm bd): This layer consisted primarily of a white (10YR 8/2) storm deposited marine sand. This deposit was either related to a very violent cyclone or tectonically influenced marine activity which may have occurred over some time period, evidence for which is seen in the layer below. The presence of frequent pumice within the matrix lends support to tectonic influence, potentially in the form of a series of tidal waves.

Layer 4b (110–140cm bd): Directly beneath the marine sand deposit appeared concentrated coral cobbles and boulders set within a very pale brown (10YR 7/3) sand matrix. The nature of this deposit lends further support to the tidal wave scenario. Pumice was again present. Midden materials were recovered from this layer but have been invariably heavily mixed amongst the marine deposited matrix. A dispersed (110–130cm bd) charcoal sample from this layer returned a completely anomalous date of 106.5±2.4%M (ANU–10641) Modern*.

Layer 3b (140–180cm bd): Rather suddenly the concentrated coral cobbles and boulders disappeared to be replaced by a black (10YR 2/1) sandy soil. This represented an occupation horizon sandwiched between the marine deposited material above and the former foreshore deposit below. Midden material was more frequent and included shell, charcoal and pottery, all of which was Mangaasi-style. Two radiocarbon dates were gleaned from the lower level of this layer. One date of 3040±140 BP (ANU-10642) 3551-2790 BP* from a dispersed charcoal sample (160–170cm bd) can be rejected as being clearly inconsistent with the ceramic remains. A marine shell sample date from the same level returned a much more realistic date of 2480±60 BP (ANU-10643) 2300–1975 BP.

Layer 11 (180–265cm bd): The former foreshore comprising of marine sand and coral cobbles and branches was reached at 180cm bd. Midden remains immediately became very sparse. Excavation continued to a depth of 265cm bd with little change in the matrix. Occasional very water worn sherds were recovered to the bottom of the excavation. The few rim forms recovered could be identified as Erueti-style indicating that an earlier settlement had once been located further inland.

TP 2 (10.13m asl) was located up-slope (almost 1m higher) and 25m east of TP 1 with which it demonstrated a broadly similar stratigraphic record (Fig. 3.15).

Layer 1 (10–20cm bd): This layer consisted of the black (10YR 2/1) humic topsoil with concentrations of coral pebbles providing evidence of former house floors. This layer graded into the layer below with an increasing tephra content with depth.

Layer 2i (20–50cm bd): At this depth the very dark gray (10YR 3/1) layer became increasingly rich in tephra. It represented a developed soil with a major component comprising of weathered Kuwae tephra. This layer graded into the layer below.

Layer 2ii (50–70cm bd): This layer consisted primarily of weathered Kuwae, although somewhat less weathered than the layer above as indicated both by its dark gray (10 YR 4/1) colour and less soil content. An abrupt change in the makeup of the layer below provided a clear boundary.

Layer 3a (70–170cm bd): At this depth an *in situ* cultural layer associated with Mangaasi-style ceramics appeared. It consisted of very dark gray (10YR 3/1) charcoal-rich silt with frequent medium to large coral cobbles, many of which were fire-blackened. The layer was somewhat mixed in the upper level due to later gardening activity. Recovered midden materials included pottery, shellfish, and bone. Charcoal and ash lenses were recorded throughout much of the layer. The lower levels of the layer became increasingly sandy and branch coral also became more frequent. A dispersed charcoal sample from the mid level (120–130cm bd) of this layer returned a radiocarbon determination of 1660±90 BP (ANU–10646) 1816–1349 BP.

Layer 4 (170–210cm bd): An abrupt change at 170cm bd revealed the remains of marine deposited debris. The matrix consisted of unconsolidated branch coral, light gray (10YR 7/1) marine sand and coral cobbles and boulders. This accumulation was the result of tidal wave or similarly violent cyclonic activity. Water worn pottery which had been churned up from the foreshore deposits was recovered throughout this layer.

Layer 3b (210–225cm bd): This layer consisted of a very dark gray (10YR 3/1) charcoal rich sediment with an increased midden content. The matrix was very similar to Layer 4 above. A dispersed charcoal sample from the upper level (210–220cm bd) of the layer returned an AMS date of 2340±50 BP (OZC829) 2467–2208 BP which corresponds well with a date of 2300–1975 BP from the same layer in TP 1. Little of the recovered pottery was diagnostic.

Layer 11 (260–290cm bd): From 260cm bd the former foreshore was revealed. This layer was excavated to a depth of 290cm bd at which point the excavation was terminated. Occasional water worn sherds of pottery were recovered throughout this layer. All rim forms were Erueti-style, indicating again that an earlier settlement had been located further inland.

TP 3 (10.55m asl) was another 25m east of TP 2 and at a similar elevation. Excavation demonstrated that although the stratigraphy of TP 3 could be directly correlated with that of TPs 1 and 2, it was largely outside the area of settlement associated with ceramic use (Fig. 3.10). Recovered midden materials were either in secondary deposition or represented ephemeral *in situ* occupation.

Layer 1 (10–40cm bd): This layer consisted of the ubiquitous black (10YR 2/1) humic topsoil with concentrated coral pebbles. A dispersed (25–30cm bd) charcoal sample returned a radiocarbon determination of 490±60 BP (ANU–10647) 629–463 BP. This layer graded into the layer below with an increasing tephra content with depth.

Layer 2i (40–80cm bd): From 40cm bd a very dark gray (10YR 3/1) developed soil, largely derived from weathered Kuwae tephra, was in evidence. A dispersed (30–50cm bd) charcoal sample returned a radiocarbon determination of 330±60 BP (ANU-10648) 510–160 BP.

Layer 2ii (80–140cm bd): This layer was a dark gray (10YR 4/1) developed soil, again primarily comprising weathered Kuwae tephra, although its colour and decreased soil content were an indication of less weathering than seen in the layer above. Two dispersed charcoal samples from 90–110 and 130–140cm bd respectively returned radiocarbon determinations of 98.5±1.5%M BP (ANU-10649) 246–0 BP* and 2220±130 BP (ANU–10650) 2708–1897 BP*. Neither were particularly useful for the dating of the layer. The first can clearly be rejected as anomalous while ANU–10650 may relate to the lower (Layer 3) cultural horizon.

Layer 4 (140–160cm bd): This layer consisted substantially of branch coral and worn coral cobbles and appears to represent the edge of the tidal wave deposit. A dispersed (150–160cm bd) charcoal sample from amongst this material returned a completely anomalous date of 6190±80 BP (ANU-10651) 7268–6807 BP. The charcoal was perhaps derived from driftwood of some kind.

Layer 3 (160–175cm bd): Very dark gray (10YR 3/1) sediment with concentrated coral cobbles, patches of ash and charcoal along with other midden materials characterised this Mangaasi cultural horizon. The layer however clearly had been somewhat disturbed and mixed by the later marine activity above along with some mixing of foreshore deposits below. This mixing was reflected in the problematic dating of the layer. A single marine shell (170cm bd) and a dispersed (170–180cm bd) charcoal sample respectively returned radiocarbon determinations of 2850±50 BP (ANU-10652) 2732–2442 BP and 130.0±2.4%M (ANU-10653) Modern*. The charcoal date presumably reflects modern tree root contamination and the marine shell seems likely to relate to the Layer 11 deposits.

Layer 11 (175–290cm bd): Abruptly at this level the former foreshore appeared. It has been divided up into a number of sub-layers to illustrate the stratified nature of the deposit and directly position a number of radiocarbon determinations. Layer 11a (175–200cm bd) consisted principally of semi-cemented coarse coral sediment. Layer 11b (200–220cm bd) consisted of more loosely compacted cobble pebbles. A dispersed (210–230cm bd) charcoal sample from this layer returned a date of 2290±100 BP (ANU–10654) 2710–2044 BP. Finally Layer 11c (220–290cm bd) consisted of more coarse coral sediment such as cobbles and boulders. A shell from the upper level (220cm bd) of this layer submitted for dating returned a radiocarbon determination of 3690±60 BP (ANU-10655) 3752–3449 BP, a date clearly un-related to the human occupation of the area. The dispersed charcoal sample however taken from the lowest level (270–290cm bd) of the layer returned a more realistic date of 2450±50 BP (OZC830) 2736–2348 BP for the initial human occupation of this area of the site assuming the charcoal had a cultural origin. Occasional water worn pottery (all rims were indicative of Erueti-style pottery) was found throughout the foreshore deposit again illustrating earlier settlement further inland.

Layer 13 (290cm bd): The former reef was revealed at a depth of 7.65m asl.

TP 4 (11.23m asl) was located further inland (35m south east) from TP 1 and the area originally excavated by Garanger. The ground surface was some 2m higher in elevation. It was the first test pit (1996) to reveal *in situ* Erueti-style pottery (Fig. 3.11).

Layer 1 (10–55cm bd): This layer was a somewhat thicker version of the humic black (10YR 2/1) topsoil layer (as per all test pits) found across the site. No midden materials were recovered.

In the upper level of the test pit, concentrations of coral pebbles relating to former house floors were noted. This layer graded into the layer below with an increasing tephra content with depth.

Layer 2i (55–100cm bd): This very dark gray (10YR 3/1) developing soil layer appeared to be principally made up of weathered Kuwae tephra. Much of the layer was disturbed however by a 20cm deep pit feature cut from the upper level of the layer. This was largely filled with coral pebbles along with a sparse component of black soil. Midden remains recovered from the pit feature included shellfish, bone and charcoal. A dispersed charcoal sample from the lowest level of the pit feature returned a date of 220±60BP (ANU–10656) 431–0 BP, indicating it was related to relatively late activity in the area. This layer graded into the one below.

Layer 5i (100–140cm bd): This dark gray (10YR 4/1) silty sediment layer was again made up largely of weathered tephra. Two radiocarbon dates on shell and charcoal respectively from the middle (110–130cm bd) of the layer (2410±60BP [ANU-10657] 2179–1889 BP and 2090±50 BP [OZC831] 2297–1928 BP) strongly suggest that the weathered tephra in this case is related to the Nguna eruption. Midden remains were sparse and seem most likely to be the result of secondary deposition and disturbance.

Layer 5ii (140–170cm bd): This layer again consisted largely of a grayish brown (10YR 5/2) developing soil derived from the weathering of the Nguna tephra. At this depth it was somewhat less weathered and patches of more pure tephra were noted across the square. More frequent midden began to appear in the lower levels of this layer but it clearly derived from the layer below. This layer represented the interface of the developing soils and the *in situ* cultural remains below.

Layer 9 (170–230cm bd): This black (10YR 2/1) greasy silty sediment layer represented an *in situ* cultural deposit which included concentrated coral and basalt cobbles, ash and charcoal, shellfish, bone, pottery and a several shell artefacts. All the pottery from this layer was of Erueti-style. A dispersed charcoal sample from the upper level (170–190cm bd) of this layer returned a date of 2380±60 BP (ANU–10658) 2711–2213 BP and from the lower level another dispersed charcoal sample (210–230cm bd) returned a date of 2520±60 BP (ANU–10659) 2754–2357 BP.

Layer 11 (230–250cm bd): Abruptly at a depth of 230cm bd the former foreshore was encountered. Several water worn pottery sherds were recovered from the 20cm deep sondage excavated into this layer indicating earlier settlement was located further inland from this test pit. Excavation was terminated at this level.

TP 5 (11.61m asl) was positioned 25m south of TP 4 in order to further define the limits of the Erueti phase of the settlement. A thin lens of *in situ* cultural material was identified at a depth of 140cm bd (Fig. 3.11). This test pit marked the landward edge of the early settlement.

Layer 1 (5–40cm bd): This layer consisted of a black (10YR 2/1) humic topsoil (as per all test pits). Deposits of concentrated coral pebbles, remains of former house floors, were recorded throughout the upper level of the layer.

Layer 2ii (40–80cm bd): An increasing tephra (Kuwae) content and a decrease in coral pebble content defined this very dark gray (10YR 3/1) developed soil. Midden remains were sparse and were in secondary deposition. This layer became increasingly lighter coloured and tephra-rich with depth, grading into the layer below.

Layer 2iii (80–100cm bd): This layer consisted primarily of light brownish gray (10YR 6/2) Kuwae tephra. Midden remains were, not surprisingly, somewhat sparse.

Layer 5i (100–140cm bd): From 100cm bd another dark gray (10YR 4/1) developed soil was encountered. It consisted largely of weathered Nguna tephra. Increased midden materials were recovered from the bottom level of the layer. These appeared to be derived both from the layer below and later dispersed remains.

Layer 9 (140–170cm bd): This layer was characterised by a black (10YR 2/1) greasy sediment with concentrated coral and basalt cobbles and midden materials. It represented a thin layer of an *in situ* cultural deposit associated with Erueti-style pottery.

Layer 11 (170–417cm bd): The cultural horizon abruptly terminated at 170cm bd on top of the former foreshore. Concentrated coral sand, pebbles, cobbles and occasional boulders comprised this layer. Midden materials immediately became very sparse and were only recovered from the upper most level (170–180cm bd) of the layer. The test pit was excavated down to the former reef (Layer 13) at a depth of 417cm bd (7.44m asl), a similar height above sea level to the reef identified in TP 3 (7.65m asl).

TP 6 (12.48m asl), located some 25m east of TP 5, and almost 1m higher, confirmed that the area of concentrated settlement lay toward the creek to the west. The stratigraphy of this test pit consisted primarily of a series of weathered and more pure tephras (Fig. 3.13). The dramatic accumulation of tephras in this area of the site is due to its location at the bottom of the nearby slope. Scattered midden material, largely ceramics, recovered from amongst the tephras was clearly in secondary deposition.

Layer 1 (10–50cm bd): This layer consisted of a black (10YR 2/1) humic topsoil with patches of coral pebbles (as per all test pits). At the interface of this layer and Layer 2I, patches of unweathered Kuwae tephra were recorded. This material appears to be related to a nearby slope failure event which is seen more clearly in the stratigraphy of TP 7.

Layer 2i (50–90cm bd): From 50cm bd a tephra-rich, very dark gray (10YR 3/1) developed soil was in evidence. This layer was derived primarily from weathered Kuwae tephra. Midden remains were sparse.

Layer 2iii (90–160cm bd): This largely sterile layer consisted primarily of more pure, light brownish gray (10YR 6/2) Kuwae tephra. At a depth of 160cm bd the tephra appeared more finely sorted and more gray coloured. This appeared to simply represent some stratification of the tephra deposit. There was a clear distinction between this layer and the layer below.

Layer 5i (180–230cm bd): This dark gray (10YR 4/1) developed soil consisted largely of weathered Nguna tephra. Occasional coral cobbles were noted throughout. Midden content increased at the lower level of this layer but the worn nature of the pottery suggested it was largely in secondary deposition or had been exposed on the surface for a long period. All pottery was Mangaasi ware.

Layer 11 (230–260cm bd): At a depth of 230cm bd very cemented coral boulders and cobbles were encountered. This signalled the appearance of the upper levels of the former foreshore. Excavation terminated at this point.

TP 7 (13.75m asl) was located at the far east of the site at the bottom of the rapidly rising slope. Due to its location near the base of the slope much thicker deposits of developed soils from weathered tephras along with more pure tephra layers were in evidence. Some evidence of slope failure in the form of later deposits of pure Kuwae tephra, was also recorded (Fig. 3.13).

Layer 1 (+10–45cm bd): This layer consisted of a very dark gray (10YR 3/1) humic soil with occasional coral cobbles. The patches of concentrated coral pebbles found across all other test pits on more level ground were absent in this test pit.

Layer 2iii (45–60cm bd): At a depth of 45cm bd patches of Kuwae tephra were recorded. As further excavation revealed this tephra, somewhat enigmatic stratigraphically, appeared to be the result of slope failure.

Layer 2i (60–100cm bd): This layer was a very dark gray (10YR 3/1) developed soil largely comprising weathered Kuwae tephra. Midden remains were completely absent. A very well defined layer change existed between this weathered tephra and the more pure Kuwae tephra below.

Layer 2iii (100–140cm bd): This layer consisted purely of light brownish gray (10YR 6/2) unweathered Kuwae tephra. Not surprisingly midden remains were non-existent. The tephra provided a very clearly defined layer separation.

Layer 5i (140–200cm bd): A dark gray (10YR 4/1) developed soil characterised this layer which consisted largely of weathered Nguna tephra. Mangaasi-style pottery appeared at the top level of the layer, indication of ephemeral settlement in the area. A dispersed charcoal sample from between 145–170cm bd returned an AMS date of 1670±60 BP (OZD580) 1710–1413 BP. The layer became increasingly lighter coloured with depth and graded into the layer below which consisted of less weathered Nguna tephra.

Layer 5ii (200–320cm bd): This layer consisted primarily of gray (10YR 5/1) Nguna tephra, its thickness due to rapid buildup at the base of the slope. Midden remains were very sparse and indicate that this part of the site was outside the area of concentrated settlement. A dispersed charcoal sample recovered from the lower level (270–290cm bd) of the layer returned an AMS date of 2540±50 BP (OZD581) 2753–2363 BP. This pre-dates the Nguna caldera forming event and appears more likely to be associated with mixed ephemeral midden remains once sandwiched between the tephra and the former foreshore below. The recovered rim forms were all indicative of Erueti-style pottery.

Layer 11 (from 320cm bd): From 320cm bd large coral boulders marking the top of the former foreshore were encountered. A very similar matrix at a correspondingly similar level (10.55cm asl) was identified in TPs 6 and 8 (TP 6, Layer 5 [10.18m asl]; TP 8, Layer 5 [9.83m asl]).

TP 8 (11.43m asl) was located 25m south of TP 5 in order to both further define the limits of the Erueti phase of settlement and test further inland for any evidence of an earlier occupation. The test pit proved to be largely outside the area where any evidence of early settlement was located. The stratigraphy consisted primarily of sterile soils developed from weathered tephras (Fig. 3.17). Midden remains were sparse throughout and appeared to be in secondary deposition.

Layer 1 (10–35cm bd): This layer comprised the black (10YR 2/1) humic topsoil with patches of concentrated coral pebbles indicative of former house floors. Scattered human bone was retrieved from the layer which was clearly associated with a burial cut feature recorded in the layer below.

Layer 2i (35–55cm bd): A decrease in the level of coral pebble content and a corresponding increase in tephra content characterised this layer of developing soil. The very dark gray (10YR 3/1) matrix was derived primarily from weathered Kuwae tephra. A burial feature had been cut into the layer which related to the post Kuwae (i.e. post dates 600 BP) settlement of the area.

Layer 2ii (55–100cm bd): From a depth of 55cm bd the stratigraphy consisted primarily of less weathered, dark gray (10YR 4/1) Kuwae tephra. Later settlement appears to have caused some disturbance to the layer which is further highlighted by the presence of patches of pure Kuwae tephra amongst the matrix.

Layer 5i (100–160cm bd): This dark gray (10YR 4/1) silty matrix consisted of a developed soil derived from weathered Nguna tephra. It became increasingly lighter brown with depth along with a corresponding increased content of less degraded tephra.

Layer 11 (160–195cm bd): This layer consisted of cemented coral boulders and cobbles laying on top of the former foreshore deposit. No midden materials were recovered. Excavation was terminated at a depth of 195cm bd.

TP 9 (10.23m asl) was located some 50m south, and over 1m higher in elevation, than Garanger's original 1967 excavations. The excavation of TP 4 (1996) suggested that earlier settlement lay further inland and TP 9 confirmed this. Its lowest layers marked the initial settlement in the Mangaasi area associated with Erueti style ceramics (Fig. 3.12). These demonstrated change over time within the over 150cm of cultural deposit.

Layer 1 (5–20cm bd): This layer consisted of the ubiquitous black (10YR 2/1) humic topsoil (as per all test pits) with patches of concentrated coral pebbles indicating the presence of former house floors. It graded into the layer below.

Layer 2i (20–40cm bd): This very dark gray (10YR 3/1) developed soil was largely derived from weathered Kuwae tephra. The concentrated coral pebbles seen in the layer above had disappeared and the tephra content had increased. Midden remains were sparse.

Layer 9a (40–90cm bd): Rather abruptly at 40cm bd a black (10YR 2/1) gritty sediment was encountered. Amongst the matrix were concentrations of cobbles, ash, charcoal and frequent midden remains. This represented a cultural horizon associated with late Erueti-style ceramics. A dispersed charcoal sample from the lower level (60–80cm bd) of the layer returned a radiocarbon determination of 2180±130 BP (ANU–10801) 2467–1871 BP. A lens of ash was recorded at a depth of 90–95cm bd. Midden materials from the layers above and below were mixed amongst the ash. A continuation of this lens was recorded in TP 12 (Fig. 3.12).

Layer 9b (95–105cm bd): This layer consisted of a similar matrix to the cultural layer above but was separated by the ash. It was a very dark gray (10YR 3/1) silty sediment with concentrated cobbles, ash, charcoal and frequent midden materials, such as shell, bone and pottery. The pottery was all of Erueti-style. The layer was sandwiched between an ash lens above and a storm-deposited marine sand below. One shell sample and two dispersed charcoal samples were dated from this layer. The marine shell (80–100cm bd) returned a date of 2820±50 BP (Wk–6602) 2716–2359 BP and the two charcoal samples (90–100cm bd and 95–105cm bd respectively) returned dates of 2960±140 BP (ANU-10802) 3468–2776 and 2550±110 BP (ANU-10803) 2855–2347 BP.

Layer 8a (105–115cm bd): This layer consisted primarily of a white (10YR 8/1) storm deposited marine sand. Water-rolled pottery was the only midden material recovered from the layer. This deposit demonstrated the close proximity of the earlier settlement to the former foreshore.

Layer 9c (115–155cm bd): This layer was similar to the upper cultural horizons comprising of a very dark gray (10YR 3/1) silty sediment including concentrations of cobbles, ash, charcoal and midden materials. The pottery consisted solely of Erueti-style sherds. One marine shell sample (110–130cm bd) and a dispersed (155–165cm bd) charcoal sample respectively returned radiocarbon determinations of 2790±50 BP (Wk–6603) 2702–2344 BP and 2550±130 BP (ANU–10800) 2921–2336 BP.

Layer 8b (155–170cm s bd): A further white (10YR 8/1) storm-deposited marine sand along with water worn coral branches comprised this layer.

Layer 9d (170–205cm bd): This layer was similar to the above cultural horizon and was characterised by an identical matrix. It lay directly on top of the former foreshore. A series of dates were determined from samples taken from different levels of the layer. They included dispersed (175–185cm bd and 195–205cm bd) charcoal samples which returned dates of 2420±80 BP (ANU–10798) 2744–2214 BP and 2790±110 BP (ANU–10799) 3236–2741 BP. A marine shell sample collected from the 195–205cm bd level returned a date of 3050±80 BP (ANU–10796) 3000–2700 BP while one from the 180–190 level returned an AMS date of 2580±50 BP (OZD582) 2776–2492 BP. Several Arapus-style rims were recovered from this lowest cultural layer but they were very much a minor component amongst a cultural horizon dominated by Erueti-style ceramics.

Layer 11 (205–230cm bd): At this depth the former foreshore was revealed. It consisted largely of water-rolled coral cobbles and branches and sloped slightly both towards the creek and the sea. At a depth of 230cm bd excavation was terminated. At this depth the layer was largely sterile apart from occasional sherds and scattered charcoal which had clearly filtered down from the layer above. A dispersed charcoal sample from the interface of the cultural horizon and the former foreshore returned an anomalous radiocarbon determination of 1350±180 BP (ANU–10797) 1685–927 BP*.

TP 10 (10.02m asl) was located between TP 9 and TP 1 in an attempt to locate a section of stratigraphy that demonstrated the transition from the Erueti to Mangaasi cultural horizons. The test pit proved to be well positioned (Fig. 3.12).

Layer 1 (15–40cm bd): This layer consisted of a black (10YR 2/1) humic topsoil with patches of concentrated coral gravel and pebbles indicating the presence of former house floors. It graded into the layer below.

Layer 2i (40–60cm bd): This very dark gray (10YR 3/1) developing soil displayed an increasing tephra content which coincided with a decrease in coral gravel and pebble content. It consisted primarily of weathered Kuwae tephra. Midden remains were sparse and clearly in secondary deposition. The layer graded into the layer below with a decreasing tephra content with depth.

Layer 3a (60–100cm bd): This layer comprised a black (10YR 2/1) silty charcoal rich sediment. It represented an occupational horizon associated with Mangaasi-style ceramics. From the bottom (90–100cm bd) of the layer a dispersed charcoal sample returned an AMS date of 2250±60 BP (OZD578) 2353–2119 BP. Frequent coral cobbles and midden remains were recovered from throughout the layer.

Layer 3b (100–140cm bd): This layer was similar to above but with a greatly increased coral cobble and midden content. Patches of ash and charcoal were also noted. Although there is clearly some mixing of deposits, this layer and the layer below demonstrate the transition from Mangaasi to Erueti cultural horizons. Mangaasi pottery dominated in this layer although occasional Erueti-style sherds were encountered. Erueti-style pottery dominated from Layer 9a below. A radiocarbon determination on a marine shell from this layer (110–120cm bd) returned a date of 2670±50 BP (Wk-6598) 2473–2280 BP.

Layer 9a (140–160cm bd): There was no identifiable change in the matrix of this layer to that which had been recorded above but the ceramic remains were dominated by Erueti-style ceramics. Two marine shell samples taken from two different levels of the layer were submitted for dating. They returned radiocarbon determinations of 2740±50 BP (Wk-6599) 2665–2321 BP (140–150cm bd) and 2910±45 BP (Wk-6600) 2755–2601 BP (150–160cm bd).

Layer 8 (160–170cm bd): At a depth of 160cm bd the cultural horizon was abruptly interrupted by a marine deposited white (10YR 8/1) sand related to cyclonic activity. This layer further highlighted the proximity of the settlement to the former foreshore.

Layer 9b (170–200cm bd): Sandwiched between the marine deposited sand and the former foreshore is this lowest cultural horizon consisting of black (10YR 2/1) sandy sediment with concentrated coral cobbles and midden materials. The recovered ceramics were exclusively Erueti-style sherds. A marine shell from the lower level (180–190cm bd) returned a radiocarbon determination of 3160±50 BP (Wk-6601) 3077–2805 BP and a dispersed charcoal sample from 190–200cm bd returned an AMS date of 2620±110 BP (OZD579) 2950–2357 BP.

Layer 11 (200–225cm bd): The former foreshore was reached at a depth of 200cm bd and excavation continued to a depth of 225cm bd. Sparse water worn sherds were recovered from throughout this level of the foreshore.

TP 11 (9.55m asl) was the most northern of all the test pits some 25m north of TP 3 but located on the same north-east south-west aligned beach terrace where Garanger had carried out his earlier excavations (Fig. 3.10). This area of the site proved to be on the periphery of the settlement associated with ceramic use. The stratigraphy was somewhat mixed due to more recent gardening and settlement activities. Recovered midden remains were sparse and included Mangaasi-style pottery.

Layer 1 (10–30cm bd): This layer consisted of a black (10YR 2/1) humic topsoil (as per all test pits). Concentrated patches of coral gravel and pebbles, indicative of former house floors, were recorded in the upper level of the layer.

Layer 2i (30–90cm bd): Much of this very dark gray (10YR 3/1) developed soil was derived from weathered Kuwae tephra. However as noted above, the layer was somewhat mixed and an increased midden content with depth hinted at the presence of a cultural horizon. It was not possible to define whether the midden remains were derived from more intact deposits nearby or were *in situ* but mixed. The lower level (90–110cm bd) of this layer became increasingly sandy until the former foreshore was revealed.

Layer 11a (110–120cm bd): This layer was made up primarily of a light gray (10YR 7/1) marine deposited sand which lay directly on top of the former foreshore.

Layer 11b (120–170cm bd): The former foreshore was reached at a depth of 120cm bd. It consisted of loosely compacted coral branches, cobbles and boulders amongst a gritty sand. Worn pottery found was found throughout this layer, indicating that earlier settlement had occurred further inland. Excavation was terminated at a depth of 170cm bd.

TP 12 (10.14m asl) was located between TP 9 and 10 in order to further define the relationship of the Erueti and Mangaasi cultural horizons. The excavation revealed a somewhat similar stratigraphy (Fig. 3.12) to that in TP 9. A greater sample of Erueti-style ceramics and artefacts were recovered.

Layer 1 (10–30cm bd): This layer consisted of the ubiquitous black (10YR 2/1) humic topsoil with concentrations of coral pebbles from former house floors appearing in the upper level.

Layer 2i (30–60cm bd): This very dark brown gray (10YR 3/1) developed soil was derived primarily from weathered Kuwae tephra. Increased midden materials were recovered with depth but these had been clearly disturbed from the lower cultural horizon.

Layer 9a (60–90cm bd): Rather abruptly concentrated burnt coral and basalt cobbles along with concentrated midden signalled the appearance of an *in situ* cultural horizon. The matrix consisted of a black (10YR 2/1) sandy sediment with patches of ash and frequent charcoal. All of the recovered ceramics were Erueti-style sherds. At a depth of 90–105cm bd an ash lens was recorded in the southern half of the test pit. It would seem to correspond with a similar fire ash recorded in TP 9. Midden from the adjacent layers had been mixed amongst it. It provided a boundary between this layer and the layer below.

Layer 9b (105–160cm bd): This layer represented a continuation of the cultural horizon above (Layer 9a) and below (Layer 9c). The matrix again consisted of a black (10YR 2/1) charcoal-rich sandy sediment with concentrated coral and basalt cobbles from fire-rakeout. Midden remains were frequent including shellfish, bone and pottery (all Erueti-style sherds).

Layer 8 (160–165cm bd): This thin layer consisted principally of a light gray (10YR 7/1) marine deposited sand related to a similar cyclonic episode as seen in TP 9 (Layers 8 and 8a). This is further confirmation of the proximity of the settlement to the original foreshore.

Layer 9c (165–185cm bd): This layer represented the earliest *in situ* cultural deposit found in this test pit. The matrix consisted of a black (10YR 2/1) charcoal-rich sandy sediment with frequent coral and basalt cobbles. Midden remains were concentrated and included shellfish, bone, pottery and other artefacts. All pottery was Erueti-style except for one sherd which was a distinctive Arapus-style rim.

Layer 11 (185–406cm bd): The marine deposited former foreshore was encountered at a depth of 185cm bd. It consisted of light gray (10YR 7/1) sand and coral debris such as branches, cobbles and boulders. Sparse water worn midden materials were found throughout this layer down to the reef platform below, indicating that earlier settlement had been located further inland (i.e. in the area of TP 9). Layer 13 (406cm bd): The top of the former reef was reached at a depth of 406cm bd (6.08m asl).

TP 13 (9.91m asl) was located 25m further south of TP 9 in order to further delineate the area of early settlement. Much of the stratigraphy however consisted of alluvial sediments deposited by the nearby Pwanmwou Creek (Fig. 3.14). This part of the site was clearly outside the area of settlement associated with ceramic use. Sparse midden remains, including pottery were found throughout the stratigraphy but were clearly in secondary deposition. Decorated sherds demonstrated an affinity with Erueti-style ceramics and may have derived from localised disturbances.

Layer 1 (0–35cm bd): This layer consisted of a black (10YR 2/1) humic topsoil. Patches of coral gravel/pebbles were noted. These are the remains of former house floors which relate to more recent settlement of the area. Sparse scattered midden, in secondary deposition, was recovered.

Layer 2i (35–65cm bd): An increased tephra content and corresponding disappearance of coral pebbles differentiated this layer from the layer above. The very dark gray (10YR 3/1) developed soil was derived largely from weathered Kuwae tephra. Several patches of more pure tephra were also noted within the matrix.

Layer 5i (65–90cm bd): At a depth of 65cm the developed soil changed to a dark gray (10YR 4/1) which was primarily derived from weathered Nguna tephra. This layer graded into the layer below with an increasing non-weathered tephra content.

Layer 5ii (90–150cm bd): This layer was characterised by a grayish brown (10YR 5/2) sediment comprising principally of less weathered Nguna tephra. Midden materials became increasingly sparse. The stratigraphy below this layer, to the bottom of the excavation at 250cm bd, was consisted largely of alluvial deposits. They were characterised by layers of re-deposited former foreshore and fine silty sands. Midden materials throughout this lower deposit were sparse and were clearly in secondary deposition. All the recovered sherds were heavily worn.

Layer 7a (150–200cm bd): This layer consisted principally of cemented coral cobbles and pebbles in a fine white (10YR 8/1) silty sand. A colour change in the sand was noted mid way through the layer (7b) where from 180cm bd it was iron stained. The sparse midden consisted solely of heavily water rolled pottery.

Layer 7c (200–220cm bd): At 200cm bd the coral debris was replaced by a sterile white (10YR 8/1) sand, a further indication of the alluvial nature of the deposit.

Layer 7d (220–240cm bd): Further fine alluvial sand was recorded at this depth, but it was finer and more light gray (10YR 7/1) in colour.

Layer 7e (from 240cm bd): At this depth white (10YR 8/1) coral sand, pebbles and cobbles were encountered. This may be evidence of a marine deposit marking the former foreshore or a further alluvially derived layer. Excavation was terminated at 250cm bd.

TP 14 (8.6m asl) was the most western test pit excavated at the Mangaasi site. Designed to further delineate the boundaries of the early settlement associated with ceramic use it was located down slope and some 25m west of TP 10. However much of the stratigraphy consisted of alluvial deposits (from 80cm bd) due to its proximity to the Pwanmwou Creek (Fig. 3.14). Sparse water worn midden remains were found throughout this alluvial deposit, indicative of earlier settlement further inland.

Layer 1 (10–30cm bd): This layer comprised a black (10YR 2/1) humic topsoil with little evidence of former settlement in the form of coral pebble concentrations associated with house floors. It graded into the layer below with an increasing tephra content

Layer 2i (30–50cm bd): This very dark gray (10YR 3/1) developed soil derived primarily from weathered Kuwae tephra. Midden remains were sparse and in secondary deposition.

Layer 2ii (50–80cm bd): This dark gray (10YR 4/1) layer consisted of less weathered Kuwae. Within the matrix several patches of light brownish gray (10YR 6/2), more pure tehpra were noted. From below the bottom of this layer the stratigraphy clearly consisted of alluvially deposited sediments.

Layer 7a (80–170cm bd): This layer consisted of alluvially re-deposited tephra with coral cobbles and boulders. Part of the layer may have consisted of the tidal wave deposit as seen in other test pits (e.g. TP 1, 2, 3, 16) but alluvial activity has thoroughly transformed the stratigraphy.

Layer 7b (170–205cm bd): At a depth of 170cm bd the stratigraphy comprised an increasing light gray (10YR 7/1) sand content with a corresponding decrease in tephra content. Branch coral and pebbles were also noted.

Layer 7c (205–220cm bd): An increasing coral pebble content provided the distinction between this layer and the one above. A distinct change in the matrix of the layer below again provided a distinct delineation of layers.

Layer 7d (220–240cm bd): This layer consisted of an alluvially deposited, compacted grayish brown (10YR 5/2) sand with concentrated coral pebbles.

Layer 7e (240–255cm bd): This layer was yet another alluvial deposit with a similar matrix to the layer above but the sand was white (10YR 8/1).

Layer 7f (255–265cm bd): From a depth of 255cm bd the stratigraphy was characterised by large coral boulders set within a cemented white (10YR 8/1) sand. Excavation terminated at a depth of 265cm bd.

TP 15 (9.25m asl) was located immediately adjacent to TP 1 and Garanger's areal excavation. Detailed descriptions of the stratigraphy are presented above (see TP 1 and Fig. 3.12).

TP 16 (9.6m asl) was located between TP14 and TP 10. It was designed to further define the stratigraphic relationships between the alluvially influenced deposits found in TP14 and those devoid of alluvial interference in TP 10. The stratigraphy of TP 16 (Fig. 3.14) was similar to TP 17 but with a greater marine deposited component.

Layer 1 (10–20cm bd): This layer comprised a black (10YR 2/1) humic topsoil with patches of concentrated coral pebbles from former house floors.

Layer 2i (20–35cm bd): The disappearance of the concentrated coral pebbles and a corresponding increased tephra content characterised this very dark gray (10YR 3/1) developed soil. It derived primarily from weathered Kuwae tephra. Midden remains were sparse.

Layer 3 (35–110cm bd): This layer consisted of a black (10YR 2/1) gritty matrix with concentrated coral and basalt pebbles and frequent midden remains. It represented a occupational horizon associated with Mangaasi style ceramics.

Layer 4 (110–150cm bd): This layer comprised a marine sand, large coral blocks and concentrated coral cobbles. It appeared to relate to the same tidal wave event recorded in TPs 1/15, 2, 3, and 17. It graded into the layer below, with a diminishing coral block content.

Layer 8 (150–190cm bd): This layer was also derived largely from marine deposits and consisted of sand and less frequent coral debris. Sparse water worn midden remains were recovered from throughout the layer but were clearly derived from foreshore deposits.

Layer 5iii (190–200cm bd): A further decrease in coral debris content and the presence of a tephra characterised this marine deposited light gray (10YR 7/1) silty sand layer. The tephra content seems likely to have derived from an eruption of Nguna c. 2200 BP.

Layer 11 (200–230cm bd): The former foreshore comprising of marine sand and coral cobbles and branches was reached at a depth of 200cm bd. Water worn pottery only was recovered to a depth of 230cm bd at which point the excavation was terminated.

TP 17 (9.72m asl) was located between TP 15 and 10, two test pits which displayed somewhat varied stratigraphies. TP17 was excavated to further clarify the stratigraphic relationship (Fig. 3.9 and 3.12) across this 25m gap.

Layer 1 (10–20cm bd): This layer comprised the black (10YR 2/1) humic topsoil with patches of coral pebbles indicative of earlier house floors. It graded into the layer below with an increasing tephra content with depth and a corresponding decrease in the coral pebble content.

Layer 2i (20–50cm bd): A very dark gray (10YR 3/1) developed soil characterised this layer which was derived from weathered Kuwae tephra. Midden remains were sparse and worn and clearly in secondary deposition.

Layer 3 (50–125cm bd): This layer consisted of black (10YR 2/1) sandy, charcoal rich sediment. Also included amongst the matrix were concentrated coral and basalt cobbles, shellfish, bone and ash. The layer represented an occupation horizon associated with Mangaasi-style ceramics. It was further subdivided into three layers based on concentrations of coral and basalt cobbles. The matrix of Layer 3a (50–75cm bd) included concentrated cobbles while a lesser frequency was noted in Layer 3b (75–100cm bd) with a return to more concentrated cobbles being noted in Layer 3c (100–125cm bd). A clearly defined boundary between this layer and the layer below was provided by a change to more sterile marine deposited sediments.

Layer 4a (125–180cm bd): This layer consisted primarily of a series of stratified marine deposited sediments which are described below. The same deposit is seen in a number of other test pits (e.g. TP1/15, TP 2, TP 3 and TP 16). As already suggested it appears to have been either related to a very violent cyclonic event or tectonically influenced marine activity (i.e. tidal wave). There was no evidence of any *in situ* cultural horizons or humic accumulation separating the deposits which suggests the deposit built up over a short time period. The stratified deposits included Layer 4a (125–140cm bd) comprising concentrated coral pebbles, Layer 4b (140–155cm bd) a sterile white (10YR 8/1) marine sand, Layer 4c (155–165cm bd) included the marine sand but with a substantial coral pebbles and cobbles content, and finally Layer 4d (165–180cm bd) was a white (10YR 8/1) coarse marine sand. Sparse, heavily water worn, midden remains were recovered from throughout the deposit. Diagnostic sherds were all Erueti-style but in secondary deposition having been churned up from the foreshore deposits.

Layer 3d (180–190cm bd): Beneath the marine deposited sediments a further cultural horizon was identified. It consisted of a very dark gray (10YR 3/1) charcoal rich sediment with frequent coral cobbles. Shellfish, bone and pottery were recovered. The ceramics were all plain sherds but the layer would seem to correspond to Layer 6 in TP1/15 where the recovered sherds were overwhelmingly Mangaasi-style ceramics. Water worn Erueti-style sherds were recovered above and below this layer but appear to have derived from disturbed marine deposits.

Layer 5iii (190–205cm bd): A layer of tephra was recorded lying directly on top of the former foreshore. Gray (10YR 5/1) in colour this tephra seems most likely to relate to an eruption of Nguna c. 2200 BP. This is the northern most part of the site where it was recorded.

Layer 11 (205–230cm bd): The former foreshore comprising of coral gravel, cobbles and boulders was encountered at a depth of 205cm bd. Excavation continued to a depth of 230cm bd where very occasional heavily water worn sherds were recovered.

TP 18 (11.18m asl) was the only test pit excavated at Mangaasi in 1999. Located some 25m east of TP 4 it was designed to further define the boundaries of the early settlement associated with Erueti style ceramics. Excavation demonstrated that it largely lay outside the area of settlement associated with ceramic use (Fig. 3.10).

Layer 1 (10–40cm bd): This layer comprised a black (10YR 2/1) humic soil. It graded into the layer below, becoming lighter coloured along with an increasing tephra content.

Layer 2i (40–70cm bd): This very dark gray (10YR 3/1) developed soil layer was primarily derived from weathered Kuwae tephra. Midden remains were sparse. A clearly defined boundary could be seen between this layer and the layer below.

Layer 2iii (70–110cm bd): This light brownish gray (10YR 6/2) layer consisted of more pure Kuwae tephra, devoid of midden remains.

Layer 5i (110–180cm bd): This dark gray (10YR 4/1) developed soil derived primarily from weathered Nguna tephra. Midden remains were sparse which suggested this area of the site was on the periphery of the main settlement.

Layer 11 (180–230cm bd): At a depth of 180cm bd the former foreshore was encountered. It consisted of worn branch coral, cobbles and pebbles. Excavation continued to a depth of 230cm bd at which point midden remains were no longer in evidence. Further indication that this part of the site was outside the area of early settlement.

Arapus Test pits

A total of 24, 1 by 1m test pits were excavated at the Arapus site in 1999. Only three of those are described in any detail here, primarily to illustrate the stratigraphic relationships of the various ceramic traditions. It was at the Arapus site that an early phase of settlement, pre-dating anything identified at Mangaasi, was located. It is this earlier phase of settlement that is associated with Arapus-style pottery (now known as Arapus Ware). Arapus-style pottery was associated with initial settlement in the area and in a number of the test pits further inland (e.g. TP 14 and 17) the ceramics are almost exclusively Arapus-style sherds which date to c. 2900 BP. In other test pits (e.g. TP 4) the basal cultural layers included Arapus-style sherds which were sealed by later layers comprising exclusively of Erueti-style sherds. The three test pits (Fig. 3.15) below illustrate both the *in situ* Arapus cultural horizons and the transition to Erueti cultural horizons.

TP 4 (10.21m asl) was located on a terrace some 10m above the live coral datum and revealed the transitional phase from Arapus to Erueti-style ceramics.

Layer 1 (0–25cm bd): This uppermost layer consisted of the ubiquitous black (10YR 2/1) humic topsoil found across the entire site. It graded into the layer below with an increasing tephra content.

Layer 6 (25–75cm bd): Layer 2 consisted of a very dark gray (10YR 3/1) developed soil, a large percentage of which was made up of weathered tephras. A patch of un-weathered Kuwae tephra was recorded in section. Occasional coral cobbles and sherds of Erueti-style pottery were recorded. This material clearly derived from the layer below.

Layer 9 (75–135cm bd): This thick layer comprised a silty black (10YR 2/1) sediment with concentrated cobbles and midden materials and represented an *in situ* cultural deposit. The pottery remains were almost exclusively Erueti-style sherds.

Layer 10a (135–160cm bd): The matrix of this layer was identical to the layer above but the ceramics displayed a marked change. Arapus-style sherds only was recovered from this depth (e.g. TP 14, Layer 10 and TP 17, Layer 10).

Layer 8 (160–165cm bd): A thin lens of light gray (10YR 7/1) storm deposited sand was recorded between 160–165cm bd, an indication that the settlement was located near the sea. There was no evidence of any such storm deposited sand in TP 14 or 17, most likely due to their location further inland.

Layer 10b (165–175cm bd): A thin lens of cultural deposit (165–175cm bd) was recorded sandwiched between the sand lens and the former foreshore below. All recovered ceramics were in the Arapus-style.

Layer 11 (175cm bd): The sterile former foreshore comprising primarily water-worn coral cobbles, pebbles and branches was located at this level. There was no sign of any tephra content within the matrix as seen in TP 14, Layer 12.

TP 14 (11.57m asl) was located on the edge of a terrace, which marks the southern boundary of the site, some 12m above the live coral datum. Recovered cultural material was largely restricted to midden associated with the initial settlement of the area, including Arapus ware pottery.

Layer 1 (0–10cm bd): Layer 1 was a black (10YR 2/1) humic topsoil found across the site which graded into the lower layer with an increasing tephra content.

Layer 2i (10–30cm bd): This layer, a very dark gray (10YR 3/1) matrix, consisted primarily of weathered Kuwae tephra. No midden materials were recovered. Some evidence of gardening activity in the form of pit features were in evidence.

Layer 2iii (30–60cm bd): Remains of the distinctive light brownish gray (10YR 6/2) Kuwae tephra comprised this layer. Again some evidence of gardening activities were recorded. No midden materials were recovered.

Layer 5i (60–100cm bd): This somewhat sterile dark gray (10YR 4/1) developed soil layer was substantially comprised weathered Nguna tephra. The striking difference between this layer and the layer below provided a clear boundary.

Layer 10 (100–160cm bd): This layer represented the concentrated *in situ* cultural deposit which consisted principally of a black (10YR 2/1) silty sediment with concentrated coral and basalt cobbles, and midden materials. Several Erueti-style rims were recovered from the upper level of this layer but from 110cm bd all the pottery consisted solely of Arapus-style sherds (as per TP 4, Layer 10 and TP 17, Layer 10). At the lower level of this layer (140–160cm bd) concentrated large shellfish were encountered along with frequent bone remains. Pottery and the black silt content became increasingly sparse. A large *Trochus niloticus* from this lower level returned a radiocarbon determination of 3010±50 BP (ANU-11160) 2868–2706 BP. Large water worn coral boulders lay at the interface of this layer and the increasingly sterile layer below.

Layer 12 (160–230cm bd): This basal layer consisted of the uplifted former foreshore with a high content of an as yet unidentified tephra. At a depth of 230m the tephra content had completely disappeared (as per TP 4, Layer 11 and TP 17, Layer 11b).

ST 17 (11.24m asl) was located 25m west of ST 14 and further revealed the nature and extent of the initial settlement site. All recovered ceramics were Arapus-style sherds.

Layer 1 (0–15cm bd): This layer again consisted of black (10YR 2/1) humic topsoil, bereft of midden material.

Layer 6 (15–70cm bd): A greatly increased tephra content characterised this very dark gray (10YR 3/1) developed soil indicating that much of it consisted of weathered tephras. No pure tephra was in evidence. An abrupt change in the matrix marked the appearance of Layer 10. No midden material was in evidence.

Layer 10 (70–120cm bd): Black (10YR 2/1) gritty soil with concentrated midden remains characterised this layer. Frequent basalt and coral cobbles were also recorded. The pottery consisted completely of Arapus-style sherds (as in TP 14, Layer 10 and TP 4, Layer 10). At a depth of 110–130cm bd concentrated large shellfish were again noted in association with frequent bone remains (as per TP 14, Layer 10) coinciding with a decrease in the black soil content and pottery remains.

Layer 11a (120–160cm bd): This layer was primarily made up of concentrated water worn coral cobbles and boulders which seem most likely to represent a natural storm beach deposit (as per TP 14, Layer 10/11 interface). Midden material was sparse and any that was recovered appeared to have filtered down from the layer above. Again a large *Trochus niloticus* was selected for dating from this layer and returned a radiocarbon determination of 3200±50 BP (ANU-11159) 3152–2853 BP.

Layer 11b (160–180cm bd): This level marked the appearance of the former foreshore which in this test pit did not include any tephra content (as in TP 14, Layer 12). This layer was completely bereft of midden remains.

Appendix Three Petrographic reports compiled by William Dickinson

Petrographic Report WRD-147 (2 July 1997)[revised 22 Nov 1998]

Sand Tempers in Prehistoric Sherds from Erromango in Vanuatu

Petrographic examination of sand tempers in thin sections of prehistoric sherds from Erromango in southern Vanuatu (sherds and thin sections provided by Matthew Spriggs and Stuart Bedford of Australian National University) was undertaken to address three key issues: (1) comparison of Erromango sherd tempers with tempers in sherds from other islands strung along the New Hebrides island arc chain (Petro Rpt WRD-117); (2) comparison of tempers in sherds from Ifo on the southeast coast with those from Ponamla on the northwest coast of Erromango; (3) comparison of tempers in rare Lapita sherds from both Ifo and Ponamla with those in abundant undecorated and presumably younger sherds from both sites.

The following materials were available for study: (1) new thin sections of four sherds (#61-1 through #61-4) from Ponamla, including one (#61-4) decorated Lapita sherd; (2) new thin sections of four sherds (#62-1 thru #62-4) from Ifo, including one (#62-1) decorated Lapita sherd; (3) three old thin sections of non-Lapita Ifo sherds that were originally examined by T.S. Dye for Matthew Spriggs.

The new non-Lapita thin sections were made from sherds personally selected, from collections at ANU, both as representative of non-calcareous Ponamla and Ifo tempers and to provide the most informative temper display (least weathered sherds with abundant sand temper). The Lapita sherds were provided separately by Spriggs and Bedford. The older thin sections from Ifo were culled at ANU from a set of sixteen (16) examined by Dye previously, and were selected to be the most informative regarding presumably indigenous placer tempers (F287, M203) and an anomalous temper (M402) that is probably non-local (exotic to Erromango). Two other older thin sections (LF84, C197) examined at ANU also contain especially informative temper sands because they include abundant calcareous limeclasts clearly representing fragments of indurated carbonate rock (rather than modern reef detritus), presumably derived from erosion of the uplifted coastal terraces prevalent along the coasts of Erromango.

Geologic Context

In southern Vanuatu, the Tafea district [acronym of Tanna-Anatom (Aneityum) Futuna-Erromango-Aniwa] is geologically a southern continuation of the central island chain of central Vanuatu (Mitchell and Warden 1971; Carney *et al.* 1985), and owes its existence to the construction of stratovolcanoes and lava shields erupted along a magmatic arc in response to late Neogene subduction at the New Hebrides trench lying to the west of Vanuatu. The volcanogenic bedrock of Tafea is entirely latest Miocene or younger (<6 Ma) in age (MacFarlane et al. 1988). Futuna has been displaced eastward from other islands of Tafea by extensional backarc rifting along the Coriolis Trough (Carney and MacFarlane, 1979; MacFarlane et al. 1988), but was formed by Tafea arc volcanism prior to its separation from the axis of the present volcanic arc.

Several Quaternary volcanic cones rising individually to elevations of 760-885 m on Erromango were constructed atop a topographically more subdued volcanogenic base underlain by the Miocene-Pliocene Plateau Formation that has yielded a maximum K-Ar age of 5.8 Ma (Colley and Ash 1971). The Plateau Formation is composed mainly of varied polygenetic volcanic breccias and reworked volcanic sediments, with pyroclastic and epiclastic components difficult either to distinguish or to interpret, sedimentologically. The younger volcanic units, named for the various edifices they form, include both lavas and polygenetic fragmental rocks in variable proportions. The island is almost entirely ringed by limestone terraces representing uplifted Pliocene-Quaternary fringing reefs tilted gently downward to the ENE, away from the New Hebrides trench.

Despite the heterogeneity of the volcanogenic breccias and lavas of Erromango, the basic nature of Neogene volcanism did not vary appreciably through time, and the petrology of the rocks is broadly consistent (Colley and Ash 1971). About 90% of the assemblage is basalt and basaltic andesite, with the remainder somewhat less mafic andesite. The rocks are composed mainly of plagioclase and clinopyroxene, with orthopyroxene and olivine each also present in perhaps 20% of the rocks, but in a different 20% in each case. Rare hornblende occurs sparingly in some andesites.

On geologic grounds there is thus every reason to suppose that volcanic sands throughout Erromango might be similar in overall composition. As the minerals in the volcanic rocks all occur at least locally as phenocryts of sand size, derivative sands are expected to contain varying proportions of the constituent mineral grains, together with volcanic rock fragments (VRF) of generally mafic character and microlitic to intergranular internal textures (microcrystalline with little or no interstitial volcanic glass except for glassy grains derived from quenched fragments in pyroclastic deposits). As olivine weathers readily, and both orthopyroxene and hornblende are volumetrically minor in any Erromango volcanic rocks, plagioclase and clinopyroxene are expected to be the principal mineral grains in derivative volcanic sands. Admixtures of limeclasts derived from the uplifted reef terraces may enter streams draining to the coast, and calcareous reef detritus from modern fringing reefs that surround the island but are widest off the east coast, may appear in coastal sand aggregates. Opaque iron oxide mineral grains, ubiquitous in subordinate amounts in nearly all volcanic sands, can also be expected.

The Ifo River, draining to the southeast coast near Ifo, heads in uplands underlain by the Plateau Formation, which underpins the southeast cape lacking in younger volcanic edifices (Colley and Ash 1971:Fig. 2). The Ponamla River, entering the sea at Ponamla on the northwest coast, drains the slopes of Mt. William, a cone of Pleistocene age built largely of pyroclastic deposits. Provided all or most of the pottery found at each site was made locally or nearby, the Ifo and Ponamla tempers should reflect some approximation of the full range of sands available for temper on Erromango. The empirical differences noted between Ifo and Ponamla tempers are thus not surprising, but not readily predictable or explicable in terms of the known areal geology of Erromango.

Indigenous Erromango Tempers

Tempers interpreted as indigenous to Erromango, and indicative of locally made pottery, include the same array of expected grain types, although in variable proportions. Most are moderately sorted, subangular to subrounded aggregates, medium sand in median size, collected from local streams or beaches, with the former favored by the limited rounding and sorting of the grain aggregates. Three variants of the temper sands inferred to be of stream origin can be distinguished: (1) a plagioclase-poor and pyroxene-rich temper type present in three non-Lapita sherds from Ponamla (Table 147-1); (2) a plagioclase-rich non-placer temper type in three sherds, one of which is Lapita ware, from Ifo (Table 147-2); (3) a pyroxene-rich (plagioclase-poor and VRF-poor) placer temper type in three non-Lapita sherds from Ifo (Table 147-3). The proportion of temper is variable from sherd to sherd, but distinct contrasts in grain size between the coarsest silt in the clay pastes and the finest temper grains implies manual addition of temper to generally sand-free clay bodies. A well sorted pyroxene-rich placer temper of beach origin in the Lapita sherd (#61-4) from Ponamla contains the following subrounded to subangular grain types of medium sand size (frequency percentages based on an areal count of 550 grains in thin section): plagioclase, 6; pyroxene, 68; olivine, 6; opaque iron oxide, 17; volcanic rock fragments, 3. The olivine content of the placer beach sand is higher than in the stream sands of the other nine sherds (Tables 147-1, 2, 3).

Three key parameters serve to set the main temper variants apart: (1) the ratio of pyroxene to plagioclase is only 1:1 to 2:1 in non-placer Ifo tempers but 5:1 to 20:1 in Ponamla tempers and 10:1 to 50:1 in Ifo placer tempers; (2) bulk percentages of volcanic rock fragments in Ponamla and non-placer Ifo tempers exceed 25%, but total less than 25% in Ifo placer tempers; (3) pyroxene grains consistently represent more than half the grain aggregates in Ifo placer tempers, but generally less than half in Ponamla and non placer Ifo tempers.

All or nearly all the calcareous grains in the Erromango tempers appear to be reworked limeclasts, rather than modern reef detritus. Their morphology and internal fabric are compatible with derivation from erosion of uplifted limestone terraces and with interpretation of the three main temper types as stream placers, because beach placering would not only tend to round the heavy mineral grains more, but also to admix calcareous reef detritus in at least minor amounts. Proportions of different kinds of volcanic rock fragments overlap too broadly, and vary too much within each temper type, to be helpful in distinguishing between the different temper variants. Bulk percentages of opaque iron oxide grains are also highly variable over similar ranges within the different temper types. Empirically, the distinctions drawn between the different temper variants seem robust, but the differences noted are not great in a generic sense and further work might reveal the presence of intermediate temper types across a spectrum of indigenous Erromango wares.

The well sorted beach sand temper in the Lapita sherd from Ponamla contains no grain types that could not have been derived from Erromango, but bears an overall generic resemblance to placer sand tempers in Lapita sherds from Nendo and the Reef Islands in the Santa Cruz Group at the northern end of the New Hebrides island arc (Table 117-5, Petro Rpt WRD-117). In particular, the olivine content is comparable. Santa Cruz/Reef placer tempers also contain traces of hornblende, however, which is absent from the Ponamla Lapita sherd. Moreover, pyroxene grains in the Ponamla sherd display distinctly more greenish tints than the nearly colorless (in thin section) pyroxenes of Nendo-Reef placer tempers. Subordinate populations of polyminerallic volcanic rock fragments in the Ponamla Lapita sherd and in Nendo-Reef Lapita sherds include similar tachylitic grains (semi-opaque volcanic glass heavily charged with opaque iron oxides), with pilotaxitic to hyalopilitic textures (plagioclase microlites set in glassy groundmasses), but the Nendo Reef tempers also contain prominent brownish to reddish, glassy to hyalopilitic grains not present in the Ponamla Lapita or other Erromango sherds. On balance, therefore, placer temper in the Ponamla Lapita sherd is inferred to represent an indigenous coastal beach sand derived at least in part from olivine-bearing Erromango bedrock not sampled by the other Erromango tempers. Olivine microphenocrysts, although subordinate to plagioclase and clinopyroxene microphenocrysts, occur in most Erromango basalts, and basalt is the commonest rock type of the local volcanogenic assemblage (Colley and Ash 1971:70-71). The gross similarity of Ponamla and Nendo-Reef Lapita tempers can be ascribed to sedimentological convergence of sand types from similar environmental settings on different islands along an island arc of generally uniform geotectonic character.

Exotic Erromango Temper

One Erromango sherd (M402) contains an anomalous temper, wholly unlike the indigenous temper variants in significant generic respects. The temper sand is moderately sorted, of medium sand median size, composed of subangular to subrounded quartz bearing sand admixed with subordinate well sorted, subrounded to rounded calcareous grains (21%) of probable reef detritus. The paste is markedly silty clay containing tiny detrital grains of the same general type that form the terrigenous sand fraction of the sherd. The paste texture suggests that the clay body may have been naturally tempered, without requiring manual addition of temper sand. The admixture of reef detritus implies collection at a coastal setting, perhaps of deltaic, estuarine, or lagoonal mud.

Frequency percentages of grain types, summed to 100% exclusive of the admixed calcareous grains, are as follows (based on an areal count of 380 terrigenous grains and 100 calcareous grains):

24 plagioclase feldspar grains

20 monocrystalline quartz grains

20 amphibole grains, including green-brown and nearly colorless hornblende 10 pyroxene grains, dominantly clinopyroxene with a trace of orthopyroxene 10 quartz-mica 'tectonite' fragments, varied metasedimentary schist and phyllite

8 quartzite rock fragments, including foliated (metamorphic) varieties

4 microgranular phaneritic rock fragments, dominantly diorite-gabbro

4 opaque iron oxide grains

The mineralogy of the temper almost certainly indicates derivation of the sherd from New Caledonia. The high proportion (more than a third) of combined quartz-rich grains, including metasedimentary quartzite and phyllite-schist fragments, could not have been derived from bedrock on any other island southeast of New Guinea. The fragments of intrusive igneous rocks (diorite-gabbro) are compatible with New Caledonian sources, which include an overthrust mafic and ultramafic massif as a prominent geologic feature of the island.

The significance of the imported sherd for culture history is beyond the scope of petrographic analysis to assess, but isolated sherds of New Caledonian origin have also been recovered on Malo and Santo in central Vanuatu (Petro Rpt WRD-118). Prehistoric contacts between Vanuatu and New Caledonia are not surprising in principle, because each is closer to the other than to any other island groups in Oceania. The temper in the exotic Erromango sherd is not, however, identical to any tempers in the exotic Malo and Santo sherds. The sand temper in the Malo sherd was derived from the ultramafic massif of New Caledonia and consequently contains prominent grains of chrome spinel, a mineral similar to magnetite but translucent (in thin section) in deep red hues, rather than being fully opaque in microscopic view (Dickinson 1971). Two sherds from Bubit Maras on Santo contain temper sands that are approximately 85% quartz and quartzite grains, similarly abundant in sherds from New Caledonia (Galipaud 1990). Quartzose tempers are otherwise unknown to date from Pacific Island sherd suites. Importation of ceramics and perhaps other goods from New Caledonia to Vanuatu, or perhaps exchange between the two, seems indicated for some part of prehistory, and broad contact involving more than one part of New Caledonia is implied by the variety of New Caledonian tempers represented by the exotic sherds recovered on different islands in Vanuatu.

Summary Conclusions

In terms of the main questions that petrographic study was intended to address, the following are the key inferences to be made:

1. Erromango temper sands are similar in a broad generic sense, as andesitic arc tempers derived from an active to dormant island-arc volcanic chain, to other Vanuatu tempers (Petro Rpt WRD-117), but are dissimilar in detail to all other Vanuatu tempers studied to date. A single sherd examined from Tanna, the only other island within Tafea for which petrographic data on temper is available, contains a volcanic sand dominated by vitric volcanic rock fragments (>75%) of yellow-brown glass unlike the darker brown glass in Erromango temper sands. Tempers from Efate and the nearby Shepherd Islands contain characteristic pumiceous grains of a type not observed in any Erromango temper variants, and the volcanic rock fragments of felsitic internal texture characteristic of the tempers in sherds from Malekula and its neighboring offshore islets are also wholly lacking in Erromango sherds. Tempers from Santo and Malo, as well as most Malekula tempers, contain subordinate hornblende grains, which do not occur in any of the Erromango temper variants. Available petrographic data thus seemingly preclude importation of any of the Erromango sherds from anywhere within central Vanuatu.

2. Ifo and Ponamla tempers, in the sherds examined, are distinguishable from details of grain proportions, but the same generic array of grain types is present in sherds from both sites, and proportions of grain types in placer and non-placer variants of Ifo tempers can be viewed as bracketing the range of grain proportions in Ponamla tempers. There is no certainty that all Ponamla and Ifo sherds would be distinguishable in terms of the grain parameters derived here, nor for that matter that all the sherds studied were made with local sands at the sites where they were recovered. The overall generic similarity of the temper variants suggests, however, that each sherd contains indigenous temper collected somewhere on Erromango. Trying to distinguish with full confidence between tempers collected at different sites on Erromango might require extensive petrographic analysis, with no guarantee of ultimate success.

3. There is no hint from the petrography of the Erromango non-Lapita sherds that they represent other than local wares made somewhere on the island, and the single Lapita sherd from Ifo contains a temper similar both texturally and mineralogically to the non placer type of non-Lapita Ifo temper. The single Lapita sherd from Ponamla contains better sorted and more olivine-rich temper sand than other Erromango sherds, but also probably contains indigenous Erromango sand derived from some nearby coastal locale.

4. One anomalous non-Lapita sherd from Ifo contains a wholly exotic temper sand almost certainly indicative of importation from New Caledonia.

References Cited

Carney, J.N., and A. MacFarlane 1979. Geology of Tanna, Aneityum, Futuna, and Aniwa: Port Vila, Vanuatu, New Hebrides Condominium Geological Survey Regional Report Series, 71.

Carney, J.N., A. MacFarlane, and D.I.J. Mallick 1985. The Vanuatu island arc: an outline of the stratigraphy, structure, and petrology, in A.E.M. Nairn, F.G. Stehii, and S. Uyeda (eds.) *The Ocean basins and margins: Vol. 7A, The Pacific Ocean*: New York, Plenum Press, p. 683–718.

Colley, H., and R.P. Ash 1971. The Geology of Erromango: Port Vila, Vanuatu, New Hebrides Condominium Geological Survey Regional Report Series, 112.

Dickinson, W.R., 1971. Temper sands in Lapita-style potsherds on Malo. *Journal of the Polynesian Society*, 80:244–246.

Galipaud, J.-C., 1990. The physico-chemical analysis of ancient pottery from New Caledonia, in M. Spriggs (ed), *Lapita design, form, and composition*: Canberra, Australian National University, Research School of Pacific Studies, Department of Prehistory, Occasional Papers in Prehistory No. 19, pp. 134–142.

MacFarlane, A., J.N. Carney, A.J. Crawford, and H.G. Greene 1988. Vanuatu - a review of the onshore geology, in H.G. Greene and F. L. Wong (eds). *Geology and offshore resources of Pacific island arcs - Vanuatu region*: Houston, Texas, Circum-Pacific Council for Energy and Mineral Resources Earth Science Series, Vol. 8:45–91.

Mitchell, A.H.G. and A.J. Warden 1971. Geological evolution of the New Hebrides island arc. *Geological Society of London Journal*, 127:501–529.

Table 147-1. Frequency Percentages of Grain Types in Non-Lapita Ponamla Tempers
(based on areal or traverse counts of n grains per thin section)

sherd no.	#61-1 (n=400)	#61-2 (n=125)	#61-3 (n=290)	range (n=125-400)	mean (n=272)
grain type					
plagioclase	7	3	5	3-7	5
pyroxene[1]	44	57	32	32-57	44
olivine[2]	tr	tr	tr	tr	tr
intergran[3]	12	12	11	11-12	12
microlitic[4]	5	11	5	5-11	7
vitric[5]	24	7	26	7-26	19
(tot VRF)[6]	(41)	(30)	(42)	(30-42)	(38)
opa Fe Ox[7]	8	4	18	4-18	10
calcareous	tr	6	3	tr-6	3

1 predominantly clinopyroxene, with trace amounts (~l grain per slide) of orthopyroxene

2 trace amounts reported imply ~1 grain per slide

3 intergranular crystalline groundmass of plagioclase laths with interstitial clinopyroxene

4 Varied internal textures with visible plagioclase microlites present

5 abraded fragments of brown (to reddish) volcanic glass (of mafic andesitic to basaltic composition), some with tiny crystallites included

6 sum of intergranular, microlitic, and vitric volcanic rock fragments

7 detrital grains of opaque iron oxides, probably mainly magnetite

8 calcareous grains (dominantly or exclusively detrital limeclasts reworked from carbonate rock rather than modern reef detritus)

Table 147-2. Frequency Percentages of Grain Types in Non-Placer Ifo Tempers (#62-1 is a Lapita Sherd)
[based on areal or traverse counts of n grains per thin section] (footnotes same as for Table 147-1)

sherd no.	#62-1 (n=230)	#62-2 (n=85)	#62-3 (n=180)	range (n=85-230)	mean (n=165)
grain type					
plagioclase	32	15	22	15-32	23
pyroxene[1]	26	32	18	18-32	25
olivine[2]	tr	tr	tr	tr	tr
intergran[3]	4	12	3	3-13	7
microlitic[4]	9	8	9	8-9	9
vitric[5]	12	5	27	5-27	15
(tot VRF)[6]	(25)	(26)	(39)	(25-39)	(30)
opa Fe Ox[7]	1	27	9	1-27	12
calcareous	16	tr	12	tr-16	9

Table 147-3. Frequency Percentages of Grain Types in Placer Ifo Tempers (all non Lapita) [based on areal or traverse counts of n grains per thin section] (footnotes same as for Table 147-1)

sherd no.	#F287 (n=325)	#M203 (n=395)	#62–4 (n=315)	range (n=315–395)	mean (n=345)
grain type					
plagioclase	4	5	2	2–5	4
pyroxene[1]	55	68	76	55–75	66
olivine[2]	tr	tr	tr	tr–4	1
intergran[3]	9	12	4	4–12	8
microlitic[4]	7	5	1	1–7	4
vitric[5]	8	3	1	1–8	4
(tot VRF)[6]	(24)	(20)	(6)	(6–24)	(17)
opa Fe Ox[7]	16	6	12	6–16	11
calcareous	1	1	–	<2	1

Petrographic Report WRD-138 (10 June 1997)

Petrography of deep sherds and stream sediment from Efate, Vanuatu

Two fragments of pottery from an unusual provenience on Efate in Vanuatu, and some stream sediment from a nearby ravine, were provided for petrographic analysis by Matthew Spriggs and Stuart Bedford. The sherds came from beneath a cemented horizon at the Mangaasi site below the depth of excavation achieved by Garanger. The purposes of my study were to assess the nature of the temper in the two deep sherds and to further test for the nature of sediment available for collection as temper on Efate. Megascopically, the sherds resemble typical indigenous Efate sherds, rounded and thick with scattered outsized pumiceous clasts visible with a hand lens.

Based on frequency percentages of sand grain types (derived from counts of 400 grains in each sherd), the tempers in the two sherds are more feldspathic (plagioclase 82–83 vs 46–73) than the standard Efate temper type established by past work (Petro Rpt WRD-117), but in other respects are closely similar texturally and composed of the same subordinate grain types (parentheses show contrasts in percentage frequency for each with respect to standard Efate temper): clinopyroxene grains (1 vs 2–10), pale brown volcanic glass fragments (10–14 vs 19–37), opaque iron oxide grains (2–5 vs 2–10), and microlitic volcanic rock fragments (1 vs 0–1). The lower percentages of pyroxene grains and glass fragments are clearly just a reciprocal function of the greater abundance of plagioclase. The predominance of pale brownish glass fragments among the volcanic rock fragments, and the absence of non-opaque mineral grains other than plagioclase and clinopyroxene, with the former more abundant, are the key characteristics of indigenous Efate temper. There seems no reason to doubt that the two deep sherds were made locally, either at Mangaasi or nearby on Efate.

The stream sediment is actually fine gravel, rather than sand, composed dominantly of pumiceous volcanic rock fragments, composed dominantly of pale volcanic glass but containing sparse microphenocrysts of plagioclase feldspar. Also present are ferruginous pedogenic grains of irregular shape evidently reworked from indurated lateritic soil. The latter variously contain pumiceous clasts and either plagioclase or clinopyroxene crystals. The character of the stream sediment is fully compatible with the nature of temper sands interpreted as indigenous to Efate (Petro Rpt WRD-117).

Petrographic Report WRD-155 (12 July 1997)

Temper sands in early plainware sherds from Malekula, Vanuatu
The following two (2) early plainware sherds from Malekula were sent for petrographic examination by Stuart Bedford:

NHM-47: sherd from lowest cultural level (~2500 BP), Navaprah cave site [bag 47, TP3 130-150] NHM-320: sherd from lowest cultural level (~2700 BP), Malua Bay site [bag 320, TP 3, 60-80]

Both contain tempers that are partly calcareous but are otherwise volcanic sand, rich in volcanic rock fragments and probably indigenous to Malekula. Admixture of calcareous grains has not been observed in previously studied Malekula sherds (Petro Rpt WRD-117), but volcanic rock fragments in the temper aggregates include felsitic types (microcrystalline feldspar-quartz mosaic fabric) characteristic of other Malekula tempers. There is no petrographic reason to doubt origin of the tempers from Malekula, and no non-Malekula tempers known to date from Vanuatu or nearby island groups closely resemble the early Malekula plainware tempers. Felsitic internal texture is typical of volcanic rock fragments in Malekula temper sands, and is interpreted provisionally as positive evidence for indigenous origin.

The tempers are moderately (NHM-320) to well (NHM-47) sorted, fine-to medium-grained sands composed largely of subrounded grains, and are probably of beach derivation. In most instances, calcareous grains in Pacific islands temper sands are reef detritus that provides conclusive evidence for coastal derivation, but in this case the calcareous grains are probably detrital limeclasts reworked from uplifted limestone strata of Pliocene-Pleistocene age forming extensive coastal terraces and plateaus, especially on northern Malekula (Mitchell 1966, 1971). Evidence for limeclast origin of the calcareous grains includes uniformly sugary internal textures, with no hint of skeletal structures, and non-calcareous impurities not expected for modern reef debris but commonly imbedded in uplifted carbonate strata.

The most straightforward interpretation of the tempers is derivation from a coastal locale, or stream near the coast, on Malekula itself. Absence of such calcareous temper in younger Malekula wares may have stemmed from bad experiences with the calcareous aggregates during firing, or a shift to non-calcareous tempers as improving technology achieved higher firing temperatures, or just fortuitously from vagaries of sampling.

Table 155-1 shows the general composition of the two tempers, which are different sands in detail, but contain the same overall grain types.

References Cited
Mitchell, A.H.G. 1966. Geology of South Malekula: Port Vila, Vanuatu. New Hebrides Condominium Geological Survey Report No. 3,41 pp.

Mitchell, A.H.G. 1971. Geology of Northern Malekula: Port Vila, Vanuatu, New Hebrides Condominium Geological Survey, Regional Report Series, 56 pp.

Table 155-1. Frequency Percentages of Grain Types in Temper Sands of Early Plainware Potsherds from Malekula (based on counts of n grains in thin section)

sherd number => grain type	NHM-47 (n=305)	NHM-320 (n=225)
plagioclase feldspar	15	43
clinopyroxene	18	3
hornblende	tr	1
opaque iron oxide	5	2
volcanic rock fragments	(45)	(34)
felsitic	12	13
microlitic	22	15
glassy	8	4
microgranular	3	2
calcareous limeclasts	17	17

Petrographic Report WRD-180 (5 January 1999)

Petrography of sand tempers in selected Late-style sherds from Malekula in Vanuatu and comparison with other Malekula sherds

Thin sections of ten late-style sherds from excavations and surface collections on northwest Malekula (n=9) and neighboring Wala Is. (n=l, #3), offshore from northeast Malekula, were sent to me by Stuart Bedford for petrographic examination. Five of the Malekula sherds are from coastal sites (Nuas, #1; Waal, #5; Navaprah, #6; Malua Bay, #7 & #10), with the other four from inland villages, one from Metkhun (#8) and the other three from Chachara (#2, #4, #9). Stylistically, three are from ribbed pots (#1, #2, #3), two from incised 'bullet' pots (#4, #7), four from large pots (#6, #9, #10) or bowls (#8), and one decorated in Mangaasi style (#5).

Available for comparison were thin sections of twenty sherds collected previously by Richard Shutler from various locales on Malekula and nearby islets (Petro Rpt WRD117, Table 117-2): six from Malekula (one each from Amok, Bethel, Lakatoro, Litzlitz, Tuato, and Port Stanley), and the remainder from offshore islets off northeast Malekula (Vao, n=1; Atchin, n=5; Walarono, n=l; Rano, n=2; Uripiv, n=2; Uri, n=l; and an islet in Port Stanley between Lakatoro and Uri, n=2).

Two sherds of early plainware sherds from the lowest cultural levels at Navaprah (#NMH-47) and Malua Bay (#NHM-320) have also been examined in thin section (Petro Rpt WRD-155). Although each contains hybrid temper sand with a significant proportion (~17%) of calcareous grains, probably detrital limeclasts from uplifted limestone outcrops, the terrigenous sand fraction is volcanic sand similar to other Malekula temper sands.

General Character of Malekula Temper Sands

Despite the prevalence of calcareous sand on Malekula beaches away from river mouths (Mitchell 1966:23), and the widespread distribution of post-Pliocene limestone plateaus and terraces parallel to many coasts (Mitchell 1972), especially around northern Malekula, no calcareous sand grains are present in any except the two early Malekula sherds. The tempers are typically moderately sorted, subangular to subrounded volcanic sands, and probably represent stream alluvium or colluvial detritus derived from Miocene volcanic and volcaniclastic bedrock units of the island interior (Mitchell 1966, 1971). As the calcareous grains in the hybrid sand tempers of the two early Malekula sherds are apparently detrital limeclasts, the sedimentology of the hybrid tempers might not differ greatly from that of the presumably interior sands lacking calcareous grains.

Textures of temper and paste leave open the question of whether temper sands were added manually to silty clay bodies, or whether the volcanic sand is natural temper imbedded in clay-rich alluvial or fluvial sediment. In about a third of the sherds (with like proportions in both Shutler and Bedford collections), there appears to be a break in grain size between the smallest sand grains and the coarsest silt grains imbedded in the clay pastes (suggesting manual addition of temper), but in the other sherds, sand grains pass gradationally downward in size toward the size of silt grains imbedded in the clay pastes (suggesting natural temper). In neither case, however, are the textural relations conclusive, and it may be that ancient potters added sand in varying proportions to silty and sandy clay bodies containing varying proportions of natural temper.

Geographic Origins of Malekula Temper Sands

In no two of the sherds are the volcanic temper sands identical in either texture or composition, with the possible exception of closely comparable individual sherds from Tuato on the mainland (#12-6) and Uripiv Island offshore (#12-12). One sherd (#2) from Chachara (inland) and one (#10) from Malua Bay (coastal), however, contain closely related temper sands rich in feldspar (plagioclase) grains, whereas lithic fragments of polycrystalline igneous rocks are the most abundant grains in all the other temper types (except that plagioclase is marginally more abundant than lithic grains in one of the hybrid sand tempers). Although differences in pyroxene and hornblende content and ratio are the most quantitatively reliable measures of variation, the internal textures of volcanic rock fragments provide a readily observable measure that is qualitatively robust. The wide range of lithic grain types characteristic of different sherds, coupled with the comparative uniformity of lithic grain types in each individual sherd, suggests that multiple local alluvial or colluvial sources of temper sand were used. The occurrence of one sherd in each collection (sherds from Port Stanley and Nuas) in which the dominant rock fragments are grains of intrusive microdiorite, with granular internal textures of non-volcanic aspect, supports that inference. Although hundreds of microdiorite dikes and sills and plugs are widespread on Malekula (Mitchell 1966:14-17; Mitchell 1971:18-20), they are nowhere the dominant bedrock type except over small areas. Sands in which the dominant rock fragments are intrusive rock were doubtless derived, therefore, from small local drainages. As shown by Table 180-1, subordinate microdiorite grains are present in a number of the volcanic sand tempers as well.

Volcanic rock fragments in Malekula sherds include dominant or abundant felsitic grains with groundmasses formed by intergrown mosaics of quartz and feldspar (Petro Rpt WRD-117). Table 180-1 indicates the relative prevalence of felsite grains among volcanic rock fragments in different Malekula sherds, other types of volcanic or associated igneous rock fragments present in each case, and the relative abundance of clinopyroxene and hornblende in general terms. The microlitic volcanic rock fragments of the table display microscopic plagioclase microlites, commonly of pilotaxitic (felted) internal texture. No areal or stylistic patterns in composition emerge from the tabulation, apart from the admixture of calcareous grains in both early sherds (NMH-47 & NMH-320), leading to the conclusion that the variability of Malekula bedrock in detail precludes any correlation of temper type with specific areas of ceramic manufacture on a broad scale. Although the sherd tempers as a group form a spectrum of generically related sands, occurrences of similar tempers in sherds from widely separated locales suggests that local variability in available temper sands is repetitive in many areas of the island. Although petrographic analysis thus seems satisfactory for distinguishing between sherds from Malekula, as opposed to other islands in Vanuatu (Petro Rpt 117), it apparently cannot achieve effective distinctions among tempers from different parts of Malekula.

Santo-Malo feldspathic tempers (Petro Rpt WRD-117, Table 117-1) differ sharply from the feldspathic tempers of Malekula (Chachara #2 and Malua Bay #10) because the plagioclase in the Santo-Malo tempers is almost exclusively fresh (transparent) and unaltered, evidently representing phenocrystic debris from volcanic rocks, whereas the plagioclase in the Malekula feldspathic tempers is almost exclusively altered plagioclase (albitic composition), clouded optically by myriad microscopic inclusions. The cloudy altered plagioclase in the Malekula tempers was presumably derived from metavolcanic wallrocks adjacent to some one or more of the many small subvolcanic intrusions of microdiorite on Malekula, or from altered portions of the intrusions themselves.

Summary of Malekula Temper Studies

Examination of 32 prehistoric Malekula sherds in thin section shows that Malekula tempers can be successfully distinguished petrographically from other Vanuatu tempers, but that the spectrum of variation in temper composition is essentially as great locally as for the island as a whole, suggesting (a) that temper sands were collected from local drainages tapping restricted samples of heterogeneous island bedrock, which does not vary regionally in any systematic way (Mitchell 1966, 1971), and (b) that petrographic study is not a fruitful method for addressing questions of the places of manufacture of indigenous Malekula pottery.

References Cited

Mitchell, A.H.G. 1966. Geology of Southern Malekula: British Service, New Hebrides: New Hebrides Condominium Geological Survey Report No. 3, 41.

Mitchell, A.H.G. 1971. Geology of Northern Malekula: British Service, New Hebrides: New Hebrides Condominium Geological Survey Regional Report, 56.

Mitchell, A.H.G. 1972. Geology of Malekula: New Hebrides Geological Survey Sheet 7, scale 1:100 000

Table 180-1. Mineralogical-Petrological Character of Malekula Temper Sands [S = Shutler sherds; B = Bedford sherds]

Sherd No.	Locale	No Nature of Lithic Fragments and type			Pyribole Content		
		mainly felsite	felsitic & microlothic	felsite & microdio	traces only	clinopyx> hornblend	hornblend only
S12-2	Lakatoro	X			X		
S12-4	Pt Stanley			X			X
S12-5	Bethel		X		X		
S12-6	Tuato	X			X		
S12-8	Litzlitz	X					X
S635	Amok		X		X		
S12-10	Vao	X				X	
S12-11	Atchin	X				X	
S631	Atchin	X					X
S639	Atchin		X			X	
S641	Atchin		X			X	
S643	Atchin			X		X	
S601	Wala	X			X		
B3	Wala	X				X	
S632	Rano			X	X		
S642	Rano			X			X
S12-12	Uripiv	X			X		
S13-1	Uripiv			X	X		
S12-9	Uripiv	X				X	
S633	islet		X			X	
S633X	islet	X			X		
B1	Nuas			X		X	
B5	Waal			X		X	
B6	Navaprah		X		X		
B7	Malua Bay	X			X		
B10	Malua Bay		X		X		
NMH-47	Navaprah		X			X	
NMH-320	Malua Bay		X			X	
B8	Metkrun		X		X		
B2	Chachara		X		X		
B4	Chachara	X					X
B9	Chachara		X			X	

Petrographic Report WRD-184 (15 March 1999) [revised 30 March 1999]

Petrography of Two Unusual Sherds from Erromango and Malekula
Two unusual sherds, one from Erromango and one from Malekula, were sent for petrographic examination by Stuart Bedford: (a) PON-4, from Unvoriu cave (site ER-L-59) in the Dillons Bay area of the west coast of Erromango; (b) TEN-5, a Tenmaru surface sherd of possible Santo origin from northwest Malekula.

Unvoriu Cave Sherd (PON-4)
The Unvoriu cave sherd is difficult to source because it contains no discernible temper apart from deeply weathered volcanic rock fragments and ferruginous particles of pedogenic origin. Both are probably inherent constituents of residual soil clay collected as the ceramic body that was worked and fired. The sherd thus differs from the tempered Ponamla and Ifo sherds examined previously from Erromango (Petro Rpt WRD-147), but the clay body does not differ optically from the clay bodies of other Erromango sherds, and an indigenous origin seems likely. Perhaps some early or late ceramic technology on Erromango did not involve deliberate tempering of clay. Unless there are archaeological reasons to suspect the sherd might be exotic, there is no reason from the petrography to infer ceramic transfer from any other island to explain its occurrence on Erromango.

Tenmaru Surface Sherd (TEN-5)
The temper sand in the Tenmaru sherd is poorly sorted, subrounded to subangular, fine to medium sand of probable stream origin, and the lack of any significant size contrast between the finest sand grains and the silt imbedded in the clay body suggests that the latter may have been a naturally tempered alluvial clay. The silt content of the clay body is similar to clay bodies in Santo-Malo sherds, but Malekula clay bodies are comparably silty.

The volcanic sand temper in the Tenmaru surface sherd differs both compositionally and texturally, however, from typical Malekula tempers, but falls within the spectrum of feldspathic to lithic temper sands in sherds from Santo (Petro Rpts WRD-117, 157). Because Malekula tempers are so variable in detail (Petro Rpt WRD-180), it would be risky on petrographic grounds alone to conclude definitively that the sherd was not made on Malekula, but derivation from Santo is apparently suspected from its archaeological occurrence and typology, and seems fully compatible with temper analysis. Features of the temper sand particularly suggestive of origin on Santo, rather than Malekula, are the lack of felsitic volcanic rock fragments (intergrown groundmass mosaic of microcrystalline quartz-feldspar), the high ratio of pyroxene to hornblende, the prevalence of fresh and unaltered plagioclase (as opposed to inclusion-charged altered feldspar), and the lack of visible microdiorite grains. Although individual Malekula tempers fail to display one or more of the properties characteristic of Malekula tempers, none are uncharacteristic in all those respects. The texture (sorting and rounding) of the sand grains of the temper is also indistinguishable from comparable Santo temper sands. On balance, therefore, derivation of the Tenmaru sherd from Santo is logical, whereas origin from Malekula seems distinctly less likely. Overall similarities among the volcanic and volcaniclastic bedrock units of Santo and Malekula mean, however, that clearcut generic contrasts cannot be drawn between derivative temper sands, with empirical differences the only reliable criteria for distinguishing Malekula and Santo tempers.

Table 184-1 compares a frequency count of the sand grains with an overall mean for non-placer Santo-Malo tempers, which vary considerably in detail. The only significant compositional anomaly is the higher content of clinopyroxene grains in the Tenmaru sherd (with accordingly fewer plagioclase grains and volcanic rock fragments), as no known non-placer Santo tempers contain more than 22% clinopyroxene, but placer tempers in Malo sherds that were probably derived from Santo contain up to 5 % clinopyroxene. Given the technical difficulty in performing accurate frequency counts for poorly sorted sands, the clinopyroxene content higher than normal does not preclude derivation from Santo.

Table 184-1. Frequency percentages of grain types in sherd TEN-5 from Tenmaru (based on traverse count of 515 grains in thin section), and comparison with mean temper composition in 24 non-placer Santo-Malo tempers (Petro Rpts WRD-117, 157)

grain type	Tenmaru sherd TEN-5	Average Santo Temper
plagioclase feldspar	35	40
clinopyroxene	25	12
hornblende	2	2
opaque iron oxides	6	7
volcanic rock fragments	32	39

Appendix Four Inventory of design motifs

The following is an inventory of design motifs that were assigned a numeric. Often design motifs were clearly not complete and comprised only elements of much larger motifs but they were seen as being sufficiently distinctive to be separately listed. The inventory is restricted to motifs identified from excavated and surface collected sherds related to this more recent research only. This does not claim or aspire to be a complete inventory of all Vanuatu ceramic motifs. Many more can be identified, particularly from the surface sherds recovered by Garanger from Efate and the later ceramic traditions of the north. Each description lists the sites where the motifs occur along with their frequency and illustrated figure.

Erromango motifs

E-motif 1. Horizontal parallel continuous vertical fingernail pinch (Figs. 5.3a–c). Ponamla (21), Ifo (10).

E-motif 2. Horizontal parallel non-continuous vertical fingernail pinch (Figs. 5.3d–f, 5.15d). Ponamla (3).

E-motif 3. Oblique and perpendicular non-continuous fingernail pinch (Fig. 5.3g). Ponamla (1)

E-motif 4. Oblique parallel rows of vertical fingernail pinch (Figs. 5.4a,b, 5.15e–i). Ponamla (35), Ifo (11).

E-motif 5. Horizontal parallel non-continuous rows (2) of fingernail pinch replaced by more random vertical fingernail pinch (Fig. 5.4c). Ponamla (2).

E-motif 6. Horizontal parallel continuous rows of horizontal fingernail pinch creating horizontal rows (Fig. 5.4d,e). Ponamla (2).

E-motif 7. Tightly packed parallel and perpendicular rows of deep fingernail pinch (perpendicular to the pot surface) creating a pronounced central ridge (Fig. 5.4f). Ponamla (13).

E-motif 8. Tightly packed parallel rows of fingernail pinch (angled to the pot surface) creating a much less pronounced central ridge (Fig. 5.4g,h). Ponamla (13), Ifo (11).

E-motif 9. Tightly packed parallel rows of fingernail pinch (perpendicular to pot surface) deeply cut leaving a flat central ridge (Fig. 5.4i). Ponamla (7).

E-motif 10. Butterfly-like fingernail pinch (Fig. 5.5f). Orientation and larger design undetermined. Ponamla (2).

E-motif 11. Curvilinear rows of lopsided fingernail pinch with emphasis more on one finger than the other (Fig. 5.5g). Orientation and larger design undetermined. Ponamla (2).

E-motif 12. Very large random fingernail pinch (Fig. 5.5h). Orientation and design undetermined. Ponamla (1).

E-motif 13. Parallel rows of continuous single vertical fingernail gouge (Fig. 5.5i). Ponamla (8).

E-motif 14. Random single fingernail gouge (Fig. 5.5j). Ponamla (3).

E-motif 15. Loosely parallel rows of discontinuous vertical fingernail gouge (Fig. 5.5k). Ponamla (1), Ifo (2).

E-motif 16. Uniform parallel rows of heavy discontinuous vertical fingernail gouge (Fig. 5.5e). Ponamla (6).

E-motif 17. Random horizontal fine fingernail impression (Figs. 5.5a, 5.6a). Ponamla (14), Ifo (2).

E-motif 18. Oblique and perpendicular rows of fine fingernail impression (Figs. 5.5b, 5.6c,d). Ponamla (3).

E-motif 19. Widely spaced vertical to oblique parallel rows of fine fingernail impression (Figs. 5.5c, 5.6e). Ponamla (8), Ifo (1).

E-motif 20. Vertical to oblique rows of small tightly packed fingernail impression (Fig. 5.5d). Ponamla (1).

E-motif 21. Tight parallel rows of fingernail impression (Fig. 5.6f-l). Although the orientation and larger design represented by these motifs could not be determined they are distinctive enough and prevalent in the site to warrant a separate motif category. Ponamla (30), Ifo (1).

E-motif 22. Random fingernail impression (Fig. 5.6m). Ponamla (3).

E-motif 23. Continuous horizontal fingernail impression creating horizontal lines (Figs. 5.6n, 5.7a). Ponamla (2), Ifo (2).

E-motif 24. Horizontal parallel rows of vertical fingernail impression (Fig. 5.7c-e). Ponamla (12).

E-motif 25. Complex incision, including rectilinear pattern infilled with oblique parallel lines and curvilinear incision (Fig. 5.8a). Ponamla (1).

E-motif 26. Complex incision, including checkerboard incision with alternate squares filled with tight cross-hatch (Fig. 5.8b). Ponamla (1).

E-motif 27. Geometric incision, possibly diamond shaped cross-hatch (Fig. 5.8c). Ponamla (1).

E-motif 28. Regularly spaced parallel vertical incision infilled with oblique parallel incision (Fig. 5.8d,e). Ponamla (8).

E-motif 29. Herringbone incision (Fig. 5.8f-h). A less well defined motif but a distinctive component of a larger design. Ponamla (3).

E-motif 30. Zigzag incision, including in some cases an incised boundary (Fig. 5.9a-c). Ponamla (3)

E-motif 31. Possible anthropomorphic figure (Fig. 5.9j), made up of curvilinear incision and incised triangle. Ponamla (1).

E-motif 32. Punctate with linear incised border (Fig. 5.10c). Ponamla (1).

E-motif 33. Punctate bounded by linear incision, plus further oblique parallel linear incision and fingernail impression (Fig. 5.10d). Ponamla (1).

E-motif 34. Fingernail impression bounded by linear incision plus cross-hatch incision (Fig. 5.10f). Ponamla (1).

E-motif 35. Fingernail gouge plus linear incision (Fig. 5.10h). Ponamla (1).

E-motif 36. Tight vertical rows of spaced vertical fingernail pinch creating a central ridge (Fig. 5.14d). Ifo (2).

E-motif 37. Continuous vertical rows of tightly spaced fingernail pinch creating pronounced ridges (Fig. 5.13e). Ifo (5).

E-motif 38. Widely spaced parallel vertical rows of vertical fingernail pinch (Fig. 5.14c). Ifo (1).

E-motif 39. A row of continuous fingernail pinch with flat central ridge plus random fingernail pinch (Fig. 5.13k). Ifo (1).

E-motif 40. Horizontal rows of vertical fingernail impression, alternating with vertical parallel incision (Fig. 5.14a). Ifo (3).

E-motif 41. Oblique parallel incision (Fig. 5.11a). Ifo (1).

E-motif 42. Cross-hatch incision enclosed within a triangular (incised) boundary (Fig.5.11c). Ifo (1).

Erromango lip modification.
E-lip motif 1. Incised lip (Figs. 5.3a, b). Ponamla (10), Ifo (1).

E-lip motif 2. Notched lip (Fig. 5.14a-c). Ifo (4).

E-lip motif 3. Fingernail impression (Fig. 5.15d). Ifo (2)

E-lip motif 4. Fingernail pinch (Fig. 5.14e). Ifo (4)

E-lip motif 5. Deep fingernail pinch creating crenellated effect (Fig. 5.14i). Ifo (1).

Efate motifs
Ef-motif 1. Spaced vertical parallel linear incision infilled with oblique parallel incision alternating in orientation (Fig. 6.11b). Mangaasi (11).

Ef-motif 2. Spaced vertical parallel linear incision infilled with assorted oblique parallel incision (Fig. 6.9a). Mangaasi (1), Arapus (1).

Ef-motif 3. Spaced vertical linear incision infilled with parallel single orientation linear incision (Fig.6.11d). Mangaasi (2).

Ef-motif 4. Spaced vertical linear incision infilled with vertical and horizontal incised detached chevrons and oblique incision (Fig. 6.10e). Mangaasi (1).

Ef-motif 5. Spaced vertical linear incision infilled with single orientation oblique parallel incision plus parallel incised detached 'V' motif (Fig. 6.11c). Mangaasi (3).

Ef-motif 6. Single horizontal linear incision above discrete parallel vertical incision infilled with detached chevrons (Fig. 6.7f). Arapus (1).

Ef-motif 7. Vertical, parallel incised zigzag bordered with vertical linear incision (Fig. 6.13f). Mangaasi (1).

Ef-motif 8. Spaced single parallel linear incision infilled with oblique multi-orientated parallel incision (Fig. 6.4b). Mangaasi (4).

Ef-motif 9. Discrete vertical linear incision infilled with oblique parallel incision (Fig. 6.10b, c, d). Mangaasi (11).

Ef-motif 10. Discrete spaced parallel vertical linear incision infilled with oblique parallel incision and separated by a single vertical linear incision (Fig. 6.10f). Mangaasi (4).

Ef-motif 11. Widely spaced vertical incision infilled with wide parallel incised chevrons (Fig. 6.4a). Mangaasi (3).

Ef-motif 12. Spaced vertical linear incision creating columns which are infilled with parallel oblique incision and separated by longer oblique parallel linear incision (Fig. 6.9b). Arapus (1).

Ef-motif 13. Horizontal rows of vertical incised gashes plus spined and non-spined chevrons (Fig. 6.12c). Mangaasi (1).

Ef-motif 14. Horizontal rows of vertical incised gashes (Fig. 6.12a). Mangaasi (6).

Ef-motif 15. Spaced parallel vertical linear incision infilled with horizontal rows of vertical gashes, single orientation oblique parallel or multi-orientated parallel linear incision (Fig. 6.13d). Mangaasi (3).

Ef-motif 16. Horizontal rows of vertical incised gashes bordered by vertical or oblique linear incision (Fig. 6.7h). Arapus (1).

Ef-motif 17. Horizontal rows of vertical incised gashes bordered by single vertical linear incision and oblique gashes. Mangaasi (1).

Ef-motif 18. Horizontal rows of vertical gashes bordered by single vertical linear incision (Fig. 6.12b, 6.13m). Mangaasi (10), Arapus (1).

Ef-motif 19. Complex incised motif comprising a series of discrete motifs including parallel incised diamond pattern, cross-hatch, half chevrons with central spine, full chevrons with central spine and further cross-hatch (Fig. 6.5b). Mangaasi (1)

Ef-motif 20. Incised geometric pattern comprising vertical incision infilled with parallel oblique incision plus a central panel of squares which are in turn infilled with concentric squares or partitioned with oblique incision (Fig. 6.5c). Mangaasi (1).

Ef-motif 21. Complex incised decoration comprised a series of distinct and separate motifs including oblique parallel incision bordering a panel of horizontal parallel rows of vertical gashes, incised chevrons, vertical parallel linear incision infilled with oblique parallel incision, further incised chevrons with central spine and vertical linear incision infilled with cross-hatch (Fig. 6.6b). Arapus (1).

Ef-motif 22. Complex incised motif comprising a row of horizontal gashes above a horizontal line which borders a series of cross-hatched areas, incised ladder-like motifs and circular incision (Fig. 6.21a). Mangaasi (2).

Ef-motif 23. Horizontal parallel lines infilled with incised chevrons and parallel vertical lines (Fig. 6.3a). Mangaasi (1).

Ef-motif 24. Incised square grid pattern partitioned by oblique incision and partly infilled by oblique parallel incision (Fig. 6.5d). Mangaasi (1).

Ef-motif 25. Single horizontal zigzag incision bordered by horizontal linear incision (Fig. 6.7d). Mangaasi (1).

Ef-motif 26. Spaced horizontal parallel incision infilled with incised diamond pattern (Fig. 6.7j). Arapus (1).

Ef-motif 27. Horizontal row of continuous parallel incised triangles partitioned by a single vertical linear incision and bounded by a curvilinear wave-like incision (Fig. 6.6c). Arapus (1).

Ef-motif 28. Vertical parallel linear incision (Fig. 6.3c,d). Mangaasi (3).

Ef-motif 29. Vertical and horizontal parallel incision surrounding a central square which is partitioned by oblique parallel incision (Fig. 6.13c). Mangaasi (1).

Ef-motif 30. Oblique parallel incision in association with occasional incised vertical or oblique gashes (Figs. 6.3e, 6.5a). Mangaasi (16)

Ef-motif 31. Oblique parallel incision creating large horizontal zigzag ('V') motif (Fig. 6.6a). Mangaasi (54).

Ef-motif 32. Horizontal and oblique incised ladder-like patterns (Fig. 6.3b). Mangaasi (1).

Ef-motif 33. Spaced vertical parallel incision infilled with oblique parallel linear incisions creating a diamond motif (Fig. 6.13b). Mangaasi (1).

Ef-motif 34. Spaced vertical and horizontal parallel linear incision infilled with oblique parallel incision (Fig. 6.13a). Mangaasi (1).

Ef-motif 35. Incised squares partitioned with single oblique incisions and further infilled with oblique gashes (Fig. 6.13k). Arapus (1).

Ef-motif 36. Incised chevrons with central spine plus oblique gashes and vertical linear incision (Fig. 6.12e). Mangaasi (2).

Ef-motif 37. Horizontal rows of vertical gashes bordered by vertical linear incision adjacent to incised spined chevrons (Fig. 6.12f). Mangaasi (2).

Ef-motif 38. Spined chevrons (Fig. 6.12d,h). Mangaasi (4).

Ef-motif 39. Not utilised.

Ef-motif 40. Geometric incision plus fine detached chevron (Fig. 6.20c). Mangaasi (1).

Ef-motif 41. Not utilised.

Ef-motif 42. Not utilised.

Ef-motif 43. Branch-like incision bordered by linear incision (Fig. 6.12i). Mangaasi (1).

Ef-motif 44. Horizontal parallel rows of punctation (Fig. 6.14a-c). Mangaasi (17).

Ef-motif 45. Horizontal rows of punctation above a series of oblique parallel linear incisions (Fig. 6.16a,b). Arapus (2)

Ef-motif 46. Horizontal parallel rows of fine punctation above a band of short vertical parallel linear incisions above oblique parallel incision embellished with occasional vertical gashes (Fig. 6.16e). Arapus (1).

Ef-motif 47. Horizontal parallel rows of punctation above a band of short vertical linear incisions above parallel oblique incisions (Fig. 6.16d). Arapus (1).

Ef-motif 48. Horizontal parallel rows of punctate above spaced columns comprised vertical rows of punctation (Fig. 6.14c). Mangaasi (1).

Ef-motif 49. Horizontal parallel rows of punctate above spaced single vertical lines of punctation (Fig. 6.14d). Mangaasi (1).

Ef-motif 50. Horizontal rows of detached chevrons creating vertical parallel zigzag motif bordered by oblique parallel linear incision and single gashes (Fig. 6.16c). Mangaasi (55).

Ef-motif 51. Horizontal rows of alternating vertical and oblique gashes bordered by parallel linear incision (Fig. 6.18c). Mangaasi (4).

Ef-motif 52. Horizontal rows of uniform oblique gashes bordered by oblique parallel linear incision (Fig. 6.18b). Mangaasi (2).

Ef-motif 53. Horizontal rows of uniform oblique gashes bordered by oblique parallel linear incision and gashes plus an applied nubbin (Fig. 6.18e). Mangaasi (4).

Ef-motif 54. Horizontal rows of punctate above oblique parallel incision forming a V motif which in turn infilled with punctation (Garanger 1972:Fig 129). Mangaasi (1).

Ef-motif 55. Horizontal rows of punctate above oblique parallel incision forming V motifs infilled with further parallel incision and bordered by punctation (Garanger 1972: Fig.126). Mangaasi (1).

Ef-motif 56. Horizontal rows of oblique gashes (Fig. 6.7i). Mangaasi (3), Arapus (1).

Ef-motif 57. Horizontal rows of punctate above oblique parallel linear incision bordered by a single row of punctate (Fig. 6.18a). Mangaasi (1).

Ef-motif 58. Single horizontal row of braided nubbins above vertical linear incision infilled with parallel oblique linear incision (Fig. 6.18i) which can also appear above the applied relief (Fig. 6.18h). Mangaasi (2).

Ef-motif 59. Single horizontal row of nubbins above oblique parallel linear incision and gashes (Fig. 6.18f). Mangaasi (2).

Ef-motif 60. Single horizontal row of nubbins above horizontal rows of punctate above parallel oblique incision (Garanger 1972:Fig.132, 26) Mangaasi (1).

Ef-motif 61. Horizontal row of oval nubbins below parallel oblique incision surrounded by random punctation (Fig. 6.19d). Mangaasi (1).

Ef-motif 62. Series of V shaped designs created with plain applied bands which are infilled with a combination of punctation and geometric incision (Garanger 1972:Fig 139). Mangaasi (6).

Ef-motif 63. Incised square grid pattern partitioned by single diagonal linear incisions (Fig. 6.15a). Mangaasi (2).

Ef-motif 64. Incised diamond grid pattern divided by a single vertical linear incision surrounded by vertical linear incision (Fig. 6.17b). Mangaasi (1).

Ef-motif 65. Incised triangular grid alternately infilled with vertical parallel incision (Fig. 6.9c, 6.14f-h). Mangaasi (57).

Ef-motif 66. Incised triangular grid pattern infilled with incised cross-hatching (Fig. 6.17a). Mangaasi (7).

Ef-motif 67. Incised geometric grid with squares alternately infilled with fine cross-hatch (Garanger 1972:Fig 87, 19). Mangaasi (3).

Ef-motif 68. Incised geometric grid with adjacent triangular partitions infilled with vertical linear incision (Fig. 6.17c). Mangaasi (1).

Ef-motif 69. Incised checkerboard pattern alternately infilled with parallel vertical incision (Fig. 6.21b-f). Mangaasi (10).

Ef-motif 70. Spaced vertical incision partitioned with a single horizontal line and infilled with oblique parallel incision. Recorded by Garanger at Mangaasi (1972:Fig. 115).

Ef-motif 71. Incised grid infilled with oblique parallel incision (Fig. 6.20g). Mangaasi (1).

Ef-motif 72. Incised grid infilled with oblique parallel incision plus adjacent punctate (Fig. 6.20e). Mangaasi (1).

Ef-motif 73. Short oblique parallel incisions forming diamond grid pattern (Fig. 6.20d). Mangaasi (2).

Ef-motif 74. Incised diamond grid pattern partitioned by horizontal incision (Fig. 6.20f). Mangaasi (1).

Ef-motif 75. Panels of vertical linear incision beneath cross-hatch incision bordered by incised gashes (Fig. 6.20i). Mangaasi (1).

Ef-motif 76. Panels of vertical parallel linear incision separated by discontinuous horizontal zigzag (Fig. 6.20a). Mangaasi (2).

Ef-motif 77. Curvilinear incised anthropomorphic (?) figure (Fig. 6.20h). Mangaasi (1).

Ef-motif 78. Single horizontal notched applied band bordering paired parallel oblique linear incisions (Fig. 6.21h). Mangaasi (2).

Ef-motif 79. Parallel notched applied bands (2) plus incised ladder-like design (Fig. 6.21l). Mangaasi (1).

Ef-motif 80. Horizontal notched applied band plus triangular incised designs infilled with cross-hatch (Fig. 6.21k). Mangaasi (2).

Ef-motif 81. Notched applied band plus parallel rows of punctate (Fig. 6.21n). Mangaasi (1).

Ef-motif 82. Three parallel horizontal applied notched bands plus geometric incision infilled with parallel vertical incision (Fig.6.21i). Mangaasi (4).

Ef-motif 83. Parallel and perpendicular linear incision infilled with rows of punctation (Fig. 6.19g, h). Mangaasi (26).

Ef-motif 84. Complex incised and punctate motif combining parallel linear incision infilled and punctate with parallel zigzag incision and triangular incised patterns infilled with parallel linear incision (Fig. 6.19f). Mangaasi (1).

Ef-motif 85. Incised ladder-like motif (Fig. 6.19e). Mangaasi (2).

Ef-motif 86. Geometric notched applied relief pattern infilled with punctation and fine cross-hatch (Fig. 6.22a). Mangaasi (3).

Ef-motif 87. Continuous notched applied relief bordering discrete incised triangles infilled with parallel linear incision (Fig. 6.22b). Mangaasi (1).

Ef-motif 88. Parallel rows of notched applied relief with incised ladder-like design and incised triangular grid infilled with parallel linear incision (Fig. 6.22c). Mangaasi (2).

Ef-motif 89. Parallel oblique linear notched applied bands in association with parallel linear incision infilled with oblique parallel incision (Fig. 6.22f). Mangaasi (2).

Ef-motif 90. Linear notched applied relief band plus fine cross-hatching (Fig. 6.22g). Mangaasi (1).

Ef-motif 91. Central incised ladder-like design bordered by fine incised gashes and notched applied bands (Fig. 6.22h). Mangaasi (4).

Ef-motif 92. Parallel linear incision bordered by fine gashes and notched applied bands (Fig. 6.22i). Mangaasi (3).

Ef-motif 93. Single horizontal linear notched applied band plus two parallel linear oblique notched applied bands and zigzag incision crossed by incised gashes (Fig. 6.21g). Mangaasi (1).

Ef-motif 94. Curvilinear notched applied band plus cross-hatch incision (Fig. 6.21m). Mangaasi (1).

Ef-motif 95. Horizontal detached chevron plus parallel linear incision infilled with punctation (Fig. 6.23a). Mangaasi (1).

Ef-motif 96. Series of discrete oblique parallel linear incisions above a single horizontal linear incision above cross-hatch incision (Fig. 6.23b). Mangaasi (1).

Ef-motif 97. Geometric incision infilled with incised cross-hatch plus a horizontal row of punctation (Fig. 6.23c). Mangaasi (1).

Ef-motif 98. Vertical parallel single incised zigzag above horizontal linear incision infilled with punctation (Fig. 6.23d). Mangaasi (1).

Efate lip motifs

Ef-lip motif 1. Notching on the lip or in the case of wide flat lips on the exterior edge of the lip (Fig. 6.2a). Mangaasi (see tables)

Ef-lip motif 2. Incision on the lip or in the case of wide flat lips on the exterior edge of the lip (Fig. 6.2b).

Ef-lip motif 3. Single line of punctate plus notching on the exterior edge of the lip. Restricted to wide flat lips (Fig. 6.2c-e). Mangaasi (11).

Ef-lip motif 4. Notching on interior and exterior edges of wide flat lips (Fig. 6.2f). Mangaasi (1).

Ef-lip motif 5. Oblique parallel incision on wide flat lips plus notched lip exterior (Fig. 6.2g). Mangaasi (1).

Ef-lip motif 6. Zigzag incision and associated single linear incision on wide flat lips. Includes notching on the lip exterior (Fig. 6.2h). Mangaasi (1).

Ef-lip motif 7. Incision on wide flat lip (Fig. 6.2i). Mangaasi (1).

Malekula motifs

M-motif 1. Horizontal ribbed surface (Fig. 7.4a), widespread across Malekula and other northern islands including Pentecost, Malo, Santo and Ambae.

M-motif 2. Horizontal ribbed surface with addition of scaling (Fig. 4.2, 1ii). Recorded across Malekula and Pentecost.

M-motif 3. Horizontal ribbed surface with parallel vertical incision (Fig. 7.4d). Found across Malekula, also recorded from Pentecost, Malo and Santo.

M-motif 4. Panels of oblique parallel incision divided by horizontal incision (Fig. 7.14a). Found across Malekula.

M-motif 5. Horizontal parallel linear incision bordering parallel horizontal rows of linked crescents (Fig. 7.14c). Found across Malekula.

M-motif 6. Roughly parallel incised stretched zigzag (Fig. 7.14d). Recorded at Chachara, Malekula.

M-motif 7. Horizontal parallel linear incision bordering parallel oblique linear incision which in turn borders parallel rows of linked crescents (Fig. 7.14e). Recorded at Chachara, Malekula.

M-motif 8. Horizontal parallel linear incision bordering discontinuous rows of parallel zigzag incision (Fig. 7.14f). Recorded at Chachara, Malekula.

M-motif 9. Spaced parallel linear incision infilled with punctation (Fig. 7.15a). Recorded at Chachara, Malekula.

M-motif 10. Vertical and oblique parallel linear incision bordering random punctation (Fig. 7.15b). Recorded at Chachara, Malekula.

M-motif 11. Spaced parallel vertical incision infilled with oblique parallel incision plus parallel rows of punctation (Fig. 7.15g). Recorded at Chachara, Malekula.

M-motif 12. Wide shallow herringbone incision (Fig. 7.16a). Recorded at Chachara and Espeigle Bay, Malekula.

M-motif 13. Vertical and horizontal parallel linear incision (Fig. 7.16b, c). Recorded at Chachara, Malekula.

M-motif 14. Horizontal parallel linear incision interspersed with sections of vertical linear incision (Fig. 7.16g). Recorded at Chachara, Malekula.

M-motif 15. Oblique parallel linear incision partitioned by vertical incision (Fig. 7.16h). Recorded at Chachara, Malekula.

M-motif 16. Spaced vertical parallel linear incision (3-4) separated by horizontal rows of vertical fingernail (?) impression (Fig. 7.17a, b). Recorded at Chachara, Malekula.

M-motif 17. Horizontal discontinuous rows of fingernail (?) impression bordering herringbone incision plus horizontal parallel linear incision (Fig. 7.17c). Recorded at Chachara, Malekula.

M-motif 18. Horizontal parallel linear incision interspersed with incised parallel joined crescents (Fig. 7.17d, f). Recorded across Malekula and also recorded on Maewo, Malo and Pentecost.

M-motif 19. Parallel single vertical rows of fingernail gouge (Fig. 7.17e). Recorded at Chachara and Norsup, Malekula and on Malo.

M-motif 20. Parallel linear and curvilinear incision (Fig. 7.6a-c). Recorded at Nuas, Malekula and on Malo.

M-motif 21. Parallel curvilinear incision plus attached perpendicular short incised parallel gashes (Fig. 7.6d). Recorded at Navepule B and Albalak, Malekula.

M-motif 22. Oblique parallel incision bordered by single horizontal linear incision (Fig. 7.6e). Recorded at Eldru, Malekula.

M-motif 23. Geometric motif comprising parallel linear incision plus radiating parallel vertical incision (Fig. 7.6f). Recorded at Malua Bay, Malekula.

M-motif 24. Geometric motif comprising incised triangular (?) grid pattern infilled with oblique parallel incision (Fig. 7.6g). Recorded at Pitar, Malekula.

M-motif 25. Incised crosshatch with horizontal incised linear border (Fig. 7.6h). Recorded at Espeigle Bay and Eldru, Malekula.

M-motif 26. Vertical linear incision infilled with oblique parallel incision (Fig. 7.6j). Recorded at Tenmiel and Malua Bay, Malekula.

M-motif 27. Parallel comb incised linear incision (Fig. 7.6k). Recorded at Lakatoro, Malekula.

M-motif 28. Herringbone type motif comprising parallel rows of linear incision with parallel oblique attached gashes (Fig. 7.6l). Recorded at Albalak, Malekula.

M-motif 29. Geometric motif comprising spaced vertical incision partitioned with oblique linear incision forming four triangular sections which in turn are infilled with horizontal parallel incision (Fig. 7.7a). Recorded at Malua and Espeigle Bay, Malekula.

M-motif 30. Complex incised and excised motif alternating with parallel excised multi-angled channels decorated with parallel excision creating ladder effect and parallel remnant surface sections. The motif is partly bordered by vertical incision (Fig. 7.7f). Recorded at Albalak, Malekula.

M-motif 31. Vertical parallel wavy comb incision (Fig 7.8c). Recorded at Navepule B and South West Bay, Malekula.

M-motif 32. Spaced paired horizontal linear incision infilled with discontinuous pairs of crescents (Fig. 7.8d). Recorded at Wala, Malekula and Pentecost

M-motif 33. Spaced vertical rows of punctate (Fig. 7.8e). Recorded at Nuas and Lakatoro, Malekula and Malo.

M-motif 34. Wavy comb incision plus geometric pattern (Fig. 7.8g). Recorded at Eldru, Malekula.

M-motif 35. Thick exposed coils. A variant of motif 1 (Fig. 7.8h). Recorded at Tenmiel and Wala, Malekula.

M-motif 36. Complex incised motif comprising spaced vertical incision infilled with oblique incision creating a geometric pattern which in turn is partly infilled with parallel incision. This is bordered on at least two sides by linear incision crossed by parallel gashes (Fig. 7.9a). Recorded at inland North West Malekula.

M-motif 37. Spaced vertical parallel applied bands (Fig. 7.9b). Recorded at Tenmaru, Malekula.

M-motif 38. Single row of horizontal punctate bordering geometric incised pattern (Fig 7.9c). Recorded at Tenmiel, Malekula.

M-motif 39. Oblique parallel incision bordered with single horizontal incision (Fig. 7.9d). Recorded at Tenmiel, Malekula.

M-motif 40. Parallel incised gashes forming a diamond motif (Fig. 7.9e). Recorded at Malua Bay, Malekula.

M-motif 41. Parallel linear incision radiating from a central diamond shape which is alternately blank and infilled with single incised horizontal and vertical lines (Fig. 7.9f). Spaced nubbins also present. Recorded at Lakatoro, Tenmiel and South West Bay, Malekula.

M-motif 42. Complex incised motif includes central blank incised diamond shape surrounded by vertical parallel incision plus adjoining incised zigzag (Fig. 7.9g). Recorded at Tenmiel, Malekula.

M-motif 43. Geometric incision including triangular (?) grid pattern infilled with oblique parallel incision (Fig. 7.9h). Recorded at Eldru, Lakatoro, Malekula.

M-motif 44. Adjoining panels of oblique parallel incision (Fig. 7.9i). Recorded at Tenmiel, Malekula.

M-motif 45. Complex incised and excised motif including multi-partitioned panels (Fig. 7.18a). Some comprise parallel curvilinear incision infilled with perpendicular parallel excision creating a ladder-like effect. Others are bounded by linear excision creating triangular (?) panels which are in turn infilled with parallel linear excision and further perpendicular excision. Recorded at North West area and South West Bay, Malekula.

M-motif 46. Spaced parallel linear excision infilled with either punctation (Fig. 7.18d) or further excision creating ladder-like motif (Fig. 7.18b-c). Recorded at Yalo South, Navepule B, Malua Bay and Uripiv, Malekula.

M-motif 47. Complex excised motif including concentric circles partitioned by vertical excision (Fig. 7.10a). Recorded at Norsup, Malekula.

M-motif 48. Parallel rows of horizontal punctate plus horizontal row of discontinuous nubbins (Fig. 7.12e). Recorded at Lakatoro, Malekula.

M-motif 49. Geometric motif including V shaped incision surrounded with vertical incision. Partly bordered by single horizontal linear excision (Fig. 7.12f). Recorded at Tenmiel, Malekula.

M-motif 50. Geometric motif comprising parallel vertical and horizontal incision bordered by horizontal linear ridge (Fig. 7.12i). Recorded at Tenmiel, Malekula.

M-motif 51. Heavily excised motif with horizontal chevron pattern in relief plus curvilinear and linear patterns in relief (Fig. 7.12j). Recorded at Tenmiel, Malekula.

M-motif 52. Parallel rows of horizontal punctation bordered by single horizontal linear incision (Fig. 7.12k). Recorded at Malua Bay, Malekula.

M-motif 53. Bands of horizontal linear incision (3) interspersed with impressed paired parallel 'V' pattern (Fig. 7.13e). Recorded at Espeigle Bay, Malekula.

M-motif 54. Horizontal parallel (2) rows of notched continuous applied relief plus panels of cross-hatch (Fig. 7.13f). Recorded at Waal, Malekula.

M-motif 55. Horizontal parallel rows (3) of notched continuous applied relief plus geometric incised pattern infilled with oblique parallel incision (Fig. 7.13g). Recorded at Lakatoro, Malekula.

M-motif 56. Single horizontal row of notched continuous applied relief plus star-like pattern of ladder-like incision (Fig. 7.13h). Recorded at Lakatoro, Malekula.

M-motif 57. Single vertical, horizontal and oblique rows of notched continuous applied relief connected by parallel ladder-like incision (Fig. 7.13i). Recorded at southern Malekula.

M-motif 58. Incised cross-hatch bordered by linear incision infilled with parallel perpendicular gashes creating ladder-like effect (Fig. 7.13j). Recorded at Lakatoro, Malekula.

M-motif 59. Miscellaneous incised cross-hatch (Fig. 7.13k) Recorded at Espeigle Bay, Malekula

M-motif 60. Geometric incised interconnected diamond pattern. Diamonds are alternately infilled with incised cross-hatch (Fig. 7.13 l). Recorded at Lakatoro, Malekula.

www.ingramcontent.com/pod-product-compliance
Lightning Source LLC
Chambersburg PA
CBHW061306270326

41935CB00026B/1847